Advancements in Sustainable Architecture and Energy Efficiency

Roberto A. González-Lezcano
Universidad CEU San Pablo, Spain

A volume in the Practice, Progress, and Proficiency in Sustainability (PPPS) Book Series

IGI Global
PUBLISHER of TIMELY KNOWLEDGE

Published in the United States of America by
IGI Global
Engineering Science Reference (an imprint of IGI Global)
701 E. Chocolate Avenue
Hershey PA, USA 17033
Tel: 717-533-8845
Fax: 717-533-8661
E-mail: cust@igi-global.com
Web site: http://www.igi-global.com

Library of Congress Cataloging-in-Publication Data

Names: González-Lezcano, Roberto Alonso, 1971- editor.
Title: Advancements in sustainable architecture and energy efficiency / Roberto Alonso González-Lezcano, editior.
Description: Hershey, PA : Engineering Science Reference, [2021] | Includes bibliographical references and index. | Summary: "This book addresses the different perspectives of energy consumption and demand to ensure sustainable energy, increased energy efficiency, improved energy policies and reasonable energy costs"-- Provided by publisher.
Identifiers: LCCN 2021005797 (print) | LCCN 2021005798 (ebook) | ISBN 9781799870234 (h/c) | ISBN 9781799870241 (s/c) | ISBN 9781799870258 (ebook)
Subjects: LCSH: Buildings--Energy conservation. | Sustainable buildings.
Classification: LCC TJ163.5.B84 A278 2021 (print) | LCC TJ163.5.B84 (ebook) | DDC 720/.47--dc23
LC record available at https://lccn.loc.gov/2021005797
LC ebook record available at https://lccn.loc.gov/2021005798

This book is published in the IGI Global book series Practice, Progress, and Proficiency in Sustainability (PPPS) (ISSN: 2330-3271; eISSN: 2330-328X)

British Cataloguing in Publication Data
A Cataloguing in Publication record for this book is available from the British Library.

All work contributed to this book is new, previously-unpublished material. The views expressed in this book are those of the authors, but not necessarily of the publisher.

For electronic access to this publication, please contact: eresources@igi-global.com.

Practice, Progress, and Proficiency in Sustainability (PPPS) Book Series

Ayman Batisha
International Sustainability Institute, Egypt

ISSN:2330-3271
EISSN:2330-328X

Mission

In a world where traditional business practices are reconsidered and economic activity is performed in a global context, new areas of economic developments are recognized as the key enablers of wealth and income production. This knowledge of information technologies provides infrastructures, systems, and services towards sustainable development.

The **Practices, Progress, and Proficiency in Sustainability (PPPS) Book Series** focuses on the local and global challenges, business opportunities, and societal needs surrounding international collaboration and sustainable development of technology. This series brings together academics, researchers, entrepreneurs, policy makers and government officers aiming to contribute to the progress and proficiency in sustainability.

Coverage

- Knowledge clusters
- Intellectual Capital
- Outsourcing
- Socio-Economic
- ICT and knowledge for development
- Global Business
- Innovation Networks
- E-Development
- Environmental informatics
- Sustainable Development

IGI Global is currently accepting manuscripts for publication within this series. To submit a proposal for a volume in this series, please contact our Acquisition Editors at Acquisitions@igi-global.com or visit: http://www.igi-global.com/publish/.

The Practice, Progress, and Proficiency in Sustainability (PPPS) Book Series (ISSN 2330-3271) is published by IGI Global, 701 E. Chocolate Avenue, Hershey, PA 17033-1240, USA, www.igi-global.com. This series is composed of titles available for purchase individually; each title is edited to be contextually exclusive from any other title within the series. For pricing and ordering information please visit http://www.igi-global.com/book-series/practice-progress-proficiency-sustainability/73810. Postmaster: Send all address changes to above address.

Titles in this Series

For a list of additional titles in this series, please visit: www.igi-global.com/book-series

Handbook of Research on Novel Practices and Current Successes in Achieving the Sustainable Development Goals
Cristina Raluca Gh. Popescu (University of Bucharest, Romania & The Bucharest University of Economic Studies, Romania & University of Craiova, Romania & The National Institute for Research and Development in Environmental Protection (I.N.C.D.P.M.), Romania & National Research and Development Institute for Gas Turbines COMOTI, Roania)
Engineering Science Reference • © 2021 • 461pp • H/C (ISBN: 9781799884262) • US $295.00

Management and Conservation of Mediterranean Environments
Rui Alexandre Castanho (WSB University, Poland) and José Martín Gallardo (Universidad de Extremadura, Spain)
Engineering Science Reference • © 2021 • 336pp • H/C (ISBN: 9781799873914) • US $195.00

Handbook of Research on Institution Development for Sustainable and Inclusive Economic Growth in Africa
Evans S. Osabuohien (Covenant University, Nigeria) Emmanuel A. Oduntan (Covenant University, Nigeria) Obindah Gershon (Covenant University, Nigeria) Olaronke Onanuga (Covenant University, Nigeria) and Oluyomi Ola-David (Covenant University, Nigeria)
Business Science Reference • © 2021 • 507pp • H/C (ISBN: 9781799848172) • US $235.00

Digital Solutions and the Case for Africa's Sustainable Development
Albert Ong'uti Maake (University of Lay Adventist of Kigali, Rwanda) Benard Magara Maake (Kisii University, Kenya) and Fredrick Mzee Awuor (Kisii University, Kenya)
Information Science Reference • © 2021 • 200pp • H/C (ISBN: 9781799829676) • US $195.00

Examining Biophilia and Societal Indifference to Environmental Protection
Mary Ann Markey (Grand Canyon University, USA) and Lonny Douglas Meinecke (King University, USA)
Engineering Science Reference • © 2021 • 279pp • H/C (ISBN: 9781799844082) • US $195.00

Strategies for Business Sustainability in a Collaborative Economy
Ramona-Diana Leon (National School of Political and Administrative Studies, Romania)
Business Science Reference • © 2020 • 440pp • H/C (ISBN: 9781799845430) • US $225.00

IGI Global
PUBLISHER of TIMELY KNOWLEDGE

701 East Chocolate Avenue, Hershey, PA 17033, USA
Tel: 717-533-8845 x100 • Fax: 717-533-8661
E-Mail: cust@igi-global.com • www.igi-global.com

Editorial Advisory Board

Table of Contents

Preface xv

Chapter 1
Application of Phase Change Materials in Construction Materials for Thermal Energy Storage Systems in Buildings 1
Eray Mertkan Meric, Istanbul University-Cerrahpasa, Turkey
Savas Erdem, Istanbul University-Cerrahpasa, Turkey
Ezgi Gurbuz, Istanbul University-Cerrahpasa, Turkey

Chapter 2
Contribution of Water Flow Glazing to Net-Zero Energy Buildings: Validation of Analytical and Numerical Models Through Experimental Data 21
Fernando del Ama Gonzalo, Keene State College, USA
Belen Moreno, Technical School of Architecture of Madrid, Technical University of Madrid, Spain
Matthew Griffin, Keene State College, USA
Juan Antonio Hernandez Ramos, School of Aeronautical and Space Engineering, Technical University of Madrid, Spain

Chapter 3
Energy-Efficient Homes: A Heaven for Respiratory Illnesses 49
Alejandro Moreno-Rangel, Lancaster University, UK
Genny Carrillo, Texas A&M University, USA

Chapter 4
Forecasting Energy Impact in Multifamily Buildings Through Airtightness Models 72
Jesica Fernández-Agüera, Universidad de Sevilla, Spain
Samuel Domínguez-Amarillo, Universidad de Sevilla, Spain
Juan José Sendra, Universidad de Sevilla, Spain

Chapter 5
Interactive Exhibition as a Form of Participation in the Future of Sustainable Housing 96
Juan Bautista Echeverría, Universidad de Navarra, Spain
Iosu Gabilondo, COAVN, Spain
Teresa Meana Rodríguez, COAVN, Spain
Juana Otxoa-Errarte, COAVN, Spain
Claudia Pennese, COAVN, Spain
Montserrat Puig Tramunt, COAVN, Spain
Judith Ubarrechena, COAVN, Spain

Chapter 6
Nursing Homes in Spain and Their High Number of Deaths by COVID-19 as an Alarm in the Study of Indoor Air Quality 118
Gastón Sanglier Contreras, Universidad CEU San Pablo, Spain
Jose Amílcar Rizzo Sierra, Universidad Politécnica de Querétaro, Mexico
Eduardo José López Fernández, Universidad CEU San Pablo, Spain
Roberto Alonso González Lezcano, Universidad CEU San Pablo, Spain

Chapter 7
Occupants' Habits and Natural Ventilation in a Hot Arid Climate 146
Hanan Al-Khatri, Sultan Qaboos University, Oman
Farah Al-Atrash, German Jordanian University, Jordan

Chapter 8
Smart Power Microgrid Impact on Sustainable Building 169
Hussain Attia, American University of Ras Al Khaimah, UAE

Chapter 9
Statistical Understanding and Optimization of Building Energy Consumption and Climate Change Consequences 195
José Antonio Orosa, University of A Coruña, Spain
Diego Vergara, University Catholic of Ávila, Spain
Feliciano Fraguela, University of A Coruña, Spain
Antonio Masdías-Bonome, University of A Coruña, Spain

Chapter 10
Studying Physical and Chemical Properties of Graphene Oxide and Reduced Graphene Oxide and Their Applications in Sustainable Building Materials 221
Omar Ibrahim, Istanbul University-Cerrahpasa, Turkey
Savaş Erdem, Istanbul University-Cerrahpasa, Turkey
Ezgi Gurbuz, Istanbul University-Cerrahpasa, Turkey

Chapter 11
The Importance of Light in Our Lives: Towards New Lighting in Schools 239
David Baeza Moyano, Universidad CEU San Pablo, Spain
Roberto Alonso González Lezcano, Universidad CEU San Pablo, Spain

Chapter 12
The Uses of Renewable Energy in Buildings: A Case Study of Australia.......... 257
Paul C. Okonkwo, Dhofar University, Salalah, Oman
El Manaa Barhoumib, Dhofar University, Salalah, Oman
Wesam H. Beitelmal, Dhofar University, Salalah, Oman
Israr I. U. Hassan, Dhofar University, Salalah, Oman
Michael Nnamdi Azubuike, Federal University of Technology, Owerri, Nigeria
Fadhil Khadoum Al Housnia, Dhofar University, Salalah, Oman
Florence Jonathan, The Gateway Polytechnic Saapade, Nigeria

Chapter 13
Water Efficiency Evaluation Analysis Among Environmental Certification Methods: LEED, BREEAM, DGNB, HQE, EDGE, and BONO VERDE 275
Estela Karem Samamé Zegarra, Navarra University, Spain

Compilation of References 292

About the Contributors 331

Index 335

Detailed Table of Contents

Preface xv

Chapter 1
Application of Phase Change Materials in Construction Materials for Thermal Energy Storage Systems in Buildings 1
Eray Mertkan Meric, Istanbul University-Cerrahpasa, Turkey
Savas Erdem, Istanbul University-Cerrahpasa, Turkey
Ezgi Gurbuz, Istanbul University-Cerrahpasa, Turkey

There has been a continuous increase in world population and industrialization. In parallel, energy consumption grew substantially in the world. The residential buildings constitute the majority of the total consumption in the developed world. Therefore, energy saving and reducing heat losses of buildings are of major concern for our society. At this point, phase changing materials (PCMs) show up as one of the most useful materials in building design. In this study, PCMs, which are used at places such as facade of buildings, bricks, and inside of concrete in order to be able to reduce the energy consumption due to heating and cooling, to provide comfort temperature inside buildings, are extensively reviewed.

Chapter 2
Contribution of Water Flow Glazing to Net-Zero Energy Buildings: Validation of Analytical and Numerical Models Through Experimental Data 21
Fernando del Ama Gonzalo, Keene State College, USA
Belen Moreno, Technical School of Architecture of Madrid, Technical University of Madrid, Spain
Matthew Griffin, Keene State College, USA
Juan Antonio Hernandez Ramos, School of Aeronautical and Space Engineering, Technical University of Madrid, Spain

Net-zero energy buildings (NetZEBs) are of a building typology designed to combine energy efficiency and renewable energy generation to consume only as much energy as produced onsite through renewable resources over a specified time. The successful creation of NetZEBs is crucial to combating the current climate crisis. Water flow glazing (WFG) is a key technology that will assist in achieving this goal. Several experimental facilities have been designed and constructed to collect data based on WFG technology. These experimental facilities demonstrate that the successful implementation of WFG will allow reducing heating and cooling loads, primary energy consumption, and CO2 emissions. However, a wrong WFG selection can lead to failure in NetZEBs design. The goal of this text was to assess WFG performance through key performance indicators to understand the need of other renewable energies so that the construction of NetZEBs becomes a realistic target.

Chapter 3
Energy-Efficient Homes: A Heaven for Respiratory Illnesses 49
Alejandro Moreno-Rangel, Lancaster University, UK
Genny Carrillo, Texas A&M University, USA

The built environment has changed dramatically due to the increased interest in mitigating climate change. Homes are becoming more energy-efficient, responding to energy issues, and reducing carbon emissions primarily. Nevertheless, we started to realize the unintended consequences of these changes that impact a home's indoor environment and occupants' health. Indoor air quality is a critical aspect as indoor pollutants are increasing in homes. More than ever, it is crucial to adhere to the best ventilation practices, building materials, and cleaning products. Additionally, behaviour changes, such as those for healthy homes, can prevent their health impact. Interdisciplinary research between public health and building professionals needs to educate citizens and present evidence for legislative changes and recommendations to spur change to reduce indoor air pollution and protect vulnerable populations preventing harmful effects on future generations' health.

Chapter 4
Forecasting Energy Impact in Multifamily Buildings Through Airtightness Models 72
Jesica Fernández-Agüera, Universidad de Sevilla, Spain
Samuel Domínguez-Amarillo, Universidad de Sevilla, Spain
Juan José Sendra, Universidad de Sevilla, Spain

Southern European areas are traditionally considered to have a benign climate and therefore most homes often do not have controlled ventilation systems or air conditioning. Our society faces challenging situations with global pandemics and health emergencies, while housing provides the basic elements of refuge and activity. A profound transformation of social and labor dynamics is underway, increasing at-home presence, hence the capital importance of improving building envelopes to reduce energy demand and to increase occupant comfort levels in order to ensure their protective function. Air leakage is one of the key factors both in indoor air quality and energy demand. This study aims to explore the sensitivity of energy demand to airtightness. A representative set of multifamily buildings, built in the last 20 years in Seville, is analyzed. The results show that air permeability has a significant effect on energy demand in the sample studied. Although impact is greater in severe climates, it needs to be considered in temperate climates, especially when more time is spent at home.

Chapter 5
Interactive Exhibition as a Form of Participation in the Future of Sustainable Housing 96
Juan Bautista Echeverría, Universidad de Navarra, Spain
Iosu Gabilondo, COAVN, Spain
Teresa Meana Rodríguez, COAVN, Spain
Juana Otxoa-Errarte, COAVN, Spain
Claudia Pennese, COAVN, Spain
Montserrat Puig Tramunt, COAVN, Spain
Judith Ubarrechena, COAVN, Spain

The Gipuzkoa branch of the Basque and Navarre College of Architects organized, within the MUGAK Architecture Biennial, the exposition "The Transgenerational House." It took place in a pavilion specially built for the purpose in a public space in the city of San Sebastian (Spain). In it, both a conventional furnished home and an alternative one, with the possibility of allowing free spatial divisions and furnishing distribution, were recreated. Some architectural teams showed their experiences on housing. A set of components with a color code was developed to link the two homes and the work of the architects. The pavilion was opened to the public, which had access to the contained information in a partially directed way and participated answering to posed specific questions. Additionally, 10 structured workshops with different collectives were organized, making specific proposals on the alternative home. The overall exposition is shown, reflecting on the advantages and limitations of citizen participation as an instrument of sustainable development.

Chapter 6
Nursing Homes in Spain and Their High Number of Deaths by COVID-19 as an Alarm in the Study of Indoor Air Quality 118
Gastón Sanglier Contreras, Universidad CEU San Pablo, Spain
Jose Amílcar Rizzo Sierra, Universidad Politécnica de Querétaro, Mexico
Eduardo José López Fernández, Universidad CEU San Pablo, Spain
Roberto Alonso González Lezcano, Universidad CEU San Pablo, Spain

Nursing homes have been one of the most prominent targets of the COVID-19 coronavirus in Spain. The factors that have determined that this is the case are very diverse. In this study, physical agents and chemical pollutants, population density, and different capacities of the residences are analyzed to see their influence on the number of elderly people who have died in the geriatric centres in the different autonomous communities (AACC) of Spain. A statistical analysis has been carried out on the variables observed. The conclusions show that in many places where this overflow of deaths has occurred, the residences were private, with some exceptions. The influence of physical agents and pollutants has been shown to be a determining factor, especially for the communities of Extremadura and Castilla-La Mancha, although it is true that the large number of factors makes the study complicated. The dilemma between air quality and energy efficiency is of great importance, especially when human health is at stake.

Chapter 7
Occupants' Habits and Natural Ventilation in a Hot Arid Climate 146
Hanan Al-Khatri, Sultan Qaboos University, Oman
Farah Al-Atrash, German Jordanian University, Jordan

This chapter presents the findings of a questionnaire distributed in the Arabian Gulf region to explore the potential of utilising natural ventilation to moderate thermal conditions in residential buildings. It, additionally, explores the occupants' acceptance of the idea of applying natural ventilation when outdoor thermal environments are acceptable. Natural ventilation is a key sustainable solution to improve thermal conditions in the region considering its extreme climate, huge consumption of cooling energy, and the cultural attitude to depend on mechanical ventilation. The chapter discusses the thermal adaptation including physical or behavioural adaptation, and it sheds light on selected studies discussing similar issues. A detailed climatic analysis is presented with reference to Muscat city. Discussing the questionnaire's findings revealed high acceptance to depend on natural ventilation of around 93.7% of the participants. In addition, the possibility to depend on natural ventilation in the period from November to March, especially during daytime, was revealed.

Chapter 8
Smart Power Microgrid Impact on Sustainable Building 169
Hussain Attia, American University of Ras Al Khaimah, UAE

This chapter explains the positive impacts of the smart microgrid with respect to the sustainable building performance. The role of the renewable energy sources within the microgrid is also demonstrated. The solar photovoltaic energy source is explained in terms of the working principle and positive effects on reducing CO2 emission through reducing the needs of the traditional power stations' electricity. This chapter explains the advantages of the smart microgrid power system and how reflects on the performance of a sustainable building. The differences between the components of a DC smart grid system and of an AC smart grid system are shown. Different algorithms for maximum power point tracking (MPPT) to improve the PV systems have reviewed in this chapter. A step-by-step overall design of any desired single-phase or three-phase alternating power of any capacity for a PV matrix-based microgrid system, in addition to the role and the importance of inserting DC-DC converter in the photovoltaic systems, is discussed.

Chapter 9
Statistical Understanding and Optimization of Building Energy Consumption and Climate Change Consequences... 195
José Antonio Orosa, University of A Coruña, Spain
Diego Vergara, University Catholic of Ávila, Spain
Feliciano Fraguela, University of A Coruña, Spain
Antonio Masdías-Bonome, University of A Coruña, Spain

In the present chapter, a new tool was designed to find a better alternative for improving building energy consumption in the next years. In this sense, in the first stage of this calculation procedure, ISO Standard 13790 calculation procedure was developed in accordance with Monte Carlo method and results showed the probability of energy consumption as a Weibull model. Furthermore, a map of different Weibull models in accordance with different input parameters and future climate change effect was developed as a future building design guide. This tool defines the probability of energy consumption of an existing building, or a building that is being designed today and in the near future, preventing the climate change effect. More applications at the time of building retrofitting and healthy indoor ambiences are proposed.

Chapter 10
Studying Physical and Chemical Properties of Graphene Oxide and Reduced Graphene Oxide and Their Applications in Sustainable Building Materials.. 221
Omar Ibrahim, Istanbul University-Cerrahpasa, Turkey
Savaş Erdem, Istanbul University-Cerrahpasa, Turkey
Ezgi Gurbuz, Istanbul University-Cerrahpasa, Turkey

In this chapter, the authors identified graphene oxide (OG)/reduced graphene oxide (RGO) as nano composites by studying its nanoparticles' properties physically, chemically, and mechanically. In this study, they mentioned the photo catalyst materials PC regarding carbon nanostructures such as GO and RGO, which have excellent oxygen functionalities, efficient adsorption areas, and considerable surface area. The compositions of GO and RGO exceed electron-holes pair reinstallation time and minimize energy hiatus by adjusting valence band level (VBL) with conducting band level (CBL) bringing high suction of the exist radiance, which improves photo degeneration achievement of material oxides and composites made from polymers. They also studied the main applications of GO and RGO in engineering fields and summarized the usefulness of intercalation of GO and RGO in construction sectors. Moreover, many synthesis techniques lead to many types of GO. Therefore, in this chapter, the authors tried to collect most GO and RGO properties, structures, and applications.

Chapter 11
The Importance of Light in Our Lives: Towards New Lighting in Schools 239
David Baeza Moyano, Universidad CEU San Pablo, Spain
Roberto Alonso González Lezcano, Universidad CEU San Pablo, Spain

The light that enters through our eyes is not only for vision. The human circadian system responds to light differently than the visual system. The timing of each biological function in mammals is directed by the main clock located in the Supraquiasmic Nucleus, which is regulated by light. However, until now, only the interaction of light with our visual system has been taken into account when choosing the parameters of indoor lighting sources, including those in the classroom. In the publications about school lighting, the first concern was the common parameters of indoor lighting such as horizontal workplane illuminance, illuminance uniformity, and avoiding reflections on different surfaces. In this chapter, the authors show publications about new findings on the effects of light on people, studies carried out in different countries aimed at improving classroom lighting, current regulations on lighting related to classroom lighting, and new parameters that are being considered, along with those already used for new and better lighting.

Chapter 12
The Uses of Renewable Energy in Buildings: A Case Study of Australia... 257
Paul C. Okonkwo, Dhofar University, Salalah, Oman
El Manaa Barhoumib, Dhofar University, Salalah, Oman
Wesam H. Beitelmal, Dhofar University, Salalah, Oman
Israr I. U. Hassan, Dhofar University, Salalah, Oman
Michael Nnamdi Azubuike, Federal University of Technology, Owerri, Nigeria
Fadhil Khadoum Al Housnia, Dhofar University, Salalah, Oman
Florence Jonathan, The Gateway Polytechnic Saapade, Nigeria

Only 5% of Australia's energy utilization comes from renewables, while 86.3% of the electricity is produced from fossil fuels. Nonetheless, this pattern has been disturbed by the ongoing decommissioning and closure of old coal power plants, alongside the Australian policy to reduce fossil fuel emissions. Presently, Australia is at a pivotal phase of its change to renewable energy power generation and utilization specifically in residential and commercial buildings. Sustainability in renewable energy utilization is being achieved through guided government policies, reasonable energy costs, and improved energy technology transfer approaches. To give a refreshed delineation of renewable energy, related government policy, and the route ahead in the Australian setting, this chapter presents a deliberate Australia update with renewable energy generation and utilization in Australian buildings.

Chapter 13
Water Efficiency Evaluation Analysis Among Environmental Certification Methods: LEED, BREEAM, DGNB, HQE, EDGE, and BONO VERDE 275
Estela Karem Samamé Zegarra, Navarra University, Spain

Developing countries such as Peru are not strangers in promoting sustainability in buildings through local certification programs such as My Green House funded by international organizations; or architectural design competitions for houses such as “Build to Grow”; compliance standards such as the one published in March 2018, “Climate Change Law”; in addition to other optional legal standards such as “EM.110 Thermal and Light Comfort with Energy Efficiency”: a technical code of voluntary sustainable construction; and the entry of international environmental certifications such as EDGE and LEED, which are currently very welcomed by real estate developers due to the incentives. One of them is the height bonus, which is promoted by some municipal ordinances, mostly located in the capital city of Lima as a product of a project developed and promoted by the IFC and World Bank. On the other hand, in the retail or office sector, they are promoted by green corporate policies; however, there is a long way to go.

Compilation of References 292

About the Contributors 331

Index 335

Preface

Thermal comfort and indoor air quality (IAQ) issues have gained significant interest in the scientific and technical community involved in building performance analysis and other related subjects. In terms of thermal comfort, the achievement and maintenance of a thermally acceptable indoor environment is affected by energy costs, and energy poverty is a widespread problem globally. There is a call for energy-efficient architecture for a developed and sustainable world. However, with the use of renewable energy that increased considerably in recent years, new technical challenges arose for the energy sector. Consumers are key players in this context, as flexibility in demand is crucial to cope with the intermittent nature of most renewable energy sources. Active demand-side participation is particularly important to ensure the efficient use of locally and globally available energy. Sustainability, human comfort, and healthy living environments have become top priorities.

Advancements in Sustainable Architecture and Energy Efficiency explores how housing is a key health factor for individuals and looks at factors such as air quality, ventilation, hygrothermal comfort, lighting, physical environment, building efficiency, and other areas as important pieces in healthy architecture. It discusses how the poor application of these parameters can directly affect human health and how sustainable architecture provides a solution. Beyond just labeling the important facets of architecture for healthy living, this book will look at different perspectives of energy consumption and demand to ensure sustainable energy, increased energy efficiency, improved energy policies, and reasonable energy costs for homes. This book is ideal for architects, designers, engineers, energy engineers, environmental scientists, practitioners, researchers, academicians, and students interested in architecture that is both conducive to healthy living and energy efficiency.

Energy sustainability is a challenge for a developed and sustainable world. The use of renewable energy has increased considerably in recent years, bringing with it new technical challenges for the energy sector. Consumers are key players in this context, as flexibility in demand is crucial to cope with the intermittent nature of most renewable energy sources. Active demand-side participation is particularly important to ensure the efficient use of locally and globally available energy.

This book will address the different perspectives of energy consumption and demand to ensure sustainable energy, increased energy efficiency, improved energy policies and reasonable energy costs. Careful analysis is needed to guide policies that promote energy conservation, energy justice, the growth of renewable energy and improved energy technology transfer. This book also will address a topic of great relevance today, taking into account that in developed countries, most of the time is spent indoors and, depending on each person, the presence in the home ranges from 60% to 90% of the day, and 30% of that time is spent sleeping. Taking into account these data, indoor residential environments have a direct influence on human health. Added to this in developing countries, significant levels of indoor pollution make housing unsafe, with a impact on the health of inhabitants. Housing is therefore a key health factor for people all over the world and various parameters such as air quality, ventilation, hygrothermal comfort, lighting, physical environment, building efficiency and others can contribute to healthy architecture, and the conditions that can result from the poor application of these parameters.

This research book will update the current state of knowledge on the subject, since the authors of the chapters will all be researchers and professors from prestigious universities. It will not be a theoretical book, but will be based on where current research is taking us in terms of energy saving.

The book has been divided into the following chapters:

Chapter 1. Application of Phase Change Materials in Construction Materials for Thermal Energy Storage Systems in Buildings

There has been a continuous increase in world population and industrialization. In parallel, energy consumption grew substantially in the world. The residential buildings constitute the majority of the total consumption in the developed world. Therefore, energy saving and reducing heat losses of buildings are of major concern for our society. At this point, phase changing materials (PCMs) show up being one of the most useful materials in building design. In this study, PCMs, which are used at places such as facade of buildings, bricks and inside of concrete in order to be able to reduce the energy consumption due to heating and cooling, to provide comfort temperature inside buildings, are extensively reviewed.

Chapter 2. Contribution of Water Flow Glazing to Net-Zero Energy Buildings: Validation of Analytical and Numerical Models Through Experimental Data

Net-zero energy buildings (NetZEB's) are a building typology designed to combine energy efficiency and renewable energy generation to consume only as much energy as produced onsite through renewable resources over a specified time. The successful creation of NetZEB's is crucial to combating the current climate crisis. Water Flow Glazing (WFG) is a key technology that will assist in achieving this goal. Several experimental facilities have been designed and constructed to collect data based on WFG technology. These experimental facilities demonstrate that the successful implementation of WFG will allow reducing heating and cooling loads, primary energy consumption, and CO2 emissions. However, a wrong WFG selection can lead to failure in NetZEBs design. The goal of this text was to assess WFG performance through key performance indicators to understand the need of other renewable energies so that the construction of NetZEB's becomes a realistic target.

Chapter 3. Energy-Efficient Homes: A Heaven For Respiratory Illnesses

The built environment has changed dramatically due to the increased interest in mitigating climate change. Homes are becoming more energy-efficient, responding to energy issues, and reducing carbon emissions primarily. Nevertheless, we started to realize the unintended consequences of these changes that impact a home's indoor environment and occupants' health. Indoor air quality is a critical aspect as indoor pollutants are increasing in homes. More than ever, it is crucial to adhere to the best ventilation practices, building materials, and cleaning products. Additionally, behaviour changes, such as those for Healthy Homes, can prevent their health impact. Interdisciplinary research between public health and building professionals needs to educate citizens and present evidence for legislative changes and recommendations to spur change to reduce indoor air pollution and protect vulnerable populations preventing harmful effects on future generations' health.

Chapter 4. Forecasting Energy Impact in Multifamily Buildings Through Airtightness Models

Southern European areas are traditionally considered to have a benign climate and therefore most homes often do not have controlled ventilation systems or air conditioning. Our society faces challenging situations with global pandemics and health emergencies, while housing provides the basic elements of refuge and activity. A profound transformation of social and labor dynamics is underway, increasing at-home presence. Hence the capital importance of improving building envelopes to reduce energy demand and to increase occupant comfort levels in order to ensure their protective function. Air leakage is one of the key factors both in indoor air quality and energy demand. This study aims to explore the sensitivity of energy demand to airtightness. A representative set of multifamily buildings, built in the last 20 years in Seville, is analyzed. The results show that air permeability has a significant effect on energy demand in the sample studied. Although impact is greater in severe climates, it needs to be considered in temperate climates, especially when more time is spent at home.

Chapter 5. Interactive Exhibition as a Form of Participation in the Future of Sustainable Housing

The Gipuzkoa branch of the Basque and Navarre College of Architects organized, within the MUGAK Architecture Biennial, the exposition "The transgenerational house". It took place in a pavilion specially built for the purpose in a public space in the city of San Sebastian (Spain). In it, both a conventional furnished home and an alternative one, with the possibility of allowing free spatial divisions and furnishing distribution, were recreated. Some architectural teams showed their experiences on housing. A set of components with a color code was developed to link the two homes and the work of the architects. The pavilion was opened to the public, that had access to the contained information in a partially directed way and participated answering to posed specific questions. Additionally, ten structured workshops with different collectives, were organized, making specific proposals on the alternative home. The overall exposition is shown, reflecting on the advantages and limitations of citizen participation as an instrument of sustainable development.

Chapter 6. Nursing Homes in Spain and Their High Number of Deaths by Covid-19 as an Alarm in the Study of Indoor Air Quality

Nursing homes have been one of the most prominent targets of the Covi-19 coronavirus in Spain. The factors that have determined that this is the case are very diverse. In this study, physical agents and chemical pollutants, population density and different capacities of the residences are analyzed to see their influence on the number of elderly people who have died in the geriatric centres in the different Autonomous Communities (AACC) of Spain. A statistical analysis has been carried out on the variables observed. The conclusions show that in many places where this overflow of deaths has occurred, the residences were private with some exceptions. The influence of physical agents and pollutants has been shown to be a determining factor, especially for the communities of Extremadura and Castilla-La Mancha, although it is true that the large number of factors makes the study complicated. The dilemma between air quality and energy efficiency is of great importance, especially when human health is at stake.

Chapter 7. Occupants' Habits and Natural Ventilation in a Hot Arid Climate

This chapter presents the findings of a questionnaire distributed in the Arabian Gulf region to explore the potential of utilising natural ventilation to moderate thermal conditions in residential buildings. It, additionally, explores the occupants' acceptance of the idea of applying natural ventilation when outdoor thermal environments are acceptable. Natural ventilation is a key sustainable solution to improve thermal conditions in the region considering its extreme climate, huge consumption of cooling energy, and the cultural attitude to depend on mechanical ventilation. The chapter discusses the thermal adaptation including physical or behavioural adaptation and it sheds light on selected studies discussing similar issues. A detailed climatic analysis is presented with reference to Muscat city. Discussing the questionnaire's findings revealed high acceptance to depend on natural ventilation of around 93.7% of the participants. Besides, the possibility to depend on natural ventilation in the period from November to March, especially during daytime, was revealed.

Chapter 8. Smart Power Microgrid Impact on Sustainable Building

This chapter explains the positive impacts of the smart microgrid with respect to the sustainable building performance. The role of the renewable energy sources within the microgrid is also demonstrated. The solar photovoltaic energy source is explained in terms of the working principle and positive effects on reducing CO2 emission through reducing the needs of the traditional power stations' electricity. This chapter explains the advantages of the smart microgrid power system and how reflects on the performance of a sustainable building. The differences between the components of a DC smart grid system and of an AC smart grid system are shown. Different algorithms for Maximum Power Point Tracking MPPT to improve the PV systems have reviewed in this chapter. A step-by-step overall design of any desired single-phase or three-phase alternating power of any capacity for a PV matrix-based microgrid system, in addition to the role and the importance of inserting DC-DC converter in the photovoltaic systems, all are discussed.

Chapter 9. Statistical Understanding and Optimization of Building Energy Consumption and Climate Change Consequences

In the present paper, a new tool was designed to find a better alternative for improving building energy consumption in the next years. In this sense, in the first stage of this calculation procedure, ISO Standard 13790 calculation procedure was developed in accordance with Monte Carlo method and results showed the probability of energy consumption as a Weibull model. Furthermore, a map of different Weibull models in accordance with different input parameters and future climate change effect was developed as a future building design guide. This tool defines the probability of energy consumption of an existing building, or a building that is being designed today and in the near future, preventing the climate change effect. More other applications at the time of building retrofitting and healthy indoor ambiences are proposed.

Chapter 10. Studying Physical and Chemical Properties of Graphene Oxide and Reduced Graphene Oxide and Their Applications in Sustainable Building Materials

In this chapter, we identified Graphene Oxide OG/Reduced Graphene Oxide RGO as nano composites, by studying its nanoparticles' properties physically, chemically, and mechanically. In this study, we have mentioned the photo catalyst materials PC regarding carbon nanostructures such as GO and RGO which have excellent oxygen functionalities, efficient adsorption areas, and considerable surface area. The compositions of GO and RGO exceeds electron-holes pair reinstallation time and minimize energy hiatus by adjusting Valence Band Level (VBL) with Conducting Band Level (CBL) bringing high suction of the exist radiance which improves photo degeneration achievement of material oxides and composites made from polymers. We also studied the main applications of GO and RGO in engineering fields and summarizing usefulness of intercalation of GO and RGO in construction sectors. Moreover, many synthesis techniques lead to many types of GO. Therefore, in this chapter, we tried to collect most GO and RGO properties, structures, and applications.

Chapter 11. The Importance of Light in Our Lives: Towards New Lighting in Schools

The light that enters through our eyes is not only for vision. The human circadian system responds to light differently than the visual system. The timing of each biological function in mammals is directed by the main clock located in the Supraquiasmic Nucleus, which is regulated by light. However, until now, only the interaction of light with our visual system has been taken into account when choosing the parameters of indoor lighting sources, including those in the classroom. In the publications about school lighting, the first concern was the common parameters of indoor lighting such as horizontal workplane illuminance, illuminance uniformity and avoiding reflections on different surfaces. In this article we will show publications about new findings on the effects of light on people, studies carried out in different countries aimed at improving classroom lighting, current regulations on lighting related to classroom lighting and new parameters that are being considered, along with those already used, for new and better lighting.

Chapter 12. The Uses of Renewable Energy in Buildings: A Case Study of Australia – The Uses of Renewable Energy in Buildings

Regardless of the worldwide setting on energy, only 5% of Australia's energy utilization was gotten from renewables, while 86.3% of the electricity was produced from fossil fuels. Nonetheless, this pattern has been disturbed by the ongoing decommissioning and closure of old coal power plants, alongside the Australian's policy to reduce fossil fuel emissions. Presently, Australia is at a pivotal phase of its change to renewable energy power generation and utilization specifically in residential and commercial buildings. Sustainability in renewable energy utilization is being achieved through guided government policies, reasonable energy costs, and improved energy technology transfer approaches. To give a refreshed delineation of renewable energy, related government policy, and the route ahead in the Australian setting, this exploration paper presents a deliberate Australia update with renewable energy generation and utilization in Australian buildings.

Chapter 13. Water Efficiency Evaluation Analysis Among Environmental Certification Methods: LEED, BREEAM, DGNB, HQE, EDGE, and Bono Verde

Developing countries such as Peru are not strangers in joining this with efforts to promote sustainability in buildings, promoting through local certification programs such as My Green House funded by international organizations, or architectural design competitions for houses such as "Build to Grow", compliance standards such as the one published in March 2018 "Climate Change Law", in addition to other optional legal standards such as "EM.110 Thermal and Light Comfort with Energy Efficiency", a technical code of Voluntary Sustainable Construction and the entry of international environmental certifications such as EDGE and LEED, which are currently very welcomed by real estate developers due to the incentives, one of them is the height bonus, which are promoted by some municipal ordinances, mostly located in the capital city of Lima as a product of a project developed and promoted by the IFC World Bank and on the other hand in the Retail or Office sector, by green corporate policies, however still There is a long way to go.

The chapters address issues concerning indoor environmental quality (IEQ), which are described more simply as the conditions inside the building. This includes air quality, but also access to daylight and views, pleasant acoustic conditions, and occupant control over lighting and thermal comfort. It also include the functional aspects of the space, such as whether the layout provides easy access to tools and people when needed and whether there is sufficient space for the occupants. Building managers and operators can increase building occupant satisfaction by considering all aspects of IEQ rather than focusing on temperature or air quality alone.

Within these objectives, housing ventilation becomes a challenging goal to solve as it is directly related to occupants' health and wellbeing. The biggest challenge is its strong connection to energy demand as a big share of building energy losses are due to air renovations and enclosure infiltrations.

In this book, we look for a range of innovative research and evidence-based policy ideas that demonstrate the importance of housing research and policy for health and wellbeing.

Roberto A. González Lezcano
Universidad CEU San Pablo, Spain

Chapter 1
Application of Phase Change Materials in Construction Materials for Thermal Energy Storage Systems in Buildings

Eray Mertkan Meric
Istanbul University-Cerrahpasa, Turkey

Savas Erdem
https://orcid.org/0000-0003-0906-5233
Istanbul University-Cerrahpasa, Turkey

Ezgi Gurbuz
Istanbul University-Cerrahpasa, Turkey

ABSTRACT

There has been a continuous increase in world population and industrialization. In parallel, energy consumption grew substantially in the world. The residential buildings constitute the majority of the total consumption in the developed world. Therefore, energy saving and reducing heat losses of buildings are of major concern for our society. At this point, phase changing materials (PCMs) show up as one of the most useful materials in building design. In this study, PCMs, which are used at places such as facade of buildings, bricks, and inside of concrete in order to be able to reduce the energy consumption due to heating and cooling, to provide comfort temperature inside buildings, are extensively reviewed.

INTRODUCTION

The requirement for energy has been rising with the increasing population and developing economy. Most of the energy is consumed for heating and cooling purposes in buildings. Especially recently, with the increase of time spent at home, energy consumption levels have significantly increased. This causes

DOI: 10.4018/978-1-7998-7023-4.ch001

great environmental problems (Du et al., 2018; Cao et al., 2017). Depending on the climatic conditions, the amount of energy used in buildings for the heating and cooling purposes can constitute a remarkable portion of the whole consumed energy. It has been informed that buildings represent nearly 30% of the total amount of energy used throughout the year. Besides, 28% of the annual CO_2 emissions emerge from heating-cooling energy consumption in buildings (Hekimoğlu et al., 2021). The reduction of energy consumption can be achieved by minimizing the heating and cooling energy used in buildings. With the developing engineering technologies and innovative materials, energy consumption can be reduced considerably, and comfortable living temperature can be provided by less energy consumption in buildings (Cao et al., 2019). Energy-efficient design of the building not only enhances the indoor thermal behavior of the building, but it also increases the utilization of renewable sources of energy like solar energy and wind. Improving thermal energy storage capacity of the building is one of the efficient techniques of developing energy-efficient building. Generally, bricks, cement and concrete are used as sensible heat storage techniques to absorb, store, and release the heat energy in buildings (Rathore & Shukla, 2021).

There are three types of thermal energy storage system used for this purpose. These are latent heat storage, sensible heat storage and chemical reaction heat storage. Latent heat storage material can stock heat at almost constant temperature while undergoing a phase transition and store 5-14 times more energy per unit volume in comparison to sensible heat technique for storage. Thereby, the latent heat storage can be an influential method for thermal energy storage in buildings (Rathore & Shukla, 2021). Phase change materials based on latent heat storage are used in buildings because phase change materials (PCM) are in an advantageous position due to their high heat storage density and low level of temperature and volume changes during phase change (Xu & Li, 2014). Most of these PCMs retain their latent heat even after thousands of cycles without a change in their physical and chemical properties (Ling & Poon, 2013). Phase change materials, which have been considered as thermal storage materials since before 1980, have become a solution even for lightweight buildings due to the PCMs used in gypsum, plaster, concrete and other wall covering materials for thermal storage (Ling & Poon, 2013; Cabeza et al., 2007). Depending on their types, PCMs are compounds which absorb, store, and release the energy during the process of changing phase at certain temperature ranges. During hot times, PCM raises above the melting point and converts into liquid from solid, absorbing the heat. During cold times, when the temperature is under the PCM's melting point, PCM solidifies and the heat spreads back into the ambient. This solidifying and melting cycle proceeds depending on the temperature rates. PCMs keep buildings at ideal temperatures and reduce their heating and cooling requirements. This cycle is visualized in Figure 1.

Because of its high latent heat storage capacity, using Phase Change Material (PCM) for thermal energy storage has become a center of attraction among researchers and scientists more and more. Thus, to develop the energy efficiency of buildings, PCMs are often embedded with the existing thermal masses of the building like wallboard, bricks and concrete slab (Rathore & Shukla, 2021).

In this work, the type and characterization of Phase Change Materials for the applications of thermal energy storage in buildings have been extensively studied. This work emphasized that PCMs can have an important role over decreasing energy consuming and increasing temperature comfort in buildings.

Figure 1. Heating and cooling function for concrete wall merged with PCM to ensure agreeable temperature for people in the room (Ling & Poon, 2013)

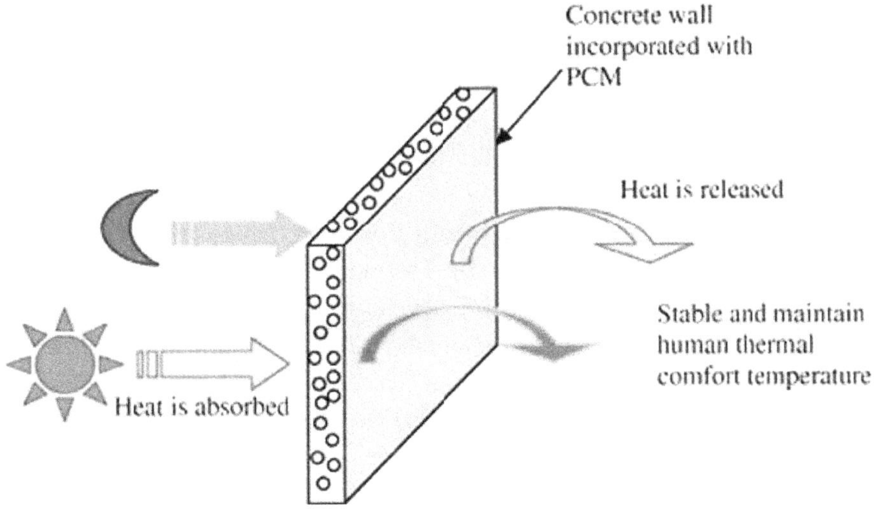

PHASE CHANGE MATERIALS AND TYPES

Phase change materials in accordance with their types are used to control temperature at certain ranges, storing heat energy. Phase change can be solid-solid, solid-liquid, solid-gas, and liquid-gas forms and vice versa. It is considered that only solid-liquid PCM can be used for heating and cooling (Faraj et al., 2020). When ambient temperature rises, the chemical bonds of the PCM which reaches its melting point break and liquefy from solid form. This is an endothermic reaction that provides energy emission. When the ambient temperature decreases, the bonds reappear, the PCM turns into solid and the heat is released (Baetens et al., 2010). As long as this process continues, a more comfortable living environment can be provided thanks to the PCM which balances internal temperature in buildings.

Among the three groups of PCMs which are solid-solid, solid-liquid and liquid-gas; solid-liquid group has the largest thermal energy storage capability. Solid-liquid PCMs can be classified as organic, inorganic, and eutectic (Zhou et al., 2012). Organic PCMs are divided into two groups as paraffin and non-paraffin. Non-paraffin ones are fatty acids, esters, alcohols, and glycols. Inorganic ones are classified as salt hydrate and metals. Eutectic PCMs are obtained by mixing organic-organic, inorganic-organic, or inorganic-inorganic PCMs. This classification is demonstrated in Figure 2.

The most commonly used inorganic PCM is salt hydrate that is both economic and a great thermal conductor. It is of very high heat storage and non-flammability property. Along with these advantages, it has some disadvantages such as super cooling, degradation in repeated phase changes, being corrosive and very high-volume change in solid-liquid state changes (da Cunha & Eames, 2016; Javadi et al., 2020; Souayfane et al., 2016). For these reasons, it is not suitable for inclusion in concrete (Ling & Poon, 2013; Gürbüz, 2018). Organic PCMs are categorized as paraffin and non- paraffin. They are generally chemically stable, non-corrosive and non-reactive. They do not cause any super cooling problem and have good nucleation property. In addition to these properties, although they are flammable and have low thermal conductivity, they are suitable for use with building materials (Chandel & Agarwal, 2017). Organic PCMs have properties like better chemical and thermal stability, suitable availability in various temperature ranges, high latent heat of fusion, also they are cheaper and non-corrosive. So, PCM is much preferred as thermal energy storage material. However, leakage and poor thermal conductivity restrict their applicability as a thermal energy storage material in buildings. Because the low thermal conductivity degrades the heat transfer rate and leakage during the phase transition leads to wastage of the PCM (Rathore & Shukla, 2021).

Non-paraffin organic PCMs have very good melting and freezing properties. These materials have very high latent heat storage and in flammability property. However, they cannot always provide the desired phase change at the comfort temperature like paraffin wax. Paraffin wax ($CH_3(CH_2)$ nCH_3) is economic, chemically stable, not reactive and is a hydrocarbon that has thermal storage intensity between 12 kj/kg and 20 kj/kg (Baetens et al., 2010). The most commonly used one is the organic PCM. Its melting point changes between 20 C and 70 C depending on the carbon atom in the chain. By this means intended melting temperatures can be obtained. It does not disintegrate in phase change cycles. It has a low thermal conductivity in solid form, and it is also inflammable (Ling & Poon, 2013; Baetens et al., 2010; Faraj et al., 2020; Chandel & Agarwal, 2017).

Eutectics are the mixtures that have sharp phase change temperature of two or more compounds with melting points that are as low as possible. Their volumetric storage intensity is relatively higher than that of organic compounds. They freeze and melt congruously without any phase exception (Baetens et al., 2010; Chandel & Agarwal, 2017; Tatsidjodoung et al., 2013).

Figure 2. Classifications of PCMs

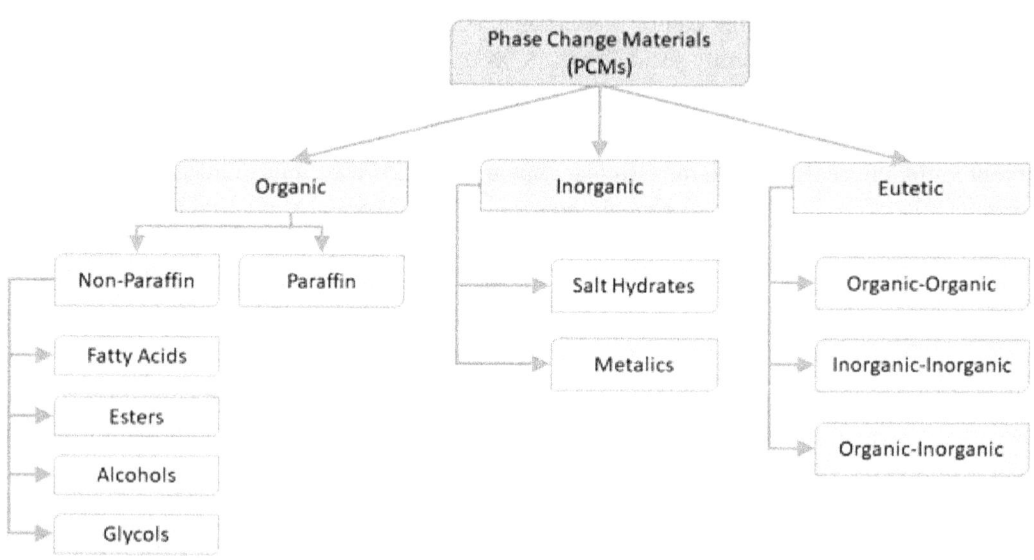

The phase transition temperatures of pure PCMs are constant. In addition, PCMs with different phase transition temperatures are required for some applications in practice. The most significant method used for this is the production of mixtures consisting of different PCMs. In other words, two or more PCMs are mixed in definite proportions, resulting in a new PCM with the desired phase transition temperature (Hekimoğlu et al., 2021). Working temperatures and melting enthalpy of various PCMs are demonstrated in Figure 3.

Figure 3. Working temperatures and melting enthalpy of various PCMs (Faraj et al., 2020)

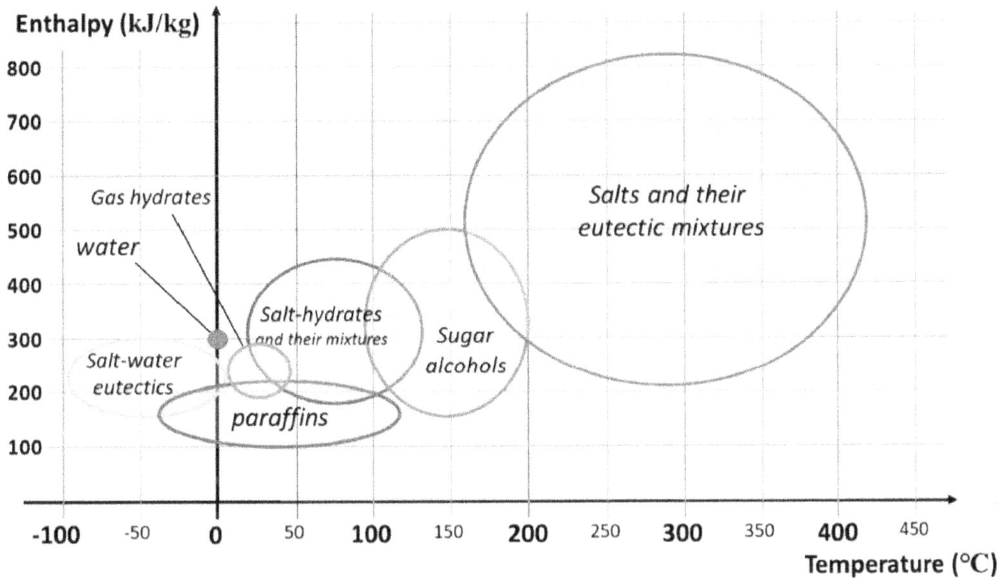

METHODS FOR PCM INCORPORATION IN COMPOSITES

After the selection of the most suitable PCM for the area where it is to be used, another important issue is how that material is added to the building material. PCM can be added to building materials in five different ways (Soares et al., 2013). These methods are as follows: direct incorporation, immersion microencapsulation and macro-encapsulation, shape-stabilization.

Direct Incorporation

The direct incorporation method is the simplest and most economical method in which liquid or powdered PCM is added directly during the production of building materials such as gypsum, concrete and plaster. The biggest problem in this method is that it may cause leakage as a result of incompatibility of some building materials and PCM (Faraj et al., 2020; Soares et al., 2013; Zhou et al., 2012).

Immersion

In immersion method porous structured materials such as concrete, brick and gypsum are immersed in melted PCM to fill the gaps with PCM through capillary effect. In this method also, leakage problem can be encountered in long term just as in direct incorporation method (Al-Yasiri & Szabó, 2021; Memon, 2013; Soares et al., 2013; Zhou et al., 2012).

Macro-Encapsulation and Micro-Encapsulation

Macro-encapsulation and microencapsulation methods, by placing PCMs into capsule, prevent some problems like leakages and interruption of interaction with the environment. This method is more suitable for long term use (Faraj et al., 2020; Soares et al., 2013; Zhou et al., 2012; Pasupathy et al., 2008).

Macro-encapsulation technology consists of capsulizing PCMs with the panels, globes, tubes that are bigger than 1 cm and other receptacles (Cabeza et al., 2011). These capsules are merged in building material to be used afterwards. Macro capsules should be designed particularly for the application. As a result of macro encapsulation, there occur organic polymers and silica in the outer layer. There is PCM in this layer (Faraj et al., 2020). It can also be used with inorganic PCM as it is resistant to abrasion (Whiffen & Riffat, 2013). It is a technology which has some disadvantages such as having low heat conductivity and solidification around the edges of the capsule (Kalnæs & Jelle, 2015; Soares et al., 2013; Zhou et al., 2012).

Microencapsulation technology emerges by covering the PCM with a thin and high molecular weighted polymeric film that prevents the leakage during the phase changing process (Figure 4). The size of these capsules ranges from a few micrometre to a few milimeter. Offering a greater surface area in proportion to its volume, it decreases low thermal conductivity effects of the PCM (Tyagi et al., 2011). The use of microencapsulated PCMs in building materials is both easier and more efficient. The PCM placed in microcapsules is used in building materials such as concrete, mortar, artificial marble, sealing materials and paints (Soares et al., 2013). Microencapsules can be created through both physical and chemical ways. Physical ways are spray drying, centrifugal extrusion, vibrational nozzle, solvent evaporation. And chemical ways are suspension polymerisation, mini-emulsion polymerisation (Soares et al., 2013; Whiffen & Riffat, 2013). However, micro-capsulation method has negative effects on mechanical properties such as compressive strength and bending strength.

Figure 4. A figuration of microencapsulated PCM

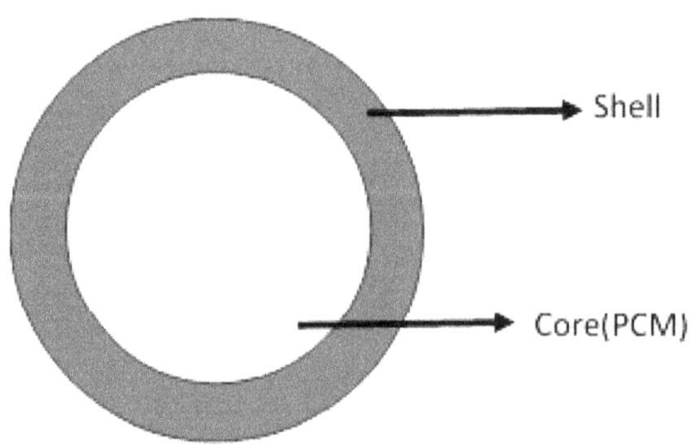

Shape Stabilization

Shape stabilization is a technique by which leakage is avoided and the energy storage ability of PCMs can be improved. Dispersed in another phase of supporting material such as high-density polyethylene, PCM forms a stabilized composite material which is called shape stabilized PCM. Having large apparent specific heat, sufficient thermal conductivity, the capability of keeping the shape of PCM stabilized in the process of phase changing and an acceptable performance of longstanding multiple thermal cycles, shape-stabilized PCMs have been studied in many researches recently (Zhou et al., 2012; Faraj et al., 2020; Kuznik et al., 2010; Soares et al., 2013; Jouhara et al., 2020).

THE USE OF PCM IN PASSIVE LATENT HEAT THERMAL ENERGY STORAGE SYSTEMS

Wallboards, Envelopes and Other Wall Applications

Wallboards suitable for use with PCMs are inexpensive and used in buildings commonly. The efficiency of these PCM-added Wallboards depends on the following factors: the type of the PCM incorporated into the wallboard, the melting temperature of the PCM, the melting temperature range of the PCM, the direction of the wall, climatic conditions, the color of the wallboard surface, latent heat capacity per unit area of the wallboard, direct solar gains etc (Baetens et al., 2010; Soares et al., 2013). Many studies have been done on wallboards containing PCM. Yu et al. (2020) combined two new shape-stabilized diatomite-based PCMs with organic alcohol PCMs to evaluate the chemical, morphological, and thermal performance properties of this mixture using Fourier transform infrared spectroscopy, scanning electron microscopy, and differential scanning calorimeter analysis. It is stated that the composite PCMs obtained had phase change temperatures between 33.1-35.1 °C and 41.7-42.7 °C and their latent temperatures are 90.5 ± 1.5 J/g and 92.6 ± 1.3 J/g. Wallboards created with these composite PCMs were placed on two different facades outside of a test room, and the importance of the position of the wallboard containing PCM and the effects of the phase transition temperature on the wallboard containing PCM were investigated in summer in order to reduce the cooling load. It has been determined that the surface temperatures of composite PCM wallboards are lower than traditional polystyrene plastic insulation wallboards. This is because traditional polystyrene plastic insulation wallboards have more favourable thermal storage capacity. It has been stated that the thermal resistance of this wallboard is too low to reduce indoor cooling energies in summer. It has been observed that the appropriate phase change temperatures for wallboards located on east and west facades are different.

Koo et al. (2011) investigated the effects of various PCM parameters such as nominal average phase change temperature, convective heat transfer coefficients and the wallboard thickness on the thermal storage performance of the wallboard. The PCM wallboard containing encapsulated organic phase change material is used. Wallboards containing PCM were observed by being placed them in a room with heaters and air conditioners. As a result, they have found that the average phase change temperature in wallboards should be close to room temperature to maximize thermal heat storage and the phase change temperature should be narrow. They have determined that thermal heat storage increases as the convective heat transfer coefficient increases.

Arıcı et al. (2020) conducted a numerical investigation to determine the place, layer thickness and melting temperature of the PCM that is used in building walls in three different locations in Turkey and different climatic conditions for maximum use of latent heat. They have stated that the melting temperature of PCM varies between 6-34 °C in layer thicknesses between 1-20 mm depending on climatic conditions.

Lee et al. (2018) examined the thermal performance by adding paraffin-based PCM into cellulose insulation in the building walls using the direct mixing method in their study. They built two identical test houses outdoor, one with PCM and the other without PCM, during the test process. At the end of the tests, it is determined that the daily average peak heat flux reduction is 25.4% for walls facing different directions and 20.1% for the total of four walls hourly.

Yang et al. (2018) investigated the thermal performance of lighweight wall panels through comparative experiments. They conducted the experiments in the readily- prepared artificial weathering room. They have placed PCM inside and outside light wall panels at different area ratios. They determined that the PCM layer placed on the inner surface can reduce the temperature fluctuation amplitude by 32.4% and increase the lag time from 2.9 hours to 4.1 hours, when compared to the PCM layer placed on the outer surface. With the thermal design method recommended for predicting the thermal performance of PCM envelopes and its early design, they were able to detect the internal surface temperature obtained from the experimental results with 7.7% relative error.

Boussaba et al. (2021) conducted an experimental and numerical study with the aim of improving the thermal comfort in buildings and to find the thermal, physico-chemical and mechanical characterization properties of the environmentally friendly and low-cost bio-PCM, which is a new composite recovered from the food industry. They included non-cocoa vegetable oil recovered from the confectionery manufacturing industry into the mineral sheet prepared from natural clay and cellulose fibers. They have found the melting freezing temperatures found with Differential Scanning Calorimetry as 34.83 °C, 22.34 °C and thermal energy storage capacity as 60.41 J / g, 62.39 J / g. In addition, the researchers who performed compressive strength tests before and after the addition of bio-PCM into the plate, observed that the added bio-PCM reduced the compressive strength. They have emphasized that the panel that has been created can be used in the buildings located in regions with hot or dry climates.

Piselli et al. (2020) researched the potential performance development of PCMs in passive cooling of natural ventilation control in buildings mentioning that using phase change materials in the building envelope is a potential passive energy strategy to increase energy efficiency. They made coupled dynamic simulation and optimization analysis to find out PCM's optimum melting temperature. They estimated the effects of PCM on the charge-discharge cycle using night-time ventilation and whole-day temperature-controlled ventilation control strategies. They also determined that adding PCM into the building envelope can provide energy saving up to 300 kWh/year in temperate climate and both natural ventilation controls can increase the efficiency in PCM's thermal energy storage charge-discharge cycle. They found that the highest cooling energy savings would be achieved in all climatic zones by using temperature-controlled natural ventilation method with optimized PCMs. They stated that energy saving will increase significantly with this method, especially in temperate climate.

Gómez et al. (2012) proposed a CFD (computational fluid dynamics)-based model to analyse the effect of phase change materials (PCMs) on the thermal behaviour of the cubicle walls exposed to the environmental conditions and on the resistance of the walls to climate changes. Walls are made of a mixture of self-compacting concrete and PCM. The cubicle is made of precast, self-compacting concrete panels, which are 2.4 m high, 2.6 m wide and 12 cm thick. They examined the effect of PCMs with a melting point of 26 ° C in the construction and experimental assembly. The concrete cell contains 5% PCM. Temperature sensors are built into concrete cells to record any temperature changes. The results of the simulation and experimental data are compared. As seen in Figure 5, the results show that the maximum temperature recorded in the normal concrete cell is 1 ° C higher than that of the concrete with PCM. In addition, the maximum temperature in the wall containing the PCM was 2 hours delayed compared to that in the non-PCM cell. Thus, it was concluded that the concrete cell containing PCM has better thermal and energy saving properties and increased thermal inertia than the one without PCM.

Figure 5. Evolution of the temperature field inside the cubicle and inside the walls (a) in the PCM (b) simulation (Gómez et al., 2012)

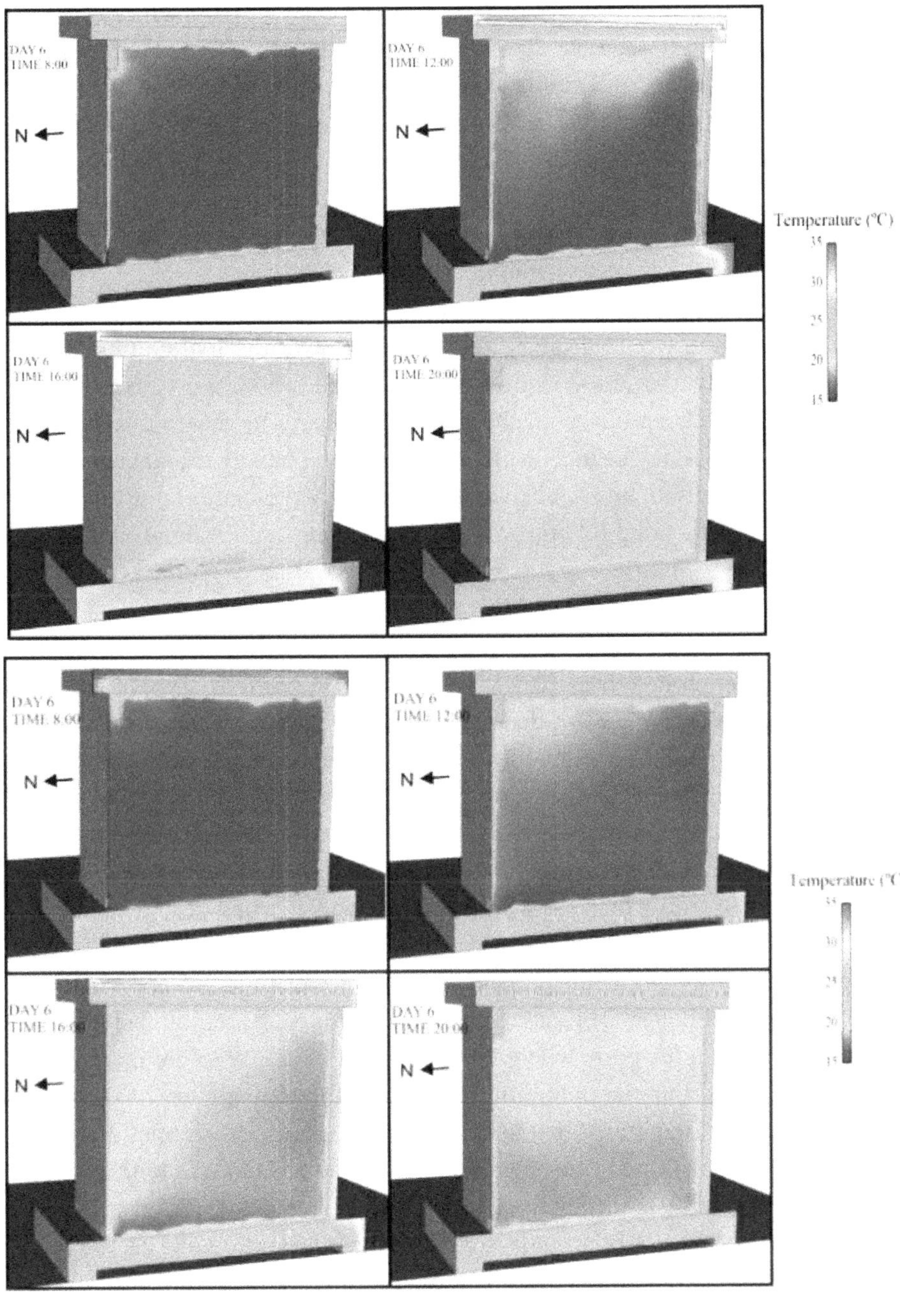

Bricks

Saxena et al. (2019) empirically examined temperature differences and heat flow changes that the bricks with PCMs would create mentioning the benefits of PCM for energy saving in Delhi. They used two different types of the PCM in the bricks. They chose eicosane and OM35 due to their suitability for

Delhi's climatic conditions. It has been found that the bricks containing PCM, tested in the hottest summer months, provide a temperature drop of 5-6 ° compared to conventional bricks. They also concluded that the heat flow decreased by 8% for eicosane and 12% for OM35 as a result of energy calculations.

Another study in a hot region was conducted by Mahdaoui et al. (2021). This study was carried out in a region of Morocco where the temperature varies between 45 °C during the day and 25 ° C at night. Physical modelling and numerical analysis of heat transfer were performed by adding n - Nonadecane, a kind of PCM, to hollow bricks. As a result, it is observed that it has increased the storage ability and insulating power of the bricks. At the same time, it was determined that the indoor temperature and temperature fluctuation were reduced. They also mentioned that the temperature comfort in buildings would increase and energy consumption would decrease.

Gao et al. (2020) mentioned that due to the use of lightweight materials in high-rise buildings, the thermal inertia of the structure would decrease, resulting in lower thermal comfort and higher energy consumption. With this motivation, they made a numerical study to improve the thermal behavior by putting PCM in bricks. It has been observed that the PCM clearly increases the thermal inertia of the wall to be created and increases its thermal insulation relatively. They found that when the phase change temperatures are at the appropriate level, it can reduce the PCM attenuation rate from 13.07% to 0.92% - 1.93%, increase the lag time from 3.83 hours to 8.83-9.83 hours and peak heat fluxes fall from 45.26 W/m2 to 19.19 W/m2-21.4 W/m2.

Li et al. (2017) created PCM bricks by adding shape stabilized PCM to cement mortar to improve thermal performance in mid-season and summer days. The wall with PCM content that has been created was observed for a long time in the 3.25m x 3.86m x 2.91m room by comparing it with the regular energy-saving wall in the specified seasons. It was observed that, when compared to the traditional wall in mid-season, between March 29-April 30, the surface temperature of the inner wall containing PCM fluctuated and the maximum value of the temperature was less. However, with the warming of the weather after April 30, the temperature begins to rise above the phase change temperature range of the PCM and they have determined that the inner wall surface temperature of the wall containing PCM is higher than the conventional wall which means it shows weaker thermal performance.

Tuncbilek et al. (2020) analysed the effect of the traditional bricks containing PCM on heating and cooling loads bricks using different fusion temperatures and also examined PCM's locations and quantities through numerical modelling in the climatic conditions of Marmara region, Turkey. They determined that the PCM samples filled in the parts where the brick gaps are close to the indoor environment will save more energy. They found that the optimum phase transition temperature of PCM changes seasonally in the range of 18-26 °C. They stated that in summer, more energy consumption occurred due to the inappropriate phase transition temperature selection. They emphasized that with the use of traditional bricks containing PCM, the annual thermal load is reduced by 17.6% and 13.2% of this reduction is due to the use of latent heat. At the end of the study, they said that conventional bricks containing PCM can reduce heating and cooling loads and increase thermal comfort.

Khedache et al. (2015) combined paraffin and red brick as PCMs with the conventional dispersion technique and focused on the determination and characterization of the thermal properties of the composite formed. They also used expanded graphite. They found that the maximum PCM mass fraction value retained in the red brick was 40% by weight. Although the PCM was heated over its melting point they did not recognize any leaks. They measured the melting temperature of the composite PCM as 52.73 °C, the freezing temperature as 55.80 °C and the latent heat as 41.39 J/g and 42.39 J/g with differential scanning calorimetry (DSC). They found that the thermal conductivity of the composite containing form-stable paraffin, red brick and 10% expanded graphite by weight was 3 times higher than pure paraffin. They concluded that form-stable paraffin and expanded graphite have significant latent heat energy storage potential.

Concrete and Mortar

Gürbüz and Erdem (2020) have worked to obtain a new material that can be used as facade or structural panels by adding MicroPCM to engineering cementitious composite mix containing 0.5% steel fiber and 1.5% polyvinyl alcohol fiber. By using MicroPCM and hybrid fiber ECC together, it is aimed to improve the thermal mass capacity, compressive strength, and tensile strength of this material. As a result of the study, it was determined that hybrid fibers significantly compensate the loss of compressive strength caused by MicroPCMs and it was also observed that the strength increased in some samples. In addition, it was concluded that with the increase in the amount of MicroPCM, there was no significant change in thermal conductivity values, but by reducing the temperature fluctuation in the building, the heating-cooling loads could be reduced and the thermal mass storage capability would increase.

In another study, Erdem and Gürbüz (2019) prepared five different ECC mixtures containing different amounts of MicroPCM and 0.5% steel fiber and 1.5% polyvinyl alcohol (PVA) to find out these samples' flexural behavior and the effect over the macro-mechanical damage behavior caused by impact load. Toughness indices, deflection at max load, flexural modulus, strength at first-cracking and post-cracking stages are determined in samples and in addition to these, they used 3D X-ray computed tomography technique in quantitative evaluation of microstructural damages caused by impact load. As a result of the impact test, it has been determined that the bond at the matrix-fiber interface was one of the most important factors in the damage behavior and also the damage was localized in the most critical area. Composite with high deformation capacity could be produced with high MicroPCM content. Then, this provided increased fracture strength under impact load. They also stated that in the sample in which the MicroPCM ratio increased from 1% to 2%, the energy absorption capability was increased up to 220%.

Hattan et al. (2021) used shape stabilized PCM (PEG 600) to provide thermal energy storage capacity and silica fume as supporting material in the cement mortar. The aim of the study is to examine the mechanical and thermal behavior of the prepared cement mortar. As a result of the experiments, they obtained positive results about using the mortars as plaster for the walls. They concluded that the PCM-containing mortar material is effective in lowering the peak temperature, would reduce the temperature fluctuations in the environment and provide thermal energy storage capacity.

Bahrar et al. (2018) designed various mixtures of new textile reinforced concrete panels containing microencapsulated PCM to improve energy efficiency in buildings. They performed multi-scale experimental characterization to determine the effect of these microencapsulated PCM added panels on the thermal behavior. To measure the thermal conductivity, they did a stable state test in the guarded hot box facility. As a result, they found that microencapsulated PCM particles lead to a decrease in thermal conductivity and as the PCM ratio increases it decreases comparatively. As a result of the experiments, they found that the melting temperature of PCM was between 23-27 ° C. They recognized an increase in the heat storage capacity and thermal inertia of the panel. They verified the results by developing a numerical model.

Hekimoğlu et al. (2021) produced cement mortars with thermal energy storing/releasing properties, containing silica fume (SF)/capric acid-stearic acid eutectic mixture (CA-SA) as form-stable composite phase change material (FSC-PCM). To achieve new type cementitious composite mortar, the FSC-PCM was replaced with ordinary cement mortar (OCM) in weight fraction of 10%, 15% and 20%. According to the data obtained from the test setup results shown in Figure 6, the maximum indoor temperature differences in the OCM based and latent heat storage-cement mortar (LHS-CM) based rooms were found as 2.48 °C during the heating stage and 1.71 °C during the cooling stage. Also, the SF/CA-SA composite produced as FSC-PCM was of relatively high TES capacity (about 66 J/g) for solar passive heating and cooling applications in buildings. Mechanical test findings of the LHS-CMs showed that sufficient mechanical properties and new type cementations composite mortars have suitable properties for regulation of indoor temperatures and reducing energy consumption in buildings.

Figure 6. The schema for thermoregulation performance test (Hekimoğlu et al., 2021)

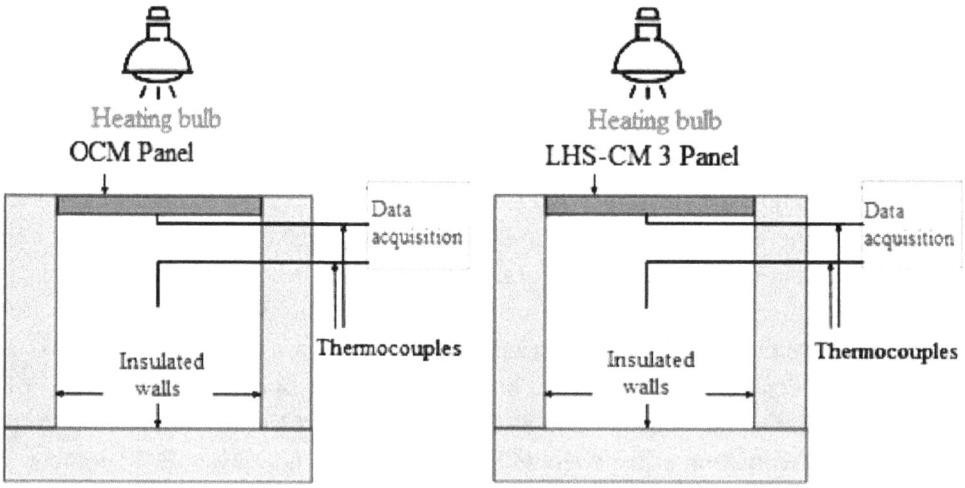

Yang et al. (2020) prepared ceremide-based shape stabilized PCM and embedded it in concrete. Six concrete based-CPCM blocks were prepared following the Chinese national standard of concrete producing with a varying mass fraction of 0%, 3%, 6%, 10%, 15%, and 20%. The size of concrete based-CPCM blocks are 70.7 x 70.7 x 70.7 mm for the stress test. According to the test results, the apparent density of the PCM added concrete blocks decreased compared with the blocks without PCM. For the bearing capacity, the maximum stress that the block can bear decreased with the increase of CPCM fraction. The low intensity of the CPCM reduced the global intensity of the block. In the continuation of the study, they made concrete panels to investigate the thermal performances of the concrete with PCM (15% of CPCM) and without PCM (normal concrete). The samples produced are as shown in Figure 7. According to the measured results, the temperature of the hot side and cold side of the concrete based CPCM panel is lower than normal concrete panel. Thus, if the concrete thickness increases, the thermal energy storage performance will increase.

Ren et al. (2021) developed a novel structural–functional integrated concrete (MPCM-UHPC) by incorporating microencapsulated phase change material (MPCM) into ultra-high-performance concrete (UHPC) for excellent mechanical properties and thermal storage capacity. According to the mechanical properties and autogenous shrinkage results, any chemically phenomenological interaction between MPCM and cementitious components occurs in the hydration process. Because of increased air voids content in cementitious matrix, harmful effect has been formed on the mechanical properties and autogenous shrinkage of MPCM-UHPC. The maximum surface temperature of UHPC with 10% MPCM was 3.9 °C lower than control mix. Also, compressive strength was obtained from the compressive strength test as 78 MPa with the reduction of 35% in 10% MPCM in UHPC. So, the MPCM-UHPC composite is shown as a promising building material.

Figure 7. Concrete based-CPCM blocks and panels (Yang et al., 2020)

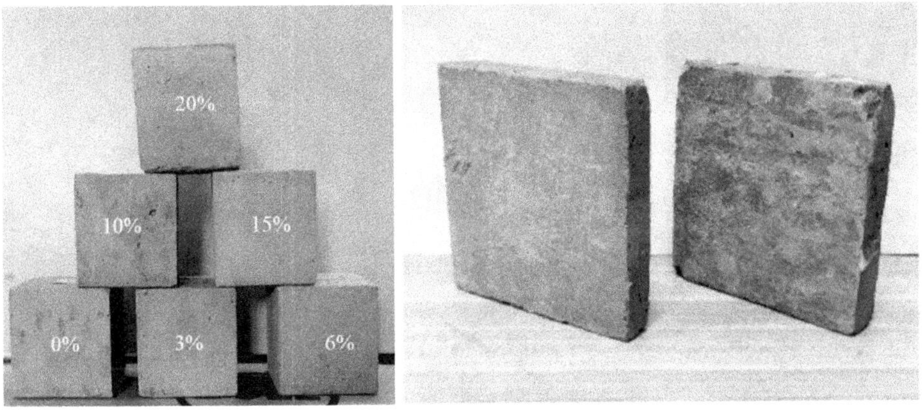

Parameshwaran et al. (2021) aims at improving a novel microencapsulated bio-based phase change material (MbP) integrated into a micro concrete composite (MbPMC) so as to store thermal energy in buildings. The surface morphology results figured that, the MbP particles were near spherical in shape with sizes ranging between 2 μm and 10 μm. It was observed that the rate of decrease in compressive strength was very marginal at rates after 0.025% MbP. The as-prepared MbPMC has displayed sufficient structural integrity with a compressive strength of 38.78 MPa at a MbP dosage of 0.075% by the weight of cementitious materials added in the mix. Ultrasonic pulse velocities (UPV), for the directions orthogonal to the direction of pour of the concrete specimens, were exhibited to be very close, thus proving that the densities, across the cross section of the specimen are conservatively uniform. When the morphology of the MbPMC powder obtained after the compressive strength test was examined (Figure 8), it was observed that the microcapsules were not broken/deteriorated even after the experiment.

Problems Caused by PCMs in Concrete

Although the use of phase change materials in concrete is beneficial for reducing energy consumption, it is a situation that affects the mechanical properties of concrete. Hunger et al. observed the reduction in the compressive strength of self-compacting concrete containing microencapsulated PCM with the increase in the amount of PCM, it is shown in Figure 9. Adding PCM to concrete through the direct incorporation and immersion method will cause leakage in the concrete with the transition of the PCM in the hollow structure concrete to the liquid phase. Thus, the amount of PCM in concrete would be reduced and when the PCM hardens again, the thermal surface area would be reduced. It should also be remembered that PCM can react chemically with concrete (Gürbüz, 2018). As previously mentioned, the leakage problems have been solved by encapsulation methods. However, there are some states that need to be considered and can cause problems while mixing concrete in the capsulation technique. The hydrophobic microcapsule can absorb water while mixing into concrete. As stated in the studies outlined, it may cause problems that would reduce the mechanical properties of concrete.

Figure 8. Surface morphology of A) Reference Specimen, B) MbPMC-1, C) MbPMC-2, D) MbPMC-3 and E) MbPMC-4 (Parameshwaran et al, 2021)

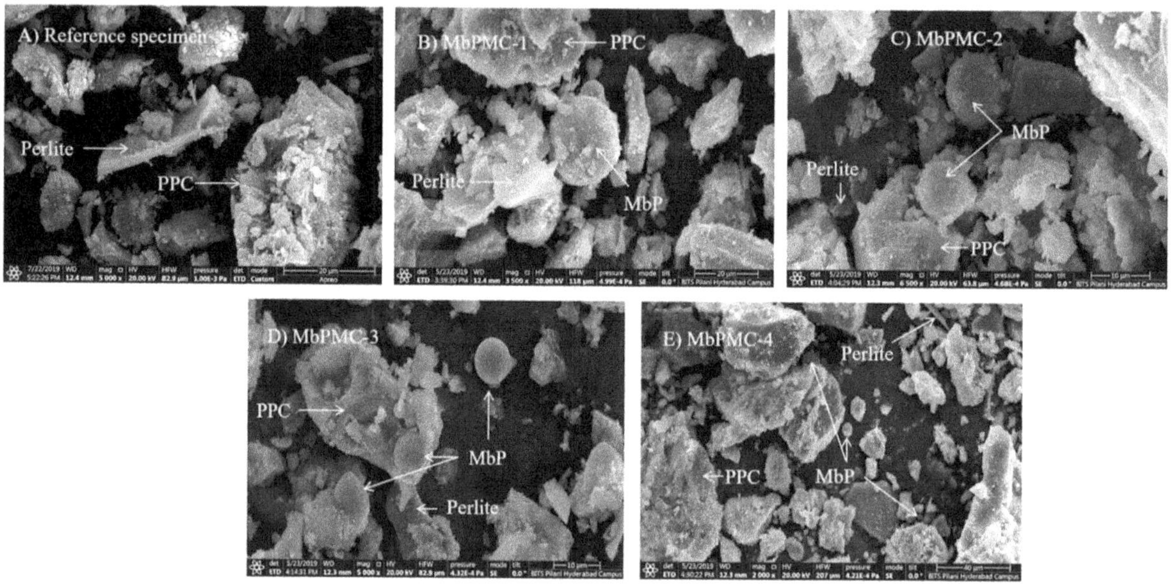

Figure 9. The compressive strengths of the concretes containing different amounts of PCM (Hunger et al., 2009)

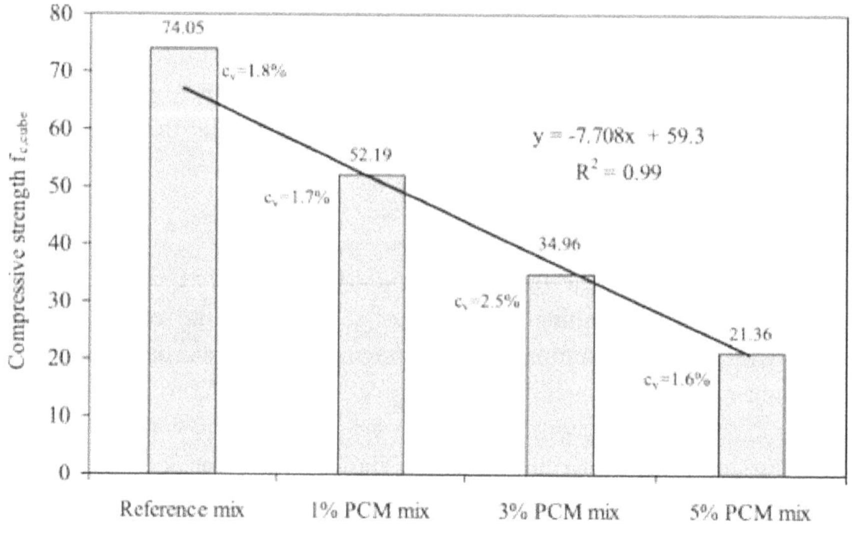

Studies have been carried out on different types of concrete. Cao et al. (2017) incorporated 3.2% and 2.7% microencapsulated phase change material (MPCM) into Portland cement concretes and geopolymer concretes. According to the results, the compressive strength decreased by 42% and 51%, respectively. However, the compressive strength even so satisfies the mechanical European regulation (EN 206-1, compressive strength class C20/25) for concrete applications. Besides, MPCM-concrete ensured a good thermal stability after subjecting the samples to 100 thermal cycles at high heating/cooling rates.

FUTURE RESEARCH DIRECTIONS

In this study, the wide range of information on PCMs and their use in construction materials for the thermal comfort and energy efficiency in buildings are culled from the literature works and summarized. As it is well-established by the literature studies, the leakage problems of PCMs during mixing or under mechanical loading limits their wider application in construction projects. In this term, a future will be conducted on the nano-encapsulation, in particular via environmentally friendly biopolymers, of PCMs to minimize the leakage and, thus increasing the effectiveness of these materials. Similarly, the hybridization of PCMs and aerogels materials would be great for reducing heat loss and improving energy efficiency on the building facades. Finally, a numerical modelling associated with real site data obtained from a range of different buildings located in different environmental conditions will be beneficial for the future research direction.

CONCLUSION

Studies on PCMs are very important nowadays when most of the energy consumption is made for heating and cooling purposes in buildings. PCMs can significantly reduce energy consumption and at the same time provide comfort temperature inside the building. This study, which refers to the types and properties of latent heat storage-based phase change materials and summarizes the numerical and experimental studies on these materials, has shown that PCMs can play an important role in reducing energy consumption in buildings. In addition, it has been emphasized in the studies that PCMs can increase the temperature comfort in buildings. It is very important that PCM, which is used in buildings with a wide variety of methods, is suitable for climatic conditions, and the type of application should be selected according to the climate. The position of the PCMs determined according to the climatic conditions during the application is also effective.

Consequently, the contribution of this study to existing knowledge is the provision of an approach to determine material's behavior when designing new buildings or when structurally evaluating or strengthening existing building elements. Further, cementitious composites produced with PCMs has a great potential for reducing the emission of greenhouse gases and the surface temperature of the structural elements. However, it should not be forgotten that PCMs may lead to the large reductions in mechanical strengths.

REFERENCES

Al-Yasiri, Q., & Szabó, M. (2021). Incorporation of phase change materials into building envelope for thermal comfort and energy saving: A comprehensive analysis. *Journal of Building Engineering*, *36*, 1–23. doi:10.1016/j.jobe.2020.102122

Arıcı, M., Bilgin, F., Nižetić, S., & Karabay, H. (2020). PCM integrated to external building walls: An optimization study on maximum activation of latent heat. *Applied Thermal Engineering*, *165*, 1–13. doi:10.1016/j.applthermaleng.2019.114560

Baetens, R., Jellea, B. P., & Gustavsend, A. (2010). Phase change materials for building applications: A state-of-the-art review. *Energy and Building*, *42*(9), 1361–1368. doi:10.1016/j.enbuild.2010.03.026

Bahrar, M., Djamai, Z. I., El Mankibi, M., Larbi, A. S., & Salvia, M. (2018). Numerical and experimental study on the use of microencapsulated phase change materials (PCMs) in textile reinforced concrete panels for energy storage. *Sustainable Cities and Society*, *41*, 455–468. doi:10.1016/j.scs.2018.06.014

Boussaba, L., Lefebvre, G., Makhlouf, S., Grados, A., & Royon, L. (2021). Investigation and properties of a novel composite bio-PCM to reduce summer energy consumptions in buildings of hot and dry climates. *Solar Energy*, *214*, 119–130. doi:10.1016/j.solener.2020.11.060

Cabeza, L. F., Castell, A., Barreneche, C., de Gracia, A., & Fernández, A. I. (2011). Materials used as PCM in thermal energy storage in buildings: A review. *Renewable & Sustainable Energy Reviews*, *15*(3), 1675–1695. doi:10.1016/j.rser.2010.11.018

Cabeza, L. F., Castellón, C., Nogués, M., Medrano, M., Leppers, R., & Zubillaga, O. (2007). Use of microencapsulated PCM in concrete walls for energy savings. *Energy and Building*, *39*(2), 113–119. doi:10.1016/j.enbuild.2006.03.030

Cao, V. D., Pilehvar, S., Salas-Bringas, C., Szczotok, A. M., Rodriguez, J. F., Carmona, M., Al-Manasir, N., & Kjøniksen, A. L. (2017). Microencapsulated phase change materials for enhancing the thermal performance of Portland cement concrete and geopolymer concrete for passive building applications. *Energy Conversion and Management*, *133*, 56–66. doi:10.1016/j.enconman.2016.11.061

Cao, V. D., Pilehvara, S., Salas-Bringas, C., Szczotok, A. M., Buia, T. Q., Carmona, M., Rodriguez, J. F., & Kjøniksen, A. L. (2019). Thermal analysis of geopolymer concrete walls containing microencapsulated phase change materials for building applications. *Solar Energy*, *178*, 295–307. doi:10.1016/j.solener.2018.12.039

Chandel, S. S., & Agarwal, T. (2017). Review of current state of research on energy storage, toxicity, health hazards and commercialization of phase changing materials. *Renewable & Sustainable Energy Reviews*, *67*, 581–596. doi:10.1016/j.rser.2016.09.070

Da Cunha, J. P., & Eames, P. (2016). Thermal energy storage for low and medium temperature applications using phase change materials – A review. *Applied Energy*, *177*, 227–238. doi:10.1016/j.apenergy.2016.05.097

Du, K., Calautit, J., Wang, Z., Wu, Y., & Liu, H. (2018). A review of the applications of phase change materials in cooling, heating and power generation in different temperature ranges. *Applied Energy*, *220*, 242–273. doi:10.1016/j.apenergy.2018.03.005

Erdem, S., & Gürbüz, E. (2019). Influence of microencapsulated phase change materials on the flexural behavior and micromechanical impact damage of hybrid fibre reinforced engineered cementitious composites. *Composites. Part B, Engineering*, *166*, 633–644. doi:10.1016/j.compositesb.2019.02.059

Faraj, K., Khaled, M., Faraj, J., Hachem, F., & Castelain, C. (2020). Phase change material thermal energy storage systems for cooling applications in buildings: A review. *Renewable & Sustainable Energy Reviews*, *119*, 79–97. doi:10.1016/j.rser.2019.109579

Gao, Y., He, F., Meng, X., Wang, Z., Zhang, M., Yu, H., & Gao, W. (2020). Thermal behavior analysis of hollow bricks filled with phase-change material (PCM). *Journal of Building Engineering*, *31*, 1–11. doi:10.1016/j.jobe.2020.101447

Gómez, M. A., Feijoo, M. A. A., Comesaña, R., & Eguía, P. (2012). CFD Simulation of a Concrete Cubicle to Analyze the Thermal Effect of Phase Change Materials in Buildings. *Energies*, *5*(7), 2093–2111. doi:10.3390/en5072093

Gürbüz, E. (2018). *Investigation of Mechanical and Impact Behavior of Engineered Cement-Based Composites Incorporating MicroPCM* (Master Thesis). Istanbul University, Turkey.

Gürbüz, E., & Erdem, S. (2020). Development and thermo-mechanical analysis of high-performance hybrid fibre engineered cementitious composites with microencapsulated phase change materials. *Construction & Building Materials*, *263*, 139–152. doi:10.1016/j.conbuildmat.2020.120139

Hattan, H. A., Madhkhan, M., & Marani, A. (2021). Thermal and mechanical properties of building external walls plastered with cement mortar incorporating shape-stabilized phase change materials (SSPCMs). *Construction & Building Materials*, *270*, 1–13. doi:10.1016/j.conbuildmat.2020.121385

Hekimoğlu, G., Nas, M., Ouikhalfan, M., Sarı, A., Tyagi, V. V., Sharma, R. K., Kurbetci, Ş., & Saleh, T. A. (2021). Silica fume/capric acid stearic acid PCM included-cementitious composite for thermal controlling of buildings: Thermal energy storage and mechanical properties. *Energy*, *219*, 1–10. doi:10.1016/j.energy.2020.119588

Hunger, M., Entrop, A. G., Mandilaras, I., Brouwers, H. J. H., & Founti, M. (2009). The behavior of self-compacting concrete containing micro-encapsulated Phase Change Materials. *Cement and Concrete Composites*, *31*(10), 731–743. doi:10.1016/j.cemconcomp.2009.08.002

Javadi, F. S., Metselaar, H. S. C., & Ganesan, P. (2020). Performance improvement of solar thermal systems integrated with phase change materials (PCM), a review. *Solar Energy*, *206*, 330–352. doi:10.1016/j.solener.2020.05.106

Jouhara, H., Żabnienska-Góra, A., Khordehgah, N., Ahmad, D., & Lipinski, T. (2020). Latent thermal energy storage technologies and applications: A review. *International Journal of Thermofluids*, *5-6*, 39–63. doi:10.1016/j.ijft.2020.100039

Kalnæs, S. E., & Jelle, B. P. (2015). Phase change materials and products for building applications: A state-of-the-art review and future research opportunities. *Energy and Building*, *94*, 150–176. doi:10.1016/j.enbuild.2015.02.023

Khedache, S., Makhlouf, S., Djefel, D., Lefebvre, G., & Royon, L. (2015). Preparation and thermal characterization of composite "Paraffin/Red Brick" as a novel form-stable of phase change material for thermal energy storage. *International Journal of Hydrogen Energy*, *40*(39), 13771–13776. doi:10.1016/j.ijhydene.2015.03.075

Koo, J., So, H., Hong, S. W., & Hong, H. (2011). Effects of wallboard design parameters on the thermal storage in buildings. *Energy and Building*, *43*(8), 1947–1951. doi:10.1016/j.enbuild.2011.03.038

Kuznik, F., David, D., Johannes, K., & Roux, J. J. (2010). A review on phase change materials integrated in building walls. *Renewable & Sustainable Energy Reviews*, *15*(1), 379–391. doi:10.1016/j.rser.2010.08.019

Lee, K. O., Medina, M. A., Sun, X., & Jin, X. (2018). Thermal performance of phase change materials (PCM)-enhanced cellulose insulation in passive solar residential building walls. *Solar Energy*, *163*, 113–121. doi:10.1016/j.solener.2018.01.086

Li, L., Yu, H., & Liu, R. (2017). Research on composite-phase change materials (PCMs)-bricks in the west wall of room-scale cubicle: Mid-season and summer day cases. *Building and Environment*, *123*, 494–503. doi:10.1016/j.buildenv.2017.07.019

Ling, T. C., & Poon, C. S. (2013). Use of phase change materials for thermal energy storage in concrete: An overview. *Construction & Building Materials*, *46*, 55–62. doi:10.1016/j.conbuildmat.2013.04.031

Mahdaoui, M., Hamdaoui, S., Msaad, A. A., Kousksou, T., El Rhafiki, T., Jamil, A., & Ahachad, M. (2021). Building bricks with phase change material (PCM): Thermal performances. *Construction & Building Materials*, *269*, 1–12. doi:10.1016/j.conbuildmat.2020.121315

Memon, S. A. (2013). Phase change materials integrated in building walls: A state of theart review. *Renewable & Sustainable Energy Reviews*, *31*, 870–906. doi:10.1016/j.rser.2013.12.042

Parameshwaran, R., Naresh, R., Ram, V.V., & Srinivas, P.V. (2021). Microencapsulated bio-based phase change material-micro concrete composite for thermal energy storage. *Journal of Building Engineering*, 1-39. doi:10.1016/j.jobe.2021.102247

Pasupathy, A., Velraj, R., & Seeniraj, R. V. (2008). Phase change material-based building architecture for thermal management in residential and commercial establishments. *Renewable & Sustainable Energy Reviews*, *12*(1), 39–64. doi:10.1016/j.rser.2006.05.010

Pilehvar, S., Sanfelix, S. G., Szczotok, A. M., Rodríguez, J. F., Valentini, L., Lanzón, M., Pamies, R., & Kjøniksen, A. L. (2020). Effect of temperature on geopolymer and Portland cement composites modified with Micro-encapsulated Phase Change materials. *Construction & Building Materials*, *252*, 55–66. doi:10.1016/j.conbuildmat.2020.119055

Piselli, C., Prabhakar, M., de Gracia, A., Saffari, M., Pisello, A. L., & Cabeza, L. F. (2020). Optimal control of natural ventilation as passive cooling strategy for improving the energy performance of building envelope with PCM integration. *Renewable Energy*, *162*, 171–181. doi:10.1016/j.renene.2020.07.043

Rathore, P. K. S., & Shukla, S. K. (2021). Enhanced thermophysical properties of organic PCM through shape stabilization for thermal energy storage in buildings: A state of the art review. *Energy and Building, 236*, 1–29. doi:10.1016/j.enbuild.2021.110799

Ren, M., Wen, X., Gao, X., & Liu, Y. (2021). Thermal and mechanical properties of ultra-high performance concrete incorporated with microencapsulated phase change material. *Construction & Building Materials, 273*, 1–10. doi:10.1016/j.conbuildmat.2020.121714

Saxena, R., Rakshit, D., & Kaushik, S. C. (2019). Phase change material (PCM) incorporated bricks for energy conservation in composite climate: A sustainable building solution. *Solar Energy, 183*, 276–284. doi:10.1016/j.solener.2019.03.035

Soares, N., Costa, J. J., Gaspar, A. R., & Santos, P. (2013). Review of passive PCM latent heat thermal energy storage systems towards buildings' energy efficiency. *Energy and Building, 59*, 82–103. doi:10.1016/j.enbuild.2012.12.042

Souayfane, F., Fardoun, F., & Biwole, P. H. (2016). Phase change materials (PCM) for cooling applications in buildings: A review. *Energy and Building, 129*, 396–431. doi:10.1016/j.enbuild.2016.04.006

Tatsidjodoung, P., Le Pierrès, N., & Luo, L. (2013). A review of potential materials for thermal energy storage in building applications. *Renewable & Sustainable Energy Reviews, 18*, 327–349. doi:10.1016/j.rser.2012.10.025

Tunçbilek, E., Arıcı, M., Bouadila, S., & Wonorahardjo, S. (2020). Seasonal and annual performance analysis of PCM-integrated building brick under the climatic conditions of Marmara region. *Journal of Thermal Analysis and Calorimetry, 141*(1), 613–624. doi:10.100710973-020-09320-8

Tyagi, V. V., Kaushik, S. C., Tyagi, S. K., & Akiyama, K. (2011). Development of phase change materials based microencapsulated technology for buildings: A review. *Renewable & Sustainable Energy Reviews, 15*(2), 1373–1391. doi:10.1016/j.rser.2010.10.006

Whiffen, T. R., & Riffat, S. B. (2013). A review of PCM technology for thermalenergy storage in the built environment:Part I. *The International Journal of Low Carbon Technologies, 8*(3), 147–158. doi:10.1093/ijlct/cts021

Xu, B., & Li, Z. (2014). Performance of novel thermal energy storage engineered cementitious composites incorporating a paraffin/diatomite composite phase change material. *Applied Energy, 121*, 114–122. doi:10.1016/j.apenergy.2014.02.007

Yang, L., Qiao, Y., Liu, Y., Zhang, X., Zhang, C., & Liu, J. (2018). A kind of PCMs-based lightweight wallboards: Artificial controlled condition experiments and thermal design method investigation. *Building and Environment, 144*, 194–207. doi:10.1016/j.buildenv.2018.08.020

Yang, Y., Wu, W., Fu, S., & Zhang, H. (2020). Study of a novel ceramsite-based shape-stabilized composite phase change material (PCM) for energy conservation in buildings. *Construction & Building Materials, 246*, 1–9. doi:10.1016/j.conbuildmat.2020.118479

Yu, H., Li, C., Zhang, K., Tang, Y., Song, Y., & Wang, M. (2020). Preparation and thermophysical performance of diatomite-based composite PCM wallboard for thermal energy storage in buildings. *Journal of Building Engineering*, *32*, 1–11. doi:10.1016/j.jobe.2020.101753

Zhou, D., Zhao, C. Y., & Tian, Y. (2012). Review on thermal energy storage with phase change materials (PCMs) in building applications. *Applied Energy*, *92*, 593–605. doi:10.1016/j.apenergy.2011.08.025

Chapter 2

Contribution of Water Flow Glazing to Net-Zero Energy Buildings:

Validation of Analytical and Numerical Models Through Experimental Data

Fernando del Ama Gonzalo
https://orcid.org/0000-0001-8366-5910
Keene State College, USA

Belen Moreno
Technical School of Architecture of Madrid, Technical University of Madrid, Spain

Matthew Griffin
Keene State College, USA

Juan Antonio Hernandez Ramos
School of Aeronautical and Space Engineering, Technical University of Madrid, Spain

ABSTRACT

Net-zero energy buildings (NetZEBs) are of a building typology designed to combine energy efficiency and renewable energy generation to consume only as much energy as produced onsite through renewable resources over a specified time. The successful creation of NetZEBs is crucial to combating the current climate crisis. Water flow glazing (WFG) is a key technology that will assist in achieving this goal. Several experimental facilities have been designed and constructed to collect data based on WFG technology. These experimental facilities demonstrate that the successful implementation of WFG will allow reducing heating and cooling loads, primary energy consumption, and CO2 emissions. However, a wrong WFG selection can lead to failure in NetZEBs design. The goal of this text was to assess WFG performance through key performance indicators to understand the need of other renewable energies so that the construction of NetZEBs becomes a realistic target.

DOI: 10.4018/978-1-7998-7023-4.ch002

INTRODUCTION

The Energy Performance Buildings Directive (EPBD) promotes policies that will produce highly energy-efficient and decarbonized structures by 2050. Starting December 31st of 2020, all new buildings will have to be Nearly Zero Energy Buildings. According to this directive, Zero Energy Buildings are defined as those that have a very low energy yearly energy consumption, which can be achieved by both highest energy efficiency and by energy from renewable sources, which shall be on-site or nearby (European Union, 2018). These balance concepts are hardly comparable, do not typically represent national standards, and differ for several reasons. The word "Net" suggests that the mission of the Zero Energy Building is to produce a balance between energy needs and energy consumption, as well as the production of energy and exportation of this energy to the greater power grid. The goal of Net-Zero Energy Buildings (NetZEB) is quite simple: to produce a neutral result for energy or emission balance, which should be created within the time span of one year. This goal can usually be achieved in buildings by implementing a two-step concept: reducing energy demand (via implementing passive means), as well as producing energy on-site (through photo-voltaic panels, wind turbines or Solar Thermal Collectors).

Several different variables must be taken into consideration when accounting for NetZEB energy. These variables could include but are not limited to heating, ventilation, and air-conditioning (HVAC) systems, domestic hot water (DHW), lighting (both indoor and outdoor), plug loads, and transportation within the building itself. Several factors have to be taken into account to set a global standard to define a NetZEB. Firstly, the choice of key performance indicators (i.e., final energy, primary energy, equivalent carbon emissions, and energy costs). Secondly, the "accounting system" used to classify the energy demands into sectors that are included in the balance (i.e., HVAC systems, domestic hot water, and lighting). Finally, we must look at "conversion factors" regarding the chosen metric (political factors, asymmetric weighting factors, and time-dependent conversion factors). The project "Industrial Development of Water Flow Glazing Systems" InDeWaG, supported by program Horizon 2020-EU.3.3.1, was committed to reducing energy consumption and carbon footprint by smart and sustainable use of the building envelope. The research (and furthermore, the project as a whole) assumed that achieving the goals of greater efficiency, aesthetic quality, and commercial renovation could only be accomplished by introducing a broad set of technologies that would encompass all of the various requirements of a building: energy, structure, and function. One of the goals of InDeWaG was to provide the building industry with a new unitized façade typology that, when coupled with a plug and play piping system, would produce a high-performance building envelope and innovative heating and cooling system.

The first goal of this chapter is to validate water flow glazing technology using real data gathered from the prototypes built within the European Project "Industrial Development of Water Flow Glazing Systems" (InDeWaG). The second goal is to develop key performance indicators to assess the performance of water flow glazing and its potential to be used in NetZEB design. The text defines some architectural issues for using Water Flow Glazing in NetZEBs design, which implies re-think the way buildings are designed.

Background

The urban planning regulations will have to analyze the energy footprint as a part of the future building design. It is a significant change since the traditional architectural design does not consider energy requirements as an input, but rather a mandatory requirement imposed by law. Sartori et al. (2012) and

Peterson et al. (2015) proposed a definition for Net-Zero Energy Buildings. This definition used the term NetZEB to refer to buildings connect to civil energy infrastructure. "Net", in this case, referred to using the energy from the electric grid (as well as other sources) to transfer excess on-site power to other locations, ensuring a balance in energy taken from the site and supplied back to the site over a time period. Andresen (2018) used previous works to present a weighting system that converts physical energy units to standard metrics (i.e. primary energy or carbon equivalent emissions). The Net-Zero Energy Building balance here is satisfied when weighted supply was equal to demand over a specific time, typically a span of a year. Boemi and Tziogas (2016) demonstrated in previous texts, existing indicators of energy efficiency, grouping these into five different parameters. Passive design strategies make NetZEB need very little energy. The design has to consider the floor area and the space required to provide for electrical and thermal energy from renewable sources and the surface necessary for placing the energy generation devices. This area is the energy footprint. The decision on the envelope design has to combine a low-energy demand passive performance, the capability of adapting to variable outdoor conditions, and the ability to produce the total energy that the building needs (Casini, 2015). Ghosh et al. (2015) showed that some dynamic envelopes react to diverse outdoor conditions by changing their opacity or varying transmission or reflection properties. The work of Abdulmohsin et al. (2019) and Allen et al. (2017) has shown polymer dispersed liquid crystal (PDLC), Electrochromic (EC) glass, and Suspended Particle Devices (SPD) as examples of dynamic technologies. The critical aspects remain in the high cost and a lack of standardization in the manufacturing process (Ghosh et al., 2016).

The main architectural implication for designing NetZEBs is the integration of renewable energies. Placing photovoltaic panels (PV) on the roof is the easiest and most reliable way to achieve the Net-Zero Energy objectives (Frattolillo et al. 2020). PV systems offer the most significant possibilities of architectural integration. The flexibility of cables transporting electricity and the characteristics of the energy produced compensate for the higher costs. The PV efficiency, the tilt, and orientation limit the energy that PV panels can generate. Therefore, the roof area and the building envelope is not enough to compensate for the building needs. Although the cost of solar panels and inverters steadily decreases, the Solar PV Balance-of-System, which refers to the components and equipment that move DC energy produced by solar panels, is a factor that matters when it comes to designing the building energy system. These components include inverters, wires, switches, enclosures, fuses, tracking systems, and ground fault detectors. They account for 10%-50% of solar purchasing and installation costs and is related to most maintenance requirements. Photo-Voltaic (PV) panels can contribute to reducing the primary energy consumption and CO_2 emissions; They also save transportation costs and losses, for the energy is consumed where it is generated. Finally, they can provide electrical energy that can be turned into thermal energy using heat pumps (Attia and Del Ama 2019). Designing Net-Zero Energy Buildings is possible when PV production can compensate for the building's energy demand.

Solar thermal collectors are mandatory in some building codes to generate domestic hot water (Buonomano et al., 2019). Thermal collectors are considered a renewable and CO_2 free energy source but have limitations in applicability. The advantage of the system is that it uses a CO_2 free energy source. The disadvantage is the investment and that the technology needs a backup energy source. The technique also depends on the availability of solar irradiation at the specific geographic location. Vertical facades generally offer large and well-sunny surfaces. However, vertical envelopes represent the building's public image, so architectural coherence is often more critical than energy performance. Thermal collectors are technical devices designed to optimize heat collection, production, and installation. Generally, marginal attention to architectural integration issues was traditionally given at the design phase of solar systems, leaving the architect's responsibility to adapt materials, dimensions, and shapes into the building. The thermal energy is intended for consumption where it is produced, but the stiffness of the pipes transporting hot water reduces the flexibility of the system (Arpino et al., 2015). New products such as the waterproof frame, the isolation systems, and the solar heater have been patented and introduced in the market without regard for their architectural impact (Frattolillo et al. 2020).

Water Flow Glazing (WFG) is a technology that can be integrated into rooftops, opaque, and transparent building envelope surfaces. In addition to this, Gil-Lopez and Gimenez-Molina (2013) showed that the implementation of WFG also allows for reduced heating and cooling costs if used as a retrofit for traditional glazing. WFG technology combines both passive and active strategies, for it employs the use of coating and Polyvinyl butyral layers, and active measures, such as variable water mass flow rate, to absorb or reject incoming infrared solar radiation to reduce the temperature of the interior glass pane (Gutai and Kheybari, 2020). By implementing flowing water into the system, WFG is able to capture a majority of solar infrared radiation, while an individual is still free to see through the visible part of the glazing (Chow et al., 2010). Radiant panels that are based on WFG technology can also be integrated into a hydronic system that heats and cools (given that the interior and exterior water temperature maintain only a small difference). Finally, this technology can also collect solar energy to provide domestic hot water to plumbing fixtures during warm seasons (Chow and Li, 2013). Lyu et al. (2018) have evaluated WFG technology and understand it as an efficient way to store cooling energy for air-conditioned structures, especially those in cold-weather climates. Although this technology is shown to store heating energy, the solar energy alone that penetrates the building may not be sufficient to maintain ideal comfort conditions during winter. Therefore, alternative heat sources should be added to reduce heat loss during cold periods (Lyu et al. 2019). The Water Flow Glazing system is primarily comprised of three constituents: glazing, a circulating device, and an aluminum frame. The glazing in the system is a compound of several different layers according to anticipated and pre-determined spectral and thermal properties that the building will experience. The circulating device is defined by including a water pump that circulates the fluid, an exchanger that can regulate heat, and several sensors (such as the water flow and water temperature) to regulate the different fluid variables involved. Finally, the aluminum assembly provides the frame with structure. The aesthetic design dictates and defines the dimensions of each WFG module. Some authors have studied WFG performance through computer simulations (Chow et al., 2010). Ji et al. (2011) have evaluated the impact of WFG components in the glazing thermal and spectral properties.

Renewable energy integrated in the building envelope might be not enough to meet the energy needs, therefore it is necessary to study different energy sources and systems to evaluate their final energy consumption and greenhouse gas emission potential. Wood chips are considered a CO_2 neutral energy source. Natural gas is a flexible, reliable and economical source to use as backup capacity in heating systems. When a detailed analysis throughout the life cycle of fossil fuel has been carried out to understand the supply chain, many different results can be obtained. The large differences are derived from not only the actual variations that exist in supply chains and processes but also from the summation of different assumptions and default values that are applied during calculations on differences in quality of source material. Since energy supply from one or other country can vary in a large amount in a very short term, there can be uncertainties in energy conversion factor calculations at the European Union level for the different fossil fuel types over their life cycle. Many nations have opted to adopt a standard typical value of 1.1 for fossil fuels, which relates to a simplification of energy inputs and distribution of primary fuels equaling 10% of energy delivered by energy carrier. In the United Kingdom or Spain, the most common values are set at 1 or 1.1. These values are calculated through a much more detailed analysis, considering the energy inputs at different phases in the energy life cycle. Baranzelli et al. (2016) and Edwards et al. (2014) have stated that it is complicated to thoroughly define a consistent primary energy factor of solid, liquid, or gas fuels in the European Union based on detailed life cycle and supply chain considerations. Table 1 demonstrates the average conversion factor per unit of electricity generated by each energy typology in EU countries on average and the values provided by the Spanish Regulation of Thermal Installations in Buildings RITE (2016). The energy conversion factors for the US are sourced from the ASHRAE Standard 105.

Table 1. Average Source Energy Conversion Factors (kWh of primary energy/kWh of electricity) in the United States and the European Union

Energy form	EPBD EU*	ASHRAE Standard 105	Spain RITE
Imported Electricity (non.renewable)	1.50-3.45	3.15	1.954-2.968
Natural Gas	1.00-1.26	1.09	1.190
Fuel Oil (1,2,4,5,6,Diesel, Kerosene)	1.00-1.20	1.19	1.179
Coal	1.00-1.46	1.05	1.082
Wood and pellets	0.01-1.26		0.085

* EU countries in average

The CO_2 factor should include all emissions related to extraction, processing, generation, storage, transport, distribution, and delivery of the energy type. The use of fuels generates carbon dioxide (CO_2) emissions as well as small quantities of other greenhouse gases, such as methane (CH_4) and nitrous oxide (N_2O). To calculate a CO_2 equivalent for a specific quantity of a gas, the "global warming potential" is measured in units of grams carbon dioxide equivalent (gCO_{2eq}). Table 2 shows the emissions in gCO_{2eq} provided by different sources. The values for the European Union are taken from Koffi et al. (2017).

Table 2. Specific CO_{2eq} conversion factors employed by different sources

Energy form	gCO_{2eq}/kWh EU*	gCO_{2eq}/kWh ASHRAE	gCO_{2eq}/kWh Spain RITE
Imported Electricity	233	324	331
Natural Gas	240	308	252
Fuel Oil	306	334	311
Coal	370	404	472
Wood and pcllcts	17	7-30	18

* EU countries in average

Previous articles have shown that WFG can be used in zero-energy building design by reducing heating and cooling loads and producing renewable thermal energy. Ji et al. (2011) have shown that, depending on the location, yearly savings of 66% can be achieved by replacing traditional double-glazing curtain walls with water flow glazing. Due to energy savings, CO2 emissions can be reduced by 70%

compared to traditional curtain walls. In addition, Moreno et al. (2020) have shown that WFG can work as a thermal collector to produce renewable thermal energy.

THERMAL PERFORMANCE OF WATER FLOW GLAZING

Water flowing between two glass panes captures the solar Near-Infrared radiation and increases its temperature through the window. The water flow transports heat and releases the energy in tanks or energy sinks, enabling energy management strategies, such as solar energy rejection or thermal energy storage. This section showed that the successful implementation of WFG could reduce heating and cooling loads, allowing the construction of NetZEB's to become a realistic target.

Mathematical Model (Steady State)

Two main parameters define the performance of traditional glazing: the thermal transmittance, U, and the g-factor. The thermal transmittance measures the glazing capability to transfer heat, whereas the g-factor is the proportion of the total solar energy which is transmitted indoors. European standards EN 673 (2011) and EN 410 (2011) determine the way to calculate these parameters in a steady state. A variable g-factor has been studied to assess the performance of transparent facade collectors showing that the g-factor depends not only on the static properties of the glass, but on the water flow of the solar collectors. The g-factor becomes a variable parameter that allows to control the incoming energy. The final thermal and spectral properties depend both on the water chamber and glass panes. This section continues the work of Sierra and Hernandez (2017) about the dynamic control of the thermal transmittance and g-factor in Water Flow glazing, and summarizes the steady-state mathematical model stated in the work of Li et al. (2011).

Figure 1 demonstrates how the glass panes in the WFG system create the circulation chamber of water. θ_{IN} represents the inlet temperature of the circulating water, while θ_{OUT} represents the outlet temperature. The g-factor is a variable that measures the amount of heat that transfers through the WFG system. *R* and *T* are the variables that measure the reflectance and transmittance of the glazing, and i_0 represents the total incoming solar radiation. The total energy absorbed by the system is separated by three different fractions: A_i, A_e, and A_v. A_i, and A_e are constants that occur from the secondary internal heat factor. A_v represents the remaining absorptance from the water chamber; this variable is further divided into $A_{v\,int}$ (which represents the energy transfer from outdoors to indoors), as well as $A_{v\,ext}$ (energy transfer from indoors to outdoors). The energy transported by the circulating water transports is represented by $A_{v\,water}$. These three variables ($A_{v\,int}$, $A_{v\,ext}$, $A_{v\,water}$) are dependent upon the water flow rate. Based on the hypothesis that the authors presented earlier in this section, the glass panes temperatures do not change over time. Finally, the spectral properties of the glazing assembly are dependent upon the variable incoming solar radiation, i_0.

U values are dependent on several variables. These include h_i (the interior heat transfer coefficient), h_e (exterior heat transfer coefficient), h_g (the air chamber heat transfer coefficient), and h_w (the water chamber heat transfer coefficient). θ_1, θ_2, and θ_3 are the temperatures at the glass panels. The variable U_i measures the heat transfer between the water chamber and indoors, while U_e measures the heat transfer between the water chamber and outdoors.

Figure 1. Temperature distribution and description of heat fluxes and temperatures of layers at a specific height of the envelope of water Flow Glazing **a)** *triple WFG* **b)** *double WFG* **c)** *Distribution of solar radiation on the glazing*

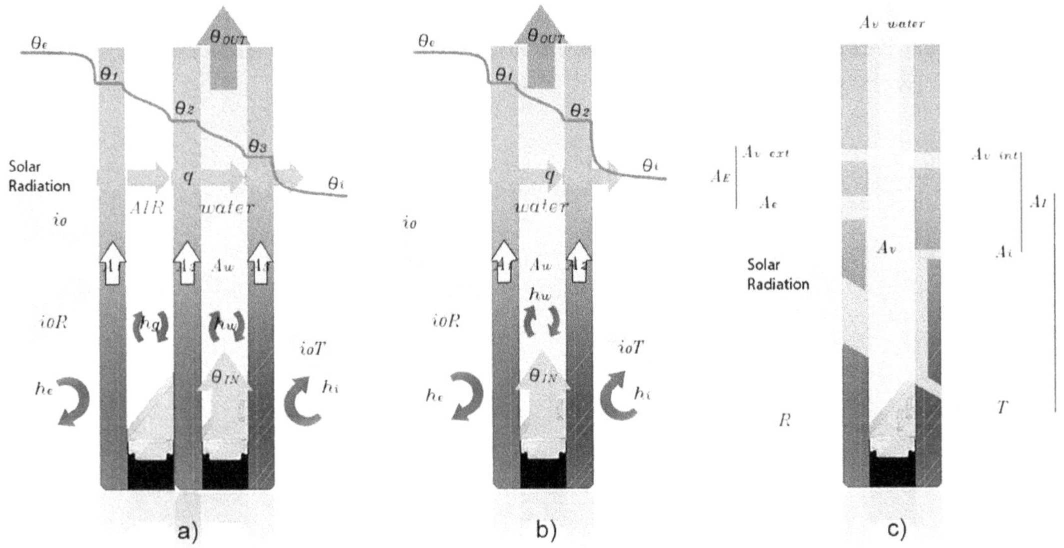

Equations (1) and (2) show U_i and U_e in a WFG module with two glass panes and a water chamber:

$$\frac{1}{U_e} = \frac{1}{h_e} + \frac{1}{h_w}, \tag{1}$$

$$\frac{1}{U_i} = \frac{1}{h_i} + \frac{1}{h_w}. \tag{2}$$

Equations (3) and (4) show U_i and U_e in a WFG module with triple glazing:

$$\frac{1}{U_e} = \frac{1}{h_e} + \frac{1}{h_g} + \frac{1}{h_w}, \tag{3}$$

$$\frac{1}{U_i} = \frac{1}{h_i} + \frac{1}{h_w}. \tag{4}$$

Equation (5) shows the total heat flux, q:

$$q = g_{i0} + U(\theta_e - \theta_i) - U_w(\theta_I - \theta_{IN}), \tag{5}$$

where g illustrates the solar energy that is transmitted into the structure, U is the thermal transmittance of the WFG, $(\theta_e-\theta_i)$ illustrates the temperature difference between the interior θ_i and exterior θ_e. U_w is considered a second thermal transmittance, due to the temperature difference between the inlet water θ_{IN} and indoor temperature θ_i. $U_w(\theta_i-\theta_{IN})$ illustrates the heat transfer between the exterior of the structure and the water chamber in the glazing assembly. Equations (6), (7), and (8) show the expressions for U_w, U, and g, respectively.

$$U_w = \frac{U_i \dot{m} c}{\dot{m}c + U_e + U_i}, \tag{6}$$

$$U = \frac{U_e U_i}{\dot{m}c + U_e + U_i}, \tag{7}$$

$$g = \left(\frac{U_i}{\dot{m}c + U_e + U_i} \right) A_v + A_i + T, \tag{8}$$

where $\dot{m}$ defines the mass flow rate, which is defined as a measurement of the total mass being moved at a point over a certain time period, c is the specific heat of the water in the glazing assembly. The absorptance, A_v, depends on the energy absorbed by the glass panes and by the water.

Equation (9) shows the expression for double WFG:

$$A_v = A_1 \left(\frac{U_e}{h_e} \right) + A_2 \left(\frac{U_i}{h_i} \right) + A_w. \tag{9}$$

Equation (10) shows the expression for triple WFG:

$$A_v = A_1 \left(\frac{U_e}{h_e} \right) + A_2 \left(\frac{1}{h_g} + \frac{1}{h_e} \right) U_e + A_3 \left(\frac{U_e}{h_i} \right) + A_w, \tag{10}$$

where A_1, A_2 and A_3 are the absorptances of the glass panes, and A_w is the absorptance of the water chamber. The Equation (11) and (12) show the absorptance A_i for double and triple glazing, respectively.

$$A_i = A_2 \left(1 - \frac{U_i}{h_i} \right), \tag{11}$$

$$A_i = A_3 \left(1 - \frac{U_i}{h_i} \right). \tag{12}$$

Heat gain in water is considered as power magnitude, which is measured in watts (W). Equation (13) is the expression used for this.

$$P = \dot{m}c\left(\theta_{OUT} - \theta_{IN}\right), \tag{13}$$

where P is the total power that is absorbed by the water, θ_{OUT} is the temperature of the water leaving the glazing assembly while θ_{IN} is the temperature of the water entering the glazing assembly. The goal of absorbing the same amount of power P can be achieved with a large mass flow rate $\dot{m}$, which will result in an increase in low temperature or a low value for $\dot{m}$, which results in a large temperature difference between the inlet and outlet. Equation (14) shows the outlet temperature from a WFG panel.

$$\theta_{OUT} = \frac{i_0 A_v + U_i\theta_i + U_e\theta_e + \dot{m}c\theta_{IN}}{\dot{m}c + U_e + U_i}. \tag{14}$$

Equation (15) results by combining Equations (13) and (14).

$$P = \frac{\dot{m}c}{\dot{m}c + U_e + U_i}\left(i_0 A_v + U_i\left(\theta_i - \theta_{IN}\right) + U_e\left(\theta_e - \theta_{IN}\right)\right). \tag{15}$$

Description of the Prototypes

Several experimental facilities have been designed and constructed to collect data based on WFG technology.

Peralveche Prototype

This experiment involves two separate prototype cabins placed in Peralveche, Spain (latitude 40°36¢42” N, longitude 2°26¢57” O, altitude 1111 MAMSL). The first one takes advantage of WFG technology, assembled with aluminum frames. The North façade and floor are constructed of sandwich panels, which have a thermal transmittance of $U = 0.6$ W/m^{2}K. This prototype is referred to as the WFG cabin. The second prototype used in this experiment is another cabin of the same dimensions, orientation, and double glazing with an air cavity. This second prototype shall be referred to as the Reference Cabin (RC). The dimensions of both prototypes are 2m x 2m x 2m. The volume is 8 m^3, the floor area is 4 m^2, and the total glass area to floor area ratio is 16m^2/4m^2. The WFG cabin has four borehole heat exchangers that are buried 50m below the Earth’s surface. The WFG cabin manages two different closed-circuit circulating systems. The primary system consists of pipes that circulate refrigerant fluid that connects from the borehole heat exchangers to the circulation pump. In addition to this, the WFG cabin manages two secondary systems, one for the WFG vertical façade and the second for the WFG horizontal roof. For this system, water connects the circulating pump to the WFG. The water in the glazing chamber circulates at a temperature near comfort conditions and the system takes advantage of the earth’s natural thermal inertia. In summer, the heat absorbed by the circulating water chamber is released in the ground through the borehole heat exchangers. Figure 2 demonstrates the schematic for the energy management system employed for the WFG cabin in this prototype. The primary circulation system circulates water from the WFG plate heat exchangers to borehole heat exchangers. The two secondary circulating systems go from the circulation pump to the WFG façade. Both secondary circuits maintain a precision flow meter and a thermometer for monitoring the inlet and outlet temperatures.

Figure 2. Peralveche prototype. Description of the cabin's glazing and layout of the energy management system: 1 heat exchanger; 2 circulation pump; 3 vertical WFG panels; 4 heat exchanger; 5 circulation pump; 6 horizontal WFG panels; 7 borehole heat exchangers.

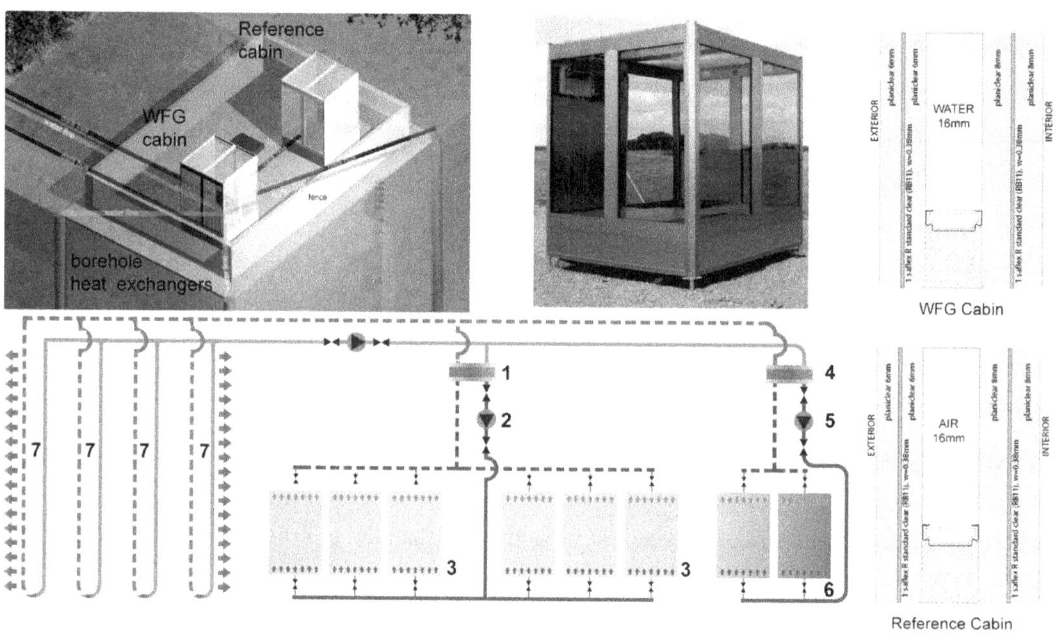

The WFG cabin comprises of a 6 + 6 mm laminated glass pane, a water chamber measuring 16mm, and another laminated glass pane, this time measuring 8 + 8 mm. The glass panes used to construct this glazing assembly are Planiclear 6mm and 8mm thick (manufactured by Saint Gobain), and the Poly-Vinyl Butyral layers are standard (1 x 0.38mm). Sierra and Hernandez (2017) give typical values for A_v, which depend on the sum of absorptances of each layer in the assembly. The external layer absorptance is 0.585, while the internal layer is 0.037. A range of values has been considered in this case study of A_v, ranging from 0.429 to 0.524. The chosen flow rate was $\dot{m}$=0.9 Liter/min m^2. Table 3 highlights the spectral and thermal characteristics of the WFG and RC. Hi and He have been referenced from the European Standard EN 673, with h_i = 100 W/m^{2}K and h_e = 25 W/m^{2}K for horizontal glazing with a vertical heat flux. The heat transfer coefficient h_w is the result of using Equation (11) in Chow at al., 2010, with a Nusselt number of 7.541. The thermal resistance of a 16 mm water layer is calculated as 0.026 m^{2}K/W. Finally, h_w = 452 W/m^{2}K. U_e is equal to 9.80 W/m^{2}K while U_i is equal to 23.69 W/m^{2}K.

Table 3. Thermal properties of the studied WFG

		Thermal properties					
		$\dot{m}$=0 (Liter/min m²)			$\dot{m}$=0.9 (Liter/min m²)		
Glazing	Glass area/floor area	U (W/m²K)	U_w (W/m²K)	g	U (W/m²K)	U_w (W/m²K)	g
Reference	4	2.6	-	0.67	-	-	-
WFG	4	4.797	0.0	0.396	0.762	5.802	0.271

Madrid Prototype

The authors constructed this prototype to compare a triple pane WFG with a double-glazing traditional solution. This prototype located near Madrid, Spain (latitude 41°22¢6² N, longitude 3°29¢57² W, altitude 1037 MAMSL). This prototype was a mobile structure, consisting of two cabins. One of these cabins had a WFG facing South, while the other had the reference glazing. The dimensions of the glazing in both cabins were 1m x 0.5m. The volume of the WFG cabin was 0.5 m^3, the floor area is 0.5 m^2, and the total glass area to floor area ratio is $0.5m^2/0.5m^2$. The design of the prototype allowed the primary and secondary circuits to be placed within independent sectors within the structure. In addition to this, the prototypes had four wheels mounted at the bottom to allow for easy orientation and transportation. Figure 3 illustrates that the prototype had two separate levels: the lower level that was 500mm high and the upper level was 1000mm high and included the two cabins. The lower level held a thermoelectric cooling device and a lithium battery. Finally, a photovoltaic panel was placed upon the roof to charge the batteries. These batteries operated a cooling device. This secondary circuit operated at a flow rate of 0.5 Liter/min and connected the circulation system to the WFG. A highly reflective WFG was chosen to reject energy as well as use the water chamber of the interior glazing to eliminate heating loads when needed. Energy absorption should be minimized if possible, therefore, a highly reflective coating (Xtreme 60.28) is oriented close to the exterior glass pane (face 2). This reflective coating allows for a reduction in the U-value of the assembly because it is also considered as a low-emissivity coating (Planitherm XM). The maximum (when mass flow rate is equal to 0) and minimum (at max flow rate) g-factor are similar and around 0.2.

Table 4 demonstrates spectral and thermal variables of the glazing assembly as shown in Figure 3. The U and g values in WFG are highly variable dependent upon the mass flow rate. U and U_w were calculated using equations (6) and (7) using $\dot{m}$ = 1 Liter/min m^2 and $\dot{m}$ = 0 Liter/min m^2, respectively.

Sofia Prototype

An experimental design was constructed in Sofia, Bulgaria (42°39'1" North, 23°23'26" East, Elevation: 590 m a.s.l.), to understand the performance of WFG modules throughout the year when exposed to real-world conditions. Unitized WFG panels were then placed in East, South, and West orientations. A pyranometer was paired with this system to measure incoming solar radiation at each façade. The heat plate exchangers of the circulation device were then all connected to the water distribution systems. The test glazing assemblies used in this prototype were triple glazing with a 16mm argon cavity and a 24mm water cavity. The circulating system was comprised of a solar water pump paired with a plate heat exchanger. The unitized WFG panels were assembled in a factory and then prepared for on-site application with an aluminum frame that held the circulating system. These panels measure 3000mm high and 1300mm in width. The main components of the circulation device are a plate heat exchanger and a solar water pump. This design allowed for hydraulic as well as electrical independence of the unitized modules with a reduced size. For this operation, the flow rate was set to 8 Liter/min for a 4m^2 module. In the Southern glazing, the following layers were employed: diamant glass (10 mm), an argon chamber (16 mm), a low-emissivity coating Planitherm XN, Planiclear (8 mm), 2 Saflex R solar (SG41), Planiclear (8 mm), water chamber (24 mm), Planiclear (8 mm), 4 Saflex R standard clear (RB11), Planiclear (8 mm). Eastern and Western glazing, on the other hand, were constructed using: diamant glass (10 mm), a highly reflective coating Xtreme 60.28, an argon chamber (16 mm), Planiclear (8 mm), 4 Saflex R standard clear (RB11), Planiclear (8 mm), water chamber (24 mm), Planiclear (8mm), 4 Saflex R standard clear (RB11), Planiclear (8 mm). Figure 4 illustrates the layout and the glazing types. This prototype is dimensioned at 7m x 7m x 3.4m and has a gross floor area of 49 m^2. In addition to this, the prototype comprises five modules that are oriented to face east, five modules that face west, and five more modules that face the southern façade. The plate heat exchangers and circulation device were connected to the water distribution systems. The total glass area to floor area ratio was 60m^2/49m^2.

Figure 3. Description of the prototype and the glazing. Energy management and monitoring system: 1 water pump; 2 thermoelectric cooling device; 3 lithium battery; 4 photovoltaic panel; 5 heat exchanger; 6 control management unit; 7 temperature sensors; pyranometer.

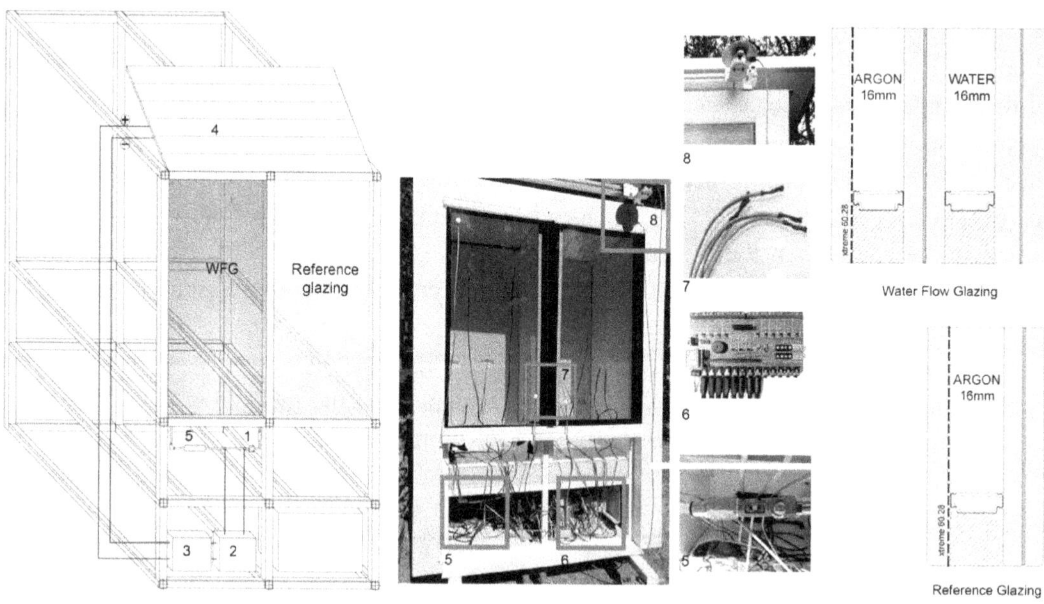

The southern glazing included a low-emissivity coating at the interior face of the argon chamber. It showed a high near-infrared (NIR) absorptance, which made this panel an ideal solution for large glazing surfaces in cold climates. Glazing in the eastern and western orientations has been designed with a solar control coating (xtreme 60 28 II) placed at the outer side of the argon chamber. Figure 4 illustrates the spectral properties with very low infrared absorptance and very high front reflectance. Although this situation would not be ideal for heating water, this effect would be better utilized to reject energy and prevent heat from entering interior space. Table 5 demonstrates the thermal properties of the glazing for when the mass flow rate was set to 0 as well as when it was set at 2 Liter/min m^2.

Table 4. Thermal properties of the studied WFG

		Thermal properties					
		$\dot{m}$=0 (Liter/min m^2)			$\dot{m}$=1 (Liter/min m^2)		
Glazing	Glass area/floor area	U (W/m^{2}K)	U_w (W/m^{2}K)	g	U (W/m^{2}K)	U_w (W/m^{2}K)	g
Reference	1	1.017	-	0.273	-	-	-
WFG	1	0.977	0.0	0.267	0.061	6.464	0.220

Data From the Prototypes

Data From Peralveche

Collected data from this prototype allowed comparing both cabins, the one with WFG and the one with the reference glazing. When solar radiation penetrated the WFG glazing, a certain amount of this energy was absorbed into the water chamber. Figure 5 demonstrates the exterior and interior temperatures for RC and WFG cabins. During the sample summer week for the WFG cabin, the exterior environment reached a temperature of 36.5 °C, while the peak interior temperature reached 38.5 °C. Meanwhile, RC reached 54.5 °C. As compared to RC, the WFG cabin had a temperature reduction of 16 °C. This was a direct consequence of the solar energy absorbed by the water chamber being transferred directly to the ground. The borehole exchangers provided the system with thermal inertia and created a cushioning effect that reduced the temperature curve of WFG compared to RC. During the sample winter week, the interior temperature curves depended on the exterior environmental temperature and the amount of incoming solar radiation. On sunny days (for example, December 04), the exterior temperature ranged between -3.5°C to 16°C. In addition to this, the interior temperature in RC ranged between -3.2°C to 34.8 °C. Simultaneously, on the same day, the interior temperature of WFG ranged from 7.7°C to 26.5 °C. On cloudy days (December 06), the exterior temperature ranged from -5°C to -2°C. In RC, the interior temperature ranged from -4.5°C to 7.8 °C, while the temperature in WFG ranged from 5°C to 8.5 °C.

Data From Madrid

The WFG envelope allowed insulating the cabin from external boundary conditions. Furthermore, the thermoelectric cooling device connected to a buffer tank produced cool water for the glazing over the summer, from 12:30 - 20:00. The goal of this prototype was to maintain the inlet water temperature for WFG within a range of 15 and 17°C, as well as measure any fluctuations in indoor air temperature. WFG cabin reached a mean maximum interior temperature of 26.5°C and the interior of RC exceeded a temperature of 38°C while the mean maximum exterior temperature reached 35°C. This means there existed a more than 11.5 °C difference between the interior of both cabins. The RC glazing mimicked the large changes and oscillations in the outside temperature. This same behavior could be observed when an analysis was done on the minimum temperatures. The minimum outdoor temperature was 10°C. During this time, the WFG cabin interior temperature reached 15 °C, and the RC cabin maintained a temperature of 11°C. It is evident that the WFG cabin was able to dampen the range of interior temperature changes compared to RC. The interior minimum and maximum values for WFG were brought closer to ideal comfort conditions when compared to the collected data for RC. Figure 6 illustrates the exterior temperature for RC and WFG cabin. During the summer, the thermoelectric cooling device used in these prototype cabins provided the necessary cooling power to keep the interior WFG cabin temperature between 16.1°C and 26.5°C. When the cabin would be experiencing colder conditions, the thermoelectric device would not be operating, allowing WFG to operate on a free-floating regime. The WFG façade system managed to reduce large interior temperature oscillations like those experienced in RC.

Figure 4. Description of Sophia's prototype glazing

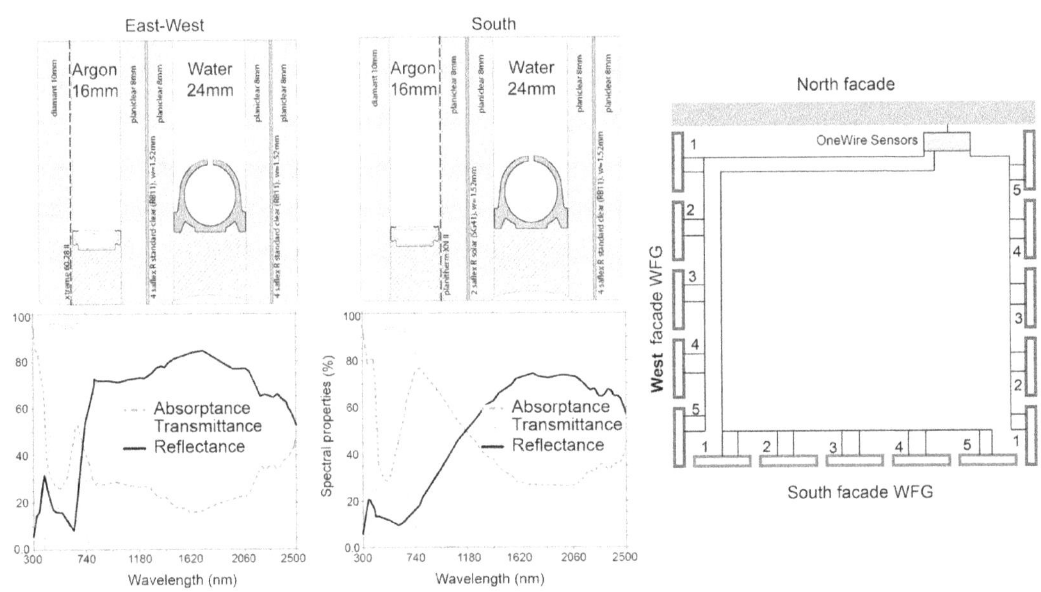

Table 5. Thermal properties of the studied WFG

		Thermal properties					
		$\dot{m}$=0 (Liter/min m²)			$\dot{m}$=2 (Liter/min m²)		
Glazing	**Glass area/floor area**	U (W/m²K)	U_w (W/m²K)	g	U (W/m²K)	U_w (W/m²K)	g
South	1.22	1.041	0.0	0.59	0.066	6.459	0.24
East-West		0.995	0.0	0.27	0.063	6.462	0.22

Figure 5. Peralveche prototype temperatures in winter and summer conditions

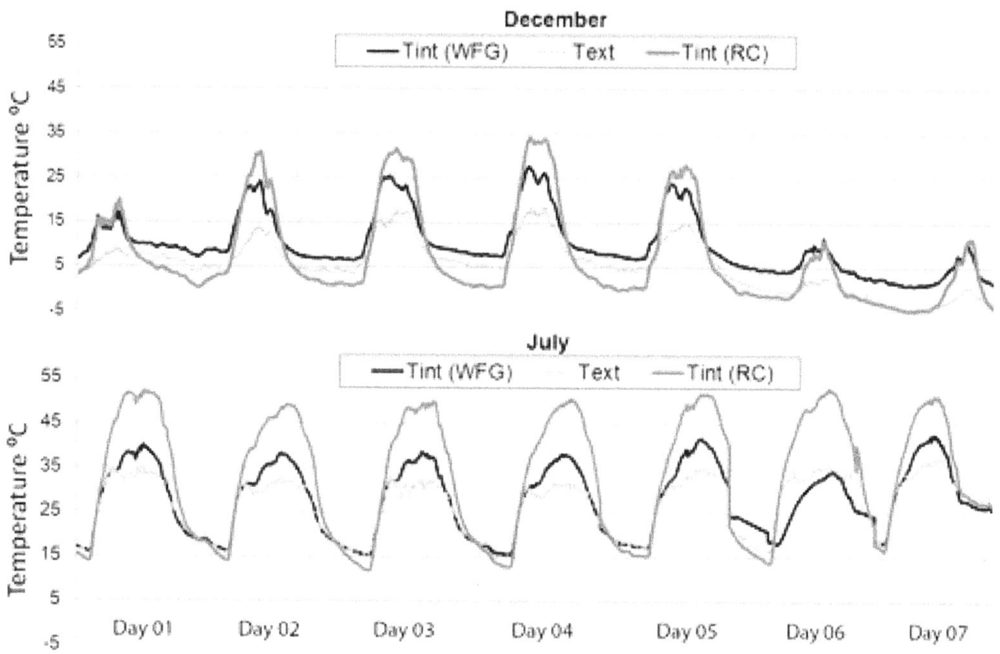

Data From Sofia

The mass flow rate and the thermal properties of glass panes have been selected to allow designers to apply energy-management strategies. Increasing the water temperature and store that energy for heating purposes was a good winter strategy. Other times it might be appropriate to reject as much solar radiation as possible without heating the water. A heat pump was employed to regulate inlet temperature. Figure 7 demonstrates outdoor and indoor air temperatures in summer and winter conditions. There occurred a maximum temperature difference from 7:00 a.m. to 10:00 a.m., when spectral radiation reached its zenith value on the eastern façade. Using values that were measured from this real-world source, the temperature difference of the WFG could reach 10°C depending on the exterior conditions. The water heat gain and the heating and cooling loads have been measured and the following sections present the thermal performance of the prototype. During winter, thermal absorption did not depend entirely upon solar radiation. The temperature difference between the interior and exterior affected energy performance more than the solar radiation on the facades.

DISCUSSION ON KEY PERFORMANCE INDICATORS

Next, the authors intended to analyze the outcomes of these experiments in terms of the potential energy savings. The key performance indicators are objective parameters defined so that external factors beyond the control of a building user cannot interfere with them. A key performance indicator must be based on actual data and must provide context and should cover a specific time divided into significant check-points. Finally, the selected KPIs should be clear so that they allow other researchers to make similar assessments. The selected KPIs were first, the absorbed energy per unit of area, second, the Primary Energy consumption and CO2 emissions per unit of area, and finally, the ratio between renewable energy production and non-renewable final energy consumption.

Figure 6. Madrid prototype temperatures in winter and summer conditions

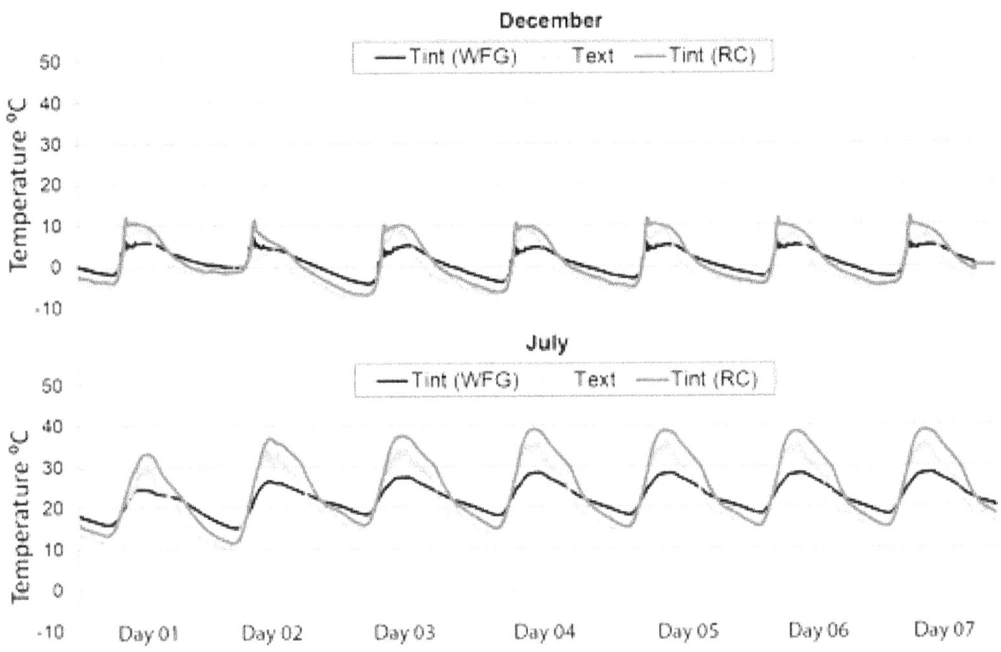

Key Performance Indicator 1: Water Heat Gain / Water Heat Gain Efficiency

In order to validate the performance of WFG, the absorbed energy per unit of area per day was calculated using Equation (15). The output of these is the water heat gain or the thermal power of WFG. Equation (13) was used to find the water heat gain of the WFG facade system. The daily water heat gain efficiency is a concept that can be used to assess thermal performance. This value is defined as the ratio of total water heat gain compared to the daily solar radiation. In this study, the daily water heat gain is accumulated if the inlet temperature is lower than outlet temperature.

Peralveche

The water chamber absorbed thermal energy during the day and then, the energy was transferred to the ground by the borehole heat exchangers. During the night, the outlet temperature was lower than the inlet temperature. For this system, the designed flow rate was 0.9 Liter/min m^2 and 0.6 Liter/min m^2 through the horizontal and vertical panels, respectively. The inlet temperature varied between 16°C and 20°C. During the daytime, the outlet temperature ranged between 24.5°C during the day, while at nighttime, it was 15°C. Figure 8 illustrates the solar irradiance and water heat gain during two sample weeks experiencing both winter and summer conditions. During July, the average daily energy absorbed by the horizontal panes was 9.47 kWh through the 3.6 m^2, meaning that the ratio of energy per area was 2.63 kWh/m^2 per day. Meanwhile, the energy absorbed through the vertical panels was 32.72 kWh in 11.4 m^2 of WFG area, meaning that the energy per area ratio was 2.87 kWh/m^2 per day. During December, it could be expected that the ratio of energy per area in the panels oriented in a horizontal direction on a sunny day would be 0.40 kWh/m^2. In a vertical orientation, this could be 0.81 kWh/m^2 per day.

During a week in December, figure 8 demonstrates that the daily water heat gain reached a peak value of 0.53 kWh/m^2 on day 4 and was 0 on day 6, due to very low incoming solar radiation. During the July week, the daily water heat gain had a maximum value of 0.72 kWh/m^2 with a minimum value of 0.56 kWh/m^2.

Figure 7. Sofia prototype temperatures in winter and summer conditions

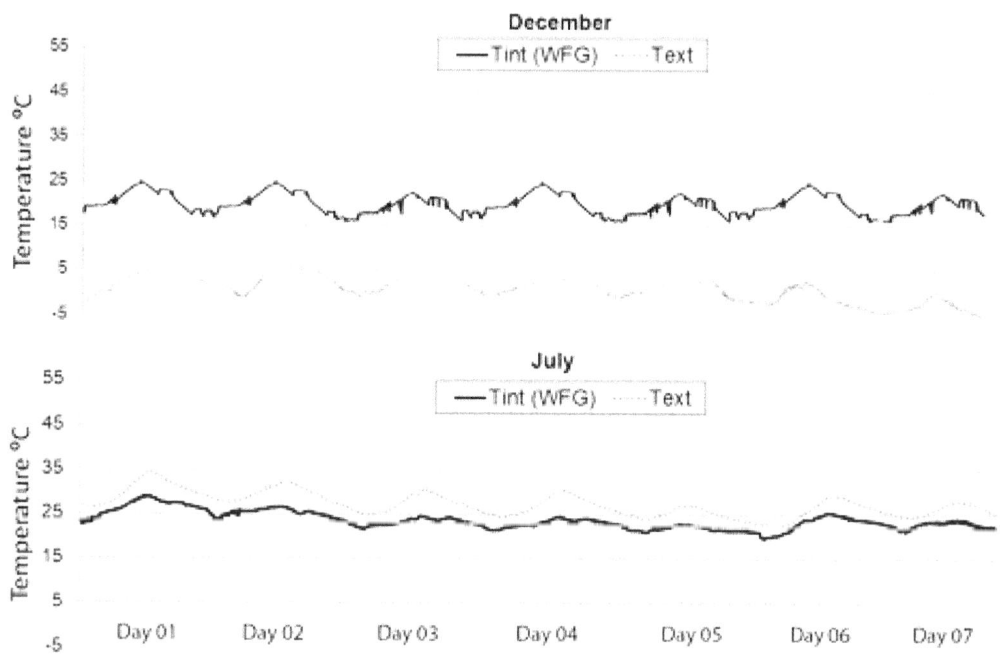

Madrid

WFG was able to reject solar energy and use the water chamber to eliminate heating loads when necessary; this was accomplished due to the highly reflective coating that was selected for the WFG prototype. In addition to this, energy absorption was minimized. Figure 9 illustrates water heat gains in winter and summer conditions when a maximum flow rate of 1 Liter/min m^2 is attained. The average daily energy per area absorbed by WFG was 0.69 kWh/m^2 in July. In December, this value is almost negligible. This is due to the weak amount of solar radiation and low ambient exterior temperature during this month. During the week in July, the daily water heat gain reached an average value of 0.14 kWh/m^2 per day.

Figure 8. Peralveche prototype. Solar irradiance and Water Heat Gain in horizontal and vertical glazing

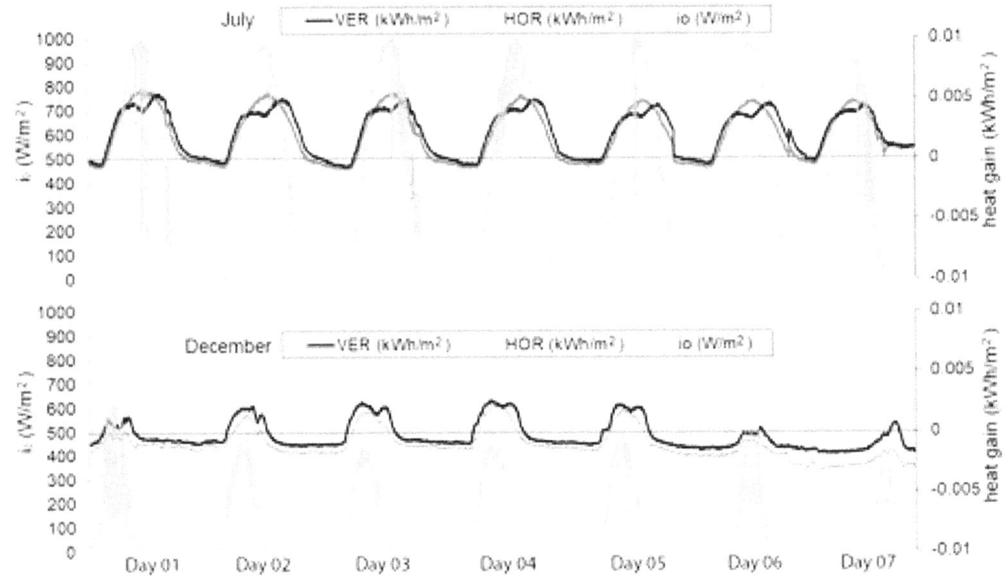

Sofia

Panels oriented in a southern direction had a daily energy absorption rate of 13.9 kWh in 3.9 m^2, meaning that the energy per area ratio was 3.57 kWh/m^2. When the maximum solar radiation was 470 W/m^2, there existed a temperature difference of 2°C. Figure 10 demonstrates the water heat gains during certain selected winter and summer time periods in three different orientations.

As expected, the energy absorption rate in the eastern and western orientations was negligible. The energy-to-area ratio was 3.81 kWh/m^2 in the southern direction, 2.12 kWh/m^2 in the western direction, and 1.77 kWh/m^2 in the east. Although there was an excess of solar radiation on eastern and western facades, the water in the circulation chamber was heated 2 °C, with most of the infrared energy being rejected. The south façade experienced similar solar energy absorption in the winter and summer, with the water heating up around 3 °C. In winter, the daily energy-to-area ratio was 2.30 kWh/m^2 in the south, 0.66 kWh/m^2 in the west, and 0.28 kWh/m^2 in the east.

Figure 11 demonstrates the monthly performance of all cases. In Sofia, the water heat gain of the southern WFG is much larger than in any other case. Two factors can explain this high heat gain. Firstly, the mass flow rate was set higher than any of the other prototypes (2 Liter/min m^2). Secondly, the southern glazing properties demonstrate the highest absorptance. The Peralveche prototype has the second-highest heat gain in the horizontal panels. In Sofia, the prototype water heat gain ranged from 60.72 kWh/m^2 in January to 97.39 kWh/m^2 in July. Meanwhile, in the horizontal panels in Peralveche, the water heat gain temperature ranged from 1.55 kWh/m^2 in January to 81.44 kWh/m^2 in July. The heat absorption rate in the southern facades is closely related to the glazing composition in the façade system. For the Sofia prototype, the goal of the authors was to make the most of the solar radiation in winter and summer months, in order to heat up the water flow. In the Madrid prototype, the goal was to reject solar energy so that the interior ambient temperature would be within a comfort range during summer months. The peak solar heat gain in the Sofia prototype was 97.39 kWh/m², while the Madrid prototype was 21.49 kWh/m².

Figure 9. Solar irradiance and Water Heat Gain in the Madrid prototype (vertical WFG)

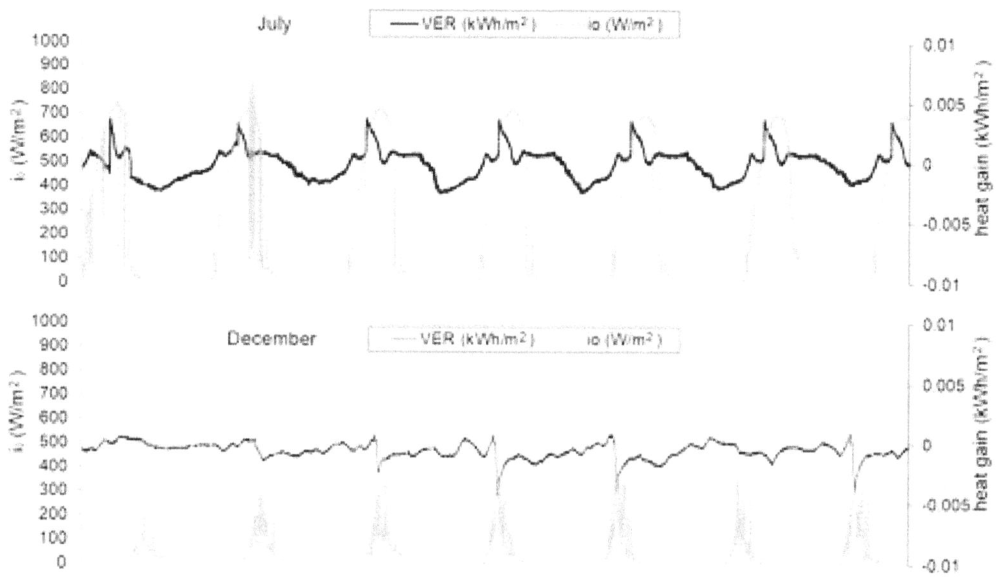

Figure 10. Solar irradiance and Water Heat Gain in Sofia prototype

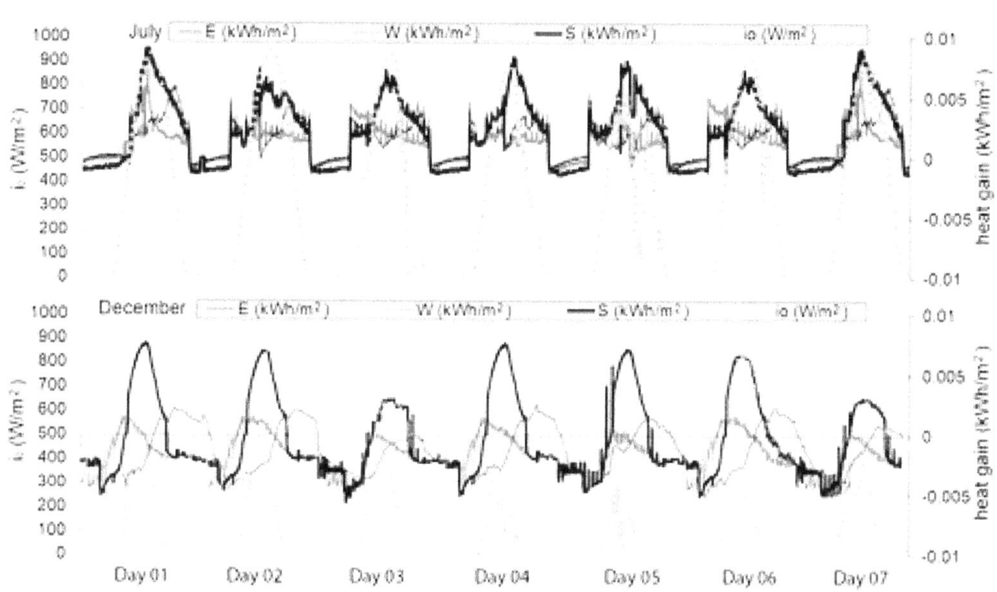

Figure 11. Water Heat Gain in the studied prototypes

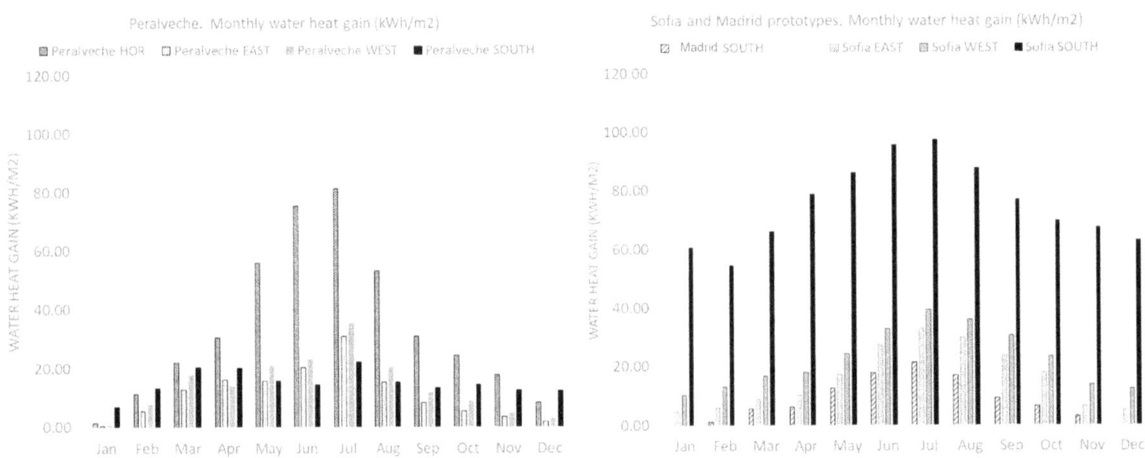

Key Performance Indicator 2: Primary Energy Consumption and CO_2 Emissions

According to the Energy Performance of Buildings Directive EPDB 2018/844 recommendations, primary energy factors (PEFs) were used to assess the energy performance. The considered energy factors were the final energy consumption (FEC in kWh/m^2), non-renewable primary energy consumption (NRPEC in kWh/m^2), and CO_2 emissions (kgCO_{2eq}/m^2). Figure 12 illustrates the monthly heating and cooling loads to maintain the indoor temperature between 22°C and 24°C throughout the year inside the prototypes with a WFG envelope.

Figure 12. Monthly heating and cooling loads in the studied prototypes

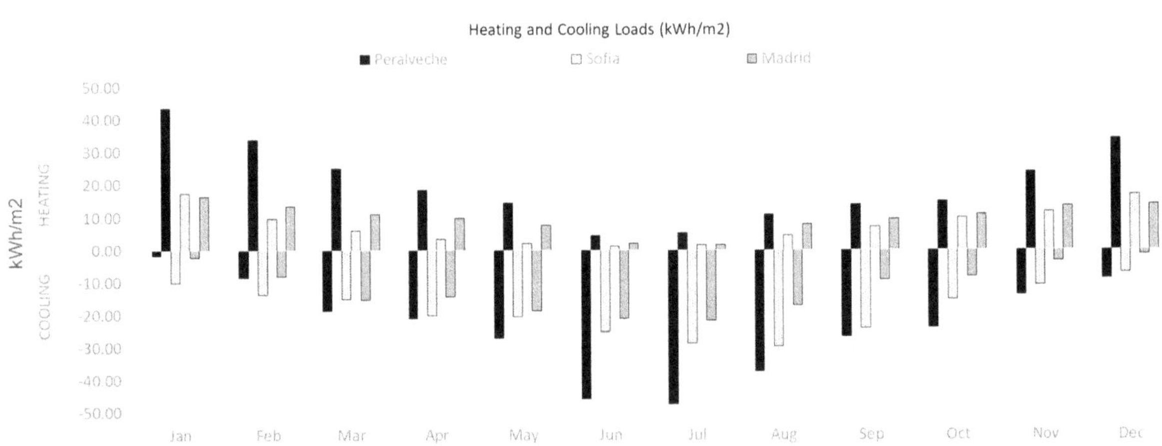

Hydronic technologies are compatible with radiant systems, such as WFG and radiant floors and walls. WFG can be a part of hydronic HVAC systems, and it is compatible with water-to-water and ground-source heat pumps. The performance of water-to-water heat pumps (WWHP) depends on the inlet temperature of the WFG (θ_{IN}), and the source inlet temperature in the heat pump. A typical value of source inlet temperature ranges from 20 °C in ground source heat pumps (GSHP) to 35 °C in other WWHPs. The source and load sides are relevant when it comes to calculating the coefficient of performance (COP) of the cooling device. Air-to-air heat pumps for heat recovery on ventilation are also analyzed. Priarone et al. (2020) have studied that the parameters that influence water-to-water and air-to-air heat pumps' performance are the dry bulb exterior air temperature and the dry bulb interior return air temperature. Tables 6 and 7 illustrate the non-renewable primary energy consumption (NRPE) with different energy generators. It takes into account the effect of the operative temperatures of the system. The conversion factor between final energy and non-renewable primary energy (kWh NRPE/kWh FE) and the factor of emitted CO_2 for electricity were taken from Spanish Regulation of Thermal Installations in Buildings (2016). The values that have been used in this research were 1.954 for the conversion factor between final energy and non-renewable primary energy (kWh NRPE/kWh FE) and 0.331 for CO_2 emissions for electricity. In the final energy balance, the accumulated energy was added to the renewable energy production.

Table 6. Final energy analysis with Water-to-Water heat pumps

	Heating			Cooling		
Location	**Peralveche**	**Madrid**	**Sofia**	**Peralveche**	**Madrid**	**Sofia**
Energy consumption (kWh/m^2)	243.82	118.29	92.25	279.70	139.77	219.10
COP [1]	6.60	6.60	6.60	5.90	5.90	5.90
FE consumption (kWh/m^2)	36.94	17.92	13.98	47.41	23.69	37.14
NRPE consumption (kWh/m^2)	72.19	35.02	27.31	92.63	46.29	72.56
CO_2 emissions (KgCO_2/m^2)	12.23	5.93	4.63	15.69	7.84	12.29

[1]COP values are taken from Appendix A inPriarone et al. (2020)

By using WFG as the heating and cooling system, it is possible to keep the cabin's indoor temperature within the comfort range by setting the inlet temperature (θ_{IN}) between 18 °C and 22 °C. Other systems, such as fan-coils, need lower operating temperatures. Table 7 shows that in the heating mode, the NRPE consumption of water-to-water heat pump is 72.19 kWh per m^2 of envelope in Peralveche, 35.02 kWh per m^2 of envelope in Madrid, and 27.31 kWh per m^2 of envelope in Sofia. In the cooling mode, the NRPE consumption of water-to-water heat pump is 92.63 kWh per m^2 of envelope in Peralveche, 46.29 kWh per m^2 of envelope in Madrid, and 72.56 kWh per m^2 of envelope in Sofia. The average NRPE use of air-to-air heat pumps is 1.57 times as much as the NRPE of water-to-water heat pumps.

When it comes to CO_2 emissions, water-to-water heat pumps generate yearly emissions of 27.92 kgCO_2 per m^2 of envelope in Peralveche, 12.77 kgCO_2 in Madrid, and 16.92 kgCO_2 in Sofia.

Key Performance Indicator 3: Renewable Energy Production vs. Non-Renewable Final Energy Consumption

The energy absorbed by the water was considered as renewable primary energy production (RPE in kWh/m^2). Another useful key performance indicator of these prototypes was to compare the amount of solar energy absorbed by the WFG with the non-renewable primary energy needed to keep the prototype's indoor temperature within a comfort range. Table 8 shows the total renewable energy production in the three prototypes compared to the yearly non-renewable primary energy considering a water-to-water heat pump.

The best production-to-consumption ratio is shown in the Sofia prototype. The primary energy consumption was 5467 kWh for heating and 12986 kWh for cooling. The total water heat gain, considering three different orientations, was 26666 kWh per year. It could be enough to cover the heating needs using an appropriate tank to store thermal energy. However, a PV array would be needed to cover the water-to-water heat pump electricity consumption. Madrid prototype showed the best performance in primary energy consumption. However, it showed the lowest production-to-consumption ratio because the goal was to minimize cooling loads. Finally, the Peralveche prototype showed a negative balance between energy production and primary energy demand and showed the maximum value for primary energy consumption.

Table 7. Final energy analysis with Air-to-Air heat pumps

	Heating			Cooling		
Location	**Peralveche**	**Madrid**	**Sofia**	**Peralveche**	**Madrid**	**Sofia**
Energy consumption (kWh/m^2)	243.82	118.29	92.25	279.70	139.77	219.10
COP [1]	4.20	4.20	4.20	3.30	3.30	3.30
FE consumption (kWh/m^2)	58.05	28.16	21.96	84.76	42.36	66.39
NRPE consumption (kWh/m^2)	113.43	55.03	42.92	165.61	82.76	129.73
CO_2 emissions (KgCO_2/m^2)	19.22	9.32	7.27	28.05	14.02	21.98

[1]COP values are taken from Appendix A inPriarone et al. (2020)

Table 8. Primary energy consumption and renewable energy production in three prototypes

	Renewable Energy Production (kWh)					NRPE Consumption (kWh)
	Horizontal	**East**	**South**	**West**	**Total**	**Total**
Peralveche	1557.26	516.69	684.03	637.23	3395.21	4091.79
Madrid			50.42		50.42	252.13
Sofia		3741.33	17629.73	5295.00	26666.05	18454.10

FUTURE RESEARCH DIRECTIONS

This chapter studies water flow glazing as a key technology to achieve Net-Zero Energy Buildings (NetZEB), reducing energy consumption and greenhouse gas emissions. The aqueous medium of WFG technology offers multiple advantages in terms of energy savings in buildings. It provides an opportunity for transparent facades to offer a real solution to climate change.

When solar radiation penetrates the WFG glazing, a certain amount of this energy is absorbed into the water chamber and later distributed to other parts of the building. Furthermore, another amount of solar energy is transmitted indoors. This energy is reflected diffusely inside the room, creating irradiance. The inner walls absorb some part of the transmitted beam radiation and some diffuse irradiance. Moreno and Hernandez (2018) demonstrated that glass panes absorb beam and diffuse solar radiation, together with diffuse irradiance created inside the room after multiple reflections. This absorbed energy contributes to increasing the indoor temperature. Hence, future work includes evaluating water flow glazing as an effective technology to alleviate overheating inside the buildings.

Energy harvesting strategies may yield considerable energy surplus stored through geothermal borehole heat exchangers or water tanks. The former produces heat with high energy efficiency but needs sufficient ground area and has a high initial cost. The latter allows a cost effective solution, although it requires a significant water volume within an insulated tank. However, Phase Change Materials (PCM) application presents a potential advantage in increasing storage capacity. Chow and Lyu (2017) highlighted that PCM has 5 to 14 times as much storage capacity per unit volume as water. Likewise, more hot water can be harvested during off-work hours for residential use in typical summer and winter weeks using PCM. A fascinating future research project is to implement PCM heat exchangers to enhance WFG energy management.

Some dynamic envelopes react to diverse outdoor conditions by changing their opacity or varying transmission or reflection properties. Gutai and Kheybari (2021) introduces the Smart Water-filled Glass (SWFG) control method, which enables the change of the façade element's opacity by coloring the fluid over a year regarding seasonal changes. A promising future work includes the evaluation of dynamic thermochromic dyes through the glazing.

CONCLUSION

Traditional curtain walls can accomplish a passive role in managing the interior environment of a structure by employing measures such as different glazing coatings and gas cavities. On the other hand, water Flow Glazing technology acts as an active element to manage the incoming solar radiation on the building and provide its occupants with comfort conditions.

The WFG prototype developed in Peralveche employed the dual utilization of renewable solar energy production and borehole heat exchangers. The most prevalent advantage is the cooling effect that is apparent during the summer months. During the sample summer week for the WFG cabin, the exterior environment reached a temperature of 36.5 °C, while the WFG cabin was able to produce a temperature reduction of 16 °C as compared to RC. Judging from yearly data, it is evident that the impact on room heat gains reduction, useful water heat gain, and thermal efficiency improvements were more significant when there was a stronger presence of solar radiation and higher ambient temperature conditions.

The first Key Performance Indicator to assess the performance of WFG is the water heat gain. The water heat gain depends on several variables, including the orientation, the glass composition, and the mass flow rate. The higher the absorptance and mass flow rate, the higher the water heat gain. For south-oriented WFG panels in summer conditions with similar solar irradiance values, the water heat gain ranged from 0.14 kWh/m2 per day in the Madrid prototype to 3.81 kWh/m2 per day in the Sofia prototype. A triple-pane system with an argon chamber outdoor and a low-emissivity coating has shown promising results in absorbing solar energy. On the other hand, if the goal is to reject solar energy, a triple pane system with an argon chamber closest to the exterior combined with a solar-control coating has shown a good performance.

The second Key Performance Indicator is the primary energy consumption and CO2 emissions of the HVAC system. WFG technology can be coupled with water-to-water heat pumps to increase efficiency. The Coefficient of Performance (COP) of water-to-water heat pumps is higher due to the lower temperature difference between the source and load sides. The Non-Renewable Primary Energy consumption (NRPE) of the water-to-water heat pump is 81.31 kWh per m2 of the envelope in Madrid, 99,87 kWh per m2 of the envelope in Sofia, and 164.82 kWh per m2 of the envelope in Peralveche. In terms of energy consumption, the best performance of these three is seen in the Madrid prototype because the triple WFG panels are combined with a low window-to-opaque envelope ratio. The worst performance of these three is the Peralveche prototype; this prototype only employs double pane WFG panels (as opposed to triple pane) and a high window-to-opaque envelope ratio. The average NRPE use of air-to-air heat pumps is 1.57 times as much as the NRPE of water-to-water heat pumps.

The third Key Performance Indicator is the renewable energy production versus non-renewable primary energy demand. The Peralveche WFG prototype showed the worst performance. The horizontal WFG produce a fair amount of hot water over the summer, but the cooling loads were too high, so the goal of achieving a NetZEB is quite far. Selecting the right glazing for each orientation and reducing the glass-to-floor area would help achieve the goal, as it was shown in the Sofia prototype.

By actively collecting solar thermal energy and reducing the unfavorable solar-related indoor heat gain, WFG helps achieve energy conservation as a high-performance building envelope. The thermal performance depends on the glass composition, the mass flow rate, and the glazing orientation. A detailed study is necessary at an early stage of the design project to select the features of WFG according to the building's energy management strategy.

The WFG system has several challenges to address. Firstly, retrieving detailed inputs is necessary to assess Water Flow Glazing operation. An advanced energy management system with smart meters and actuators should be deployed along with the WFG facade to integrate the heat pump operation and the ventilation system. These devices show limitations concerning the regularity and precision of data.

Secondly, more case studies in several different climate regions should be analyzed. The system is originally meant to operate in areas with mild winters and warm summers. When the freezing risk is high, the performance in the harsh wintertime has to be tested. Finally, a whole commissioning protocol, including design, manufacturing, and maintenance, should be explored to make construction stakeholders involved in adopting this product.

The market adoption of WFG system is limited by its high initial cost, particularly when the system is coupled with borehole heat exchangers. Production and deployment processes must be standardized to bring down initial costs and payback periods

ACKNOWLEDGMENT

This work was supported by program Horizon 2020-EU.3.3.1: Reducing energy consumption and carbon footprint by smart and sustainable use, project Ref. 680441 (InDeWaG: Industrial Development of Water Flow Glazing Systems). Special thanks to the Central Laboratory of Solar Energy and New Energy Sources of the Bulgarian Academy of Science (CL SENES–BAS) for providing measured data of the solar radiation on the different facades of the test facility in Sofia, Bulgaria. This research has been funded by a KSC Faculty Development Grant (Keene State College, New Hampshire, USA).

REFERENCES

Abdulmohsin, H., Aritra, G., Senthilarasu, S., & Tapas, K. M. (2020). Evaluation of thermal performance for a smart switchable adaptive polymer dispersed liquid crystal (PDLC) glazing. *Solar Energy*, *195*, 185–193. doi:10.1016/j.solener.2019.11.024

Allen, K., Connelly, K., Rutherford, P., & Wu, Y. (2017). Smart windows—Dynamic control of building energy performance. *Energy and Building*, *139*, 535–546. doi:10.1016/j.enbuild.2016.12.093

Andresen, I. (2018). Definitions, Targets, and Key Performance Indicators for New and Renovated Zero Emission Buildings. In R. Wang & X. Zhai (Eds.), *Handbook of Energy Systems in Green Buildings* (pp. 35–60). Springer. doi:10.1007/978-3-662-49120-1_47

Arpino, F., Cortellessa, G., & Frattolillo, A. (2015). Experimental and numerical assessment of photovoltaic collectors performance dependence on frame size and installation technique. *Solar Energy*, *118*, 7–19. doi:10.1016/j.solener.2015.05.006

Attia, H. A., & Del Ama Gonzalo, F. (2019). Stand-alone pv system with mppt function based on fuzzy logic control for remote building applications. *International Journal of Power Electronics and Drive Systems*, *10*(2), 842–851. doi:10.11591/ijpeds.v10.i2.pp842-851

Baranzelli, C., Lavalle, C., Sgobbi, A., Aurambout, J., Trombetti, M., Jacobs, C., Cristobal Garcia, J., Kancs, D., & Kavalov, B. (2016). *Regional Patterns of Energy Production and Consumption Factors in Europe—Exploratory Project EREBILAND—European Regional Energy Balance and Innovation Landscape; EUR 27697*. Publications Office of the European Union.

Boemi, S. N., & Tziogas, C. (2016). Indicators for Buildings' Energy Performance. In S. N. Boemi, O. Irulegi, & M. Santamouris (Eds.), *Energy Performance of Buildings* (pp. 79–93). Springer. doi:10.1007/978-3-319-20831-2_5

Buonomano, A., Forzano, C., Kalogirou, S. A., & Palombo, A. (2019). Building-façade integrated solar thermal collectors: Energy-economic performance and indoor comfort simulation model of a water based prototype for heating, cooling, and DHW production. *Renewable Energy*, *137*, 20–36. doi:10.1016/j.renene.2018.01.059

Casini, M. (2015). Smart windows for energy efficiency of buildings. *International Journal of Civil and Structural Engineering*, *2*, 230–238.

Chow, T. T., & Li, C. (2013). Liquid-filled solar glazing design for buoyant water-flow. *Building and Environment*, *60*, 45–55. doi:10.1016/j.buildenv.2012.11.010

Chow, T. T., Li, C., & Lin, Z. (2010). Thermal characteristics of water-flow double-pane window. *International Journal of Thermal Sciences*, *50*(2), 140–148. doi:10.1016/j.ijthermalsci.2010.10.006

Chow, T. T., & Lyu, Y. (2017). Numerical analysis on the advantage of using PCM heat exchanger in liquid-flow window. *Applied Thermal Engineering*, *125*, 1218–1227. doi:10.1016/j.applthermaleng.2017.07.098

Edwards, R., Larivé, J. F., Rickeard, D., & Weindorf, W. (2014). *Well-To-Tank Report Version 4.a. Well-to-Wheels Analysis of Future Automotive Fuels and Powertrains in the European Context*. Publications Office of the European Union.

EN 410. (2011). *Glass in Building—Determination of Luminous and Solar Characteristics of Glazing*. German Institute for Standardization.

EN 673. (2011). *Glass in Building—Determination of Thermal Transmittance (U Value)—Calculation Method*. German Institute for Standardization.

European Union. (2018). *Directive (EU) 2018/844 of the European Parliament and of the Council of 30 May 2018. Amending Directive 2010/31/EU on the Energy Performance of Buildings and Directive 2012/27/EU on Energy Efficiency* https://eur-lex.europa.eu/legal-content/EN/TXT/PDF/?uri=CELEX:32018L0844&from=EN

Frattolillo, A., Canale, L., Ficco, G., Mastino, C. C., & Dell'Isola, M. (2020). Potential for Building Façade-Integrated Solar Thermal Collectors in a Highly Urbanized Context. *Energies*, *13*(21), 5801. doi:10.3390/en13215801

Ghosh, A., Norton, B., & Duffy, A. (2016). Behaviour of a SPD switchable glazing in an outdoor test cell with heat removal under varying weather conditions. *Applied Energy*, *180*, 695–706. doi:10.1016/j.apenergy.2016.08.029

Gil-Lopez, T., & Gimenez-Molina, C. (2013). Influence of double glazing with a circulating water chamber on the thermal energy savings in buildings. *Energy and Building*, *56*, 56–65.

Gutai, M., & Kheybari, A. G. (2020). Energy consumption of water-filled glass (WFG) hybrid building envelope. *Energy and Building*, *218*, 110050.

Gutai, M., & Kheybari, A. G. (2021). Energy consumption of hybrid smart water-filled glass (SWFG) building envelope. *Energy and Building*, *230*, 110508.

Ji, J., Luo, C., Chow, T. T., Sun, W., & He, W. (2011). Thermal characteristics of a building-integrated dual-function solar collector in water heating mode with natural circulation. *Energy*, *36*(1), 566–574. doi:10.1016/j.energy.2010.10.004

Koffi, B., Cerutti, A., Duerr, M., Iancu, A., Kona, A., & Janssens-Maenhout, G. (2017). *CoM Default Emission Factors for the Member States of the European Union - Version 2017, European Commission, Joint Research Centre (JRC)* [Dataset]. https://data.europa.eu/89h/jrc-com-ef-comw-ef-2017

Li, C., & Chow, T. T. (2011). Water-filled double reflective window and its year-round performance. *Procedia Environmental Sciences*, *11*, 1039–1047. doi:10.1016/j.proenv.2011.12.158

Lyu, Y. L., Chow, T. T., & Wang, J. L. (2018). Numerical prediction of thermal performance of liquid-flow window in different climates with anti-freeze. *Energy*, *157*, 412–423. doi:10.1016/j.energy.2018.05.140

Lyu, Y. L., Liu, W. J., Su, H., & Wu, X. (2019). Numerical analysis on the advantages of evacuated gap insulation of vacuum-water flow window in building energy saving under various climates. *Energy*, *175*, 353–364. doi:10.1016/j.energy.2019.03.101

Lyu, Y. L., Wu, X., Li, C., Su, H., & He, L. (2019). Numerical analysis on the effectiveness of warm water supply in water flow window for room heating. *Solar Energy*, *177*, 347–354. doi:10.1016/j.solener.2018.11.033

Moreno, B., & Hernández, J. A. (2018). Analytical solutions to evaluate solar radiation overheating in simplified glazed rooms. *Building and Environment*, *140*, 162–172.

Moreno Santamaria, B., Ama Gonzalo, F., Lauret Aguirregabiria, B., & Hernandez Ramos, J. A. (2020). Evaluation of Thermal Comfort and Energy Consumption of Water Flow Glazing as a Radiant Heating and Cooling System: A Case Study of an Office Space. *Sustainability*, *12*, 7596.

Peterson, K., Torcellini, P., & Grant, R. (2015). *A common definition for Zero Energy Buildings. U.S. Department of energy*. Energy Efficiency and Renewable Energy.

Priarone, A., Silenzi, F., & Fossa, M. (2020). Modelling Heat Pumps with Variable EER and COP in EnergyPlus: A Case Study Applied to Ground Source and Heat Recovery Heat Pump Systems. *Energies*, *13*, 794.

Sartori, I., Napolitano, A., & Voss, K. (2012). Net zero energy buildings: A consistent definition framework. *Energy and Building*, *139*, 220–232.

Sierra, P., & Hernandez, J. A. (2017). Solar heat gain coefficient of water flow glazing. *Energy and Building*, *139*, 133–145.

Spanish Regulation of Thermal Installations in Buildings RITE. (2016). *Factores de Emisión de CO_2 y Coeficientes de paso a Energía Primaria de Diferentes Fuentes de Energía Final Consumidas en el Sector de Edificios en España. Instituto para la Diversificación y Ahorro de la Energía*. IDAE.

KEY TERMS AND DEFINITIONS

Building Envelope: The integrated elements of a building that separate its interior from the outdoor environment.

Circulating System: A group of devices that allows the water flow through the Water Flow Glazing and it is made of at least a water pump, a heat plate exchanger, and a precision flow meter and a thermometer for monitoring the inlet and outlet temperature.

Dynamic G-Factor: The coefficient used to measure the glazing's ability to transmit solar energy can change in response to an environmental, temperature, or electrical control. Water Flow Glazing can vary its dynamic g-factor by changing the mass flow rate.

Dynamic U-Value: A measure of heat loss in an element of the building envelope depending on the variable mass flow rate through the water flow glazing. It is indicated in units of Watts per meter squared per Kelvin (W/m^2K).

Mass Flow Rate: The mass of a fluid which passes through the water flow glazing cavity per unit of time. It is indicated in units of kilogram per second.

Water Flow Glazing: A double or triple glazing with a water chamber connected to a circulating device allowing the flow of water through the glass panel in a closed-circuit exchanging heat with the environment.

Water Heat Gain Efficiency: The ratio of total water heat gain compared to the daily solar radiation. In the Water Flow Glazing, the daily water heat gain is accumulated when the inlet temperature is lower than outlet temperature.

Chapter 3
Energy-Efficient Homes:
A Heaven for Respiratory Illnesses

Alejandro Moreno-Rangel
https://orcid.org/0000-0001-6405-4233
Lancaster University, UK

Genny Carrillo
https://orcid.org/0000-0002-8528-422X
Texas A&M University, USA

ABSTRACT

The built environment has changed dramatically due to the increased interest in mitigating climate change. Homes are becoming more energy-efficient, responding to energy issues, and reducing carbon emissions primarily. Nevertheless, we started to realize the unintended consequences of these changes that impact a home's indoor environment and occupants' health. Indoor air quality is a critical aspect as indoor pollutants are increasing in homes. More than ever, it is crucial to adhere to the best ventilation practices, building materials, and cleaning products. Additionally, behaviour changes, such as those for healthy homes, can prevent their health impact. Interdisciplinary research between public health and building professionals needs to educate citizens and present evidence for legislative changes and recommendations to spur change to reduce indoor air pollution and protect vulnerable populations preventing harmful effects on future generations' health.

INTRODUCTION

This chapter considers the impact of sustainable architectural design in the indoor environment and human health by looking at the built environment's energy-efficiency evolution, which has changed mainly due to climate change concerns. As homes have become more energy-efficient, unintended consequences such as poor indoor air quality and respiratory illness arise. This chapter discusses these issues within residential indoor environments, emphasizing indoor air quality (IAQ), possible health outcomes, and prevention considerations.

DOI: 10.4018/978-1-7998-7023-4.ch003

Issues, Controversies, Problems

The world's overgrowth has created challenges for many countries to have enough space to build houses, to design and develop energy-efficient homes, and reduce the cost to operate them compared to existing households. The new "energy smart" houses are 40-60% more efficient than older homes (Cooperative Energy Futures, no date). They are designed with an improved Mechanical Ventilation with Heat Recovery (MVHR) system, high-performance windows, no air filtration, and water efficiency. It is acknowledged that energy-efficient homes offer great comfort due to the combination of temperature, humidity, and air movement (Schieweck *et al.*, 2018). Part of the efficient promise is their resale value due to its high-performance standard and low utility costs. Many people are primarily interested in the body of evidence about climate change and how an energy-smart house can decrease our need for energy, water and improve our world. Nonetheless, indoor environment problems (i.e., poor indoor air quality and overheating) may arise when these aspects are not adequately addressed, causing health problems (Davies and Harvey, 2008).

The energy-efficient homes need to use non-toxic materials in construction to improve indoor air quality, which can reduce the rate of respiratory illnesses such as asthma (Institute of Medicine, 2000). The materials and products used need to be emission-free and have very little or no VOC (Volatile organic compound) content. They also need to be moisture-resistant to prevent mould spores from growing inside the house. Indoor air quality (IAQ) can improve through ventilation and materials in the house's construction that controls humidity and allow a building to breathe (Crump, Dengel and Swainson, 2009). The house needs to be maintained responsibly, and owners need to keep up to date with changing standards and policies related to energy-efficient homes.

This chapter has the following objectives (a) to review the changes for energy-efficient dwellings and their impact on the indoor environment, (b) to describe the most common indoor air pollutants problems and their implications to human health, (c) to present some considerations for prevention and (d) to discuss healthy homes practices.

RESIDENTIAL INDOOR ENVIRONMENT

Climate Change and the Built Environment

The influence of climate change is one of the most pressing matters of our time (Mcnutt, 2013). Perhaps, it is the most significant global impact caused by human activities (Treut *et al.*, 2007). Humans are polluting the Earth to a no-return point on which the natural systems cannot remain stable. In addition, the rate of the use of non-renewable resources has surpassed historic peaks to the end of jeopardizing the destabilization of Earth's carbon cycle, creating new risks, and amplifying existing ones to natural and human systems (IPCC, 2014).

The deterioration of the Earth's carbon cycle affects how natural carbon sinks absorb CO_2 from the atmosphere, causing an increase of 2°C of the planet's mean surface temperature (Schellnhuber *et al.*, 2006), unlikely to be avoided by the end of this century. In fact, between 1950 and 2010, greenhouse gas emissions have caused an increase of 0.6-0.7°C (IPCC, 2014, p. 48). The International Panel of Climate Change's (IPCC) fifth report (IPCC, 2013, p. 4) notes:

"Warming of the climate system is unequivocal, and since the 1950s, many of the observed changes are unprecedented over decades to millennia. The atmosphere and the ocean have warmed, the amounts of snow and ice have diminished, sea levels have risen, and the concentrations of greenhouse gases have increased."

According to the IPCC (2014) report, the Earth's surface temperature is projected to rise between 0.3°C to 5.4°C over all the projected scenarios by 2100 (Figure 1). They also state that heatwaves are very likely to occur more often and last longer. Extreme precipitation will become more intense and frequent, and oceans will continue to warm and acidify, and sea levels will continue to rise. Likewise, these effects are also exacerbated by air pollution. Climate change, air pollution, and human health are related (Fairweather, Hertig, and Traidl-Hoffmann, 2020) see Figure 2, particularly asthma and allergies. Air pollution and global warming are caused primarily by the combustion of gas, solid or liquid fuels during energy production and use. One of the problems from green gas emissions (i.e., nitrogen oxides, ozone) is the chemistry interactions from these air pollutants. The tropospheric ozone interacts with other gases such as CO_2 and NO_2. Furthermore, climate change impacts the tropospheric and stratospheric layers (Ramanathan and Feng, 2009).

As emphasized by Hopfe & McLeod (2015, p.3), "*society faces rising energy prices, increased resource competition, and a moral imperative to create a sustainable built environment.*" Clean growth, the sensible use of resources, as well as an increased emphasis on a reduction in energy consumption, sustainability, and resilience should move upfront on the agenda.

Drivers to Change the Way We Build

Building construction has evolved in recent years, particularly concerning housing. Environmental concerns, high energy costs, and an increasing demand for housing have stimulated the transition towards energy-efficient homes (Sadineni, Madala, and Boehm, 2011). However, these approaches primarily focus on energy consumption and carbon emissions (Anderson, Wulfhorst, and Lang, 2015), and other aspects, such as health and IAQ, had been left aside. The built environment is accountable for 40% of the global annual energy consumption (Liu, Zhao, and Tang, 2010). On their own, buildings should provide adequate indoor environmental conditions for human activities. Indoor comfort performs a vital role in building energy consumption, as cooling and heating may account for 60-70% (Pérez-Lombard, Ortiz and Pout, 2008).

A crucial concern, however, is that carbon emissions from the built environment are growing. The Paris Agreement and the Kyoto Protocol seek to mitigate climate change by reducing greenhouse emissions. In a way, they denote the first steps towards sustainable buildings. It is estimated that residential carbon emissions related to energy consumption increased at an annual rate of 1.7% between 1971 and 2004 (IPCC, 2007). In the European Union, buildings are responsible for around 40% of the total carbon emissions (Petersdorff, Boermans, and Harnisch, 2006), while in the U.K., residential carbon emissions stand at about 15% (BEIS, 2019) and 19% in the USA (EPA, 2018).

The housing demand will continue to grow in the forthcoming years, as developed regions extend—and so will related carbon emissions and energy demands. Energy-efficient homes typically use mechanical ventilation systems (Sharpe *et al.*, 2016), higher levels of airtightness, better insulation (Feist *et al.*, 2005), and reduced ventilation rates (Dimitroulopoulou, 2012) to lower their energy demands, and in doing so, the related long-term carbon emissions. Ventilation is a critical aspect that impacts thermal comfort, indoor air quality (microbial and chemical), and moisture-related allergens in homes.

Implementing a wide range of passive or active energy-efficient methodologies can help enhance energy efficiency in homes. Active methods include a range of improvements that require energy to function, such as MVHR, heating, energy-efficient electric bulbs, white goods, and other appliances. Passive technologies rely on or make the best use of natural resources, such as solar power for lighting and heating, daylight, natural ventilation, or thermal mass. As interest in energy-efficient homes is growing, new approaches to low-energy buildings, such as Passivhaus, LEED, BREEAM, have been developed. The construction process is explained in Figure 3.

Figure 1. Time series of global climate changes, risks, and impacts (2006-2100).
Source: (IPCC, 2014)

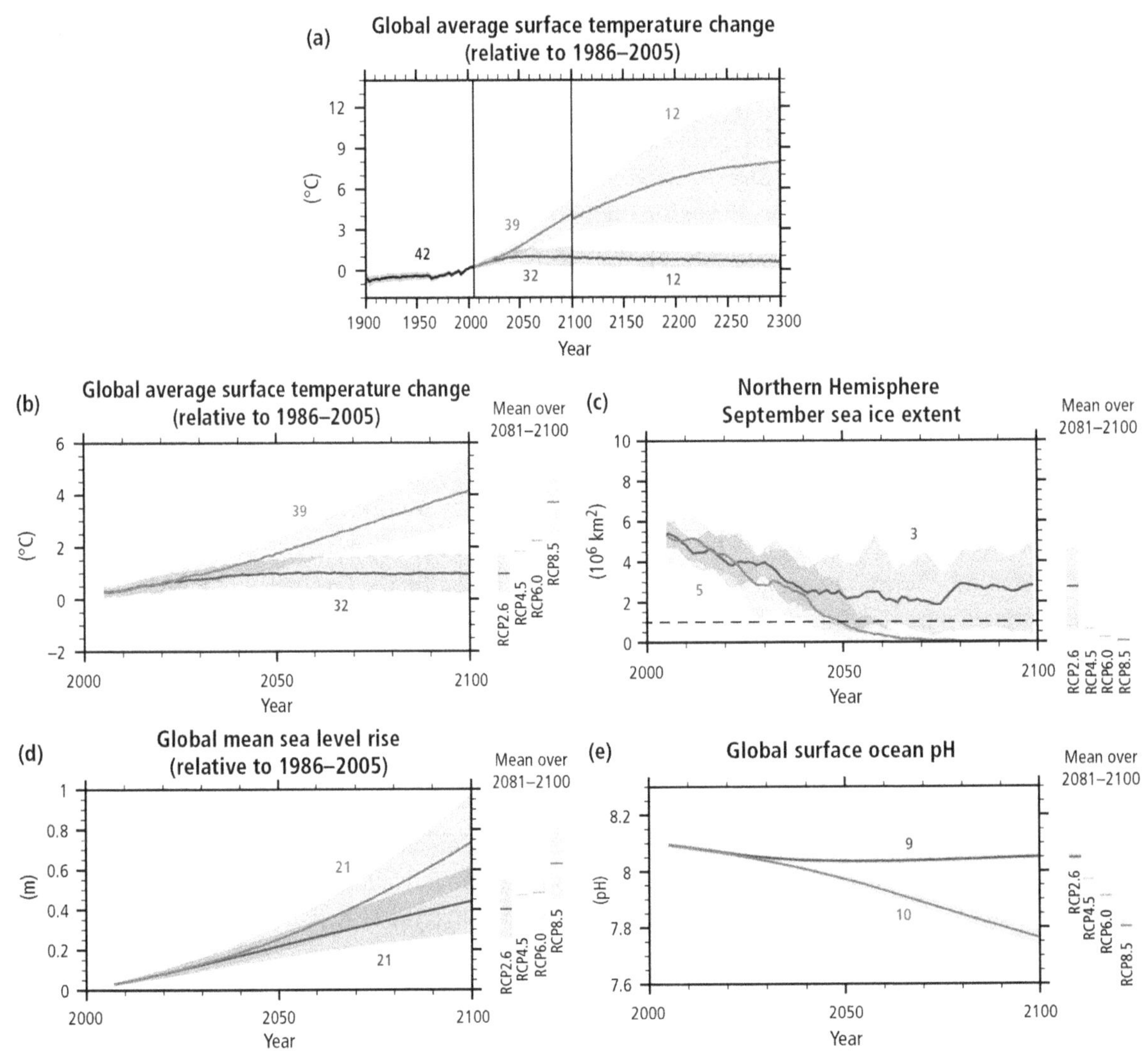

Figure 2. Air pollution trends between 1990–2040. Source: (Amann et al., 2020)

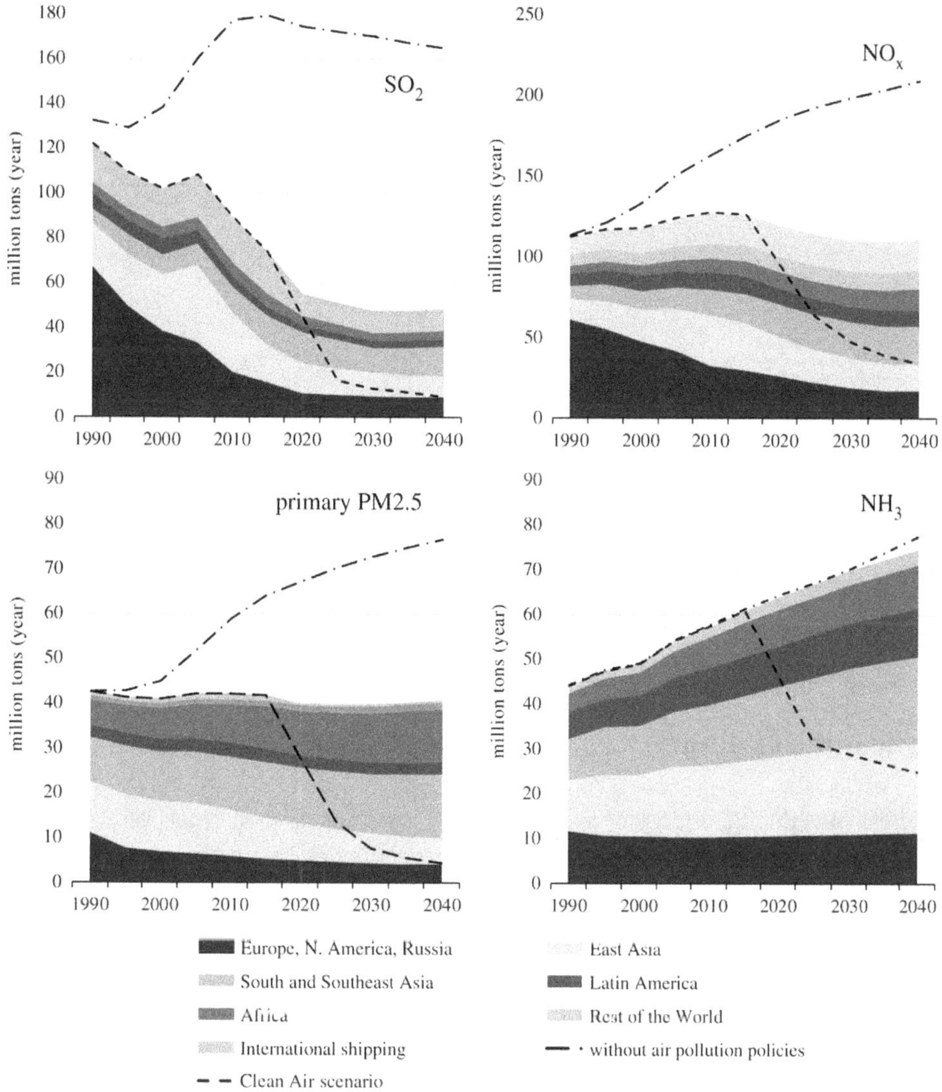

Governments have set targets for low-energy buildings to encourage the transition to energy-efficient buildings and reduce carbon emissions. For instance, the Energy Performance in Buildings Directive, established in 2010, requires all new European buildings to be near ZEB by 2020 (CEC, 2010). Some states in the USA, such as California, established similar goals for near ZEB dwellings by 2020 (CEC, 2007). The U.K. government agreed to achieve zero carbon homes by 2016 as semi-detached, detached, and end of terrace homes 46kWh/m^2, while new mid-terraced houses and apartments aimed to achieve 39kWh/m^2 for heating and cooling demands. Nevertheless, this proposal ended in 2015 (Ares, 2016), following the governments' announcement to stop the Allowable Solutions Carbon Offsetting Scheme and '*keep energy efficiency standards under review*' (H.M. Treasury, 2015, p. 46). The critical barriers for its implementation were the increased capital cost, public awareness, scheme viability, and knowledge of occupants (Heffernan *et al.*, 2016).

Figure 3. Energy-efficient construction process. Source: (Bere, 2013)

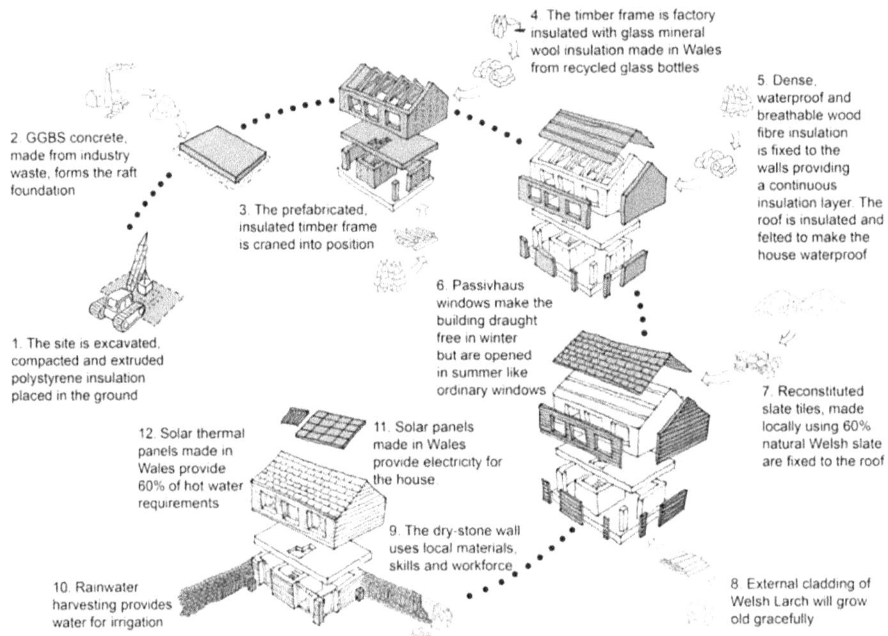

Energy-Efficient Buildings

With the interest of reducing energy consumption and lowering carbon emissions, new approaches to building design have reduced energy demands and improved indoor environmental conditions by reducing heat losses with passive and active methods. Existing building components and energy systems have improved drastically since the 1970s. Some innovative buildings achieved extraordinary heat demand reductions; nevertheless, the additional costs were so high that they could not be repaid by saving fuel costs (Feist *et al.*, 2005). We are transitioning from traditional building practices to ZEB. Such a transition has been possible through different approaches to building design and construction to meet low-energy demands. Nevertheless, most of this transition has been driven by energy and carbon emissions concerns, while others, such as health, were left aside.

The current transition to near Zero-Energy Buildings has been made possible through different building design approaches for energy demands. Such buildings provide a proven record for low/ultra-low-energy; besides, they are also economical, resource-efficient, and provide high occupant comfort and resilience to future climate changes (Hopfe and McLeod, 2015). Hence, different organizations formulated rating systems to promote low-energy-low-carbon building design and construction; and recognize and quantify such achievements through certification. Internationally recognized systems include BREEAM (Building Research Establishment Environmental Assessment Methodology), LEED (Leadership in Energy & Environmental Design), Living Building Challenge, and Passivhaus Standard, among others. However, adopting a standard or regulation does not guarantee the desired effects, as buildings still display performance gaps (Miguez *et al.*, 2006), such as poor indoor air quality (IAQ) and energy.

Some studies suggest that Belgian (Hens, Parijs and Deurinck, 2010), French (Cayre *et al.*, 2011), and British (Kelly, 2011) low-energy homes may consume more energy than expected. The common causes were related to occupant behaviour and indoor environment comfort—such as lighting use, window opening, heating expectations (Masoso and Grobler, 2010)—as the main determinants (Santin, Itard and Visscher, 2009; Gram-hanssen, 2010). For instance, the UK Standard Assessment Procedure (SAP) for new dwellings and reduced SAP (RdSAP) do not assess energy efficiency. Instead, they calculate the performance effectiveness and the additional carbon emissions (Kelly, Crawford-brown and Pollitt, 2012), which may provide misleading estimations. Most discussions on energy-efficient buildings include energy, materials, site impacts, water use, and indoor environmental quality, and IAQ, which merits urgent attention (Persily, 2014). As the building envelope becomes more and more airtight to minimize thermal—and energy—losses, we start realizing the unintended consequences to the indoor environment and our health.

AIR POLLUTION

Air pollution is a mix of hazardous substances from human-made and natural sources (National Institute of Environmental Health Sciences, 2020). Particles and aerosols produced from the combustion process comprise air pollution, and they come from vehicle emissions, burning wood, coal, oil, other fossil fuels, and manufacturing processes. Children, the elderly, and pregnant women are at higher risk of exposure if living in urban areas where they contact smog and microscopic particles. The Clean Air Act of 1970, created in the United States, required the Environmental Protection Agency (EPA) to set National Ambient Air Quality Standards for pollutants considered harmful to public health and the environment. Air quality standards are established for six common air pollutants known as "criteria pollutants" (United States, 2011): carbon monoxide, lead, nitrogen oxides, ground-level ozone, particle matter, and sulfur dioxides.

We are exposed to both indoor and outdoor pollution throughout our lives. However, studies have shown that health risks from exposure to indoor air pollution are higher than outdoor air pollution (United States Environmental Protection Agency, 2020). A mixture of pollutants from indoor (i.e., smoking, cooking, and airborne suspended particles) and outdoor activities (i.e., vehicular traffic and industrial), as well as the building-related factors (ventilation and gas emissions from building materials), influence IAQ (United States Environmental Protection Agency, 2020).

INDOOR ENVIRONMENT PROBLEMS

Thermal Comfort

Thermal comfort is an essential parameter in dwellings. It may be the one that we are more aware of and have the highest degree of control, hence impacting overall energy consumption. Although temperature, and relative humidity, to an extent, are not pollutants, they also have an exacerbating impact on indoor material emission rates (Haghighat and De Bellis, 1998), perceptions of IAQ (Fang, Clausen, and Fanger, 1998), and health in terms temperature extremes (Neill and Ebi, 2009).

Studies have established that comfort diminishes at high temperatures, causing sleep disturbance, reduced productivity, diminished attentiveness, and impaired judgment (Peacock, Jenkins, and Kane, 2010). Exposure to high temperatures increase has several impacts on health. It increases the risk of heart stroke (Bouchama and Knochel, 2002), respiratory and cardiovascular illnesses (Anderson *et al.*, 2013).

Heat-related effects can occur relatively soon after exposure. As such, it is essential to address overheating in homes. Overheating problems in energy-efficient dwellings have been measured (McGill et al., 2016) and contextualized with IAQ measurements (McGill et al., 2015). One of the most accepted definitions of overheating is "*the phenomenon of excessive or prolonged high temperatures in the home, resulting from internal or external heat gains, which may have adverse effects on the comfort, health or productivity of the occupants*" (Zero Carbon Hub, 2015, p. 11). Different criteria are utilized to evaluate the risk of overheating, either through static or dynamic criteria. However, there is no universal characterization of overheating conditions, and neither a standard for the domestic sector (Zero Carbon Hub, 2012, 2015).

Thermal comfort is not just a function of high temperature, as other factors are involved (Nicol, 2004), particularly in buildings without mechanical cooling (Nicol and Humphreys, 2002). Relative humidity also impacts thermal comfort. Relative humidity depends on temperature, as air at higher temperatures can hold higher levels of water vapour. When assessing the indoor environment, psychrometric charts provide information about the air's behaviour, revealing the relation between temperature, relative humidity, and moisture content, helping to define a comfort area. Relative humidity can also support identifying the risk of dampness, mould growth, and the proliferation of house dust mites and other invertebrates.

Indoor Air Quality

Several definitions of "acceptable/healthy IAQ" have been suggested over the years, reflecting the evolution of indoor air understanding. In the past, healthy IAQ was associated with outdoor air, building design, and indoor pollution control (WHO, 1991). It was thought that clean outdoor air guaranteed a healthy indoor air environment and that human bio-effluents were the major indoor pollutants (Mølhave *et al.*, 1997). Although these assumptions might still be relevant today, we now understand that the air quality is far more complicated than these aspects.

The number of air pollution sources and airborne contaminants found indoors is considerable; nonetheless, a few of them have been characterized (Katsouyanni *et al.*, 2004). For instance, over 900 pollutants, biological materials, and ultrafine particles associated with building materials (SCHER, 2007) are present in the air. Porteous *et al.* (2014) suggest that occupant activities such as passive indoor drying can influence IAQ as moisture levels may boost dust airborne mould spores and mite populations. Jacobs (2007, p.p. 977) defined indoor air pollution as "*chemical, physical or biological contaminants in the breathable air inside a habitable structure [...]*" with the potential to detriment the well-being of its occupants. Consequently, an IAQ definition should consider health and comfort aspects as Rosseau (2001, p.p. A-3) suggests: an acceptable IAQ should have "*absence of air contaminants which may impair the comfort on the health of building occupants*." Rosseau recognizes that air free from all contaminants is challenging to achieve. Nevertheless, it should be understood as "*the absence of pollutants that can affect typical occupants' health.*" The American Society of Heating, Refrigeration, and Air-cooling Engineers' (ASHRAE 2007, p.3) defines acceptable IAQ for high-occupancy buildings:

"air in which there are no known contaminants at harmful concentrations as determined by cognizant authorities and with a substantial majority (80% or more) of the people exposed do not express dissatisfaction."

As homes adopted more significant requirements of airtightness, ventilation, and insulation to reduce energy consumption through heat losses, the outdoor-indoor air exchange decreased. In addition, the increased use of chemicals and synthetic building materials caused high concentrations of chemical off-gassing, such as VOCs, on top of human bio-effluents (Zhang and Smith, 2003). Perhaps the most efficient way to control and reduce indoor air pollution is to limit its sources (Fanger, 2006). Sundel (2014) advocates that building factors that may influence IAQ are ventilation, dampness, building materials, and indoor air chemistry. However, "*we 'know' that building characteristics such as 'dampness' a low ventilation rate and certain building (furnishing) materials are, but we do not know how, or why*" (Sundell, 2004, p. 57).

Sources of Indoor Air Pollution

The sources of indoor air pollutants in buildings are varied. Nevertheless, we can characterize the air pollutants as they include allergens, combustion products, tobacco smoke, volatile organic compounds (VOCs), as well as gases from building materials, furnishing, cleaning, and personal care products (Jacobs, Kelly and Sobolewski, 2007). Biological particles (bacteria, fungi, pollen, and cockroach allergens) have been linked with asthma or its exacerbation (Do, 2016). A summary of sources and emissions is provided in Table 1 and Figure 4. Although sources play a vital role, the actual concentration of the pollutants also depends on other factors. According to Maroni et al. (1995), they are:

- the volume of air contained indoors
- the rate of production or release of the pollutant
- the rate of removal of the pollutant
- the rate of exchange with the outside atmosphere
- outdoor pollutant concentration.

The most common indoor air pollutants are:

Carbon dioxide (CO_2) levels are associated with human-related pollution, but they may be disassociated from other pollution sources, such as off-gassing from building materials and furniture. Levels of CO_2 are frequently used as indicators of ventilation (Porteous, 2011) and IAQ (Curwell, March and Venables, 1990). Levels below 1,000 ppm are considered adequate for these purposes (Porteous, 2011) as human bio-effluents were considered the major indoor air pollutant (Mølhave *et al.*, 1997). Wargocki (2016, p. 114) states that "*ventilation rate in homes is associated with health in particular with asthma, allergy, airway obstruction, and SBS symptoms [...] ventilation rates above 0.4h*$^{-1}$ *or* CO_2 *below 900ppm in homes seem to protect against health risk.*" Nevertheless, CO_2 itself is not an indoor air pollutant (Satish *et al.*, 2012) despite adverse effects on human well-being and productivity (Kajtar *et al.*, 2006). CO_2 is influenced by several parameters, such as its disposal, building openings, and heat (Steiger, Hellwing, and Junker, 2008). Therefore, a more prudent practise is to use CO_2 as a metric of outdoor air ventilation (Sundell *et al.*, 2011).

Indoor CO_2 concentrations above 700 ppm of the outdoor level are deemed acceptable (ASHRAE, 2007), but this is founded on the assumption that outdoor CO_2 levels are adequate (typically 300-500ppm). Mechanical ventilation is crucial to meet this target, particularly in energy-efficient homes with high airtightness levels (SCHER, 2007). The EN 13779:2007 categorizes the IAQ in occupied zones based on four IDA categories dependent on the CO_2 levels above the outdoor air with recommended outdoor air flows (Table 2). Normal outdoor CO_2 levels can reach 400ppm, although, in city centres, 500ppm is a more realistic assumption (CIBSE *et al.*, 2015). Using the IDA 2 values and 500ppm as the outdoor baseline would result in CO_2 levels between 900-1,100ppm (1,000ppm default value) for indoor spaces.

Table 1. Sources and emissions of air pollution.

Sources	Emission
Building materials and elements	
Fire retardants	Asbestos
Insulation	Asbestos, formaldehyde
Boilers	Carbon monoxide
Stoves	Carbon monoxide
Gas or kerosene heaters	Carbon monoxide
Particleboard and plywood	Formaldehyde
Furnishing	Formaldehyde
Air conditioning systems	Micro-organisms
Adhesives and solvents	Volatile organic compounds
Paint	Volatile organic compounds
Building materials (concrete, stone)	Volatile organic compounds, radon
Internal surfaces	Fungal spores
Human-related (activities and occupants)	
Respiration	Carbon dioxide
Combustion (cooking, fireplace)	Carbon dioxide, volatile organic compounds, particulates.
Fuel-burning	Carbon monoxide, nitrogen dioxide, polycyclic aromatic hydrocarbons, sulphur dioxide
Tobacco smoke	Carbon monoxide, volatile organic compounds, particulates, polycyclic aromatic hydrocarbons
People	Micro-organisms
House Dust	Allergens
Domestic animals	Allergens, micro-organisms
Cleaning products	Volatile organic compounds
Outdoors	
Motor vehicles (in garages)	Carbon dioxide, nitrogen dioxide
Outdoor air	Biological particles, benzene, nitrogen dioxide, particulates, pollens, sulphur dioxide
Trees, grass, weeds, and plants	Pollens, fungal spores
Soil	Radon, fungal spores

Adapted from Crump et al. (2009) and Spengler and Sexton (1987)

Figure 4. Sources of indoor air pollution. Source: (Holgate et al., 2020).

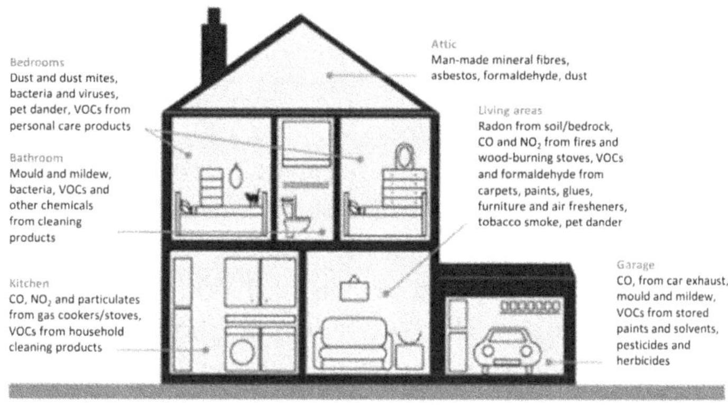

Sources and types of indoor pollution encountered in homes. VOCs = volatile organic compounds. Please note that these lists are not exhaustive and that the actual pollutants present, and their amounts, will vary from household to household.

Table 2. IDA categories based on the EN 13779:2007

Category	CO_2 levels above outdoor air (ppm)		Rate of outdoor air per person (m^3/h)			
			Non-smoking areas		Smoking areas	
	Typical range	Default value	Typical range	Default value	Typical range	Default value
IDA 1 (high IAQ)	£400	350	>54	72	>108	144
IDA 2 (medium IAQ)	400-600	500	36-54	45	72-108	95
IDA 3 (moderate IAQ)	600-1,000	800	21.6-36	28.8	43.2-72	57.6
IDA 4 (low IAQ)	>1,000	1,200	<21.6	18	<43.2	36

Particulate matter 2.5µm ($PM_{2.5}$) refers to ultrafine particles or droplets that are 2.5µm or less in diameter. Their composition fluctuates, but it comprises dust, smoke, soot (AQEG, 2012), mineral ashes (i.e., calcium carbonate, chlorides, coal, metal oxides, mould spores, oil ash, pollen, sodium), other airborne matter (i.e., fleece, fur, hair, vegetable fibres, such as cotton, flax, and hemp), silicate materials (i.e., sepiolite clays, zeolites), and textile fibres (i.e., asbestos, glass and ceramic, nylon, polypropylene, silicates, Crump *et al.*, 2002). The impact of airborne particulate matter on human health is linked directly to the size of the particles (Harrison *et al.*, 2010). Continuous exposure to $PM_{2.5}$ may impair health as respiratory disease outcomes correlate to $PM_{2.5}$ exposure and concentrations (Harrison *et al.*, 2010).

As concern for $PM_{2.5}$ effects on human health increases, particularly in residential buildings (Crump, Dengel, and Swainson, 2009), several exposure thresholds have been set. For instance, Laxen et al. (2010) suggest that there is no safe level for short- or long-term exposure. Other thresholds set daily average exposure between 8µg/m³ (National Environmental Protection Council, 2003) and 25µg/m³ (Commission, 2015). Nonetheless, it is generally accepted that levels above 25µg/m³ detriment human health (WHO, 2000).

Total Volatile Organic Compounds (tVOC) are a large, diverse, and ubiquitous group of compounds that vaporize at room temperature. The indoor Volatile Organic Compounds (VOC) mixture is often known as tVOC. In the past, VOCs were difficult to study individually as human health hazards. Still, their impact on health was investigated as a mixture (WHO, 1997), looking at associations between health and temporal exposition of tVOC and their severity (Molhave, 1991). Nowadays, individual VOCs, such as formaldehyde and benzene, are linked with significant health risks (WHO, 2010), and thresholds are set for specific VOC instead of tVOC (Teichman and Howard-reed, 2016).

The World Health Organization (WHO) published guidance for safe exposure to individual VOCs (see (WHO, 2000, 2010), and for a detailed list of individual VOC exposure limits, see CIBSE (2011) and Health and Safety Executive (2011)). There are various thresholds for tVOC concentrations in non-industrial environments, from 25μg/m^3 (Berglund *et al.*, 1997) up to 500μg/m^3 (Delia, 2012). Nevertheless, 300μg/m^3 over 8 hours is generally accepted as a maximum level (ECA, 1992).

TVOC concentrations are typically higher in new buildings, as they emanate from construction materials and building contents (Brown *et al.*, 1994). Indoor VOCs are released from various building materials (carpets, coving, linoleum, particleboard, power cables, and vinyl tiles) and construction consumer products (adhesives, caulks, cleaners, paint strippers, paint thinners, and paints). They are also related to human activities (deodorizers, dry-cleaned clothing, frying food, moulds, personal care products, pesticides, showering, and smoking (Zhang and Smith, 2003).

HUMAN HEALTH IMPACT ON ADULTS AND CHILDREN

Outdoor Air Pollution and Health.

Air pollution is a mix of hazardous substances from human-made and natural sources (National Institute of Environmental Health Sciences, 2020). Particles and aerosols produced from the combustion process comprise air pollution. They come from vehicle emissions, burning wood, coal, oil, other fossil fuels, and manufacturing processes. Children, the elderly, and pregnant women are at higher risk of exposure if living in urban areas where they contact smog and microscopic particles. The Clean Air Act of 1970, created in the United States, required the Environmental Protection Agency (EPA) to set National Ambient Air Quality Standards for pollutants considered harmful to public health and the environment. Air quality standards are established for six common air pollutants known as "criteria pollutants" (United States, 2011). Those are carbon monoxide, lead, nitrogen oxides, ground-level ozone, particle matter, and sulphur dioxides. They can be found in urban and non-urban settings and affected by geography, season, and weather conditions. Particulate matters (PM) are believed to play a primary role in air pollution's systemic health effects. PM is classified according to its aerodynamic diameter. Particle size is of utmost importance since respirable particles of less than 2.5 μm (particulate matter or $PM_{2.5)}$ can penetrate deep into the lungs and cause damage (Xing, 2016). The chemical composition of these particles varies based on source and secondary atmospheric reactions. Metals and polycyclic aromatic hydrocarbons (PAHs), a large class of fused aromatic rings, are found adsorbed to these small particles. Many individual PAHs present in complex mixtures are mutagenic and carcinogenic to humans. Air pollution affects human health in diverse manners– one-third of deaths are due to stroke, lung cancer, and heart disease. Manisalidis (2020) has shown that air pollution is considered the major environmental risk factor in incidences of asthma, lung cancer, ventricular hypertrophy, Alzheimer's and Parkinson's diseases, psychological complications, autism, and low birth weight.

Indoor Air Pollution and Health

Indoor and outdoor pollution contributes to the exposure we face in our lives. Biological particles (bacteria, fungi, pollen, and cockroach allergens) have been linked with asthma or its exacerbation (Do, 2016). Among chemicals changing IAQ are carbon monoxide (CO), radon, volatile organic compounds (VOCs), and ultrafine particulate matter ($PM_{2.5}$), asbestos, lead, nitrogen dioxide (NO_2). Figure 5 shows the different indoor air pollution sources in our houses (Martenies and Batterman, 2018). The most common sources are described in Figure 2.

Exposure to $PM_{2.5}$ has become a substantial concern in public health because it penetrates deeply into the respiratory tract and enters the circulatory system causing diverse health effects. Such diseases include lung cancer, cardiovascular diseases, respiratory diseases, and increased risk for asthma exacerbations (Xing, 2016). Published research related to asthma has shown that asthma education, coupled with home-based interventions and integrated pest management, has improved asthma outcomes (Baek et al., 2019; Moreno Rangel A, 2020). The invisible indoor air pollutants in our homes can cause allergic rhinitis, affecting up to 40% of people worldwide (World Allergy Organization, 2013). It happens when our immune system reacts to allergens in the air, such as dust mites, or moulds, pollen, causing inflammation of the nose lining (World Allergy Organization, 2013). Air pollution has a strong link with allergy prevalence and can cause headaches, runny noses, itchy or watery eyes, and problems sleeping.

Figure 5. Indoor house air pollutants. Source: (United States Environmental Protection Agency, 2020a)

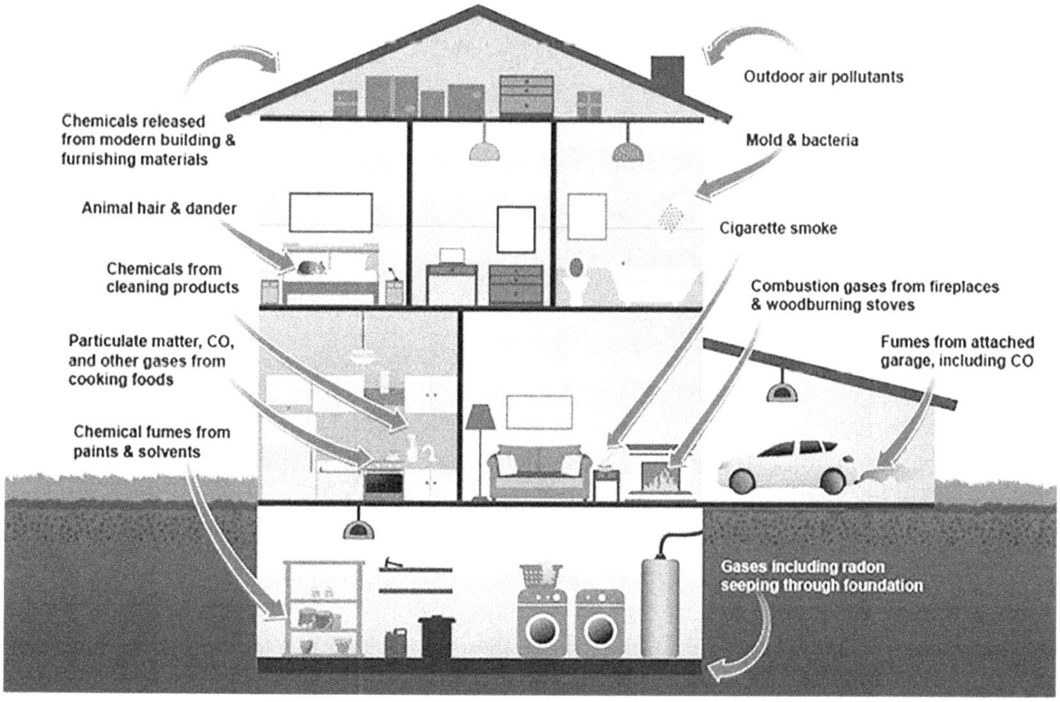

HEALTHY HOMES DEFINITION AND PRACTICES

The National Center for Healthy Housing defines a healthy home as "*housing that is designed, constructed, maintained, and rehabilitated in a manner conducive to good occupant health*" (U.S. Department of Housing and Urban Development, 2020).

We now have emerging scientific evidence associating health outcomes such as asthma, lead poisoning, and unintentional injuries to substandard housing. Applying Healthy Homes principles is now essential for residents' ongoing health; in light of this new research, its principles will help us with the health of the persons who live in the household. All of us are familiar with the challenges that lead has caused to children in different places where lead-paint-based was used for old buildings. As such, health consequences for many children included damage to the brain and nervous system, slowed growth and development, as well as learning and behavioral problems (Hauptman et al., 2017).

Houses have many hazards, not just confined to older or cheaper homes; we can identify diverse types of risks for new or expensive housing residents. When we decide to build a new house, non-toxic materials, safe for the individuals who will live in the house, must be selected. The new house smell is due to the construction chemicals released into the ambient atmosphere and chemicals emitted from new flooring, cabinets, adhesives, glues, plastics, and paint fumes. Those chemicals are off-gassed into the indoor environment (Zhu & Liu, 2014). They will pollute the indoor environment damaging our health, and endanger the air quality in our homes (Zhu & Liu, 2014).

Everyone wants and needs a healthy home since we spend most of our free time in our homes, so we need to create a space that promotes children's healthy growth and development and can save billions in health care costs. There are particular reasons to think about children: children are growing; they eat more food, drink more water, and breathe more air than adults; toddlers crawl and young children play on the ground and put their fingers frequently into their mouths, and they are dependent on their caregivers to take care of them.

The Eight Healthy Homes Principles below can help everyone make their home a healthier place to live (Ferguson & Yates, 2016; Health and Urban Department).

1. **Keep it Dry.** Prevention from water entering the home's structure and indoors through leaks in the roof, walls, faucets, or interior plumbing will keep it dry.
2. **Keep it Clean.** Controlling the dust source, reducing clutter, and using a damp towel to clean will remove the dust.
3. **Keep it Safe.** Storing all household chemicals that are dangerous to our children and placed them out of their reach is essential. Installation of smoke and carbon dioxide detectors, as well as fire extinguishers, are critical.
4. **Keep it Well-Ventilated.** Ventilation of the whole house is critical; open the windows to let fresh air circulate throughout the home, preferably through the night. Use a dehumidifier or an extraction fan in the bathroom after showering. In the kitchen, use the exhaust vent and cooking hood while and after cooking.
5. **Keep it Pest-free.** The pests are similar to humans in that they look for food, water, and shelter as we do. Therefore, look for cracks or openings in walls, windows, doors, and seal them. Store food in airtight containers and eat only in the dining room. It is best to use sticky-traps and baits in closed containers instead of pesticides since they are toxic or use the least toxic chemicals like boric acid powder with sugar, but place it out of children and pets' reach.

6. **Keep it Contaminant-free.** Test the house for lead and radon (a naturally occurring dangerous gas that enters homes through soil, crawlspaces, and foundation crack).
7. **Keep your home Maintained.** It is important to inspect, clean, and repair the home routinely, and,
8. **Thermal Controls.** Ensure the indoor temperatures in the house are maintained at an adequate temperature during cold or heat seasons. Ask a technician to check the A/C and ventilation units to avoid increased upkeep costs and possible health problems.

Asthma and Healthy Homes education were vital in improving participants' well-being in a study where participants increased their knowledge of asthma symptoms, case management, and identifying triggers and exacerbating factors (Carrillo G., 2015). Krieger et al. reported improved asthma-related health status in his Seattle-King County Healthy Homes project by reducing exposure to allergens and irritants at home (Krieger et al., 2005).

IMPLICATIONS FOR PUBLIC HEALTH

There has been an increasing recognition that air pollution extends to indoor environments. The health consequences caused by indoor air pollution are respiratory problems such as asthma, chronic pulmonary disease, myocardial arrhythmias, and many more. Research studies have shown that children and their families' education with Healthy Home's principles and asthma self-management improves health outcomes, such as decreased hospitalizations and emergency room visits (Carrillo G, 2015). Information about such principles will help healthcare leaders and paediatricians who serve children with asthma understand how their patients improve their condition and decrease hospitalization costs and Emergency Department visits. Furthermore, research has shown that following the healthy home principles in households; there is an improvement in allergies and asthma symptoms, a better quality of life, and a decreased healthcare utilization.

CONCLUSION

The chapter objectives focused on the sustainable architectural design in indoor air quality and its health impact consequences. Accumulating evidence demonstrates a significant adverse effect on a building environment's energy-efficient and the unintended consequences of the health impact they caused. Research has shown that exposure to indoor pollutants above certain thresholds could cause diverse health conditions. Both building designers and occupants must manage indoor air quality through proper ventilation, building materials, and cleaning products with low VOC emissions. Additionally, individuals living in energy-efficient homes can also make preventive behaviour changes to decrease their health impact on allergies, asthma, cardiovascular, and respiratory problems. The use of green products to clean the homes and follow the eight principles of Healthy Homes has improved health outcomes in children with asthma and allergies. Exposure to air pollution may result in academic performance deficiencies, emotional and behavioural changes and generally lead to deleterious societal consequences. While additional research is essential in larger populations to confirm recent studies, the existing data are quite compelling. Interdisciplinary research between public health and building professionals is needed to educate citizens and present evidence for legislative changes and recommendations to spur change to reduce air pollution. Such policies would protect vulnerable subgroups like children, pregnant women, the elderly population, and immunocompromised individuals to prevent these harmful effects on future generations' health.

ACKNOWLEDGMENT

This research received funding from the Research England Expanding, Excellence in England (E3) fund.

REFERENCES

Amann, M., Kiesewetter, G., Schöpp, W., Klimont, Z., Winiwarter, W., Cofala, J., Rafaj, P., Höglund-Isaksson, L., Gomez-Sabriana, A., Heyes, C., Purohit, P., Borken-Kleefeld, J., Wagner, F., Sander, R., Fagerli, H., Nyiri, A., Cozzi, L., & Pavarini, C. (2020). Reducing global air pollution: The scope for further policy interventions: Achieving clean air worldwide. *Philosophical Transactions - Royal Society. Mathematical, Physical, and Engineering Sciences*, *378*(2183), 20190331. Advance online publication. doi:10.1098/rsta.2019.0331 PMID:32981437

Anderson, G. B., Dominici, F., Wang, Y., McCormack, M. C., Bell, M. L., & Peng, R. D. (2013). Heat-related emergency hospitalizations for respiratory diseases in the medicare population. *American Journal of Respiratory and Critical Care Medicine*, *187*(10), 1098–1103. doi:10.1164/rccm.201211-1969OC PMID:23491405

Anderson, J. E., Wulfhorst, G., & Lang, W. (2015). Energy analysis of the built environment - A review and outlook. Renewable and Sustainable Energy Reviews, 44, 149–158. doi:10.1016/j.rser.2014.12.027

AQEG. (2012). *Fine Particulate Matter in the United Kingdom*. Available at: https://uk-air.defra.gov.uk/assets/documents/reports/cat11/1212141150_AQEG_Fine_Particulate_Matter_in_the_UK.pdf

Ares, B. E. (2016). *Zero Carbon Homes*. House of Commons Library.

ASHRAE. (2007). ASHRAE standard 62.1-2007 Ventilation for Acceptable Indoor Air Quality. ASHRAE.

Baek, J., Huang, K., Conner, L., Tapangan, N., Xu, X., & Carrillo, G. (2019). Effects of the home-based educational intervention on health outcomes among primarily Hispanic children with asthma: A quasi-experimental study. *BMC Public Health*, *19*(1), 912. doi:10.118612889-019-7272-5 PMID:31288792

BEIS. (2019). *Digest of United Kingdom Energy Statistics 2018*. Available at: https://assets.publishing.service.gov.uk/government/uploads/system/uploads/attachment_data/file/736148/DUKES_2018.pdf

Bere, J. (2013) *Introduction to passive house*. RIBA Publishing.

Berglund, B., Clausen, G., Kettrup, A., Lindvall, T., & Maroni, M. (1997). *Total Volatile Organic Compounds (TVOC) in Indoor Air Quality Investigation*. European Communities. doi:10.1111/j.1600-0668.1997.00002.x

Bouchama, A., & Knochel, J. P. (2002). Heat Stroke. *The New England Journal of Medicine*, *346*(25), 1978–1988. doi:10.1056/NEJMra011089 PMID:12075060

Brown, S. K., Sim, M. R., Abramson, M. J., & Gray, C. N. (1994). Concentrations of volatile organic compounds in indoor air–a review. *Indoor Air, 4*(2), 123–134. doi:10.1111/j.1600-0668.1994.t01-2-00007.x

Carrillo, G. S.-A. E., Lucio, R.L, Chong-Menard, B., & Smith, K. (2015, Sep 1). Improving Asthma in Hispanic Families Through a Home-Based Educational Intervention. *Pediatric Allergy Immunology and Pulmonology, 28*(3), 165-171. doi:10.1089/ped.2015.0523

Cayre, E., Allibe, B., Laurent, M., & Osso, D. (2011). There are people in the house! How the results of purely technical analysis of residential energy consumption are misleading for energy policies. *Proceedings of the ECEEE 2011 Summer Study on Energy Efficiency First: The Foundation of a Low-Carbon Society*, 1675–1683.

CEC. (2007). *California Energy Commission*. Energy Policy.

CEC. (2010). *Directive 2010/31/E.U. of the European Parliament and of the Council of 19 May 2010 on the energy performance of buildings, Official Journal of the European Union*. Commission of the European Communities. doi:10.1039/ap9842100196

CIBSE. (2011). *CIBSE KS17 Indoor Air Quality and Ventilation*. Available at: https://www.cibse.org/Knowledge/CIBSE-KS/KS17-Indoor-Air-Quality-Ventilation

CIBSE. (2015). CIBSE Guide A: Environmental Design (8th ed.). Norwich: CIBSE Publications.

Commission, E. (2015). *Air Quality Standards, Air Quality*. Available at: https://ec.europa.eu/environment/air/quality/standards.htm

Cooperative Energy Futures. (n.d.). *Energy Smart Homes*. Available at: https://www.cooperativeenergyfutures.com/energy-smart-homes

Crump, D., Dengel, A., & Swainson, M. (2009). *Indoor air quality in highly energy-efficient homes—a review, Report NF18*. Milton Keynes: National House.

Crump, D., Raw, G. J., Upton, S., Scivyer, C., Hunter, C., & Hartless, R. (2002). A protocol for the assessment of indoor air quality in homes and office buildings. London: BRE Bookshop.

Curwell, S. R., March, C., & Venables, R. (1990). *Buildings and Health: The Rosehaugh Guide to the Design, Construction, Use and Management of Buildings*. RIBA Publ.

Davies, I., & Harvey, V. (2008). *Zero carbon: what does it mean to homeowners and housebuilders?* IHS BRE Press.

Delia, A. (2012). Total Volatile Organic Compounds (Total VOCs) in the air. *State of Knowledge Report*, (March), 1–4. Available at: http://www.environment.gov.au/atmosphere/airquality/publications/sok/vocs.html

Dimitroulopoulou, C. (2012). Ventilation in European dwellings: A review. Building and Environment, 47, 109–125. doi:10.1016/j.buildenv.2011.07.016

Do, D. C., Zhao, Y., & Gao, P. (2016). Cockroach allergen exposure and risk of asthma. *Allergy, 71*(4), 463–474. doi:10.1111/all.12827 PMID:26706467

ECA. (1992). *European collaborative action Indoor air quality & its impact on man: Report No.11: Guidelines for ventilation requirements in buildings*. Available at: https://www.inive.org/medias/ECA/ECA_Report11.pdf

EPA. (2018). *U.S. Greenhouse Gas Emissions and Sinks, 1990-2016, Epa 430-R-18-003*. Doi: EPA 430-R-12-001.

Fairweather, V., Hertig, E., & Traidl-Hoffmann, C. (2020). A brief introduction to climate change and health. *Allergy. European Journal of Allergy and Clinical Immunology*, *75*(9), 2352–2354. doi:10.1111/all.14511 PMID:32779793

Fang, L., Clausen, G., & Fanger, P. O. (1998). Impact of temperature and humidity on the perception of indoor air quality. *Indoor Air*, *8*(2), 80–90. doi:10.1111/j.1600-0668.1998.t01-2-00003.x

Fanger, P. O. (2006). What is IAQ? *Indoor Air*, *16*(5), 328–334. doi:10.1111/j.1600-0668.2006.00437.x PMID:16948709

Feist, W., Schnieders, J., Dorer, V. & Haas, A. (2005). Re-inventing air heating: Convenient and comfortable within the frame of the Passive House concept. *Energy and Buildings*, *37*(11), 1186–1203. . doi:10.1016/j.enbuild.2005.06.020

Ferguson, A. C., & Yates, C. (2016). Federal Enactment of Healthy Homes Legislation in the United States to Improve Public Health. *Frontiers in Public Health*, *4*, 48–48. doi:10.3389/fpubh.2016.00048 PMID:27047913

Gram-hanssen, K. (2010). Residential heat comfort practices: Understanding users. *Building Research and Information*, *38*(5), 175–186. doi:10.1080/09613210903541527

Haghighat, F., & De Bellis, L. (1998). Material emission rates: Literature review, and the impact of indoor air temperature and relative humidity. *Building and Environment*, *33*(5), 261–277. doi:10.1016/S0360-1323(97)00060-7

Harrison, R. M. (2010). Size distribution of airborne particles controls outcome of epidemiological studies. *Science of the Total Environment. Elsevier B.*, *409*(2), 289–293. doi:10.1016/j.scitotenv.2010.09.04 PMID:21109288

Hauptman, M., Bruccoleri, R., & Woolf, A. D. (2017). An Update on Childhood Lead Poisoning. *Clinical Pediatric Emergency Medicine*, *18*(3), 181–192. doi:10.1016/j.cpem.2017.07.010 PMID:29056870

Health and Safety Executive. (2011). *EH4O / 2005 Workplace exposure limits EH4O / 2005 Workplace exposure limits* (2nd ed.). HSE Books.

Health and Urban Department. (n.d.). *The Healthy Homes Program*. Retrieved March 26, 2020, from https://www.hud.gov/program_offices/healthy_homes/hhi

Heffernan, E., Pan, W., Liang, X. & Wilde, P. D. (2016). Zero carbon homes : Perceptions from the U.K. construction industry. *Energy Policy*, *79*(2015), 23–36. doi:10.1016/j.enpol.2015.01.005

Hens, H., Parijs, W., & Deurinck, M. (2010). Energy consumption for heating and rebound effects. *Energy and Building*, *42*(1), 105–110. doi:10.1016/j.enbuild.2009.07.017

Holgate, S. (2020). *The inside story: Health effects of indoor air quality on children and young people*. doi:10.1039/9781839160431-00151

Hopfe, C. J., & McLeod, R. S. (2015). *The PassivHaus designer's manual: A technical guide to low and zero energy buildings* (C. J. Hopfe & R. S. McLeod, Eds.; 1st ed.). Routledge. doi:10.4324/9781315726434

Institute of Medicine. (2000). Clearing the Air: Asthma and Indoor Air Exposures. *Indoor Air*. doi: NBK224477

IPCC. (2007). *Mitigation of climate change: Contribution of working group III to the fourth assessment report of the Intergovernmental Panel on Climate Change*. Intergovernmental Panel on Climate Change. https://www.ipcc.ch/publications_and_data/.htm

IPCC. (2013). *Climatic Change 2013*. Advance online publication. doi:10.1017/CBO9781107415324. Summary

IPCC. (2014). Climate Change 2014: Synthesis Report. Contribution of Working Groups I, II and III to the Fifth Assessment Report of the Intergovernmental Panel on Climate Change. IPCC.

Jacobs, D. E., Kelly, T., & Sobolewski, J. (2007). Linking Public Health, Housing, and Indoor Environmental Policy: Successes and Challenges at Local and Federal Agencies in the United States. *Environmental Health Perspectives, 115*(6), 976–982. doi:10.1289/ehp.8990 PMID:17589610

Kajtar, L. (2006). Influence of Carbon-dioxide Pollutant on Human Well-being and Work Intensity. In *H.B. 2006 - Healthy Buildings: Creating a Healthy Indoor Environment for People* (pp. 85–90). Available at https://www.scopus.com/inward/record.url?eid=2-s2.0-84871535246&partnerID=40&md5=0b0e1ad8a48325274153bf610d7761ea

Katsouyanni, K., Part, P., Andersson, I., Autrup, H., Baños de Guisasola, E., Bjerregaard, P., Mette, B., Bert, B., Calheiros, J., Casse, F. F., Castaño, A., Cochet, C. & Eisenreich, S. (2004) *Baseline Report on Needs in the framework of the European Environment and Helath Strategy ((COM 2003)338 final)*. Academic Press.

Kelly, S. (2011). *Do Homes that are More Energy Efficient Consume Less Energy?: A Structural Equation Model for England's Residential Sector*. EPRG Working Paper No. 1117. doi:10.17863/CAM.5540

Kelly, S., Crawford-brown, D., & Pollitt, M. G. (2012). Building performance evaluation and certification in the U.K.: Is SAP fit for purpose? *Renewable and Sustainable Energy Reviews. Elsevier, 16*(9), 6861–6878. doi:10.1016/j.rser.2012.07.018

Krieger, J. W., Takaro, T. K., Song, L., & Weaver, M. (2005, April). The Seattle-King County Healthy Homes Project: A randomized, controlled trial of a community health worker intervention to decrease exposure to indoor asthma triggers. *American Journal of Public Health, 95*(4), 652–659. doi:10.2105/AJPH.2004.042994 PMID:15798126

Laxen, D., Moorcroft, S., Marner, B., Laxen, K., Boulter, P., Barlow, T., Harrison, R., & Heal, M. (2010). *Final Report - PM 2.5 in the U.K.* SINFER.

Liu, D., Zhao, F. Y., & Tang, G. F. (2010). Active low-grade energy recovery potential for building energy conservation. *Renewable and Sustainable Energy Reviews. Elsevier Ltd*, *14*(9), 2736–2747. doi:10.1016/j.rser.2010.06.005

Manisalidis, I., Stavropoulou, E., Stavropoulos, A., & Bezirtzoglou, E. (2020). Environmental and Health Impacts of Air Pollution: A Review. *Frontiers in Public Health*, *8*(14), 14. Advance online publication. doi:10.3389/fpubh.2020.00014 PMID:32154200

Maroni, M., Seifert, B., & Lindvall, T. (1995). *Indoor air quality a comprehensive reference book*. Amsterdam: Elsevier Science B.V. Available at: https://books.google.co.uk/books?hl=en&lr=&id=qsyLaKnn-nYC&oi=fnd&pg=PP2&dq=Indoor+Air+Quality+A+Comprehensive+Reference+Book,&ots=m DAkLMqddR&sig=RWCLF99Kto-MIWMGV8kLp2RUolM#v=onepage&q=Indoor Air Quality A Comprehensive Reference Book%2C&f=false

Martenies, S. E., & Batterman, S. A. (2018). Effectiveness of Using Enhanced Filters in Schools and Homes to Reduce Indoor Exposures to PM 2.5 from Outdoor Sources and Subsequent Health Benefits for Children with Asthma. *Environmental Science & Technology*, *52*(18), 10767–10776. doi:10.1021/acs.est.8b02053 PMID:30141330

Masoso, O. T., & Grobler, L. J. (2010). 'The dark side of occupants' behaviour on building energy use'. *Energy and Building*, *42*(2), 173–177. doi:10.1016/j.enbuild.2009.08.009

McGill, G. (2016). Meta-analysis of indoor temperatures in new-build housing. *Building Research & Information. Taylor & Francis*, *0*(0), 1–21. doi:10.1080/09613218.2016.1226610

McGill, G., Oyedele, L. O., & McAllister, K. (2015). Case study investigation of indoor air quality in mechanically ventilated and naturally ventilated U.K. social housing. *International Journal of Sustainable Built Environment*, *4*(1), 58–77. doi:10.1016/j.ijsbe.2015.03.002

Mcnutt, M. (2013). *Climate Change Impacts*. Science., doi:10.1126cience.1243256

Miguez, J. L., Porteiro, J., López-González, L. M., Vicuña, J. E., Murillo, S., Morán, J. C., & Granada, E. (2006). Review of the energy rating of dwellings in the European Union as a mechanism for sustainable energy. *Renewable & Sustainable Energy Reviews*, *10*(1), 24–45. doi:10.1016/j.rser.2004.08.003

Molhave, L. (1991). Volatile organic compounds, indoor air quality and health. *Indoor Air*, *1*(4), 357–376. doi:10.1111/j.1600-0668.1991.00001.x

Mølhave, L., Clausen, G., Berglund, B., Ceaurriz, J., Kettrup, A., Lindvall, T., Maroni, M., Pickering, A. C., Risse, U., Rothweiler, H., Seifert, B., & Younes, M. (1997). Total Volatile Organic Compounds (TVOC) in Indoor Air Quality Investigations. *Indoor Air*, *7*(19), 225–240. doi:10.1111/j.1600-0668.1997.00002.x

Molhave, L., Jensen, J. G., & Larsen, S. (1991). Subjective Reactions To Volatile Organic Compounds As Air Pollutants. *Atmospheric Environment*, *25*(7), 1283–1293. doi:10.1016/0960-1686(91)90240-8

Moreno Rangel, A. B. J., Roh, T., Xu, X., &, Carrillo, G. (2020). Assessing Impact of Household Intervention on Indoor Air Quality and Health of Children with Asthma in the US-Mexico Border: A Pilot Study. Texas A&M University, School of Public Health.

National Environmental Protection Council. (2003). *National Environment Protection (Ambient Air Quality)*. Measure.

National Institute of Environmental Health Sciences. (2020). *Air Pollution and Your Health*. NIEHS. Retrieved November 16 from https://www.niehs.nih.gov/health/topics/agents/air-pollution/index.cfm

Neill, M. S. O., & Ebi, K. L. (2009). Temperature Extremes and Health: Impacts of Climate Variability and Change in the United States. *Journal of Occupational and Environmental Medicine*, *51*(1), 13–25. doi:10.1097/JOM.0b013e318173e122 PMID:19136869

Nicol, F. (2004). Adaptive thermal comfort standards in the hot-humid tropics. *Energy and Building*, *36*(7), 628–637. doi:10.1016/j.enbuild.2004.01.016

Nicol, J. F., & Humphreys, M. (2002). Adaptive thermal comfort and sustainable thermal standards for buildings. *Energy and Building*, *34*(6), 563–572. doi:10.1016/S0378-7788(02)00006-3

Peacock, A. D., Jenkins, D. P., & Kane, D. (2010). Investigating the potential of overheating in U.K. dwellings as a consequence of extant climate change. *Energy Policy. Elsevier*, *38*(7), 3277–3288. doi:10.1016/j.enpol.2010.01.021

Pérez-Lombard, L., Ortiz, J., & Pout, C. (2008). A review on buildings energy consumption information. *Energy and Building*, *40*(3), 394–398. doi:10.1016/j.enbuild.2007.03.007

Persily, A. K. (2014). *Indoor Air Quality in High Performance Buildings: What is and isn't in ASHRAE/IES/USGBC Standard 189.1. In 13th International Conference on Indoor Air Quality and Climate*. Indoor Air. Available at https://www.isiaq.org/docs/paper/HP1014.pdf

Petersdorff, C., Boermans, T., & Harnisch, J. (2006). Mitigation of CO2 Emissions from the EU-15 Building Stock Beyond the E.U. Directive on the Energy Performance of Buildings. *Environmental Science and Pollution Research International*, *13*(5), 350–358. doi:10.1065/espr2005.12.289 PMID:17067030

Porteous, C. D. A. (2011). Sensing a Historic Low-CO 2 Future. In D. N. Mazzeo (Ed.), *Chemistry, Emission Control, Radioactive Pollution and Indoor Air Quality* (pp. 213–246). InTech. doi:10.5772/16918

Ramanathan, V., & Feng, Y. (2009). Air pollution, greenhouse gases and climate change: Global and regional perspectives. *Atmospheric Environment. Elsevier Ltd*, *43*(1), 37–50. doi:10.1016/j.atmosenv.2008.09.063

Rosseau, D., Bowser, D., & Mattock, C. (2001). *A Guide to Mechanical Equipment for Healthy Indoor Environments*. Available at: https://www.cmhc-schl.gc.ca/odpub/pdf/62015.pdf

Sadineni, S. B., Madala, S., & Boehm, R. F. (2011). Passive building energy savings: A review of building envelope components. *Renewable and Sustainable Energy Reviews. Elsevier Ltd*, *15*(8), 3617–3631. doi:10.1016/j.rser.2011.07.014

Santin, O. G., Itard, L., & Visscher, H. (2009). The effect of occupancy and building characteristics on energy use for space and water heating in Dutch residential stock. *Energy and Building*, *41*(11), 1223–1232. doi:10.1016/j.enbuild.2009.07.002

Satish, U., Mendell, M. J., Shekhar, K., Hotchi, T., Sullivan, D., Streufert, S., & Fisk, W. J. (2012). Is CO2 an indoor pollutant? direct effects of low-to-moderate CO2 concentrations on human decision-making performance. *Environmental Health Perspectives*, *120*(12), 1671–1677. doi:10.1289/ehp.1104789 PMID:23008272

Schellnhuber, H. J., Schellnhuber, H. J., Cramer, W., Nakicenovic, N., Wigley, T., & Yohe, G. (2006). Avoiding Dangerous Climate Change. Cambridge University Press.

SCHER. (2007). *Opinion on risk assessment on indoor air quality*. SCHER.

Schieweck, A., Uhde, E., Salthammer, T., Salthammer, L. C., Morawska, L., & Mazaheri, M. (2018). Smart homes and the control of indoor air quality. Renewable and Sustainable Energy Reviews, 94, 705–718. doi:10.1016/j.rser.2018.05.057

Sharpe, T., McGill, G., Gregg, M. & Mawditt, I. (2016). *Characteristics and performance of MVHR systems: A meta study of MVHR systems used in the Innovate U.K. Building Performance Evaluation*. Academic Press.

Spengler, J., & Sexton, K. (1987). Indoor air pollution: A public health perspective. *Science*, *221*(4605), 9–17. doi:10.1126cience.6857273 PMID:6857273

Steiger, S., Hellwing, R., & Junker, E. (2008). Distribution of carbon dioxide in a naturally ventilated room with high internal heat load. In C. Rode (Ed.), *Symposium on Building Physics in the Nordic Countries* (pp. 377–384). Available at http://publica.fraunhofer.de/documents/N-97349.html

Sundell, J. (2004). On the history of indoor air quality and health. *Indoor Air*, *14*(s7, Suppl. 7), 51–58. doi:10.1111/j.1600-0668.2004.00273.x PMID:15330772

Sundell, J., Levin, H., Nazaroff, W. W., Cain, W. S., Fisk, W. J., Grimsrud, D. T., Gyntelberg, F., Li, Y., Persily, K., Pickering, C., Samet, J. M., Spengler, J. D., Taylor, S. T., & Weschler, C. J. (2011). Ventilation rates and health: Multidisciplinary review of the scientific literature. *Indoor Air*, *21*(3), 191–204. doi:10.1111/j.1600-0668.2010.00703.x PMID:21204989

Teichman, K., & Howard-reed, C. (2016). *NIST Technical Note 1868 Characterizing Indoor Air Quality Performance Using a Graphical Approach*. doi:10.6028/NIST.TN.1868

Treasury, H. M. (2015). Fixing the Foundations: Creating a more prosperous nation. *The Cambridge Law Journal*. Advance online publication. doi:10.1017/S0008197300099359

Treasury, H. M. (2015). Fixing the Foundations: Creating a more prosperous nation. *The Cambridge Law Journal*. Advance online publication. doi:10.1017/S0008197300099359

Treut, L. (2007). Historical Overview of Climate Change Science. IPCC WG1 Fourth Assessment Report, 93–127. doi:10.1016/j.soilbio.2010.04.001

United States E. P. A. (2011). *National Ambient Air Quality Standards*. https://www.epa.gov/ttn/naaqs

United States Environmental Protection Agency. (2020a). *Indoor Sources of Contaminants*. Retrieved November 27 from https://www.epa.gov/expobox/exposure-assessment-tools-media-air#indoorair

United States Environmental Protection Agency. (2020b). *Introduction to Air Quality*. Retrieved November 16 from https://www.epa.gov/indoor-air-quality-iaq/introduction-indoor-air-quality

US Department of Housing and Urban Development. (2020). *A Guide to Healthy Homes*. https://www.hud.gov/program_offices/healthy_homes/healthyhomes

Wargocki, P. (2013). The Effects of Ventilation in Homes on Health. *International Journal of Ventilation, 12*(2), 101–118. doi:10.1080/14733315.2013.11684005

WHO. (1991). *Indoor Environment: Health Aspects of Air Quality, thermal environment, light and noise*. Available at: http://whqlibdoc.who.int/hq/1990/WHO_EHE_RUD_90.2.pdf

WHO. (1997). Assessment of exposure to indoor air pollutants. *WHO Regional Publications. European Series no. 78*, 139.

WHO. (2000). *Air quality guidelines for Europe. WHO Regional Publications, European Series; No. 91* (F. Theakston, Ed.; 2nd ed.). WHO Regional Publications. Available at http://link.springer.com/10.1007/BF02986808%5Cnhttp://www.springerlink.com/index/10.1007/BF02986808

WHO. (2010). WHO guidelines for indoor air quality: selected pollutants. World Health Organization Regional Office for Europe. doi:10.1186/2041-1480-2-S2-I1

World Allergy Organization. (2013). *White Book on Allergy*. https://www.sanofi.com/en/about-us/indoor-air-pollution-the-invisible-enemy-of-allergy-sufferers

Xing, Y. F., Xu, Y. H., Shi, M. H., & Lian, Y. X. (2016). The impact of PM2.5 on the human respiratory system. *Journal of Thoracic Disease, 8*, E69–E74.

Zero Carbon Hub. (2012). *Understanding overheating - where to start: An introduction for house builders and designers*. Author.

Zero Carbon Hub. (2015). *Overheating in Homes: the big picture*. doi:10.1017/CBO9781107415324.004

Zhang, J., & Smith, K. R. (2003). Indoor air pollution: A global health concern. *British Medical Bulletin, 68*(1), 209–225. doi:10.1093/bmb/ldg029 PMID:14757719

Zhu, X., & Liu, Y. (2014). Characterization and Risk Assessment of Exposure to Volatile Organic Compounds in Apartment Buildings in Harbin, China. *Bulletin of Environmental Contamination and Toxicology, 92*(1), 96–102. doi:10.100700128-013-1129-x PMID:24158356

Chapter 4
Forecasting Energy Impact in Multifamily Buildings Through Airtightness Models

Jesica Fernández-Agüera
Universidad de Sevilla, Spain

Samuel Domínguez-Amarillo
https://orcid.org/0000-0002-9177-8736
Universidad de Sevilla, Spain

Juan José Sendra
Universidad de Sevilla, Spain

ABSTRACT

Southern European areas are traditionally considered to have a benign climate and therefore most homes often do not have controlled ventilation systems or air conditioning. Our society faces challenging situations with global pandemics and health emergencies, while housing provides the basic elements of refuge and activity. A profound transformation of social and labor dynamics is underway, increasing at-home presence, hence the capital importance of improving building envelopes to reduce energy demand and to increase occupant comfort levels in order to ensure their protective function. Air leakage is one of the key factors both in indoor air quality and energy demand. This study aims to explore the sensitivity of energy demand to airtightness. A representative set of multifamily buildings, built in the last 20 years in Seville, is analyzed. The results show that air permeability has a significant effect on energy demand in the sample studied. Although impact is greater in severe climates, it needs to be considered in temperate climates, especially when more time is spent at home.

DOI: 10.4018/978-1-7998-7023-4.ch004

INTRODUCTION

Spain is the European country with the second-highest number of deaths in winter (Healy, 2017), and equally high figures in the warm season. These figures are directly related to the environmental conditions in housing, where occupants spend much of their day. The current housing stock, most of which is over 50 years old, presents a significant degree of obsolescence with a performance capacity that is limited in many cases. In general, the oldest group of housing units has poor construction standards, especially in low-income or social housing. It is often found that the indoor environmental conditions of the dwellings are far from what are usually regarded as comfortable or healthy, even in the climatic zones of regions usually considered mild or moderate climates (Domínguez-Amarillo, Sendra, Fernández-Agüera, & Escandón, 2017; Domínguez-Amarillo, Sendra, & Oteiza San José, 2016). The combination of building envelopes of limited capacity, thermal systems that are not very suitable - inefficient in many cases - and situations of energy poverty or high energy costs are usually responsible for this situation (Domínguez-Amarillo, Fernández-Agüera, Peacock, & Acosta, 2019; Sendra, Domínguez-Amarillo, Bustamante Rojas, & Leon, 2013). In this context the uncontrolled air exchange between the outside and the inside of the buildings, characterized by envelope airtightness, is one of the key variables in residential energy consumption and indoor-ambient performance, as are comfort and IAQ.

The main aim of this work is to provide a methodological proposal for the development of predictive models of energy consumption in dwellings in relation to airtightness, presenting a model of the case of the city of Seville in southern Spain.

To do so, using calculation engine EnergyPlus this study proposes to simulate the energy behavior of a representative sample of multifamily housing buildings in one climate area, assessing the repercussion of infiltrations and the influence of urban context in the operational conditions of the housing on energy demand. Based on these energy consumption simulations, airtightness tests and the study of their morphological and typological characteristics, a model was established to predict the energy demand of the dwellings classified by typological and morphological characteristics.

Background

Since the 1990's the airtightness of buildings has been widely studied in the USA, where the models and fundamental principles were developed, while specific studies were carried out more recently in other regions including Europe. Airtightness behavior models differ significantly between areas and national contexts, mainly due to different construction methods and architectural typologies. This makes it necessary to develop specific analyses of individual housing stock given that representative values vary significantly between regions, especially as regards climate context and building tradition.

In many situations, as in the case of southern European social housing - of which Spain is a particularly interesting example - air infiltration is the only form of continuous ventilation or basic ventilation in the dwellings, except for the voluntary opening of windows. However, infiltrations contribute significantly to the total annual energy demand of the dwellings (heating and cooling). There are fewer studies on the quantification of energy demand associated with infiltration, and most of the results refer to northern Europe, where the energy impact of air infiltration on heating demand is estimated to be in the order of 10 kWh/m2 in moderately cold regions - 2500 degree-days - (Spiekman, 2010). Despite less studies on this subject in the southern area, the impact of infiltration on the energy balance of homes has been observed, both in colder and continental areas (Feijó-Muñoz et al., 2019) as well as warmer ones (Alves,

Fernández-Agüera, & Sendra, 2014; Domínguez-Amarillo, Fernández-Agüera, Campano, et al., 2019). Within an area such as the Iberian Peninsula, there is an irregular distribution of consumption, especially between winter and summer, while the cities in the south of the country present a greater impact in terms of annual energy affected by infiltration (Poza-Casado, Meiss, Padilla-Marcos, & Feijó-Muñoz, 2020).

Balancing the need to conserve heat in order to limit infiltrations and the contribution of these to the renewal of the indoor atmosphere, which therefore becomes a fundamental part of the system for controlling IAQ in existing homes, can be a major dilemma (Fernández-Agüera, Domínguez-Amarillo, Alonso, & Martín-Consuegra, 2019). It is therefore essential to establish a knowledge of the real impact of infiltration on the energy balance, as well as seizing the opportunity to identify the points of balance between the need to save energy and the possible effects of airtightness on basic ventilation (Fernández-Agüera, Dominguez-Amarillo, Fornaciari, & Orlandi, 2019).

For the prediction of the relationship between energy consumption and leakage, the approach represented in equation 1, developed by (Younes, Shdid, & Bitsuamlak, 2012) and commonly accepted in many general studies (Feijó-Muñoz et al., 2019; Meiss & Feijó-Muñoz, 2014; Meiss & Muñoz, 2013), is often used.

$$Qinf = CP \, . \, Gt \, . \, Vinf \tag{1}$$

Where:

- Qinf is the annual energy loss due to air infiltration (kW h)
- Cp is the volumetric heat capacity of the air, which is 0.34 W h/(m^3 K),
- Gt are the annual degree days (kK h), with a base comfort temperature of 21 and 25 °C for heating and cooling, respectively.
- Vinf is the air leakage rate (m^3/h)

In this approach, the forecast of energy demand by outdoor air intake, in itself inherently uncertain, shows a linear dependence both on the degree-day value and on the air leakage rate parameter - derived from the pressurization test (Blower Door Test) - although average or standardized values for zones or regions are often used. In general terms, some authors have emphasized that this method may be acceptable mainly when calculating the load due to concentrated air flows through large openings and short passages, hence its equivalence to void balance equations. However, it could entail a significant overestimation of the energy impact in the case of distributed leaks, such as those resulting from the porous materials and the small cracks typical of construction processes, where heat exchange between the air and the enclosures may occur (Younes et al., 2012). Thus, not all the energy load of the air mass becomes real demand in the indoor environment. This is perhaps the most common case in buildings with heavy or medium-heavy walls, typical of built elements, as is the case of the residential buildings in the building stock analyzed. In addition, the degree-day concept, although useful at a global estimate level, brings about a significant generalization in buildings in the same stock, without taking into account the different thermal behavior of buildings in the same geographical area. These present different responses to external thermal oscillations, with different base-temperatures for the start of the use of air-conditioning (Domínguez-Amarillo, Fernández-Agüera, Sendra, & Roaf, 2018), depending on their morphological and constructive characteristics and factors such as location and relationship to the urban context.

MATERIALS AND METHODS

In order to evaluate the thermal and energy demand behavior of buildings, energy simulation tools are needed to generate sufficiently accurate synthetic dynamic models of the building. Using these it is possible not only to define the building itself, but also its urban environment conditions. In this study the energy analysis program DesignBuilder was selected to calibrate and validate the models. Based on the simulation results a predictive model of airtightness and energy consumption in homes was developed.

Sampling Selection

Once the unit of analysis is defined, the population to be studied is outlined and the results are generalized. A population is the set of all cases that meet a series of specifications (Hernandez Sampieri, Fernandez Collado, & Baptista Lucio, 2014). The sample is usually defined as a subgroup of the population (Sudman, 1976).

Type of Sampling

For the selection of the buildings in the sample a random probability sampling was carried out. In addition, as one of the objectives of this work is the study of the differences between dwellings within the same development, several dwellings were chosen from one development. For this selection, accessible dwellings were used as a convenience sample, as they share constructive and typological characteristics with others in the same development phase.

Optimal Sample Size

Probability samples have many advantages, most notably that the error size in predictions can be measured. It can even be stated that the main objective in designing a probability sample is to minimize this error, or standard (Kish, 1965). The sample elements should have values very similar to those of the population so that measurements on the subset provide estimates of the larger set.

The sample, n, is a subset of the population N.

The sample size is calculated in relation to the population variance, the confidence level and the margin of error for the study carried out, according to equation 2:

$$n = \frac{z^2 \tilde{A}^2}{e^2} \qquad (2)$$

Where:

- n is the number of dwellings in the sample.
- z is the deviation from the mean value according to the established confidence level.
- σ^2 is the variance of the estimated population.
- e is the maximum established margin of error.

Probability samples require the determination of the sample size, as well as a random selection process to ensure that all elements of the population have the same probability of being chosen. In this case, the list was drawn up by the authors after locating the elements of the population on plans. The sample elements were selected from these, by constructing a reference framework allowing us to physically identify the elements of the population, listing them and proceeding to select the sampling elements.

A total of 6 different samples of dwelling developments was selected as representative of low-income buildings in the area. Table 1 shows the main characteristics of the dwellings selected as a study sample and Figure 1 shows the differente developments.

Table 1. Main characteristics of the models of the sample used (Floor area (AFL), Window area (AW), Façade area (AFA))

Model	No. of dwellings	Facade	Degree of exposure	A_{FL} (m^2)	A_W (m^2)	A_{FA} (m^2)
P1	1	2	2	72.8	14.62	55.9
P2	10	3	1	58.5	9.50	45.5
P3	8	3	4	60.5	9.77	47.5
P4	7	3	3	64.5	8.96	54.3
P5	5	3	3	70.4	8.24	47.2
P6	8	3	4	68.0	9.24	47.3
Total	39					

Figure 1. Energy models in the sample studied

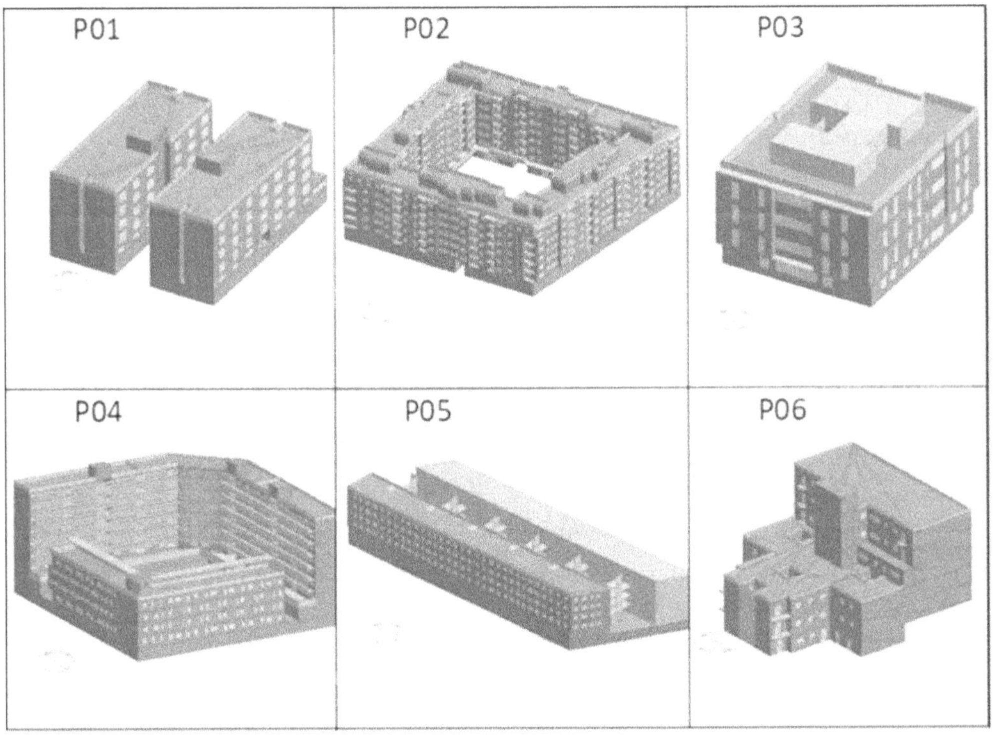

Characterization

Morphological and Constructive Assessment

In this phase the morphological and dimensional characteristics were analyzed, together with the different constructive solutions of the dwellings in the sample. A field inspection was carried out to identify the factors which may influence airtightness behavior:

- Geometric definition of the enclosure
- Orientation and situation of urban context
- Construction systems used
- Hollow-mass relationship
- State of conservation/degradation of the enclosure
- Facilities and ducts
- Building retrofittings

Airtightness

The actual airtightness behavior of dwellings is analyzed by carrying out pressurization/depressurization tests on the indoor environment, using the Blower Door Test (BDT) procedure (M. Sherman, 1995), which is generally accepted by the scientific community. The tests were carried out following UNE EN 13829:2002 (UNE-EN 13829:2002, n.d.). In order to carry out these tests, a fan was placed in an outer door of the dwelling in order to extract (depressurize) or introduce (pressurize) air, until a negative or positive pressure of about 50 Pa was reached, and then measuring the air flow. During the test the rest of the outer doors and windows of the dwelling had to remain closed, while the inner doors were open. The specific protocols applied and variations of the tests are developed in (Fernández-Agüera, Domínguez-Amarillo, Sendra, & Suárez, 2016; Fernández-Agüera, Sendra, & Domínguez, 2011).

Table 2 shows statistical descriptors for air infiltration rate at 50 Pa, a test more widely discussed in (Fernández-Agüera, Domínguez-Amarillo, Sendra, Suárez, & Oteiza, 2019).

Table 2. Sample statistical descriptors for airtihghtness

ID	N°	Med V_{50}	*std V_{50}*	Med n_{50}	*std n_{50}*	Med ELA/S	*std ELA/S*	Med n	*std n*
P1	1	2892	*0.0*	15.57	*0.00*	10.50	*0.00*	0.56	*0.00*
P2	10	1439	*202.2*	9.06	*1.39*	5.92	*1.06*	0.59	*0.01*
P3	8	826	*82.9*	5.30	*0.10*	3.52	*0.39*	0.56	*0.01*
P4	7	719	*201.8*	4.17	*0.87*	2.84	*0.70*	0.58	*0.02*
P5	5	1531	*172.9*	8.37	*0.09*	5.41	*0.09*	0.58	*0.00*
P6	8	886	*70.4*	4.95	*0.28*	3.32	*0.49*	0.56	*0.03*
Tot	159	978	*452.0*	6.52	*2.59*	4.30	*1.71*	0.58	*0.03*

Energy Assessment

Virtual models of the buildings were generated from the study sample, selected from the analysis of the building stock, in order to recreate their fundamental characteristics: geometric-morphological, construction systems, urban environment and operational conditions. For this purpose, the EnergyPlus (E+) engine was used in combination with a 3D model generator on an Open-GL basis.

Sample Modeling

The models developed are nodal, adopting one node per dwelling as a working hypothesis in order to refer the results to individual average values. The relations between each housing unit - calculation node - are dynamic (no simplifications of adiabatic elements were applied), and introduced the partition masses into the set of equations assessing heat accumulation in the dwellings, together with a proportional distribution of the estimated weight of the furnishings to calculate internal thermal mass.

Data collection process:

- Design of the experimental group through the selection made in the statistical analysis of the stock.
- Geometric, constructive definition and configuration of the urban environment. This information was modeled from the field survey data set.

The same dwelling was simulated for different levels of airtightness in order to obtain the increases of energy demand improvement as a function of airtightness, evaluating whether there was a linear relationship between these. Four different ranges were simulated: 0.35 h^{-1}; 0.45 h^{-1}; 0.55 h^{-1} and 0.65 h^{-1}.

The models were generated in such a way as to allow for evaluation closer to real situations. For this purpose, the basic geometric and morphological configuration of the buildings was generated, incorporating the modeling of surrounding conditions (urban implementation) in order to record the interference in solar radiation generated by complex urban environments, as well as the shadows typical of the built complex itself. The terrain on which the building was built, when acting as a connecting element, was represented by its main material and thermal properties.

The opaque enclosures were determined by their different components, including the number of layers, materials and thicknesses. The thermo-energetic properties of the different components of the enclosure were determined, including heat transfer parameters (convective and radiant) and thermal transmittance (U) values. This included a definition of the thermal properties of the different layers that made up the enclosure: thickness of the layer, density of the layer (kg/m^3); specific heat of the layer (J/kgK) and conductivity of the layer (W/mK). The optical properties of the surface layers of the enclosure were also provided. For semi-transparent materials, the relative amount of solar radiation passing through the material was represented, compared to 3 mm float glass. The operation of the solar protection was standardized in accordance with the Spanish Energy Labeling procedure (PBCEE) (Ministerio de la Presidencia. Gobierno de España, 2013) assuming a 30% continuous performance of the protective elements.

Natural Infiltration and Ventilation

Uncontrolled outside air inlets (not secured by mechanical means) were considered. Infiltrations were first considered as a continuous flow not associated with the voluntary actions of the occupants. Modeling was based on the BDT tests carried out at field level, using these performance values and a performance model based on the EnergyPlus Airflow Network model (Lixing, 2007) and taking into account the climatic conditions and urban configuration, as well as the characteristics of the building itself (geometry, position, orientation and air porosity of its enclosures). Secondly, natural ventilation was examined as the flow of external air let in by occupants opening windows. As it was not guaranteed by the existence of fans or other elements this flow was not initially known, but could be estimated in certain circumstances, following the implementation of the EPBD in Spain:

- The existence of nocturnal ventilation during summer, imposing a value of 4 ACH of the interior volume, emulating the process usually carried out by users for the passive cooling of dwellings.
- A base flow during the winter of 0.63 h^{-1} when infiltrations are not sufficient, in order to ensure minimum hygienic ventilation conditions.

Location and Climate

The case study examines the city of Seville, as it is not only one of the largest cities in the region - with a very extensive building stock - but is especially representative of other southern European cities due to its climate and socio-cultural context. Located in the south of Spain, the city has a typical Mediterranean climate (Köppen: Csa), with mild winters from December to March and very hot summers from June to September, as well as periods of variable moderate temperatures from April to May and October to November. Seville has a large number of sunny days, with slightly cloudy or clear skies (Kottek, Grieser, Beck, Rudolf, & Rubel, 2006). For the energy analysis, climate data from the standard meteorological year collected in the national energy certification procedure were used (de la Flor, Domínguez, Félix, & Falcón, 2008).

The wind profile in the city is mild, with speeds usually around 5 to 12 km/h and normally below 20 km/h, and no significant presence of strong winds. The winds mainly move from SW to NE and vice versa. Therefore, the dynamic action of the wind on the envelopes is relatively limited, also given the moderate profile of the height of the buildings in the city, where residential buildings normally have 4 to 8 floors.

Operating Profiles

For an actual assessment of the running performance of the dwellings and the infiltration impact on the balance, a comprehensive operating profile was adopted based on the national energy labeling procedure (IDAE, 2009; Ministerio de Fomento del Gobierno de España, 2017). The home profiles were thus standardized for a suitable comparison to emulate real housing units. Table 3 shows the indoor profiles for set-point temperature: heating and cooling, occupancy schedule, indoor loads, lighting, and ventilation. Summer regime was set from April to September, while the rest of the year was considered winter.

Table 3. Operating conditions in simulations (IDAE, 2009; Ministerio de Fomento del Gobierno de España, 2017). Maximun Set-point temperatura in OC (MT), Minimun set-point temperature in OC (Mt), Sensible heat due to occupation in W/m2 (OSH), Latent heat due to occupation in W/m2 (OLH), Lighting loads in W/m2 (LL), Equipment loads in W/m2 (E), Ventilation in ACH (V), Summer (S), Winter (W), Weekend (WE), Weekday(WD), Annual (A) and renewals by DB-HS3(*)

	MT	mT	OSH		OLH		LL	E	V	
Hour	S	W	WD	WE	WD	WE	A	A	S	W
1	27	17	2.15	2.15	1.36	1.36	0.44	0.44	4	*
2	27	17	2.15	2.15	1.36	1.36	0.44	0.44	4	*
3	27	17	2.15	2.15	1.36	1.36	0.44	0.44	4	*
4	27	17	2.15	2.15	1.36	1.36	0.44	0.44	4	*
5	27	17	2.15	2.15	1.36	1.36	0.44	0.44	4	*
6	27	17	2.15	2.15	1.36	1.36	0.44	0.44	4	*
7	27	17	2.15	2.15	1.36	1.36	0.44	0.44	4	*
8	-	20	0.54	2.15	0.34	1.36	1.32	1.32	4	*
9	-	20	0.54	2.15	0.34	1.36	1.32	1.32	*	*
10	-	20	0.54	2.15	0.34	1.36	1.32	1.32	*	*
11	-	20	0.54	2.15	0.34	1.36	1.32	1.32	*	*
12	-	20	0.54	2.15	0.34	1.36	1.32	1.32	*	*
13	-	20	0.54	2.15	0.34	1.36	1.32	1.32	*	*
14	-	20	0.54	2.15	0.34	1.36	1.32	1.32	*	*
15	-	20	0.54	2.15	0.34	1.36	1.32	1.32	*	*
16	25	20	1.08	2.15	0.68	1.36	1.32	1.32	*	*
17	25	20	1.08	2.15	0.68	1.36	1.32	1.32	*	*
18	25	20	1.08	2.15	0.68	1.36	1.32	1.32	*	*
19	25	20	1.08	2.15	0.68	1.36	2.2	2.2	*	*
20	25	20	1.08	2.15	0.68	1.36	4.4	4.4	*	*
21	25	20	1.08	2.15	0.68	1.36	4.4	4.4	*	*
22	25	20	1.08	2.15	0.68	1.36	4.4	4.4	*	*
23	25	20	1.08	2.15	0.68	1.36	4.4	4.4	*	*
24	27	17	2.15	2.15	1.36	1.36	2.2	2.2	*	*

Calibration and Validation

The phenomena to be simulated, associated with the energy behavior of the buildings, are complex and interrelated. This makes it difficult to generate models whose behavior coincides with the actual object studied. Both in research and in the development of tools, a great effort was made to optimize the modeling and calibration processes (Reddy, 2006). Although energy analysis programs currently present a high degree of precision and are useful tools with a suitable predictive capacity, there are still two fundamental barriers to the accurate simulation of buildings. There are different approaches for adjusting the simulated models to real behavior which combine reading the project documentation on the one hand, while also visiting and inspecting the constructed building. This involved collecting data on construction parameters, tests, etc., in addition to measuring thermal and electrical loads and applying sensitivity analysis (Cuerda, Guerra-Santin, Sendra, & Neila, 2020; Raftery, Keane, & O'Donnell, 2011).

The adjustment of the simulation model was proposed through the use of successive sensitivity analyses, working linearly with variables (interactions between variables and their non-linear behavior were not taken into account for simplicity of approach).

Sensitivity analysis was used to estimate the error in the simulation results as a consequence of the uncertainty of the data input. Despite the specificities of each method, these techniques basically varied the input values and checked result variations. The analysis of these techniques made it possible to locate the variables which can generate significant variations (Saltelli, Tarantola, Campolongo, & Ratto, 2004) and others which are irrelevant to individual models, thus helping to simplify the work of generating the model by adopting standard or default values.

Other studies (Raftery et al., 2011) indicate that the introduction of unnecessary details into the building model can contribute to increased uncertainty in the simulation results. A suitable physical knowledge of the behavior and experience in the analysis is therefore essential, since the tools provide a great degree of freedom in the initial definition.

The uncertainty profile can be structured in three fundamental blocks, although this grouping is better suited to behavior in cold weather (Tamburrini, Palmer, & Macdonald, 2003):

- Thermo-physical properties of materials.
- Internal gains.
- Actual infiltration rates

For this approach proposal, the fundamental action lies in the simulation of multiple cases, applying systematic perturbation in the predefined data inputs.

The first step of the calibration was to replace the standard climate profile, in this case the meteoro logical type year (EPW, based on the SWEC for the city of Seville), with the information recorded for the actual meteorology corresponding to the period used for the calibration. Subsequently, the general programs of occupation, use and operation of the dwelling were replaced by specific profiles designed for the cases of contrast for this period of analysis. These data were collected monitoring the use of the housing equipment and user surveys through a data mining process. This process allows the calibration of the uses and contributions made by the different equipment and appliances in the dwelling (including lighting). For the calibration process, periods where dwellings were unoccupied were recorded in order to reduce the uncertainty of the analysis.

The acceptance criteria for the data resulting from the calibration are based on those defined by ASHRAE 14-2002 (Ashrae, 2002) for energy modeling of buildings. In this document it is established that hourly data are suitable for the calibration of these models. For this purpose, the Mean Bias Error (MBE) and the Coefficient of Variation of the Root Mean Square Error (CVRMSE) were used as evaluation indices as were the parameters for hourly data. These indices, together with the adopted acceptance criteria established for hourly data, are shown in Table 4.

Table 4. Calibration evaluation criteria according to ASHRAE 14-2002

Evaluation index		Acceptance criteria
Mean bias error	$MBE = \frac{\sum_{i=1}^{N_s}\left(y_i - \hat{y}_i\right)}{\sum_{i=1}^{N_s} y_i}$ $\widehat{Y_S} = \frac{\sum_{i=1}^{N_s} y_i}{N_s}$	<10%
Coefficient of Variation of the Root Mean Square Error	$CVRMSE_{(S)} = \frac{\sqrt{\sum_{i=1}^{N_s} \frac{\left(y_i - \hat{y}_i\right)^2}{N_s}}}{\widehat{Y}_S}$	<30%
	• y_i: recorded data • $\hat{y}_i$: simulated data • N_s: sample size • $\widehat{Y}_S$: sample mean for recorded data	

SENSITIVITY TO THE DEGREE OF AIRTIGHTNESS OF THE BUILDING ENVELOPE

The fundamental objective of this section is to construct a set of analysis scenarios based on different degrees of airtightness for the buildings in the sample set.

Prediction Models

The use of linear regression models was chosen as these were preferred for their greater simplicity and robustness in the event of stochastic behavior. The linear regression model analyzed the relationship between the dependent variable (n_{50}) and a set of independent variables, or predictor variables, different for each model. This relationship was expressed as an equation predicting the response variable as a linear function of the parameters.

The procedure used was based on stepwise regression, which enables the model with the least number of variables to be identified in order to avoid the inflation of the model variance due to the number of variables introduced, while maximizing the information it can provide. Following the hypothesis that the relation between the initial prediction variables and the response variable is linear (or close to it), the procedure selected the subset of variables which best defined the model. Subsequently, the goodness of fit of the regression model was contrasted.

The following procedure was used to generate the multiple regression model:

1. group setting
2. choice of parameters affecting airtightness
3. testing of the multiple collinearities of the parameters considered

4. dummy variable classification and entry in keeping with the classified parameters
5. implementation of the stepwise method to identify the model with the lowest number of variables which best explained the dependent variable or determined the criterion of the goodness of fit of the data to the multiple regression model
6. estimation of equation or predictive model parameters.

A complementary set of statistical tests was performed to determine the existence of correlations between airtightness - represented by n_{50} - and the defining parameters of the sample. The tests were designed to confirm (null hypothesis) or discard (alternative hypothesis) the existence of a single population whose behavior could be explained by aspects including the geometric parameters, the design of its envelopes, and climatic zones.

In this study, the null hypothesis was adopted initially, as the sample consisted of a series of units with the same conceptual design and construction systems and similar surface areas: it was initially assumed that the data had been extracted from different samples of the same population. In addition, it was proposed that the airtightness analysis should present consistent results and comparable behavior for the different parameters.

Clustering

One of the main tasks when building the predictive model is to check that all the samples selected belong to the same population. In this way it is possible to divide these into groups in which common behaviors in certain aspects can be expected (Hennig, Meila, Murtagh, & Rocci, 2015).

The initial assumption is that the determining factors of airtightness must be fundamentally related to the morphological and construction factors defining the housing envelope. Although this relationship as analyzed above is not presented linearly or univocally, the existence of groups of patterns that determine the behavior of the groups of dwellings is proposed as a basis for identifying clusters of dwellings with a high degree of internal homogeneity and maximum external heterogeneity.

In order to define which of the variables that describe the dwellings in the sample has the most weight in this morphological grouping, a main component analysis (MCA) was performed to establish that all dwellings in this sample were part of the same building population.

RESULTS AND DISCUSSION

Energy Balance

The energy behavior of residential buildings depends on the action of multiple factors with complex interrelations. Traditionally, a distinction has been made between the elements that make up the thermal envelope and the internal loads. The energy exchanged through the enclosure is recorded through the opaque section (Floor -QFL, Roof –QR-, Partitions –QP- and Façade –QFA-) and through the openings (Window –QW-, solar gains troght the windows -QSW-). The internal loads: occupation (QO), lighting (QL) and internal equipment (QE) were measured, as were the energy associated with movements and mass exchanges between the indoor and outdoor atmosphere: night ventilation (QV) and air infiltration (QI). In the graphic representation of these energy balances, increases of energy which contributed to the indoor environment were considered to be positive while decreases of energy in the same environment were taken to be negative. All loads were standardized by dividing their value by the useful area of the dwelling analyzed in order to compare dwellings of different sizes.

An initial base state of the thermal load analysis was adopted in order to establish a distinction between the infiltration loads in the energy balance in the cooling and heating periods. This used an average global rate of air infiltration ACH = 0.35 h^{-1}, corresponding to the value of n_{50}=7, representative of the building stock of the area (Fernández-Agüera, Domínguez-Amarillo, Sendra, et al., 2019), using the Persily-Kronvall (M. H. Sherman, 1987) formula.

The results derived from the set of simulated scenarios indicate that infiltrations are one of the basic factors responsible for the thermal load of the dwellings during the cold period. However, this impact is less influential in the hot season where the action of solar radiation on the built masses - indoors and outdoors - becomes the basic factor. It should be pointed out that, despite being a temperate climate zone, the thermal differentials between the indoor and outdoor of the dwelling are much more pronounced during the cold period, so that there is generally a greater exchange of energy.

The values of monthly energy demands associated with infiltrations range from -2.68 kWh/m^2 to -3.58 kWh/m^2 for the typical winter months (Figure 2), representing around 25% of the negative demand for this period. During the summer period, the accumulated impact of air permeability on monthly energy demand ranges from 0.04 kWh/m^2 to 0.85 kWh/m^2, values lower than those reached in winter (Figure 2). It is interesting to highlight the role of the infiltrations themselves in acting as a compensation load in certain periods of the year, thus generating beneficial situations from the point of view of environmental control of dwellings. For much of the time in the shoulder seasons, the energy associated with the infiltrations is able to compensate the energy balances, reducing the need for heating in residential buildings, just as the need for cooling is reduced for a significant number of hours in summer, mainly at night.

The average net sample demand caused by infiltration for the winter is -9.2 kWh/m^2 (Figure 3), with a standard deviation of 0.83 kWh/m^2 and a range between -7.6 and -9.1 kWh/m^2. The maximum value of heat demand due to infiltration occurs in a building with multiple orientations (type P3), which is relatively exposed. For the summer the average net demand for infiltrations is in a range almost 7 times lower, 1.3 kWh/m^2 (STD 0.1 kWh/m^2 ; range 0.7 kWh/m^2 and 1.8 kWh/m^2). In this case, the maximum demand is found in dwellings with a double south-east-north-west orientation (type P5). In general, there is no clear trend in the distribution of values between groups within buildings or between buildings, and in most cases dispersion can be associated with construction processes and types of elements used in the enclosures.

This can be broken down into different structuring elements, including air changes. A distinction was established between the total energy demand due exclusively to an air infiltration rate ACH = 0.35 h^{-1} and that corresponding to the rest of the thermal loads (Figure 3). The adoption of a scenario adapted to the average airtightness measured in social housing, ACH = 0.35 h^{-1}, permits the comparison of energy demands between different building groups, defining bands of expected behavior.

The average value of annual demand due to infiltration is 9.78 kWh/m^2 (Figure 4), with a standard deviation of 1.51 kWh/m^2 and a range from 8.60 kWh/m^2 to 10.92 kWh/m^2. The maximum value was reached by model P4, corresponding to a model with intermediate heating demand and high cooling demand. The demand associated with infiltrations for ACH = 0.35 h^{-1} accounts for an average of 24% of total demand.

Figure 2. Net monthly energy balance of housing located in Seville

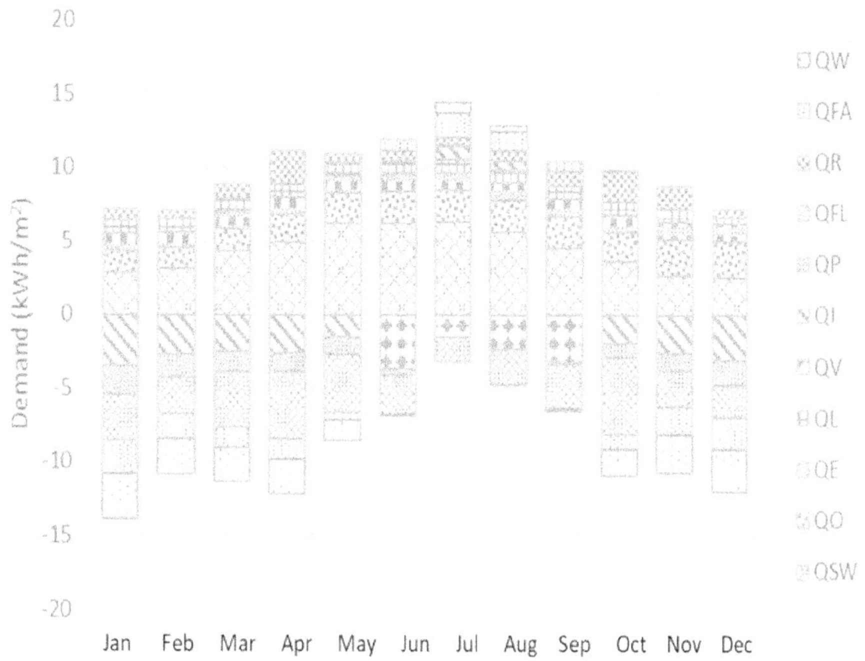

Figure 3. Average net demand due to infiltration for the winter and summer season of the Seville sample promotions [ACH=0.35 h^{-1}]

Sensitivity to Infiltration of Building Types

This analysis starts with the creation of scenarios comparing the variation of the average seasonal ACH value from the population average of 0.35 h^{-1} to lower and higher infiltration rates (ACH = 0.15 to 0.8 h^{-1}). Figure 6 shows the percentage increase in total demand for each variation in the infiltration rate for each 0.1 h^{-1} increase.

Once the scenarios for the different infiltration rates were established, it was confirmed that in the usual values found in the residential stock the demand for heating and cooling responds to a behavior that can resemble a linear one. This allows the slope of the regression equation to be associated as an indicator of the sensitivity of the energy demand to the variation of the parameter: heating and cooling demands of housing for different degrees of airtightness (equations 3 and 4).

For the heating and cooling period a regression can be established taking the value of global hourly renewals of the dwellings (ACH) as a predictor variable:

$$D.cal = PHeat \times ACH + cte \tag{3}$$

$$D.ref = PCool \times ACH + cte \tag{4}$$

Figure 4. Average annual demand due to infiltration versus other demands (set of demands not associated with infiltration) of the development phases in the Seville sample [ACH=0.35 h^{-1}]

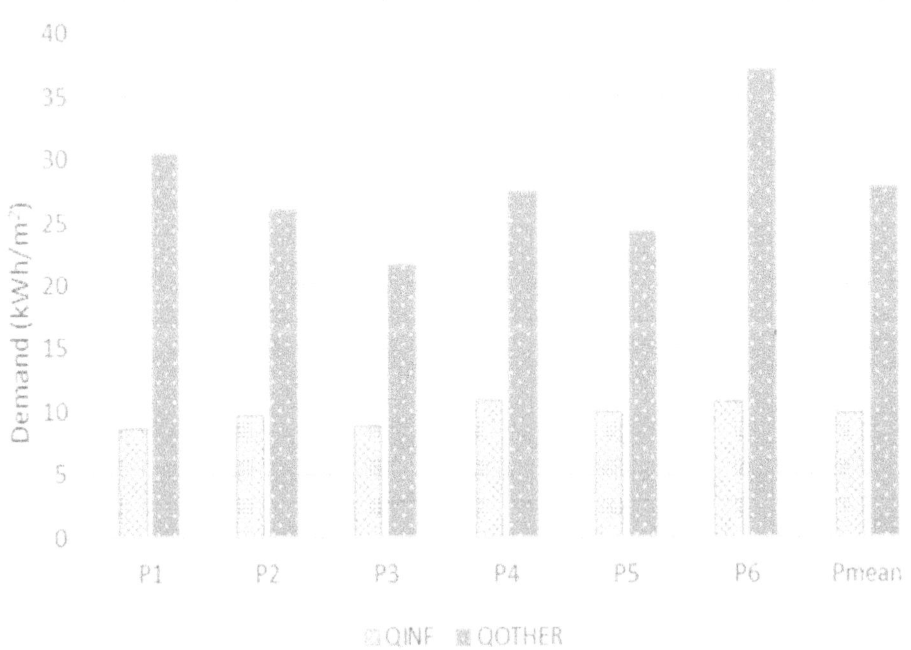

For the heating season, the average demand slope (heating) is 23.87 (Figure 5a), (STD 3.41; ranging from 17.19 to 28.07). The types with lower demand values at 0.35 seem to be more sensitive to permeability, as is the case of type P4 with multiple orientations (southwest or northwest). The least reactive type (P5) has a double orientation south-east-north-west. In contrast, in the summer cases, the sensitivity of these models is substantially reduced to one tenth of the winter value, presenting a mean slope (cool) of 2.36 (Figure 5b) [STD: 1.12 ranging from 0.75 to 3.63]. This value represents an average percentage of 9.5%. The P4 and P6 models present the highest cooling demand. The dwellings that make up the P4 sample have different typologies and orientations (southwest, northwest, northeast or north). The dwellings that make up the P6 sample display two typologies and orientations (west or north). The P5 model has the lowest cooling slope; furthermore, it is one of those with the lowest demand for cooling due to its good orientation.

Figure 5. Evolution of total annual demand for heating (a) and cooling (b) according to hourly renewals (ACH) per promotion in Seville

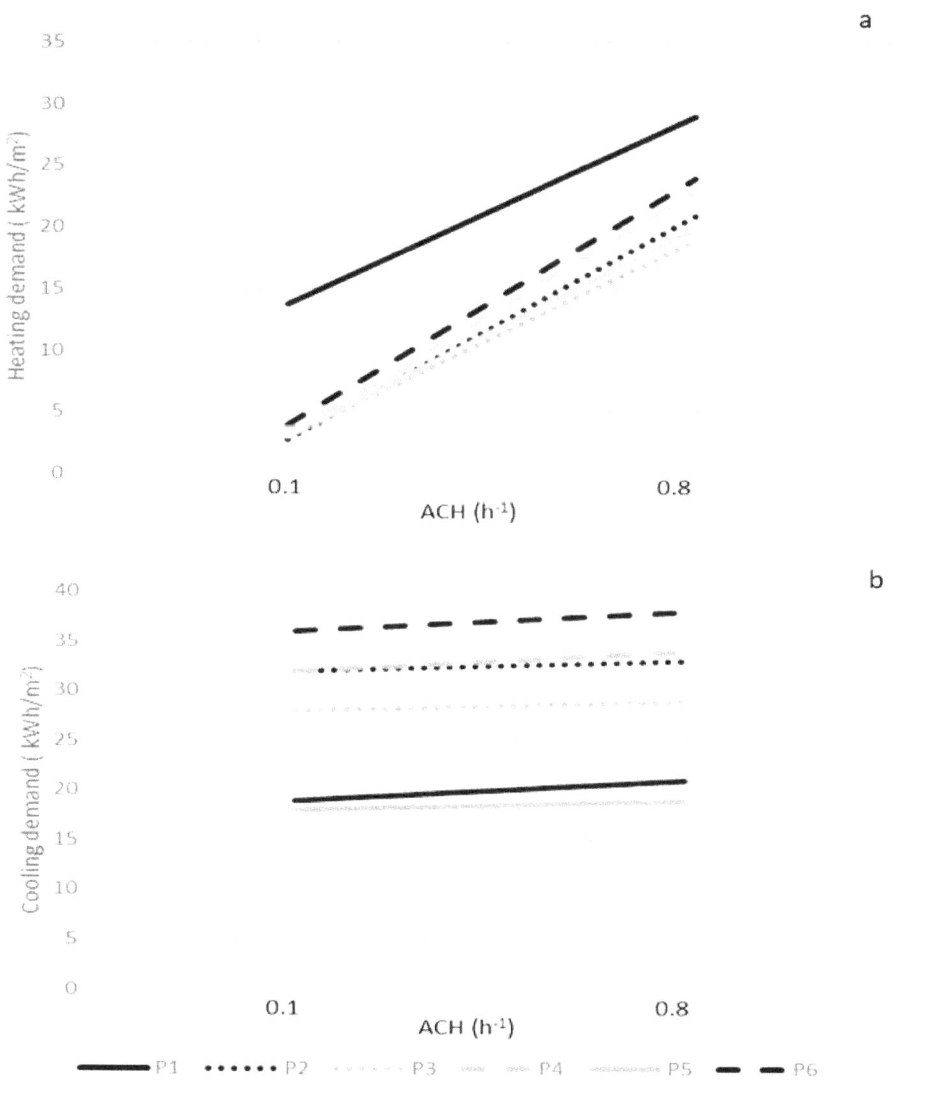

For the Seville sample, for every increase in the infiltration rate of 0.1 h^{-1}, there is an increase in demand ranging from 5.5 to 8.3%, averaging 6.3%. In terms of total energy demand, this means a variation of between 2 and 3.1 kW/hm^2 for each increase of 0.1 h^{-1} (Figure 6).

Figure 6. Total annual demand for different ACHs and relative increase in demand for a differential ACH = 0.1 h-1 in Seville

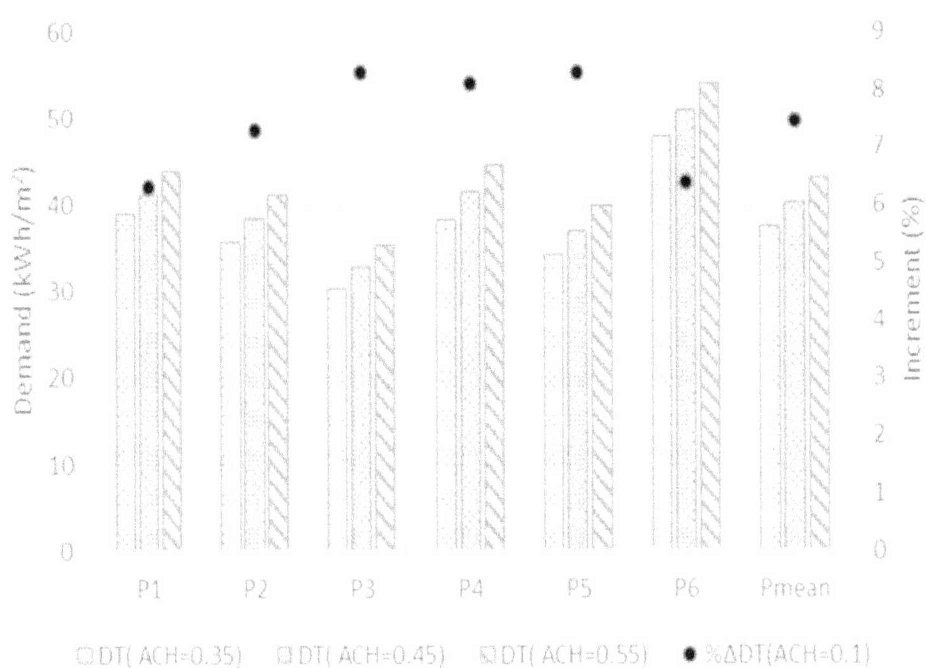

Derived Predictive Model for Energy Demand

The model was based on the parameters which could have an impact on the quantification of the degree of airtightness: typology and position of the dwelling, constructive characterization and morphology, together with dimensional aspects. The degree of exposure to wind action in the different dwellings was analyzed. The influence of the relative position of the dwelling under study within the block on airtightness was analyzed: on the floor immediately below the roof (RF) or on intermediate floors (IF). In the case analyzed, 15.7% of the dwellings were located immediately below the roof and 84.3% of the dwellings were located on intermediate floors. The model included continuous factors, geometric, and categorical, including the composition of the envelope, type of openings, presence of solar protection and singular elements.

A more in-depth description of the generation of airtightness models in multifamily buildings can be found in (Fernández-Agüera, Domínguez-Amarillo, Sendra, et al., 2019).

The factors which proved most influential in the construction method of the stepwise predictive model incorporate situational, morphological, constructive and geometric characteristics such as window surface, window perimeter, existence of blinds, existence of windows in bathrooms, type of window, degree of exposure and type of facade.

Table 5 shows the different factors of the built model. The γ_{SW} and γ_{PW} parameters represent the coefficients of the quantitative variables of window area and window perimeter respectively. The β_P, β_{WB}, β_W, β_{GE}, and β_{FA} parameters represent the constants of the dichotomous or categorical variables, existence of blinds, existence of window in bathroom, type of window, degree of exposure and type of facade respectively, according to the category to which they belong.

The categorical variables established in this sample were as follows:

- Facade type (F3: ½ foot brick (or one-brick thick) outer wall + insulation layer + hollow brick inner wall; F4: ½ foot brick (or one-brick thick) outer wall + air cavity + plasterboard inner wall; F5: fired clay paneling + air cavity + insulation + fired clay block).
- Window type (W1: hinge opening windows; W2: sliding windows; W3: hinged and sliding windows).
- Blind type (B1: no blinds; B2: external blinds; B3: roller shutter in splayed openings; B4: roller shutter in compact window blinds)
- Exposure (E1: semi-detached, linearly aligned buildings with four units per story; E2: semi-detached, linearly aligned buildings with two units per story; E3: open gallery buildings; E4: stand-alone high rises; E5: semi-detached, linearly aligned buildings with two units per story and building, located at the corner of the compound or in stand-alone buildings with H, T- or X-shaped ground plans).

The predictive model can be defined by means of the probability function found in the following equation, whose coefficients are defined in Table 5.

$$n_{50} = \gamma_{SW} \cdot SW + \gamma_{PW} \cdot PW + \beta_K + \beta_P + \beta_{WB} + \beta_{SCI} + \beta_W + \beta_{GE} + \beta_{FA} \quad (5)$$

The model (equation 5), established with 11 independent variables, indicates that 88.7% of the variability of the dependent variable is due to the independent variables. The standard error of the prediction has a value of 0.849 h^{-1}, which represents an adjustment of 35% of the standard deviation of cluster 1 (Table 6), a very limited value in relation to the variability of the group. In order to discard the dependence between variables, which even if present would not necessarily demonstrate causality, a set of ANOVA tests was established to determine whether the infiltration in the group of dwellings studied could be predicted with a model based on their classification characteristics, as the initial hypothesis seemed to indicate.

The predictive models were subject to a series of limitations imposed by the sample. The units had to be social, multifamily housing in one of the climate zones defined in Seville, with a net floor area < 105 m^2 and a window area <17 m^2.

Equation 6 represents the model for predicting demand based on infiltration for the case of housing built from 1979 onwards in Seville.

Table 7 represents the values to be included in equation 6 above the mean of the consumption slope established for Seville and the constant for the Seville models. Regionalized energy demand prediction models can be generated using the proposed method at either building or housing complex scale.

$$D_{TOT} = m\cdot\left(\frac{\gamma_{SW}\cdot SW + \gamma_{PW}\cdot PW + \beta_K + \beta_P + \beta_{WB} + \beta_{SCI} + \beta_W + \beta_{GE} + \beta_{FAC}}{20\cdot}\right) + \varepsilon \quad (6)$$

Table 5. Coefficients of the predictive model

Coefficient	Category	Value	Standard error
Constant (α)		9.058	1.624
Window area (i_{SW})		1.265	0.218
Window perimeter (i_{PW})		-0.387	0.070
Presence of Blinds (β_P)	Yes	0.000	0.000
	No	-1.957	1,051
Bathroom window (β_{WB})	Yes	1.498	0.315
	No	0.000	0.000
Window type (β_W)	V1	-5.154	0.600
	V2	-2.445	0.493
	V3	*	*
Degree of exposure (β_{GE})	T1	-0.075	0.473
	T2	-0.620	0.665
	T3	-1.558	0.588
	T4	-2.570	0.664
	T5	*	*
Façade type (β_{FA})			
	F3	1.179	0.619
	F4	-0.838	0.668
	F5	*	*

Table 6. Statistical description of the n50 predictive model of cluster 1

Model	R	R square	Adjusted R square	Standard error in calculation
12	0.943	0.887	0.865	0.849

Table 7. Parameters of the total demand prediction equation as a function of air leakage rate

Building Group	M	ε	*SD*
Building date >1979	27.37	25.95	3.28

The modeling is equally applicable at the level of stock-area for populations of buildings that share constructive and typological characteristics, such as multifamily housing stock of a region or geographical area, assuming the variability of the average values as well as the uncertainty that this introduces by widening the prediction band. For this, it is appropriate to use the representative values of the stock, central values of the population distribution for each set of variables. In order to apply the procedure a representative sample of a set of dwellings that define building groups, or buildings that describe the population of a regional group, is needed. In this way, the model can be fundamentally adjusted to the construction and morphological characteristics.

Discuss solutions and recommendations in dealing with the issues, controversies, or problems presented in the preceding section.

The following sentence is an example of a figure callout. Figure 1 is an example of a figure caption within a chapter.

FUTURE RESEARCH DIRECTIONS

In spite of the uncertainty generated by the construction methods of the enclosure, which follow more traditional building methods, especially in the regional case analyzed, the proposed models are capable of producing an acceptably accurate prediction of the energy impact of airtightness in the dwellings, particularly at the level of building stock or complexes. This is achieved by using general defining parameters requiring little effort and by increasing the usefulness of the proposed method as an analysis and work tool for planning interventions in built sectors or developments in urban areas.

CONCLUSION

The proposed method provides regional prediction models as an alternative to the generalist models developed to date. These are adapted to construction systems and architectural types, in this case, the stock of social housing in southern Europe.

The analysis presented allows the development of energy sensitivity analysis scenarios to airtightness of the building sets, generating an efficient tool for the processes for assigning objectives in retrofitting interventions.

In general, the air permeability in the study sample presents an energetic effect mainly during cold periods. Although this is usually higher in cold areas, it is still relevant in temperate areas. During the warm period air permeability has a limited effect on the energy balance of the dwellings and may even be beneficial during the periods of dissipation of the internal load. The average potential for reducing the annual energy demand associated with improving the airtightness of the enclosure is around 5% for every 0.1 h^{-1} decrease in the annual air renewal change (ARC). The usual airtightness values, although consistent with those found in other regions, differ somewhat from the norms in central and northern European countries, in some cases increasing fivefold. This means that, although indoor-outdoor temperature differentials are usually lower than those in central and northern Europe, the impact on energy demand, and in parallel on indoor environmental conditions, is particularly significant. This makes it necessary to analyze and consider this impact, especially given its relationship with energy poverty conditions and distortions of housing health and comfort conditions.

The permeability through the envelope is due to the combination of different constructive and morphological factors to which the stochastic component of the construction and the temporal degradation should be added. Therefore, models predicting the energy associated with infiltrations should introduce this set of factors. In the case of multifamily buildings in southern Europe, the model with the most information and the least uncertainty introduced is linked to the openings and construction interface, specified in the parameters of:

- Window area and perimeter
- Window opening mechanism
- Existence of solar protection
- Degree of exposure of the dwellings
- Constitution of the construction system of the facade

In the case of other typologies or specific aspects of construction, the model should be adjusted to reflect these factors. The models should be optimized to introduce as few variables as possible to ensure correct adjustment while reducing the interference introduced by an excess of variables.

ACKNOWLEDGMENT

The authors want to thank to the VI Plan Propio de Investigación de la Universidad de Sevilla.

REFERENCES

Alves, S., Fernández-Agüera, J., & Sendra, J. J. (2014). Infiltration rate performance of buildings in the historic centre of Oporto. *Informes de la Construcción*, *66*(535), e033. Advance online publication. doi:10.3989/ic.13.009

Ashrae. (2002). *Ashrae Guide and Data Book (Fundamentals)*. Ashrae. Retrieved from https://www.ashrae.org/File Library/docLib/Public/2003627132256_326.pdf

Cuerda, E., Guerra-Santin, O., Sendra, J. J., & Neila, F. J. (2020). Understanding the performance gap in energy retrofitting: Measured input data for adjusting building simulation models. *Energy and Building*, *209*, 109688. doi:10.1016/j.enbuild.2019.109688

de la Flor, F. J. S., Domínguez, S. Á., Félix, J. L. M., & Falcón, R. G. (2008). Climatic zoning and its application to Spanish building energy performance regulations. *Energy and Building*, *40*(10), 1984–1990. doi:10.1016/j.enbuild.2008.05.006

Domínguez-Amarillo, S., Fernández-Agüera, J., Campano, M. Á., Acosta, I., Domínguez-Amarillo, S., Fernández-Agüera, J., ... Acosta, I. (2019). Effect of Airtightness on Thermal Loads in Legacy Low-Income Housing. *Energies*, *12*(9), 1677. doi:10.3390/en12091677

Domínguez-Amarillo, S., Fernández-Agüera, J., Peacock, A., & Acosta, I. (2019). Energy related practices in Mediterranean low-income housing. *Building Research and Information*. Advance online publication. doi:10.1080/09613218.2019.1661764

Domínguez-Amarillo, S., Fernández-Agüera, J., Sendra, J. J., & Roaf, S. (2018). Rethinking User Behaviour Comfort Patterns in the South of Spain—What Users Really Do. *Sustainability*, *10*(12), 4448. doi:10.3390u10124448

Domínguez-Amarillo, S., Sendra, J. J., Fernández-Agüera, J., & Escandón, R. (2017). *La construcción de la vivienda social en Sevilla y su catalogación 1939-1979* (U. de Sevilla, Ed.).

Domínguez-Amarillo, S., Sendra, J. J., & Oteiza San José, I. (2016). *La envolvente térmica de la vivienda social: el caso de Sevilla, 1939 a 1979 Title* (1st ed.). Editorial Consejo Superior de Investigaciones Científicas.

Feijó-Muñoz, J., Pardal, C., Echarri, V., Fernández-Agüera, J., Assiego de Larriva, R., Montesdeoca Calderín, M., Poza-Casado, I., Padilla-Marcos, M. Á., & Meiss, A. (2019). Energy impact of the air infiltration in residential buildings in the Mediterranean area of Spain and the Canary islands. *Energy and Building*, *188–189*, 226–238. doi:10.1016/j.enbuild.2019.02.023

Fernández-Agüera, J., Domínguez-Amarillo, S., Alonso, C., & Martín-Consuegra, F. (2019). Thermal comfort and indoor air quality in low-income housing in Spain: The influence of airtightness and occupant behaviour. *Energy and Building*, *199*, 102–114. doi:10.1016/j.enbuild.2019.06.052

Fernández-Agüera, J., Dominguez-Amarillo, S., Fornaciari, M., & Orlandi, F. (2019). TVOCs and PM 2.5 in Naturally Ventilated Homes: Three Case Studies in a Mild Climate. *Sustainability*, *11*(22), 6225. doi:10.3390u11226225

Fernández-Agüera, J., Domínguez-Amarillo, S., Sendra, J. J., & Suárez, R. (2016). An approach to modelling envelope airtightness in multi-family social housing in Mediterranean Europe based on the situation in Spain. *Energy and Building*, *128*, 236–253. doi:10.1016/j.enbuild.2016.06.074

Fernández-Agüera, J., Domínguez-Amarillo, S., Sendra, J. J., Suárez, R., & Oteiza, I. (2019). Social housing airtightness in Southern Europe. *Energy and Building*, *183*, 377–391. doi:10.1016/j.enbuild.2018.10.041

Fernández-Agüera, J., Sendra, J. J., & Domínguez, S. (2011). Protocols for measuring the airtightness of multi-dwelling units in Southern Europe. In Procedia Engineering (Vol. 21). doi:10.1016/j.proeng.2011.11.1992

Healy, J. D. (2017). *Housing, Fuel Poverty and Health : A Pan-European Analysis*. Routledge.

Hennig, C., Meila, M., Murtagh, F., & Rocci, R. (2015). *Handbook of Cluster Analysis*. CRC Press. doi:10.1201/b19706

Hernandez Sampieri, R., Fernandez Collado, C., & Baptista Lucio, M. del P. (2014). *Metodología de la investigación. Metodología de la investigación*. Retrieved from https://www.casadellibro.com/libro-metodologia-de-la-investigacion-5-ed-incluye-cd-rom/9786071502919/1960006

IDAE. (2009). *Condiciones de aceptación de Procedimientos alternativos a LIDER y CALENER*. Instituto para la Diversificación y Ahorro de la Energía.

Kish, L. (1965). Survey Sampling. John Wiley and Sons.

Kottek, M., Grieser, J., Beck, C., Rudolf, B., & Rubel, F. (2006). World Map of the Köppen-Geiger climate classification updated. *Meteorologische Zeitschrift (Berlin)*, *15*(3), 259–263. doi:10.1127/0941-2948/2006/0130

Lixing, G. (2007). *Airflow Network Modeling in EnergyPlus | Simulation Research*. International Building Performance Simulation Association. Retrieved from https://simulationresearch.lbl.gov/publications/airflow-network-modeling-energyplus

Meiss, A., & Feijó-Muñoz, J. (2014). The energy impact of infiltration: A study on buildings located in north central Spain. *Energy Efficiency*, *8*(1), 51–64. doi:10.100712053-014-9270-x

Meiss, A., & Muñoz, F. (2013). La Necesidad De Estudiar Las Infiltraciones : Estudio De Caso. In *Viviendas Sociales De La Urbanización Zabalgana*. Boletín Académico, Revista de Investiga., doi:10.17979/bac.2013.3.0.997

Ministerio de Fomento del Gobierno de España. (2017). Código técnico de la Edificación (CTE). Author.

Ministerio de la Presidencia, Gobierno de España. (2013). Real Decreto 235/2013, de 5 de abril, por el que se aprueba el procedimiento básico para la certificación de la eficiencia energética de los edificios. Boletín Ofical del Estado. Author.

Poza-Casado, I., Meiss, A., Padilla-Marcos, M. Á., & Feijó-Muñoz, J. (2020). Airtightness and energy impact of air infiltration in residential buildings in Spain. *International Journal of Ventilation*, 1–7. doi:10.1080/14733315.2020.1777029

Raftery, P., Keane, M., & O'Donnell, J. (2011). Calibrating whole building energy models: An evidence-based methodology. *Energy and Building*, *43*(9), 2356–2364. doi:10.1016/j.enbuild.2011.05.020

Reddy, T. A. (2006). Literature Review on Calibration of Building Energy Simulation Programs. *ASHRAE Transactions*, *112*(1), 226–240.

Saltelli, A., Tarantola, S., Campolongo, F., & Ratto, M. (2004). *Sensitivity Analysis in Practice: A Guide to Assessing Scientific Models*. Wiley. Retrieved from https://www.wiley.com/en-us/Sensitivity+Analysis+in+Practice%3A+A+Guide+to+Assessing+Scientific+Models-p-9780470870945

Sendra, J. J., Domínguez-Amarillo, S., Bustamante Rojas, P., & Leon, A. L. (2013). Energy Intervention in the residential sector in the south of Spain: Current challenges. *Informes de la Construcción*, *65*(532), 457–464. doi:10.3989/ic.13.074

Sherman, M. (1995). The Use of Blower-Door Data. *Indoor Air*, *5*(3), 215–224. doi:10.1111/j.1600-0668.1995.t01-1-00008.x

Sherman, M. H. (1987). Estimation of infiltration from leakage and climate indicators. *Energy and Building*, *10*(1), 81–86. doi:10.1016/0378-7788(87)90008-9

Spiekman, M. (2010). *ASIEPI. The final recommendations of the ASIEPI project: How to make EPB-regulations more effective?* Retrieved from https://ec.europa.eu/energy/intelligent/projects/sites/ieeprojects/%0Afiles/projects/documents/asiepi_access_the_results._en.pdf

Sudman, S. (1976). *Applied Sampling*. A. Press.

Tamburrini, M., Palmer, D., & Macdonald, I. A. (2003). *Calibrating models for simulation use in design practices*. Retrieved from http://www.sesg.strath.ac.uk

UNE-EN 13829:2002. (n.d.). *Thermal performance of buildings - Determination of air permeability of buildings - Fan pressurization method. (ISO 9972:1996, modified)*.

Younes, C., Shdid, C. A., & Bitsuamlak, G. (2012). Air infiltration through building envelopes: A review. *Journal of Building Physics*, *35*(3), 267–302. doi:10.1177/1744259111423085

KEY TERMS AND DEFINITIONS

Air Exchange Rate: Number of interior volume air changes that occur per hour, and has units of 1/h.

Air Infiltration: Unintentional flow of air from outdoors to indoors through leaks (unplanned openings) in a building envelope.

Airtightness: Uncontrolled inward and outward leakage of outdoor air through cracks, interstices, or other unintentional openings of a building, caused by pressure effects of the wind and/or stack effect

Building Envelope: All of the elements of the outer shell that maintain a dry, heated, or cooled indoor environment and facilitate its climate control.

Mechanical Ventilation: A building ventilation system that uses powered fans or blowers to provide fresh air to rooms when the natural forces of air pressure and gravity are not enough to circulate air through a building.

Natural Ventilation: Intentional passive flow of outdoor air into a building through planned openings (such as louvers, doors, and windows).

Predictive Modeling: The process of using known results to create, process, and validate a model that can be used to forecast future outcomes.

Ventilation: Intentional introduction of outdoor air into a space that is mainly used to control indoor air quality.

Chapter 5
Interactive Exhibition as a Form of Participation in the Future of Sustainable Housing

Juan Bautista Echeverría
https://orcid.org/0000-0002-4438-6449
Universidad de Navarra, Spain

Iosu Gabilondo
COAVN, Spain

Teresa Meana Rodríguez
COAVN, Spain

Juana Otxoa-Errarte
COAVN, Spain

Claudia Pennese
COAVN, Spain

Montserrat Puig Tramunt
COAVN, Spain

Judith Ubarrechena
COAVN, Spain

ABSTRACT

The Gipuzkoa branch of the Basque and Navarre College of Architects organized, within the MUGAK Architecture Biennial, the exposition "The Transgenerational House." It took place in a pavilion specially built for the purpose in a public space in the city of San Sebastian (Spain). In it, both a conventional furnished home and an alternative one, with the possibility of allowing free spatial divisions and furnishing distribution, were recreated. Some architectural teams showed their experiences on housing. A set of components with a color code was developed to link the two homes and the work of the architects. The pavilion was opened to the public, which had access to the contained information in a partially directed way and participated answering to posed specific questions. Additionally, 10 structured workshops with different collectives were organized, making specific proposals on the alternative home. The overall exposition is shown, reflecting on the advantages and limitations of citizen participation as an instrument of sustainable development.

DOI: 10.4018/978-1-7998-7023-4.ch005

INTRODUCTION

Citizen participation is broadly recognized as a key instrument on the way to sustainable development. Different administrations and organizations play an important role in mobilizing and educating citizens because information, sensibilization an education are previous steps that need to be carefully planned (Meadowcroft 2014). Recent studies show how cities that are committed to enhancing sustainability policies tend to be more participatory (Lyons, Smuts and Stephens 2001; Portney & Berry 2010) and reinforce the idea that citizen participation is an indicator of the type and quality of democracy (de Poza-Vilches, Gutiérrez-Pérez and López-Alcarria 2019).

The second Basque Country International Architecture Biennial, MUGAK, was held between the 10th of October and the 12th of December 2019 at different venues around the Basque Country (Spain). Organized by the Department of Environment, Territorial Planning and Housing of the Basque Government, with the collaboration of the regional government of Gipuzkoa and the City Council of San Sebastian, according to data gathered over the event, it brought together more than 45,000 people in hundreds of activities and exhibitions (MUGAK 2019a). Architecture and urbanism were understood within the Biennial as tools for change, which must be redefined and rethought together to produce new paths towards the design of spaces for coexistence which materialize a collective yearning to live in more balanced and just cities and societies. (MUGAK 2019b).

With the objective of participating actively in the Biennial, the Gipuzkoa branch of the College of Architects, COAVN, created a team that applied for a grant from a funding program of the Basque Country.

The grant proposal covered the development of an interactive exhibition that would stimulate a reflection on existing housing stock present in urban space, and of possibilities to adapt this over the lifetime of residents. This concept could be defined as Transgenerational Housing. The use of this term is aligned with shifts in industrial design that emerged from 1986, when James J. Pirkl (1994, 2008) began to link the design of objects with gerontology. Transgeneracional Housing is, therefore, housing that is adapted to people's needs, whatever their ages or physical abilities.

Participation, as the basis of collaboration between a number of diverse agents and the creation of networks, was taken as the principle objective of the exhibition. To encourage participation, the design of the pavilion and the content of the exhibition were developed in parallel, so that the physically constructed outside was related to the message communicated inside. The quality of space, a key aspect in housing, had to be reflected in the physical space of the exhibition itself.

To give the exhibition an image that was clearly different to that of generic provisional installations, a decision was taken to develop a pavilion with emphatic geometry, carefully designed to host the content of the exhibition coherently. As the proposal was written by an architectural college, the prefabrication and reusability of the pavilion were also issues that were considered. The assembly of the pavilion, a completely dry construction process, was completed in just a few days, as was the disassembly. All components were reused.

Great importance was placed on graphic design, which underpinned a connection with both the activities of the Biennial and visitors. Flyers, reusable paper bags and information panels printed on fabric with details about specific projects and questions directed at visitors were printed. Black was the color chosen for the finished exterior of the pavilion, signage lettering and the plastic balls that were part of the public participation system used in the exhibition.

Both the contents and methodology that were deployed facilitated the inclusion of a large number of people, including both workshop participants and casual visits by members of the public in general.

Thanks to this, data of immediate relevance to important decisions to be made on the future of the housing was gathered over the course of the Biennial.

The exhibition ran between the18th and 27th of October of 2019 and welcomed a total of 2355 visitors. Six architectural teams presented innovative housing projects that served as a reference for the visitors. Ten directed workshops involving 178 people of different cultural, gender and age profiles were arranged, resulting in ten housing proposals.

With the objective of ensuring that event was cost neutral for COAVN, the organisational team collaborated with a range of public institutions and private companies, which made the exhibition possible.

METHODOLOGY

The design of a working methodology oriented to the collective construction of an innovative and transformative knowledge was a first step. To this end, it was suggested that a review of socially constructed knowledge through a process of shared reflection would lead to a collective process of reflection or deliberation. This means, defining possible scenarios as a basis for technical and political decisions, able to, with ethics and wisdom, materialise the outcomes of the reflective process (Cohen 1989, 1996).

A meeting space was constructed that facilitated the effective participation of a diversity of agents. Citizens themselves were centered as subjects directly affected by housing, which plays an important part in their daily lives throughout their entire lifespans.

As Patrizia Nanz and Miriam Fritsche emphasize (2014), for this type of process to be effective and produce acceptable results, specific methodologies applied in each case must meet certain standards. While this is not an exact scientific process, under common theoretical principles, the following criteria of quality are respected: **Information, Dialogue, Reflection, Empowerment,** and **Inclusion**. The authors note that, to guarantee the quality standards mentioned, the dialogue-deliberative processes must be structured according to specific formulas, generally codified, that aim to guarantee a two-way dialog in a space safe for all participants.

These five principles drove the whole exposition, and the pavilion itself took a form to allow a journey through them. The four first ones, conforming a not interchangeable sequence and Inclusion as the key aspect of a real wide participation.

Information

A quality participative process based on the dialogue is not possible if information is not provided to all participants. This information should be focused on the specific task and aim to build the capacity of participants not necessarily familiar with the area, so that their participation can be active and productive. In this case, due to the extent of the technical data involved, it was very important that this information was presented in language accessible to a diverse public. The objective was to generate understanding without reducing the technical content to banal generalities.

The next step was to select examples of innovative housing models, covering housing units but also working at a building and urban scale, because of the implications that these have for individual dwellings. The most attractive proposals at international level were chosen, though later this selection was limited to a national level in order to offer a higher degree of geographic and cultural proximity so that the models could be more easily assimilated by local citizens. The architectural teams invited were selected on the basis of their professional careers. These teams were contacted and asked to choose one of their projects that would best fit with the theme exhibition. Some of them turned down the proposal due to lack of available time or simply lack of interest. The final selection of participants was considered to be, in any case, very representative of the initial shortlist of candidates.

The commitment of the participating teams was very high and was an essential starting point for very close collaboration over the event. This is a key aspect for the realization of the goal of a collective construction of a "common" (Bailiff et al. 2011).

All the architectural teams were asked to present their projects using a variety of media (drawings, texts, scale models and videos), while using shared and accessible terminology, previously defined by the event organizers. This facilitated the construction of simple and direct mechanisms for communication, and subsequently an understanding of the similarities and differences between the proposals.

Information panels, videos and books were available in addition to this material, multiplying the available channels of communication and deepening in the level of information available to provide a quality response to the diversity of the participating public.

Dialogue

Participation is not achieved just by bringing people together and hoping for a result. (Ganuza et al. 2010). It consists of constructing physical and emotional spaces capable of facilitating the constructive interchange of knowledge and collective reflection around this, establishing an egalitarian dialogue between diverse actors.

Dialogue is a key aspect, as it makes the generation of natural, endogenous, and authentic thought processes possible. Processes that are able to provoke a substantial change in the position of all social actors who, by nature, tend to express themselves on the basis of individual and preconceived points of view (Allegretti and Torquato Luiz 2007). Dialogue makes it possible to escape these mechanisms, open wider horizons and reveal hidden necessities. It can place collective interests before individual ones as part of a shared journey of valuing diversity as a key value in socio-spatial design.

With this objective in mind, in addition to the content and the design of the space itself, a series of workshops and working groups with groups differentiated by age and origin were organized. The working dynamic was designed to encourage reflection, interaction and the generation of unexpected ideas from the participants who, besides contributing to the dialogue, could learn from it.

Reflection

Participation does not depend exclusively on the presence of target audiences, but also on the creation of a deliberative process that allows these actors to reflect collectively both on an issue and alternative courses of action around it. Reflection emerges as key to guaranteeing the effectiveness of citizen active participation processes with respect to all areas of public policy, and especially those related to the design of living spaces, be they public or private as in this case.

As Arstein (1969) comments, the participative processes that are "sold" to citizens who are called on to make decisions are manipulative, as these decisions, especially in relation to urban development, are inevitably of a technical and political nature. Methodology must facilitate reflection, which is to say, the definition of possible alternatives, prioritizing and identifying pros and cons. Alternatives should not be abstract ideas, but the result of knowledge and reflection specific to the issue under discussion, in order to guarantee the social value of all interventions, urban or architectural.

The dynamics put into practice facilitated this process, and results were not taken at any time to be "the end", but as a knowledge base essential for reflection and decision making on the part of the relevant agents defined from technical and political points of view. In any case, these processes must also generate a commitment on the part of officials, to try to build their capacity to ethically interpret the results of participative processes and not always satisfy themselves with more "standard" and "relaxing" proposals.

Empowerment

This aspect has two facets. On the one hand, it is about engaging the active participation of actors in the definition of design guidelines and, on the other, to try to change their perspectives and conventional points of view. Alternatives must be made visible before they can be considered or adopted.

This raises a fundamental question in terms of citizen motivation when getting involved in the process, determining, in addition, the quality of this involvement. It is vital that the time, energy and knowledge that people share is not perceived as a wasted effort, but that the work and the opinions they have shared are noted and exert an effective impact on decision making. This demands an effective systematization and communication of results, and also it calls for sticking to the timeline for the proposed change and insisting on steps and mechanisms that make implementation of the results, which are opportunely interpreted and formalized, possible.

As to the empowerment process for citizens themselves, as it has already been commented, the materials used for the dialogue process and dynamics used have a strategic importance.

Inclusion

The plurality and representativeness of participants is a key area, not only to guarantee the quality of the results but, above all, to provide legitimacy. With the objective of guaranteeing inclusion and as an expression of active participation, 10 workshops were organized, directed at different social agents. Prior to selecting participants, a demographic analysis was carried out identifying different collectives on the basis of age, sex and origin.

To these were added casual visitors to the pavilion, enriching the process in uncontrolled and unforeseeable ways. Nevertheless, as it is seen in point 4,4, a rigorous register of people attending linked to the variables used to identify specific collectives was maintained.

The methodology contributed, in addition to what has already been described, the integration of the languages used, for each design aspect, in a unified, if polyphonic, discourse. In this sense, the design of the pavilion itself was especially important. As is detailed in point three, it addressed more than just criteria of architectural composition and construction. The shape, the materials and constructive techniques used and the designed lifespan of the pavilion were adapted to the overall ambition of the project: to harness the communicative capacity of the overall message and its socio-cognitive contents, and facilitate dynamic dialogue.

DESIGN AND CONSTRUCTION OF THE PAVILION

The architectural image of the pavilion had a decisive importance for the exhibition. It was considered that it should be both a pole of attraction in the context of the Architecture Biennial, and a suitable container for the content it sought to display and communicate.

Its image, with a geometry and simplicity that evoked the iconography of a house, transmitted a simple relation between vision-distant outer volume and a tactile-near interior. It defined the home as refuge, a place where space is contained through material. The materiality and the scale responded to complex questions such as orientation, hierarchical structuring of uses, optimization of resources and adaptation to climate.

Figure 1. Pavilion, site (Design Team)

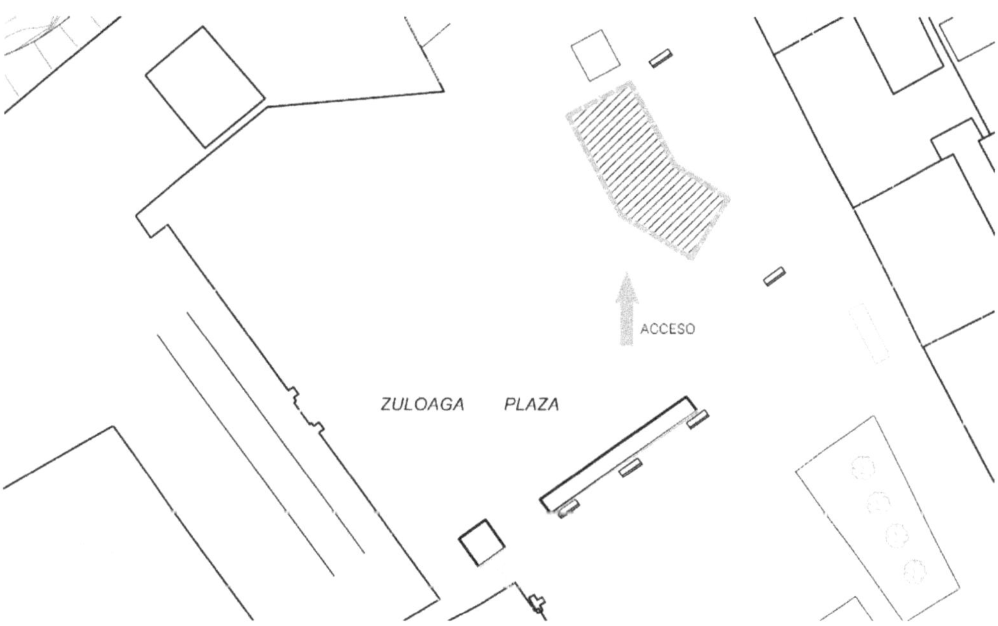

The pavilion was placed in a public space to the northeast of the historical centre of the city, in Zuloaga Plaza next to the San Telmo Museum (Figure 1). The placement of the pavilion respected the urban characteristics of the square and the buildings that surrounded it, and did not interfere with the view of the San Telmo Museum or the San Vicente Church, but maximized the attraction of pedestrian traffic towards the museum and San Juan Street. It is worth emphasizing that not a single urban element of the square was modified nor used as an anchor, and after disassembly of the pavilion the square was left exactly in its original state.

The choice of the shape and the place were not accidental (Figure 2), as the goal was to achieve maximum visibility of both the constriction and the contents of the exhibition and to take a reflection and questioning on the crisis of the current housing model into the heart of the city. The general context of the MUGAK Biennial, dedicated mainly to housing, also contributed to a deepening reflection on how homes will have to be adapted to the real necessities of XXI century society.

Figure 2. Pavilion, exterior (Design Team)

Figure 3. Pavilion, interior (Design Team)

The interior (Figure 3) was carefully designed both to meet the exhibition requirements and to reflect its sustainable character.

Characteristics and Construction System

The pavilion occupied a total floorplan of 122.35 m^2 and had a peak altitude of 5 meters along its central longitudinal axis. A one simple entrance facilitated its installation in the square, providing maximum respect to the natural circulation of pedestrians in the immediate surroundings. This entrance, which coincided with access from the square, clearly articulated the exhibition's content and allowed visual control from the entrance counter (Figure 4). In addition, it invited visitors through a planned route, in the form of a figure eight, that guided visitors through the exhibition's contents: general information, info points around hands-on models, an experimental model and a participating zone next to an empathy scale model.

The color selected for the exterior, black, was the result of a deliberate intention to link the volume directly with the graphic style of the biennial. The volume was intentionally mute from the outside, so that the interior content could surprise. This strategy encouraged people to discover what the artefact held. The smooth, shiny, black exterior contrasted with the colorful interior in which raw material was visible, achieving impact without overstatement.

In terms of construction, it was built using a wooden laminate (CLT) base of local radiata pine, 120 mm thick. Seven identical laminated wood gantries and a variable section with a seven-meter span were also screwed on to the base. The gantries were supported with eight braces and the structure was completed with 160x100 mm wooden strapping. The lateral facade and gabled roof were screwed onto the structural elements, as well as an unbroken black lacquered (mini wave) cladding. In the walls, fixed to a wooden substructure, cellular polycarbonate sheets were installed. This created an interesting lighting contrast between the opaque covering and the walls which also facilitated precise control of the artificial illumination in the exhibition area.

The pavilion incorporated a wood framed door integrated in the cladding, and an entrance ramp made with OSB boards. At the join at the meeting of the two sections of the pavilion, hidden gutters, properly waterproofed, were installed to maintain the pure lines of the volume. The result was a compact architectural piece, black and shining on the outside, that gave way to a colorful, suggestive and stimulating interior space.

The overall shape, especially the transversal cross section and the shape of the lateral walls, evoked the silhouette of the stereotypical house of collective imagination, giving visitors a hint as to the content of the exhibition. The diffuse light given off by the translucent walls harnessed the atmosphere generated by this silhouette during a visit inside.

The pavilion was completely accessible and was equipped with all the indispensable security systems demanded by construction of this type, including powder and CO_2 fire extinguishers, emergency lighting and signage. Evacuation procedures were calculated using ASET/RSET (Rey et al. 2015), given a 10-minute limit for the support structure. (R10).

Figure 4. Pavilion, floor plan (Design Team)

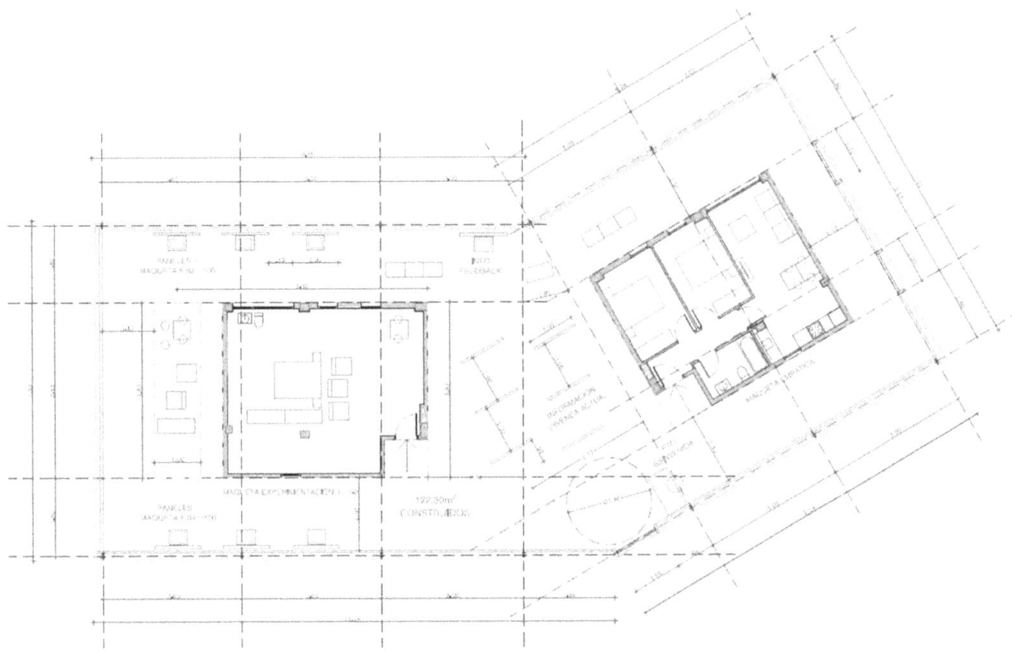

Assembly, Disassembly, and Reusability

Big importance was given to the architectural sustainability of the pavilion and the three Rs approach was followed (Dahy and Knippers 2013). **Reduction** as a consequence of the very limited number of materials used (wood, metal and polycarbonate for the container and light textile panels and EPS pieces for the exhibition). **Reuse**, because the pavilion was sold to allow a new use, the lighting is now installed in the College of Architects and the plastic balls used for participation were a gift for a kindergarten. **Recycling** was completed properly, carrying the EPS remaining pieces to a recycling point.

The assembly of the pavilion was a totally dry process, carried out in approximately 6 days. The assembly of the wooden framework was completed in 2 days, while the external cladding of polycarbonate sheets and steel flashing, along with the entrances and electrical installation, took about 4 days. The dry assembly of the pavilion was a totally totally reversible process, carried out in reverse order after the exhibition. The time needed for disassembly was notably reduced in each stage of the operation.

The reusability of the elements of the pavilion was a concern from conception for the organisers of the activity, who were conscious that one of the problems of ephemeral architecture is the endpoint of its components once its useful life is over. Amongst the design decisions taken to facilitate reuse was the deliberate division of the structure into two separate parts, revealed in its geometric breakdown. Both the duplication of the gantries in the central wedge and the join in the CLT base, allowed for a seperation of the structure into two prisms for later independent reusability. Other factors, including the provision of an open and flexible space, and the oversized structure capable of sustaining much greater loads than the exhibition contents, contributed to the reusability of the pavilion. After the closing of the exhibition, while set up in the same square, the pavilion was sold to a builder, who has given the space a new use.

CONTENTS OF THE PAVILION

The exhibition was understood, fundamentally, as a participative challenge, both in terms of content and its internal distribution inside the pavilion. To this end, the following zones were established. As has already been commented, they were arranged in a route which was not imposed but insinuated, after passing through the entryway:

1. Knowledge Zone
2. Empathy Zone
3. Experimental Zone
4. Feedback Zone

As observed in the scheme (Figure 5), the central axis of the pavilion contained two floor plans of an apartment, in a way that was harmonic with the break in the space resulting from its design. The first floor plan reflected the distribution of a house type in which elements associated with **Actions** and **Areas** were incorporated. These elements were famed in **Blocks** and color coded. The choice of colors was limited to those available in cartridges for 3D printing. The second floor plan was reduced to just the perimeter of the first apartment, creating a space for experimentation.

Figure 5. Pavilion, Zoning plan (Design Team)

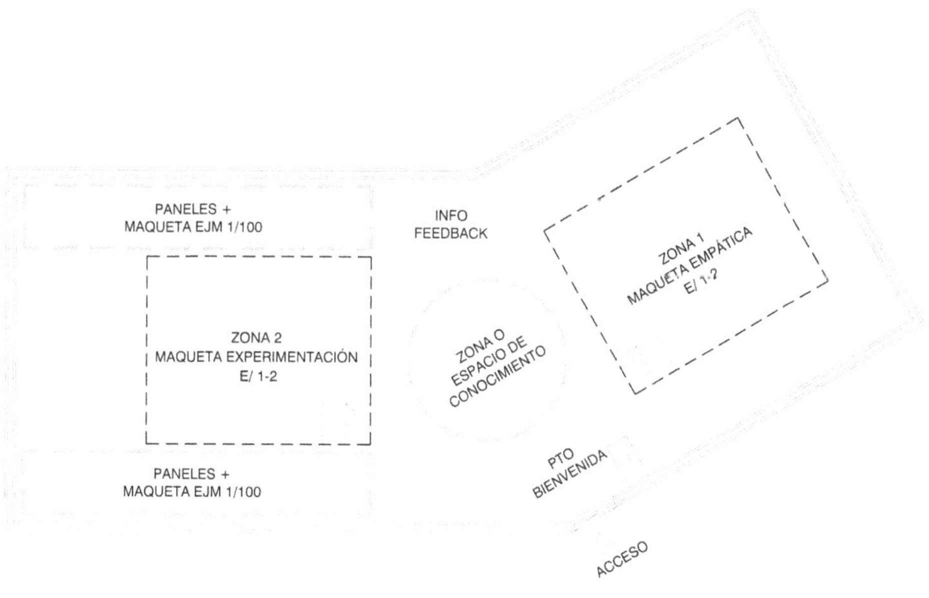

The division that was made between Blocks connected with different areas was based on recent studies on the cataloguing of areas in the house (Montaner and Muxí 2010; Paricio and Sust 2004; Loch 2006; Leupen et al. 2005) and on the relationships between supports and separable units (Habraken et al. 1979; Habraken and Mignucci 2009). The representation of the floorplans was inspired by the film *Dogville*, by Lars von Trier (2003), integrating the ideas of footprints and limits of the field of view

(Salgado 2006). Marking the perimeter of both floorplans facilitated a reflection on the possibilities of a house like one's own, one that we already know and live in, not another yet to be built.

For the two floorplans, and therefore for all other elements, a 1:2 scale was considered suitable. Its use facilitated a real relationship with the size of objects, reducing the size required of the pavilion and simultaneously offering the possibility of enriching the experience, by approaching a scale that children use in games.

The Knowledge Zone was located between the two floorplans, in the wedge that formed the intersection, next to the entrance. It extended into the Experimentation Zone. The information panels in the central zone were hung so that they could be read from any point around the floor of the pavilion, facilitating a comparison of the two floorplans and the circulation of the visitors.

The route finished in the Feedback Zone, where visitors had the opportunity to respond to questions that were raised along the route (Figure 6).

Figure 6. Exhibition opening (Design Team)

Knowledge Zone

Illustrative information was available as infopoints, with the objective that all visitors could have a virtual experience of less conventional ways of organizing their houses and living in them. In the central part of the pavilion next to the entrance, introductory information was provided by the organization Grupo Habitar, from the Catalan Polytechnical University, which consisted of three books and three videos. One of these videos, *Una conversa sobre el menjar i ciutat*, (UPC 2019) was also screened at an open session in the San Telmo Museum. Additionally, panels with information about projects by different groups of architects were placed on both sides of the Experimental Zone, accompanied by explanatory videos. This material was produced by the teams which carried out each project. The architectural firms finally selected had demonstrated professional experience the field and their initiatives contributed a theoretical and practical background to the exhibition:

- MAIO architects
- EEESTUDIO
- SATT
- ENORME estudio
- LACOL arquitectura cooperativa
- Colectivo PUNTO6

A project by each of these groups was reproduced in a 1:50 scale model, using the same coding for different colors, elements and blocks as was used in the scale models in the Empathy and Experimentation zones. Color 3D printing was possible thanks to the collaboration of the ENEDI research group at the University of the Basque Country (2020). The objective of this action was to be able to compare diverse projects, seeking an overall reflection in common terms, starting from initiatives with diametrically opposed approaches.

Finally, some time was offered to each team to present its work in person at some point over the duration of the exhibition, with a debate following their presentation.

Empathy Zone

The Empathy Zone contained the floor plan of a conventionally distributed 65 m^2 (54m^2 useful) house (Access, Living-Dining Room, Kitchen, Bathroom and 2 Bedrooms). The Empathy scale model contained the elements shown in Table 1.

Table 1. Empathy Scale Model. House type

	BLOCK	ACTION/AREA	ELEMENTS	COLOR
IN HOUSING UNIT	Hygiene	Bathroom/Individual	Washbasin-Toilet-Shower	
		Wash-drying/Unitary	Washing Machine-Dryer	
	Food	Cooking-Food/ Unitary	Fridge-Oven-Stove-Table	
	Rest	Rest/Individual	Bed	
	Relationships	Rest/Individual	Sofas-Tables	
	Learning	Study-work/Individual-Group	Tables-Chairs-Piano	
	Storage	Stored/Unitary	Everyday Items	
	Outside	Relation with the outside/ Unitary	Balcony-Terrace	
IN THE BUILDING	Support	Storage-Parking/ Unitary	Special Items	

Figure 7 shows the floor plan of the Empathy scale model of the house with elements placed and coded using the RAL color code.

Figure 7. Empathic Model (Design Team)

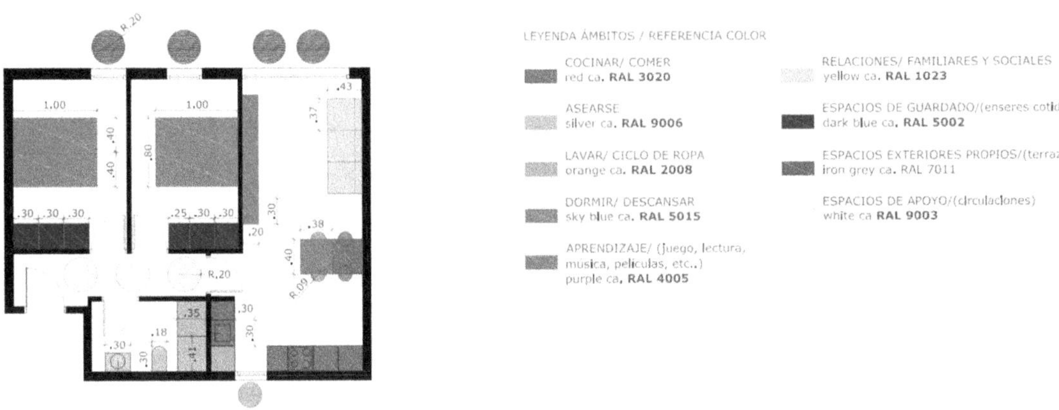

The elements were made of polystyrene, so that they could be moved easily and would not be dangerous for any member of the public. They were categorised into blocks and painted accordingly.

Experimentation Zone

The Experimentation Zone, which contained the outline of the floorplan represented in the Empathy Zone and therefore had the same surface area, facilitated the elements game, which was played by selecting and placing elements from an attached "warehouse." In this case, the placement of internal and external elements, inside the flat and in the building's common areas, was totally unrestricted, as is reflected in Table 2.

Feedback Zone

The Answer or Feedback Zone was located next to the Empathy scale model, to take advantage of the circumstance that all the other zones in the exhibition were crossed before reaching this area. At the entrance to the pavilion a paper bag which contained three black plastic balls was given to each visitor who wanted one. Questions were written with marker pen on panels covered in white fabric, and one ball could be used to answer each question by placing a ball in a corresponding basket (Figure 8).

This methodology for recording data was inspired by an installation by Stefan Sagmeister titled *The Happy Show*, at the MOCA Pacific Design Center (Sagmeister 2013). The final count of balls collated the opinions of all visitors, including both the general public and members of organized groups.

Placing the Feedback zone close to the Empathic Model, invited the public to follow a natural path through the exhibition (Figure 9) having the possibility to answer the critical questions in the close presence of the conventional housing distribution.

Table 2. Experimentation Scale Model

	ACTION	ACTION/AREA	ELEMENTS	COLOR
IN THE HOUSING UNIT (BLOCKS THAT THEY COULD BE REDISTRIBUTED)	Hygiene	Bathroom/Individual	Washbasin-Toilet-Shower	
		Wash-drying/Unitary	Washing Machine-Dryer	
	Feeding	Cooking-Food/Unitary	Fridge-Oven-Stove-Table-Chairs	
	Rest	Rest/Individual	Bed	
	Relationships	Rest/Individual	Sofa-Tables	
	Learning	Study-work/Individual-Group	Tables-Chairs-Piano	
	Storage	Stored/Unitary	Everyday Items	
	Outside	Relation with the outside/		
IN THE BUILDING (BLOCKS THAT COULD BE REMOVED)	Hygiene	Wash-drying/Collective	Washing Machine-Dryer	
	Food	Cooking-Food/Collective	Complete Kitchen and Table	
	Rest	Social Accommodation/Collective	Sofa-Tables-Beds	
	Relationships	Relationships/Collective	Sofa-Tables	
	Learning	Shared Workspaces/Collective	Tables-Connectivity-Information	
	Storage	Parking-Stored/Collective	Special Items	
	Outside	Relationship with the outside/ collective	Terraces	

Figure 8. Feedback Zone (Design Team)

Figure 9. Empathic and Model Feedback zone (Design Team)

PARTICIPATION OF ORGANIZED GROUPS

Throughout the time over which the exhibition was active, 10 structured workshops were organized. The intention was to bring together groups with different ages and profiles. The general characteristics of each group is drawn up in Table 3. The written consent of all members of the groups was required due to data protection legislation.

Table 3. Structured Workshops

Nº	Name	Date	Characteristics	Age Group	Women	Men	TOTAL
T_01	FP Bidasoa	21 October 2019	Professional training	Young people	3	13	16
T_02	EAGI	22 October 2019	Building School (Gipuzkoa)	Young people	0	9	9
T_03	H.Hitza	22 October 2019	Active Aging Association	Adults (>55)	9	2	11
T_04	Usandizaga 1	23 October 2019	High School	Young people	9	17	26
T_05	Col. Jesuits	23 October 2019	School	Young people	15	18	33
T_06	DSSLagunkoia	23 October 2019	Friendly City, San Sebastian City Council	Adults	9	3	12
T_07	Usandizaga 2	24 October 2019	High School	Young people	12	18	30
T_08	Elkartu	24 October 2019	Disabled Persons Association	Adults	2	3	5
T_09	Kunsthal	25 October 2019	Design School	Young people	9	3	12
T_10	Usandizaga 3	25 October 2019	High School	Young people	6	18	24
TOTAL					74	104	178

The work of each group was directed by at least one member of the organisational team. They were guided through the Knowledge Zone and the relationship between the Empathy and Working scale models was explained to them. The categorization of each of the Elements and their relationship with Areas and Actions was explained.

Each one of the groups was divided into smaller working groups, which discussed concrete proposals in the Experimentation Zone and transferred results to scale models on a 1:50 scale.

To close the working sessions, the smaller groups came together to agree to a shared proposal based on the work done by each group. These were formalised in the Experimentation Zone one after another until a consensus emerged around a definitive scale model.

PARTICIPATION BY THE GENERAL PUBLIC

The total number of visitors to the pavilion during the days over which the exhibition was open was **2355**, a daily average of **220**, taking into account that the inauguration took place at noon on the 18th of October. This number includes members of organised groups, but not people who were invited to specific events, including the inauguration and the members dinner. Daily records were kept at the entrance reflecting sex (M = 1113 / W= 1242) and age (Children= 325 / Teenaged= 251 / Adult = 1478 / Elderly= 301).

Casual visitors not part of organized groups, were only given a basic outline of the contents and the paper bag containing three black balls.

RESULTS

Results from Organized Groups

The result of the work done by organized groups consisted of floor plans drawn from a final consensus of the smaller working groups. It is important to indicate that complete freedom was given to groups to propose Blocks in the building's common areas, which were represented as a rectangle exterior to the perimeter of the house. The final floor plans of each group are shown in Figure 10.

The parameter considered is the number of elements used by each group in their final proposals. Table 4 shows the number of modules used for each one of the defined blocks, both in the apartment itself and the building's common areas. The total number of modules used by each group and the average is also recorded. This last number gives an idea of the overall importance given to each one of the different Blocks by the organized groups.

Table 5 shows the relationship between different pairs of blocks. The associations established link hygiene with food, and rest and relationships with food and the outside. To establish the importance of these relationships ten proposals have been analyzed and a semiquantitative criterion has been established, valuing the intensity of each relationship with a score of between 1 and 3. The averages have also been calculated.

Table 6 shows the spatial relationship between privacy and mobility of furniture studied. To establish the importance of each one, a criterion analogous to the previous section has been used.

Table 7 shows the results from the 3 circulation maps that were studied.

Figure 10. Schemes with results from the ten organized workshops. From left to right and upside down (Design Team)

Table 4. Scale Model Unit Used

PLACE	BLOCK		Workshop										TOTAL	AVERAGE
			01	**02**	**03**	**04**	**05**	**06**	**07**	**08**	**09**	**10**		
IN THE HOUSING UNIT	Hygiene	Bathroom	3	3	3	3	6	3	3	3	5	2	**34**	**3.4**
		Wash-Dry	0	0	0	0	0	0	1	0	0	0	**1**	**0.1**
	Food		10	9	8	9	5	7	12	6	12	5	**83**	**8.3**
	Rest		2	4	2	2	3	3	2	4	2	3	**27**	**2.7**
	Relationships		2	2	4	3	11	3	3	2	5	7	**42**	**4.2**
	Learning		9	1	4	4	4	6	5	2	9	3	**47**	**4.7**
	Storage		8	4	5	8	3	8	3	8	4	10	**61**	**6.1**
	Outside		10	0	0	0	4	0	0	0	0	9	**23**	**2.3**
			44	**23**	**26**	**29**	**36**	**30**	**29**	**25**	**37**	**39**		
IN THE BUILDING	Hygiene (Wash-Dry)		2	1	2	1	3	2	0	2	1	3	**17**	**1.7**
	Food		10	1	4	0	6	6	0	7	1	11	**46**	**4.6**
	Rest		1	0	0	1	0	1	0	0	2	0	**5**	**0.5**
	Relationships		2	4	0	2	1	3	1	1	0	0	**14**	**1.4**
	Learning		1	1	3	2	0	3	0	4	2	3	**19**	**1.9**
	Storage		0	2	1	0	0	3	0	1	0	0	**7**	**0.7**
	TOTAL		**16**	**9**	**12**	**6**	**10**	**18**	**1**	**15**	**6**	**17**		

Table 5. Relationships between Blocks

Relation between BLOCKS (*)	COLOR		01	02	03	04	05	06	07	08	09	10	AVERAGE
Hygiene-Food			1	3	1	1	1	1	1	1	1	1	**1.2**
Food-Relationships			3	1	3	2	3	1	1	2	2	1	**1.9**
Hygiene-Rest			2	2	1	1	2	3	1	1	1	2	**1.6**
Relationships-Outside			3	1	2	2	2	2	2	2	2	2	**2**
(*) Relationships between blocks: maximum (3), average (2), minimum (1)													

Table 6. Spatial Relationships

Spatial Relationships	01	02	03	04	05	06	07	08	09	10	AVERAGE
Common space/Private Space (*)	3	3	2	3	3	2	2	2	3	2	**2.5**
Mobility of the furniture (**)	3	2	3	1	1	2	1	2	2	2	**1.9**
(*) Common space maximum (3), average (2), minimum (1) (**) Mobility of the furniture maximum (3), average (2), minimum (1)											

Table 7. Circulation

Circulation	01	02	03	04	05	06	07	08	09	10
	M	R	R	M	M	P	R	M	R	M
Radial (R), Perimetral (P), Mixed (M)										

Results for the General Public

The questions asked, as well as the total number of answers gathered and the percentage with respect to each question is shown in Table 8.

DISCUSSION AND FUTURE WORK

If we consider that the current population of San Sebastián is **181652** (EUSTAT 2020), the total of 2355 visits can be considered a success, representing **1.3**% of the total population. It must also be taken into account that the exhibition was open for only a short period of time, in which unfortunately the weather was not conducive to being outside. The number of visitors totals, additionally, **5.2**% of participants across all activities in the Biennial.

Of the total of **2355** visitors, **1666** visitors, 71%, answered question 3. It is important to indicate that visitors were often part of small groups, and only some members of these groups participated in the survey.

The results of organized groups show that users wish to make significant changes in their current houses. First of all, they place great importance on **Blocks** that, up until now, have been linked with other rooms. This is the case with **Food**, **Learning** and **Storage**.

Table 8. Answers Submitted

Nº	Question	Answer	Total	%
1	DOES YOUR HOUSE MEET YOUR NEEDS?	Yes	791	**51.3**
		Almost	560	**36.3**
		No	191	**12.4**
TOTAL QUESTION 1			**1542**	
2	WHAT WOULD MAKE YOUR HOUSE BETTER?	Possibility of extending/ reducing	564	**34.1**
		Urban connectivity	150	**9.1**
		Light/Views/Space	414	**25.1**
		Acoustic/thermal comfort	371	**22.4**
		accessibility	153	**9.3**
TOTAL QUESTION 2			**1652**	
3	WHAT AREAS OF YOU HOUSE COULD BE SHARED SPACES?	The Kitchen/Eating	263	**15.8**
		Laundry/Clothesline	293	**17.6**
		Storerooms	102	**6.1**
		Satellite Rooms	100	**6**
		Vegetable Garden/Terrace	499	**30**
		Multiple Use Spaces	409	**24.5**
TOTAL QUESTION 3			**1666**	

The results for the general public might seem surprising due to the high degree of acceptance by users of their present houses, although this contradicts to some extent with the answers to the second question, since, basically, **bigger**, **better oriented** and **more comfortable** houses were demanded.

As has already been commented, dialogue-reflection processes are enormously complex and, therefore, it is necessary to establish achievable limits for this study. Here it is worth signaling some difficulties. One was the inclusion of a socially representative sample, while another was assessing the level of commitment by and within different working groups, which was difficult to evaluate.

Although this exhibition has demonstrated that an initiative of this type, open to all citizens, can enjoy a wide participation and a significant media presence, ongoing work with diverse selected groups is proposed. Working with the same methodology, different surfaces, forms and types of housing can be addressed.

CONCLUSION

The house, as the basic environment for human habitation, will face important challenges soon. Phenomena such as the increasing population density of cities, aging populations, cultural diversity, new structures of coexistence, new forms of communication, and situations such as that caused by the COVID_19 pandemic, demand a reframing of the design of current housing. This housing was a response to the need to house a massive exodus from rural areas into urban nuclei, and a consequence of 20th Century industrialization.

A reframing, which must include different agents (architects, engineers, designers, politicians, etc.) must not have to leave out end users of the buildings whose input is essential even though in many cases they are not specialists.

An interactive exhibition, such as the one described here, constitutes a perfect space for the integration of housing users. The success of participation and communication, as well as the significant degree of commitment taken on by both visitors and members of organized groups, demonstrates this.

The analysis of data from organized workshops show that great importance is placed on the blocks of **food**, both inside the house and in possible common areas of the building, and **storage**, fundamentally inside the house. The relationships **Food-Relationships** and **Relationships-Outside** are also prioritized.

The analysis of the collected data sample shows that, although the public in general is satisfied enough with their houses, **87.6%** if we add the two first answers for question 1, many users would like to improve the size, amount of natural light and views, and improve levels of thermal and acoustic comfort.

The difference between the results obtained in the directed workshops and the public in general demonstrates the importance of structuring dialogue-reflection processes.

Sustainability and participation are closely linked, and data show how, at different levels, encouraging participation, and therefore building a wide individual conscience of the public, is key to transit to a more sustainable built environment.

To site the exhibition in a coherent environment, a big importance was given to the reduction of the materials used for the construction, the reuse of the pavilion and the recycling of the remaining pieces.

The entire process of the exhibition was documented in a video (https://vimeo.com/379952749_)

ACKNOWLEDGMENT

The exhibition was mainly funded with a grant of 56,000 euros on Budget Line I (temporary educational exhibitions of urban planning and design and contemporary architecture in Euskadi) from the funding program for the promotion and public communication of urban design and contemporary architecture in the autonomous community of Basque Country (BOPV 2018). Other public institutions and private companies completed the funding needs, making the event possible.

REFERENCES

Alguacil, J. (2011). *Experiencia y metodología para un modelo de desarrollo comunitario. El caso de San Cristóbal de Los Ángeles*. Editorial Popular.

Allegretti, G., & Torquato Luiz, J. (2007). Quando lo spazio urbano si fa metafora di un cambiamento politico: la riscossa del territorio nel Piano di Urbanizzazione. In Viaggio in Portogallo. Dentro e fuori territori dell'architettura. Roma: ARACNE editrice S.r.l.

Arnstein, S. (1969). A Ladder of Citizen Participation. *Journal of the American Planning Association*, *35*(4), 216–224.

BOPV. (2018). Orden de 28 de noviembre de 2018, del Consejero de Medio Ambiente, Planificación Territorial y Vivienda, por la que se regulan y se convocan para el ejercicio 2019, las ayudas para la promoción y difusión del diseño urbanístico y la arquitectura contemporánea de la Comunidad Autónoma del País Vasco. *Boletín oficial del País Vasco, 236*. Accessed April 18, 2020. https://www.euskadi.eus/y22-bopv/es/bopv2/datos/2018/12/1806029a.shtml

Cohen, J. (1989). Deliberation and Democratic Legitimacy. In A. Hamlin & P. Pettit (Eds.), *The Good Polity*. Blackwell.

Cohen, J. (1996). Procedure and Substance in Deliberative Democracy. In Democracy and Difference. Princeton University Press.

Dahy, H., & Knippers, J. (2013). Influence of the three Rs on Modern Approaches in Sustainable Architecture. *The International Journal of Environmental Sustainability*, *3*(4), 43–53. doi:10.18848/2325-1077/CGP/v08i04/55063

de Poza-Vilches, M. F., Gutiérrez-Pérez, J., & López-Alcarria, A. (2019). Participation and Sustainable Development. In W. Leal Filho (Ed.), *Encyclopedia of Sustainability in Higher Education*. Springer., doi:10.1007/978-3-319-63951-2_57-1

ENEDI. (2020). *Energética en la Edificación*. Accessed April 18, 2020. https://www.ehu.eus/es/web/enedi/grupo-enedi

EUSTAT (Instituto Vasco de Estadística). (2020). Accessed April 18, 2020. https://www.eustat.eus/municipal/datos_estadisticos/donostia_san_sebastian_c.html

Ganuza. (2010). *La democracia en acción, una visión desde las metodologías participativas*. Documentos OIDP. Madrid: Antígona, procesos participativos

Habraken, N. J. (1979). El diseño de soportes. Barcelona: Gustavo Gili.

Habraken, N. J. & Mignucci, A. (2009). *Soportes: vivienda y ciudad* [Supports: Housing and City]. Barcelona: Máster Laboratorio de la Vivienda del Siglo XXI.

Leupen, B. (2005). Time-based Architecture (10th ed.). Academic Press.

Loch, S. (2006). Life transforms living transforms life. *Nordic Journal of Architectural Research*, *19*(3), 33–44.

Lyons, M., Smuts, C., & Stephens, A. (2001). Participation, Empowerment and Sustainability: (How) Do the Links Work? *Urban Studies (Edinburgh, Scotland)*, *38*(8), 1233–1251. doi:10.1080/00420980125039

Meadowcroft, J. (2004). Participation and Sustainable Development: Modes of Citizen, Community and Organisational Involvement. In W. M. Lafferty (Ed.), *Governance for Sustainable Development. The Challenge of Adapting Form and Function*. Edward Elgar Publishing, Inc. doi:10.4337/9781845421700.00014

Montaner, J. M., & Muxí, Z. (2010). Reflexiones para proyectar viviendas del siglo XXI. *Dearq*, *06*(6), 82–89. doi:10.18389/dearq6.2010.09

MUGAK. (2020a). *Noticias: MUGAK cierra su segunda edición, con una bienal más internacional y más cercana a la ciudadanía*. Accessed March 26, 2020. https://www.bienalmugak.eus/es/noticias

MUGAK. (2020b). *La Bienal: ¿Puntualizar? ¿Adaptar? ¿Repensar? ¿Volver a empezar?* Accessed March 26, 2020 https://www.bienalmugak.eus/es/bienal

Nanz, P., & Fritsche, M. (2014). La partecipazione dei cittadini: un manuale. Metodi partecipativi: protagonisti, opportunità e limiti. Assemblea legislativa della Regione Emilia-Romagna, Italia.

Paricio, I., & Sust, X. (2004). *La vivienda contemporánea.* Barcelona: Institut de Tecnología de la Construcció de Catalunya-ITeC.

Pirkl, J. J. (1994). *Transgenerational Design: Products for an Aging Population.* John Wiley & Sons.

Pirkl James, J. (2008). Transgenerational Design: A Heart Transplant for Housing. In F. Kohlbacher & C. Herstatt (Eds.), *The Silver Market Phenomenon.* Springer. doi:10.1007/978-3-540-75331-5_10

Portney, K. E., & Berry, J. M. (2010). Participation and the Pursuit of Sustainability in U.S. Cities. *Urban Affairs Review*, *46*(1), 119–139. doi:10.1177/1078087410366122

Rey, E., Aguayo, F., & Peralta, M. (2015). Integración de métodos escalares y vectoriales en la evaluación del riesgo de incendio en el ciclo de vida de una construcción. *Informes de la Construcción*, *67*(539), e110. doi:10.3989/ic.14.154

Sagmeister, S. (2013). *The Happy Show.* The Museum of Contemporary Art. Accessed April 18, 2020. https://www.moca.org/exhibition/stefan-sagmeister-the-happy-show

Salgado, M. A. (2006). Antes y después de la vivienda cinematográfica. *Revista Arquitectura*, *345*, 98–103.

UPC (Universidad Politécnica de Cataluña). (2019). *Una conversa sobre el manjar i la ciutat* (Película). Habitar-UPC.

Chapter 6
Nursing Homes in Spain and Their High Number of Deaths by COVID-19 as an Alarm in the Study of Indoor Air Quality

Gastón Sanglier Contreras
https://orcid.org/0000-0002-8981-5622
Universidad CEU San Pablo, Spain

Jose Amílcar Rizzo Sierra
Universidad Politécnica de Querétaro, Mexico

Eduardo José López Fernández
https://orcid.org/0000-0001-9103-3585
Universidad CEU San Pablo, Spain

Roberto Alonso González Lezcano
https://orcid.org/0000-0002-6185-4929
Universidad CEU San Pablo, Spain

ABSTRACT

Nursing homes have been one of the most prominent targets of the COVID-19 coronavirus in Spain. The factors that have determined that this is the case are very diverse. In this study, physical agents and chemical pollutants, population density, and different capacities of the residences are analyzed to see their influence on the number of elderly people who have died in the geriatric centres in the different autonomous communities (AACC) of Spain. A statistical analysis has been carried out on the variables observed. The conclusions show that in many places where this overflow of deaths has occurred, the residences were private, with some exceptions. The influence of physical agents and pollutants has been shown to be a determining factor, especially for the communities of Extremadura and Castilla-La Mancha, although it is true that the large number of factors makes the study complicated. The dilemma between air quality and energy efficiency is of great importance, especially when human health is at stake.

DOI: 10.4018/978-1-7998-7023-4.ch006

INTRODUCTION

Today, in developed countries, 80-90% of people's time is spent indoors (Klepeis et al., 2001), especially in the home, as the presence in the home varies between 60% and 90% of the day, and 30% of the time is spent sleeping (Wargocki, 2016; Borsboom et al., 2016). For this reason, homes contain the air that is inhaled, constituting the interior spaces of greatest exposure. The indoor environment in the home should benefit rest and recovery (Wargocki, 2000, Nathanson, 1987); therefore, poor air quality (IAQ) prevents this purpose as it has harmful effects on health).

Since the energy crisis of the 1970s, buildings have become increasingly airtight, leading to the appearance of IAQ-related diseases such as sick building syndrome (SBS) (Cao et al., 2014). In addition, the relationship between air movement in buildings due to ventilation and the spread of infectious diseases has been demonstrated (Li et al., 2007; Deisy et al., 2003). In this context, the benefits of indoor air exchange have been demonstrated, although the influence of ventilation on the spread of infectious diseases is not clear (Cao et al., 2014). Consequently, since there is no quantitative influence of airflow rates on health, ventilation rates specified in different regulations are usually set according to confort criteria (perceived conditions) (Wagner et al., 1993; Wargocki, 2015, González-Lezcano et al., 2016, WHO, 1987).

However, a healthy indoor environment can be achieved by applying the necessary strategies to improve the IAQ which are, in addition to increasing the supply of fresh air, controlling pollution from emission sources, cleaning the air and improving the efficiency of ventilation (Yocom & McCarthy, 1991; Spengler et al., 2000, Hormigos-Jiménez et al., 2018a). Therefore, indoor air quality (IAQ), especially in indoor residential spaces, has a strong influence on human health; therefore, it is essential to design adequate ventilation, which ensures good IAQ, since the main purpose of ventilation is to dilute or remove indoor contaminants by providing outdoor air (Turiel, 1986; Berenguer et al., 1994; Wadden & Schelf, 1983; Dimitroulopoulou, 2012; Van Buggenhout et al., 2006; Xu et al., 2020).

The values set for air renewal to ensure comfort and odour elimination have been modified over the course of history according to variations in ventilation theories. Nowadays, in the case of Spain and for residential buildings, the Basic Document HS 3 Indoor air quality, included in the Technical Building Code (DB-CTE-HS3, 2018), provides data on minimum ventilation rates depending on the room of the dwelling. In European countries, in addition to the EN 15251 standard (which should be used in case no national standard is available), there are also state regulations that, like the Spanish standard, show data on minimum ventilation rates based mainly on body odours with CO2 as an indicator and, to some extent, also established according to primary emissions from some building materials (Fernandez, 2008; RITE, 2020).

On the other hand, even if the regulations are complied with, indoor air quality may still be inadequate if stagnant air zones are generated, and therefore health and comfort problems may develop (Namiesnik et al., 1992; Knoppel & Wolkoff, 1992; Davidson et al., 1987; Kwon et al., 2011; Liu et al., 2010). In-depth research in this field is still needed; therefore, the debate on how much ventilation is sufficient to achieve good indoor air quality, capable of preventing both odours and the emergence and spread of diseases, is still ongoing.

In Spain, one of the countries most affected by the Covid-19 coronavirus pandemic, according to data provided by the Spanish Ministry of Health's Centre for the Coordination of Health Alerts and Emergencies (CCAyES, 2020), the number of deaths as of 16 May 2020 was 27,563 and 230,698 people were infected.

If these data are transferred to the population of elderly people living in nursing homes who have died there (EnR?, 2019), we find that the Autonomous Communities (AACC) most affected were Castilla-La Mancha (14.44%), Extremadura (14.13%), Castilla y León (13.71%) and Aragón (12.90%) with respect to the total number of deaths in these communities (elderly and non-elderly). These numbers are very similar if we reflect on the elderly who died in their community, obtaining the following data for the most affected communities: Castilla-La Mancha (17.38%), Extremadura (16.80%) and Aragón (15.36%).

According to data provided by the AACC, the number of elderly people who have died as a result of Covid-19, in the approximately 5417 Spanish homes for the elderly, whether public, subsidised or private, stands at 18,354 deaths. Only people who have died after testing positive for coronavirus have been registered as victims, i.e. the tests have not been performed post mortem, so those who have not been tested are not listed as having died of coronavirus.

Figure 1. Distribution of the number of public and private nursing homes in the different autonomous communities of Spain

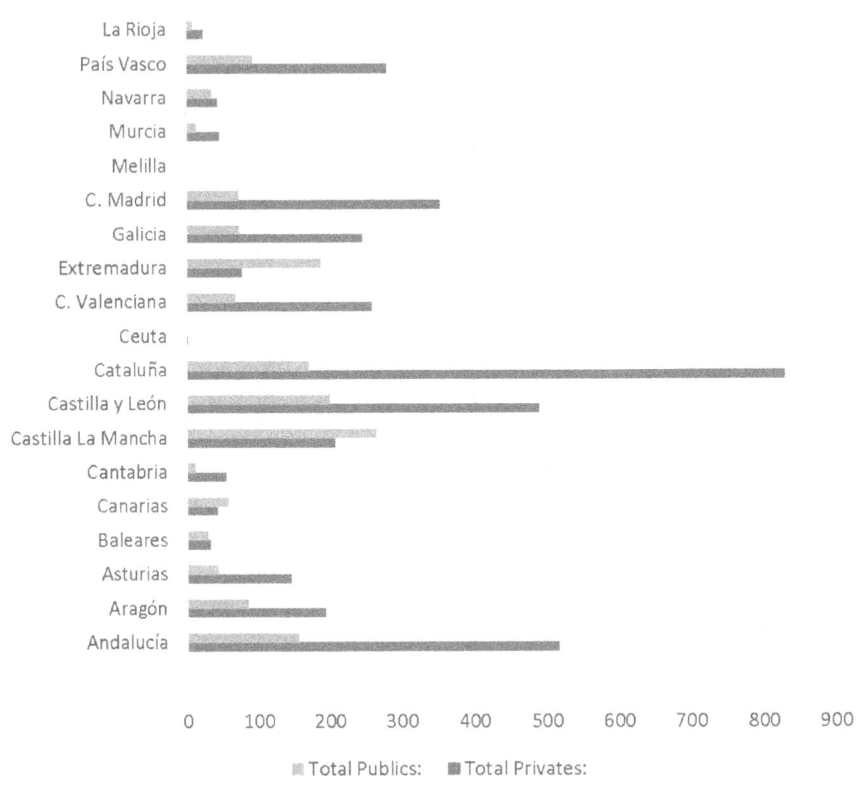

It is observed (Figure 1) that the number of private residences in the CCAA is greater than the number of public residences, in a proportion of 3 to 1, with the exception of the Autonomous Communities of Extremadura and Castilla-La Mancha, and to a lesser extent Canarias.

The population pyramid in Spain continues its ageing process, measured by the increase in the proportion of elderly persons, those aged 65 years old and over. According to the latest statistical data from the Continuous Register of the National Statistics Institute (INE), on 1 January 2019 there were 9,057,193 elderly people, 19.3% of the total population of 47,026,208 (INE, 2019) (Final data published on 27-12-2019 by the INE). This population of elderly people continues to increase, both in number and proportion. The average age of the population, which is another way of measuring this process, is 43.3 years; in 1970 it was 32.7.

Below is a comparative chart of the population over 65 years of age and the number of deaths in that age range for each AACC.

Figure 2. Graph of number of people >65 years old and deaths of people in this range by Autonomous Community

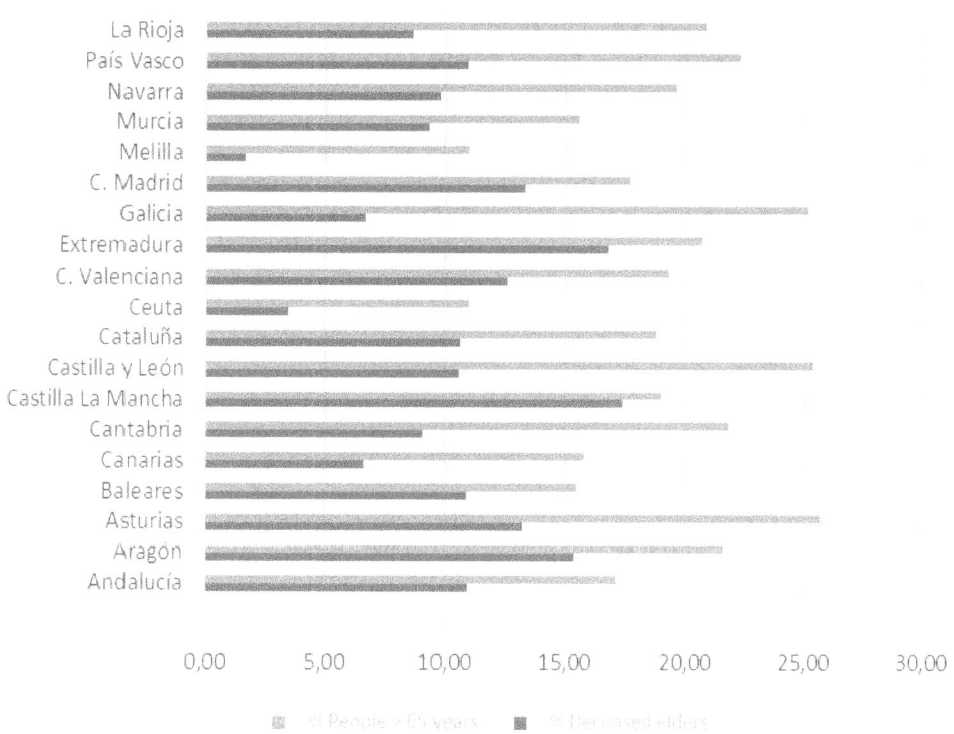

It can be observed that the communities most affected by deceased elderly persons with respect to their population of persons over 65 years of age, have been the communities of Madrid, Extremadura, Castilla-La Mancha and Baleares, on the other hand, those less affected have been Ceuta, Melilla, Canarias and Galicia.

The first report carried out in Europe on indoor air quality (Rasmussen et al., 2017) includes detailed statistical data from the Eurostat Union Statistics on Income and Living Conditions which demonstrates the correlation between people's health and the conditions of the buildings in which we live.

In March this year the European Respiratory Journal published a study that revealed the importance of maintaining adequate indoor air quality conditions in residential buildings. Based on this study it was deduced that older people are more likely than adults to suffer negative effects on lung health due to indoor air pollution in buildings.

It has been shown that city dwellers (especially the elderly and children) spend between 58 and 78% of their time in an indoor environment that is polluted to a greater or lesser degree. This is a problem that has been aggravated by the construction of buildings that are designed to be increasingly airtight and that recycle air with a smaller proportion of fresh air from outside in order to increase their energy efficiency. Pollution of the indoor environments of buildings is the cause of many health problems of various kinds, which can range from simple fatigue or discomfort to symptoms compatible with allergies, infections and cancer, among others.

The pollutants present in the indoor air of buildings (residences), whether chemical, physical or biological, vary depending on the activities that take place in such spaces, the state of health of the occupants, the physical infrastructure of the building and its material assets and the quality of the surrounding air. At present, environmental pollutants such as: environmental tobacco smoke, formaldehyde, radon, mineral fibres, isocyanates and epoxy resins, have been identified as some of the main emerging risks that can increase the risk of diseases such as: allergies, asthma, fertility disorders and cancer. Physical factors that influence comfort are mainly related to relative humidity, average air speed, temperature and noise. In addition, there are chemical pollutants including carbon dioxide (indicative of insufficient indoor air renewal), carbon monoxide, sulphur dioxide, volatile organic compounds, suspended particles, ozone, radon, etc. and various pathogens. In short, a conglomerate of pollutants that the population faces daily not only in buildings, but also in the outside air, water, food, consumer products, etc., and of which it is quite complex to know their composition, daily exposure dose, and interaction with the human body and the environment (Sanglier et al., 2020b).

Volatile organic compounds (VOCs) play an important role in the process of assessing the IAQ. They negatively affect both occupant comfort and health (Wargocki, 2013; Kim et al., 2006). Indoor spaces vary from sensory irritation at medium exposure levels to toxic effects at high exposure levels (Guo et al., 2011). Formaldehyde (HCHO) is especially important as it is known to be the highest irritant in indoor air, causing eye, nose and throat irritation (at concentrations above 0.1 mg.m3) and may even cause asthma attacks at high concentrations (WHO, 2010; INSHT; 2010).

Previous studies (Stymme, 1994; Kostiainen, 1995) provide total VOC concentration data using the concept of total volatile organic compounds (TVOC). This parameter is used to make a simpler and faster interpretation than the analysis of a high amount of VOCs, which are usually detected in indoor spaces (Tham, 2016). The concentration of any pollutant in an indoor space results from the balance between the network in that space and what is removed or added by ventilation. Therefore, if the TVOC level is high, it indicates that there are significant sources of contamination or that local ventilation is inadequate. Therefore, it is important to take measures on TVOC as they are an indication of the pollution load in the indoor environment and the sustainability of the ventilation rate provided (European Collaborative Action, Indoor Air Quality and its Impact on Man, 1992). It should be noted that TVOC does not contain all VOCs present in indoor air, as there are organic pollutants that are not included, such as low molecular weight aldehydes, like HCHO, because the method of identification differs from that of other VOCs. Therefore, in order to characterize the IAQ and determine an adequate ventilation rate, it is essential to consider the concentration of COVs and HCHs.

Therefore, it is difficult to assess the health risks (measurement, tolerance level, exposure time, effects...) in the indoor environment, and the preventive and control work of the facilities involved is relevant in order to promote healthy environments.

In this study, chemical pollutants such as formaldehyde, nitrogen dioxide, ozone, PM10 and PM2.5 particulate matter and sulphur dioxide will be analysed, as well as physical agents such as temperature, humidity, precipitation and hours of sunshine, and socio-demographic agents such as population density and the capacity of the different public or private homes for the elderly on the quality of the indoor air of the homes to determine whether they influence the health of the elderly who live there.

MATERIALS AND METHODS

If children and pregnant women are left out, there are other population groups that are vulnerable to the effects of air pollution. These include people with respiratory diseases such as asthma, chronic bronchitis or emphysema, those with cardiovascular diseases or diabetes, and elderly peple with chronic diseases, particularly those living in residential homes.

The quality of the air in the residences belongs to the IDA2 category, according to the Regulation of Thermal Installations (RITE) in Spain (RITE, 2020). In this type of building it is necessary to have a good quality of indoor environment because the elderly are a sector of the population more vulnerable to infectious agents.

Age and diseases (cardiovascular, respiratory problems) are factors that make the elderly more likely to be negatively affected by pollutants present in the quality of residential environments. They will also influence the lifestyle of those people, the food, the genetic predisposition, etc.

Confort

In indoor environments, the ability to regulate temperature is provided by the heating, ventilation and air conditioning systems. The human body has the ability to regulate its body temperature within a range of degrees. For confort in an indoor environment, most people who live there must feel a general sense of temperature and humidity.

Confort means feeling good from the point of view of the hygrothermal environment outside the person. Extreme limits, from the thermal point of view, can be harmful, and even deadly, for the human being. This is because human beings are homeothermic, i.e. they must maintain certain vital parts at approximately constant temperature. To achieve a feeling of confort the most advisable situation is that the ambient temperature is slightly higher than the air temperature and that the flow of radiant thermal energy is the same in all directions and is not excessive above the head. In the UNE EN ISO 7730 standard (confort in moderate environments), confort is given by the thermal balance of the body with the environment, i.e. a person will feel comfortable when the internal heat generated and the losses due to evaporation from the body are compensated by the losses or gains due to latent, sensitive or radiant heat with respect to the environment.

The following table (Table 1) shows data on the parameters of average temperature and humidity, precipitation, hours of sunshine (AEMet, 2020) and the number of elderly people (>=65 years) who have died in homes and in total for the different Autonomous Communities (CCAyES, 2020).

Table 1. Data on deceased elderly people and physical agents in the Autonomous Communities

CC.AA.	(RD) Residential Deads	(CD) Community Deads	Average Temperature (ºC)	Average Humidity (%)	Precipitation (mm)	Sun hours (h)
Andalucía	527	1355	19,4	53,6	344,5	3342
Aragón	704	838	14,6	61	272,7	3112
Asturias	192	313	13,8	80	908,2	1962
Baleares	84	216	19,3	73,5	466,5	3034
Canarias	18	151	22	61	78,5	3165
Cantabria	135	206	13,8	74	1157,6	1826
Castilla-La Mancha	2395	2883	16,4	49	221,8	2779
Castilla y León	2519	1940	12,1	65	384,3	2894
Cataluña	3394	5915	16,9	63,3	329,3	2731
Ceuta	0	4	19,1	74	650,2	2640
C. Valenciana	518	1365	19,1	62	335,6	2808
Extremadura	418	497	17,4	53	271,8	3326
Galicia	269	604	14,5	74	798,5	2342
C. Madrid	5909	8826	16	53	256,1	2970
Melilla	0	2	19,6	70	239,7	2778
Murcia	67	144	20	55	177,9	3348
Navarra	422	501	13,2	67	696,9	2512
País Vasco	584	1455	14,4	75,5	1456,7	1780
La Rioja	199	348	14,4	61	390,7	2708

Air Quality Factors

The quality of the environment in nursing homes is affected by indoor air pollutants such as dust, suspended particles, CO2, CO, N0x, VOCs, bacteria, fungi, viruses, as well as pollution from outside. However, nursing homes need special maintenance of the facilities and environment (air renewal) due to the following causes:

- The elderly spend practically all their time in these places (90%), since it would be their home of residence.
- The quality of the environment in the residences will be vitiated by a greater number of viruses due to the fact that their occupants usually suffer from different symptoms.
- In the residences, food and different types of drugs or medicines are preserved to care for the elderly, so these products need the air quality in the residences to be optimal.
- It is essential to use air-conditioning systems in the homes to ensure the commodity of the users and to renew the indoor air. In winter it is necessary to temper the indoor air so that cold currents that could affect the health of the elderly do not occur. In summer the opposite will occur and air conditioning systems will be used to prevent the appearance of hot flushes or heat in the occupants.

- The use of air conditioning systems in residences is as important as their cleaning and maintenance: lubrication of mechanisms, revision, change of filters, etc.
- Daily cleaning and disinfection of the building to maintain the quality of the indoor environment in the residence halls, since different groups of people live together every day. In the residents' rooms, in the reception areas and in the corridors, it will be necessary to carry out a deep cleaning or even disinfection several times a day to improve the safety of the guests and visitors.

Consequences of Poor Air Quality

The World Health Organization (WHO) warned that air pollution (outside air) kills about seven million people every year. Indoor pollution levels can be as much as 10 to 100 times higher than outdoor concentrations. And exposure of people (with the elderly being more vulnerable) to poor quality residential environments can lead to many health problems:

- Airways: dryness, itching/heartburn, nasal congestion, sneezing, sore throat.
- Lungs: chest tightness, choking sensation, dry cough and bronchitis
- Skin: redness, dryness, generalised itching.
- General malaise: headache, weakness, drowsiness/lethargy, difficulty concentrating, irritability, anxiety, nausea and dizziness.
- Diseases: hypersensitivity pneumonitis, humidification fever, asthma, rhinitis, dermatitis.
- Infections: legionellosis, Pontiac fever, tuberculosis, common cold, flu.

Chemical Pollutants

The pollutants studied in this work will be named, as well as their impact on human health.

Nitrogen Oxides (NOx)

The term 'nitrogen oxide' (NOx) is normally used to group together a set of oxidised forms of nitrogen in the atmosphere, mainly nitric oxide (NO) and nitrogen dioxide (NO2). Nitric oxide reacts with oxygen or ozone in the air to form nitrogen dioxide. When sulfur dioxide is present with nitrogen dioxide, it can have a synergistic effect increasing the damage to higher levels than those produced by the sum of the individual effects of both. The presence of NOx is related to the burning of fuels, mobile sources (vehicles), industrial processes and some natural processes (lightning, soil microorganisms). Combustion processes emit a mixture of nitric oxide (90%) and nitrogen dioxide (10%). In turn, nitric oxide reacts with other chemicals in the air to become nitrogen dioxide. In indoor environments the main source of NO2 emissions are heating systems and gas stoves, as well as tobacco smoke.

At low concentrations, nitrogen oxides are irritating to the upper respiratory tract and eyes. In prolonged exposure it can cause pulmonary edema. Excessive exposure to nitrogen oxides can cause health effects on the blood, liver, lung and spleen. Nitrogen dioxide is also one of the gases contributing to acid rain that causes damage to vegetation, buildings and contributes to the acidification of lakes and streams.

Suspended Particles

Airborne particles are a complex mixture of substances of different chemical composition and physical nature (suspensions of solids or drops of liquid) with a variable size ranging from 0.005 to 100 µm. As a whole, particles may be present as soot, dust clouds or mist; in isolation they cannot be detected by the naked eye. The elements that can be found in particles are extremely heterogeneous: coal, hydrocarbons, silica, ammonium sulphate, nitrates, metals such as lead, iron, aluminium or cadmium, pollen, micro-organisms, dioxins, pesticides, etc. Because the composition of the particles is extremely heterogeneous, an overall assessment of their toxicity cannot be made as it will depend on the type of elements that make up their composition. In general, the smaller particles have in their composition more toxic elements, such as heavy metals or organic compounds of carcinogenic capacity, such as benzo-a-pyrene. Therefore, the fine, smaller particles are those that have adverse effects on the health of the population, since they are the ones that can reach the pulmonary alveoli.

These particles are usually referred to as total suspended particles (TSP) and include all particles with diameters ranging from less than 0.1 microns to 50 microns, as larger particles are deposited by gravity. TSP is expressed as PM, particulate matter with a sub-index referring to particle diameter, in weight of particles per volume of air (mg/m3 or µg/m3). The larger the particle size, the shorter the time they remain suspended in the air and the shorter the distances they can travel. Particles larger than 10 microns fall rapidly near the source that produces them; PM10 particles (with a diameter of £ 10 microns) can remain suspended for hours and travel from 100 meters to 40 kilometers, while PM2.5 particles (with a diameter of £ 2.5 microns) can remain in the air for weeks and are capable of moving hundreds of kilometers, moving with air currents and penetrating inside the premises through ventilation systems.

The main sources of particulate matter outdoors are road traffic, especially diesel vehicles, industrial processes, incinerators, quarries, mining, stack emissions, coal heating... Other important sources of particulate matter are dust from agricultural work, road construction, or vehicle traffic on unpaved roads. On the other hand, they are present in almost all indoor environments, mainly from combustion appliances and tobacco smoke. They can also have a biological origin such as pollen, spores, bacteria and fungi. Typically, most particles of anthropogenic origin are in the range of 0.1 to 10 µ.

The size range that can be considered dangerous, in relation to causing effects on human health and affecting air quality, is between 0.1 and 10 microns in diameter, since these particles, once inhaled, generally have a greater capacity to penetrate the respiratory system. PM10 particles are deposited in the upper respiratory tract (nose) and in the trachea and bronchi, while PM2.5 particles with a smaller diameter can reach the bronchioles and alveoli of the lungs.

Formaldehyde

It is a major indoor air pollutant and due to its chemical properties and serious health effects, an individual assessment is recommended. Inside buildings, it is often present in the structure of modern building installations and furnishings, its concentrations being higher indoors than outdoors. Urea-formaldehyde, foam insulation (UFFI), was widely used in house building until the early 1980s, although its installation is now rare. The main sources of exposure to formaldehyde include: particle board, varnishes, lacquers, glues, fibreglass, carpets, non-iron fabrics, paper products and certain cleaning and disinfection products. Due to the extremely high concentrations of formaldehyde in tobacco smoke, smoking is a major source of this compound. Gas stoves and ovens and open fireplaces are also sources of formaldehyde exposure.

Studies conducted in Canada since the early 1990s indicate the presence of formaldehyde in households in concentrations ranging from 2.5 to 88 μg/m³, with an average of between 30 and 40 μg/m3 (Health Canada, 2005).

The main form of exposure is inhalation; it can also be absorbed through skin contact. The main effects of acute exposure to formaldehyde are irritation of the conjunctiva of the eye and the mucosa of the upper and lower respiratory tract. The symptoms are temporary and depend on the level and duration of exposure. Exposure to high concentrations of formaldehyde may cause burns to the eyes, nose and throat. In the long term, exposure to moderate concentrations of formaldehyde (chronic exposure) may be associated with respiratory symptoms and allergic sensitivity, especially in children. Prolonged or repeated skin contact leads to irritation and dermatitis.

Through some studies carried out on industry workers, exposed to very high concentrations of this compound, more cases of cancer of the nose and throat seem to be observed than expected, however, other studies have not confirmed these results. However, experimental studies in rodents do confirm the risk in these, to get nose cancer. For all these reasons, formaldehyde vapor has been classified as a human carcinogen by OSHA (Occupational Safety and Health Administration), IARC, and EPA (U.S. Environmental Protection Agency). An average formaldehyde concentration between 37 and 60 mg/m³ can cause pulmonary edema or pneumonia, and above 60 mg/m³ can even cause death.

Ozone

Ozone (O_3) is produced naturally in the upper layers of the atmosphere (stratosphere) from the dissociation of oxygen molecules by the action of sunlight, acting as a barrier to solar radiation. It is composed of three oxygen atoms and is a colourless gas with a penetrating and highly oxidising odour, capable of reacting with all kinds of substances, whether living tissue or inert materials.

In addition, at tropospheric level we also find ozone, which can come from the entry of ozone from the stratosphere or be originated (secondary pollutant) by the presence of other pollutants such as nitrogen oxides and volatile organic compounds (VOCs) in the presence of sunlight (local photochemical production). However, other times it can appear due to the transport from long distances of this pollutant produced in other areas.

In the indoor environment, it originates from equipment that generates a potential discharge between metal plates or with the existence of ultraviolet radiation. This occurs in photocopiers, laser printers, electrostatic equipment for air purification, electric motors and equipment with UV radiation, such as those used in disinfection.

Due to its oxidizing power, the immediate health effects are: irritation of the respiratory tract and eyes, coughing, breathing difficulties, etc. In the medium term there may be a general decrease in physical performance, as well as symptoms of general malaise such as headache, tiredness, heaviness, etc. In the long term it can produce alterations in the pulmonary function (pneumonitis and pneumonia). In general, the effects of exposure to ozone are accentuated the higher the concentration, the duration of exposure and the levels of activity during exposure, although the form of the dose-response relationship is not known. The severity of the response is strongly dependent on the sensitivity of the respiratory system and often on the health status of the exposed person.

Sulfur Dioxide

The primary source of sulphur oxides is the burning of sulphur-containing fossil fuels, such as coal, as it reacts with oxygen to produce sulphur dioxide (SO2). Therefore, it is a major pollutant outside, as this is where most of the sources are found.

Sulfur dioxide causes irritation of the mucous membranes, especially the eyes, nose and throat, and is the cause of respiratory diseases such as bronchoconstriction and bronchitis. In sensitive people exposed as asthmatics it can aggravate the symptoms, although the effects will depend on the sensitivity of the individual, being more sensitive the most vulnerable population groups such as the elderly, children and subjects with chronic respiratory diseases. These effects can be aggravated when sulphur dioxide is combined with particles.

There are very few data related to the quality of indoor air in nursing homes or geriatric residences due to the high hermiticity of the centres where these tests are carried out, especially in private centres and many concerted ones.

Data have been collected on various pollutants in the different AACCs in Spain. The limit and objective reference values that appear in this study are those established by Directive 2008/50/EC and Royal Decree 102/2011, as well as those recommended by the World Health Organization (WHO). In the following table (table 2), data have been extracted from the 127 zones and agglomerations established for the measurement of nitrogen dioxide in Spanish territory, organized by AACC, with their respective measurement stations. The exceedances of the legal and WHO references by zone or agglomeration are reflected in the table. The values that appear correspond to the average value of all the data collected by the stations that make up the zone (whether they exceed the limits or not). There are stations that are the only ones representative of their area, and therefore their data correspond to the average value of the area. The target value for the protection of human health from tropospheric ozone is set for a three-year period, in this case for the years 2017, 2018 and 2019. The remaining pollutants refer to the year 2019 (Ceballos et al, 2019).

For the interpretation of the data in table 2, the limit values of the analysed pollutants are attached:

Partículas PM_{10}:

- Daily value: No. of days during the year when the 50 has been exceeded µg/m^3 . When it is greater than 35 days, the daily limit established by the regulations is exceeded, and if it is greater than 3 days, the WHO recommendation is also exceeded.
- Annual average: Average value of PM10 during the year. The limit established by the regulation is 40 µg/m^3 per year, while the WHO recommends not to exceed 20 µg/m^3 as an annual average.

PM2.5 particles:

- Daily value: Number of days during the year when the 25 have been exceeded µg/m^3. When it is greater than 3 days, the WHO recommendation is exceeded.
- Annual average: Average value of PM2.5 during the year. The regulations do not allow exceeding 25 µg/m^3 per year. WHO recommends not to exceed 10 µg/m3 as an annual average.

Nitrogen dioxide NO2:

- Annual average: Average value of NO_2 during the year. The annual limit value set by the regulations is 40 µg/m^3, which is in line with the WHO recommendation.

Ozone O3:

- 8-hour value: Number of days over the year when the average value of 120 µg/m³ (legal) or 100 µg/m³ (WHO) ozone is exceeded during 8-hour periods (considered as the maximum daily 8-hour moving average). The regulations do not allow more than 25 days per year (averaged over three consecutive years), a threshold also adopted in this report for the WHO recommendation (in 2018).
- AOT40 May-July: sum of the difference between hourly concentrations above 80 µg/m³ and 80 µg/m³ between 8:00 and 20:00, 1 May to 31 July. The legal target is 18,000 µg/m³ h (on average over five consecutive years), and the long-term target is 6,000 µg/m³ h (in 2019).

Sulphur dioxide SO2:

- Daily value: Number of days per year when the average daily value of 125 µg/m3 (legal) or 20 µg/m³ (WHO) of SO_2 has been exceeded. The regulations do not allow more than 3 days per year, a threshold that is also adopted in this report for the WHO recommendation.

The table below shows the average chemical contaminant measurements for each AACC and the limits according to the regulations specified above:

Table 2. Table of chemical contaminants with their limits according to regulations

	PM_{10} (particles < 10 micras)		$PM_{2,5}$ (particles < 2,5 micras)		NO_2	O_3 (tropospheric ozone)			SO_2	Formal-dehyde
	Daily Value	Annual Average	Daily Value (OMS)	Annual Average	Annual Average	eighth hour (Normative)	eighth hour (OMS)	AOT40 (Normative)	Daily Value (OMS)	Annual Average
	Nº days > 50µg/m³ Normative: máx=35 OMS: máx:3	µg/m³ Normative: máx=40 OMS: máx:20	Nº days > 50µg/m³ OMS: máx:3	µg/m³ Normative: máx=25 OMS: máx:10	µg/m³ Normative: máx=40 OMS: máx:40	Nº days > 120µg/m³ Normative: máx=25	Nº days > 100µg/m³ OMS: máx:3	Normative: máx=18000	Nº days > 20µg/m³ OMS: máx:3	µg *m²
Andalucía	5,36	23,27	12,36	13,27	15,9	18,36	94,18	18767,54	0,36	59,2
Aragón	1	14,25	2,75	8,75	8	12,75	69.5	13766,5	0	60,4
Asturias	7,75	22	6,25	11,75	13,5	1,75	18,75	3223,25	8,25	54,6
Baleares	1	13,75	1,33	7,33	11,75	8,75	77	14018,5	0,75	76,1
Canarias	17	22,83	7,5	8,16	11,83	0,83	16,16	2274,66	2,66	57,4
Cantabria	4,33	19	1,66	9,3	18	1	35,33	4180,66	1,33	68
Castilla La Mancha	15,33	21,66	13,66	10,66	11,66	23,66	88	18153,66	2,33	59,3
Castilla y León	3,1	15,2	1,8	6,4	10,1	9,6	66,1	10345,5	2	57
Cataluña	2,2	19,6	5,64	10,5	15,92	14,66	77	19245,78	1,76	58,2
Ceuta	N/A	N/A	N/A	N/A	N/A	N/A	N/A	N/A	N/A	61,3
C. Valenciana	1,21	14,78	4,28	9	11,71	16,64	90,28	19060,35	0,64	61
Extremadura	5,5	15	4,5	8	6	32,5	88,5	18395	0	56,7
Galicia	5	22	10,66	12,33	20,83	6,5	33,5	4911.16	2,66	65
C. Madrid	4,71	16,28	4	10	20,42	36,14	110,71	22506,42	0,42	57,1
Melilla	N/A	N/A	N/A	N/A	N/A	N/A	N/A	N/A	N/A	60,6
Murcia	26,5	27	N/A	N/A	27	28,5	76,5	22224,5	20,5	57,8
Navarra	0,5	13,5	5	12	11,25	5	29,75	8902	1,5	59,8
País Vasco	0,87	15	1,62	8,25	15,25	5	36	7419,25	0,28	58,6
La Rioja	2	17	1	9	7	7	59	10237	0	60,1

There are various types of building which, due to their unique use and the special sensitivity of their occupants to temperature changes or indoor air pollution, among other factors, require special air-conditioning and ventilation conditions.

This is the case of homes for the elderly or geriatric centres which, without meeting the regulatory requirements for specific temperature, relative humidity and ventilation conditions in hospitals, are buildings that require appropriate air-conditioning and ventilation systems to enable the elderly people who live in them to achieve a good standard of living in terms of their personal well-being due to their special needs in terms of health and commodity.

Air conditioning and other factors related to architecture and interior design can have a profound impact on the sense of home experienced by the elderly who live in this type of centre; furthermore, the dimensions of the room and the position of various elements within a room, such as furniture, have an influence on the movement of air, and therefore on the efficiency of ventilation (Hormigos-Jiménez et al., 2018b).

If the ventilation of a dwelling is considered in general, there are research studies (Saéz, E., 2017) which show that for a type of dwelling without inhabitation, the concentration of PM2.5 particles, for conditions without ventilation and forced ventilation, the difference in concentration of particles of this type for the interior is half that of the value marked for the exterior. In the case of natural ventilation, the difference between inside and outside is almost negligible.

In the analysis of the particle size 10.0 μg/m3 (PM10), there is no appreciable difference between indoor and outdoor for conditions without ventilation or in natural ventilation, being, in both cases, the highest concentration outside. Finally, in the case of forced ventilation, the concentration sampled outside is double that sampled inside, since the filters retain the larger ones.

Geriatric centres are usually full, the people who live there are usually sedentary people who move around in small places and are usually together for a long time in different places (motor room, activity room, dining room, TV room, etc.). These are places with a high concentration of harmful agents and stale air that must be properly renovated. In order for the indoor air to be optimal for these people, good maintenance of the air conditioning equipment is necessary, as well as an adequate change of filters.

Based on all of the above, and taking into account the two ventilation systems used (natural and forced), it would be logical to use mechanical ventilation systems assisted by natural ventilation in nursing homes.

For the statistical analyses to be carried out below, and taking into account the particulate matter PM10 and PM2.5 for forced ventilation, it will be taken into account that the concentration number of particles outside is twice as high as inside for the two types of particles. However, these data have been obtained with the passenger compartment (study area) without people. In the case studied in this work, the areas are geriatric centres where there is a great deal of mass, so for the study it will be considered that the particle concentrations of both types will be identical outside and inside, using the data obtained in table 2. This hypothesis is contrasted in the Indoor Air Quality study (Guardino, 2012) carried out for the National Institute of Safety and Health (INSST) in Spain.

A statistical study will be developed with the support of the Ststgraphics Centurion v.xvi program to determine the confluence of all parameters in the number of elderly deaths. An analysis of principal components will be carried out in order to reduce the number of variables studied with respect to the number of elderly deaths in the homes of each AACC. An analysis of variance will then be carried out to see the convergence of the variables selected in the parameter studied.

RESULTS

The analysis has been carried out in two stages, firstly taking into account the sizing data of the different residences, and subsequently, with respect to the physical agents and pollutants.

For the study of the first stage, and in relation to the statistical study, statistical techniques of multivariate analysis by correlations between the variables will be applied in order to determine the relationships between them in the case of the number of deaths in the residences, as opposed to the 12 variables related to the size of the residences. Then a cluster or conglomerate analysis will be carried out to be able to determine the relationship between the nursing homes and the Autonomous Communities under study.

In Spain, a total of 3,844 private residences and 1,573 public residences have been counted (EnR?, 2019). At this stage of the study, the nursing homes have been categorised into private and public, and these in turn by their size, obtaining 12 variables: private residence with less than 25 persons (PR<25), private residence with between 25 and 49 persons (PR25_49), private residence with between 50 and 99 persons (PR50_99), residences with more than 100 persons (PR>100), residences with an unknown number of places (PR_NI) and the total number of private residences (PR) has also been included. Similarly, the same has been done for the public residences for the elderly, taking into account the following variables, PU<25, PU25_49, PU50_99, PU>100, PU_NI and PU (EnR?, 2019).

For the statistical study, a multivariate analysis has been applied, considering the number of elderly persons (old people) who have died in the homes as a dependent variable and the 12 previous variables as independent variables. The following table shows the momentary correlations produced by Pearson between each pair of variables. The range of the correlation coefficients varies from -1 to +1 and shows the strength of the linear relationship between the variables.

Table 3. Correlations between the different variables analysed

	CC.AA.	(RD) Residential Deads	(CD) Community Deads	(TD) Total Deads	PR<25 Private Residences (<25)	PR25 49 (25-49)	PR50 93 (50-93)	PR>=100 (>=100)	PR_NI No inf.	PR Total Privates:	PU<25 Public Residences (<25)	PU25_49 (25-49)	PU50_93 (50-93)	PU>=100 (>=100)	PU_NI No inf.	PU Total Publics:
CC.AA		0,0458	0,0706	0,0819	-0,0371	-0,2717	-0,2076	-0,1577	-0,4264	-0,2107	-0,1546	-0,3066	-0,278	-0,1034	-0,2121	-0,273
(RD)	0,0458		0,9699	0,9437	0,8801	0,5474	0,6624	0,7889	-0,0744	0,6228	0,2255	0,2891	0,4689	0,8157	-0,0005	0,4605
(CD)	0,0706	0,9699		0,9862	0,4415	0,5704	0,6802	0,8094	-0,0124	0,6595	0,1169	0,1946	0,5065	0,8079	0,0174	0,3872
(TD)	0,0819	0,9437	0,9862		0,545	0,6585	0,749	0,8319	-0,0041	0,7366	0,0492	0,2005	0,5835	0,7946	-0,042	0,3741
PR<25	-0,0371	0,3659	0,4415	0,545		0,8066	0,7652	0,5907	0,1824	0,8462	0,02	0,3536	0,7683	0,6352	-0,0992	0,4313
PR25_49	-0,2719	0,5474	0,57	0,65	0,8		0,96	0,79	0,33	0,97	0,12	0,58	0,92	0,69	0,11	0,6
PR50_93	-0,2	0,66	0,68	0,74	0,76	0,96		0,87	0,23	0,97	0,22	0,57	0,92	0,69	0,11	0,6
PR>=100	-0,15	0,78	0,8	0,83	0,59	0,79	0,82		0,35	0,87	0,14	0,48	0,71	0,86	0,28	0,55
PR_NI	-0,42	-0,074	-0,012	-0,004	0,18	0,33	0,23	0,35		0,34	-0,02	0,37	0,24	0,23	0,47	0,21
PR	-0,21	0,62	0,65	0,73	0,84	0,97	0,97	0,87	0,34		0,14	0,56	0,9	0,79	0,15	0,61
PU<25	-0,15	0,22	0,11	0,04	0,02	0,12	0,22	0,14	-0,02	0,14		0,71	0,2	0,39	0,6	0,83
PU25_49	-0,3	0,28	0,19	0,2	0,35	0,58	0,57	0,48	0,37	0,56	0,71		0,62	0,55	0,58	0,92
PU50_93	-0,27	0,46	0,5	0,58	0,76	0,92	0,9	0,71	0,24	0,9	0,2	0,62		0,63	0,23	0,66
PU>=100	-0,1	0,81	0,8	0,79	0,63	0,69	0,77	0,86	0,23	0,79	0,39	0,55	0,63		0,24	0,69
PU_NI	-0,21	-0,0005	0,0174	-0,042	-0,00992	0,11	0,18	0,28	0,47	0,15	0,6	0,58	0,23	0,24		0,59
PU	-0,27	0,46	0,38	0,37	0,43	0,6	0,66	0,55	0,21	0,61	0,83	0,92	0,66	0,69	0,59	

Pairs of variables with P values below 0.05 (95% confidence level) have been obtained from the analysis to prove the statistical significance of the calculated correlations. Since what is sought is the relationship of the residences with the number of deaths in them, the relationship of the RD variable with the other 12 variables must be sought, finding significant relationships with the independent variables. The most significant relationships found were the following (in brackets the P-value obtained): PR25_49 (0.0153), PR50_99 (0.0020), PR>=100 (0.0001), PU50_99 (0.0428) and PU>=100 (0.0000). The influence of private residences (three) is more significant than that of public residences (two) in the number of deceased elderly persons. It is also observed that in medium and large-sized nursing homes, the number of deaths is related to these sizes, that is, larger size plus deaths. For the same size, for example, PR50_99 and PU50_99, the significance or relationship between the number of deceased persons and the size is greater in the private residence.

Next, a cluster analysis will be performed using the Ward method and the Euclidean square distance metric, establishing four groups or clusters from 19 observations in order to be able to study the groupings of the five variables of greatest significance or relation with the number of elderly persons who have died in the residences, that is, the six variables of greatest significance to be studied are RD, PR25_49, PR50_99, PR>=100, PU50_93 and PU>=100, and they will be done with respect to the Autonomous Communities.

The graph obtained or dendogram (Figure 3) shows four groups with similar characteristics according to the variables analysed. To form the groups, the procedure starts with each observation in separate groups, then the two observations that were closest are combined to form a new group. After recalculating the distance between groups, the closest groups are combined again, so that this process is repeated until all four groups are formed.

Figure 3. Dendogram with the four final groups of analysis

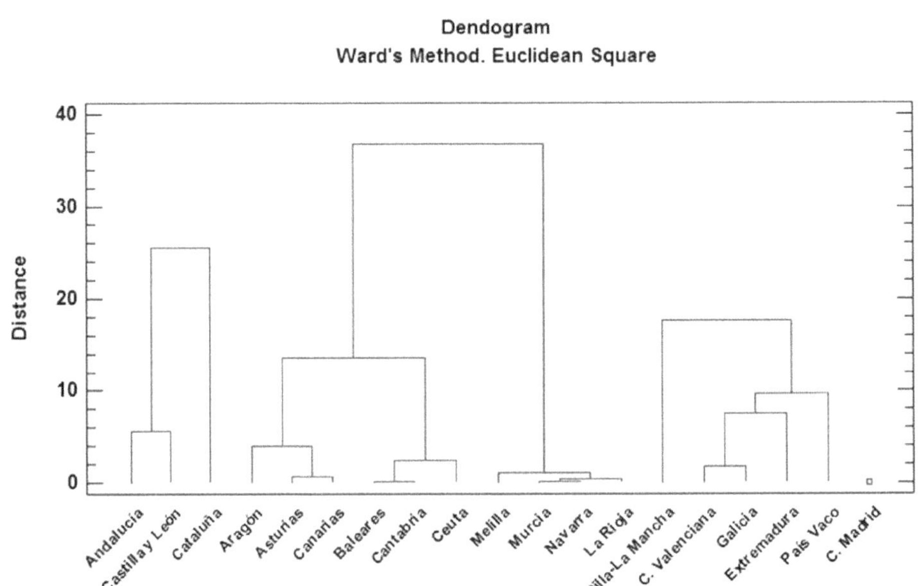

From figure 3 it can be seen that there are four very clearly defined groups, the first group (G1) made up of the communities of Catalonia, Andalucia and Castilla y León; and the second group (G2) made up of the communities of Castilla-La Mancha, C. Valenciana, Galicia, Extremadura and the País Vasco. As the third group (G3), and alone, Madrid appears with a point to the right of the lower part of the dendogram. The rest of the Autonomous Communities form the fourth group (G4).

Relating the six previous variables of influence, with the four associations or groups of Autonomous Communities, it has been found that the variable most related to the number of deaths in all of the Autonomous Communities is PR>=100, which corresponds to the private residences of more than one hundred persons (large old people's homes). This relationship is shown in the cluster dispersion diagram in figure 4.

Figure 4. Diagram of the dispersion of conglomerates of the Autonomous Communities compared to public residences with a capacity of more than 100 people

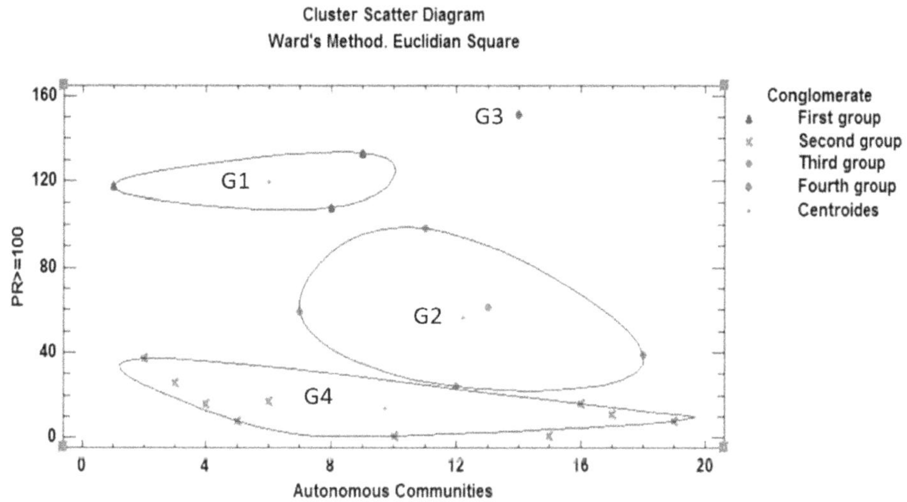

For the study of the second stage, the data corresponding to physical agents and chemical pollutants will be analysed to see a possible relationship between the quality of the indoor air in the residences and the number of elderly people who have died in the different residences in the different autonomous communities of Spain. The analysis process will be similar to the one carried out in the first stage, to analyse the influence of the size of the residences, a cluster or classification analysis will be carried out together with other techniques to relate and compare variables.

The variables to be studied are the physical agents: average temperature (AT) in °C, average humidity (AH) in %, precipitation (PREC) in mm, hours of sunshine (SH) in hours, to which we will include the population density of the communities (PD) in inhabitants/Km2. The following variables have been chosen for the pollutants: average annual particulate matter (PM10AA) and (PM2.5AA) measured in $\mu g/m^3$, average annual nitrogen dioxide (NO2) measured in $\mu g/m^3$, average annual ozone (O3OCTNOR) measured in $\mu g/m^3$, average annual sulphur dioxide (SO2) measured in $\mu g/m^3$ and formaldehyde (FOR) in $\mu g.m^2$. In total there are five variables belonging to the physical agents (including population density) and six variables belonging to the pollutants.

The cluster analysis provides the following segmentation according to the previous 11 variables around the Autonomous Communities and the number of elderly people who have died in the homes.

Figure 5. Cluster diagram of physical agents and contaminants in the AACC

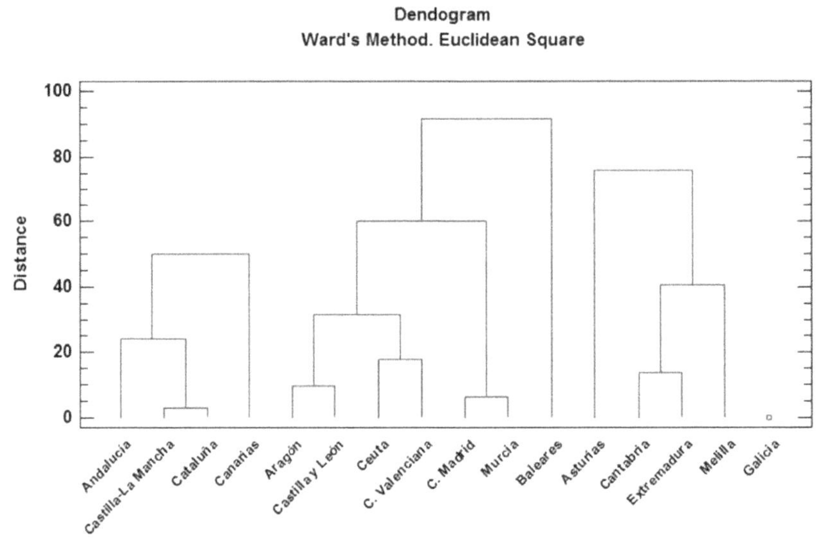

Four different groups have been obtained which will be called differently from those obtained in the analysis of the first stage in order to differentiate them, a first group (FC1) formed by the communities of Andalusia, Castilla-La Mancha, Catalonia and the Canary Islands, a second group (FC2) formed by Asturias, Cantabria, Extremadura and Melilla, a third group (FC3) where only Galicia is present and a fourth group (FC4) with the rest of the autonomous communities.

A multivariate method by factor analysis will be used using a Listwise missing value treatment, a standardization by type of factorization of main components and a varimax type rotation for the analysis of these variables and to see how they are related. The objective is to achieve a small number of factors that explain most of the variability of the 13 variables considered in the analysis (the previous 11, plus DR and AACC)

Applying the previous statistical methodology, 4 factors have been extracted, because these factors have had eigenvalues greater than or equal to the unit. These four factors explain 79.3414% of the variability of all data. When doing principal component analysis, the initial estimate of the community has been established to assume that all variability in the data is due to the common factors.

The factor load matrix will then be displayed after a Varimax rotation. It is done so that the factors have few high saturations and many almost none in the variables. This means that there are factors with high correlations with a small number of variables and zero correlations in the rest, thus redistributing the variance of factors.

Table 4. Factor analysis table

Factor Number	Eigenvalor	Variance Porcentage	Cumulative Porcentage
1	4,09141	31,472	31,472
2	2,43407	18,724	50,196
3	2,28212	17,555	67,751
4	1,50677	11,591	79,341
5	0,9121	7,016	86,358
6	0,67221	5,171	91,528
7	0,347443	2,673	94,201
8	0,315038	2,423	96,624
9	0,22916	1,763	98,387
10	0,099754	0,767	99,154
11	0,0636375	0,490	99,644
12	0,029118	0,224	99,868
13	0,0171641	0,132	100,000

The table above (Table 5) shows the equations that estimate the values of the common factors once the rotation has taken place. The rotation has been done to simplify the explanation of the factors. The first rotated factor is an equation:

F1=0,251871* AA.CC. – 0,347442*RD – 0,555084*AT + 0,912213*AH + 0,927481*PREC – 0,920873*SH -0,0857457*DP – 0,070322*PM10AA + 0,0782349*PM2,5AA + 0,383983*NO2 – 0,704236*O3OCTNOR + 0,326536*SO2 + 0,324599*FOR (1)

Table 5. Load matrix of factors after applying Varimax rotation

	Factor 1	Factor 2	Factor 3	Factor 4
Autonomous Communities	0,251871	-0,471631	0,262884	-0,52807
RD	-0,347442	-0,00841992	0,744084	-0,367933
AT	-0,555084	0,112131	0,0783305	0,662564
AH	0,912213	0,0516618	-0,145363	0,185372
PREC	0,927481	-0,00501128	0,0524455	-0,0907723
SH	-0,920873	-0,145316	-0,0741857	0,215049
DP	-0,0857457	-0,0962712	0,886502	0,076693
PM10 AA	-0,070322	0,904561	0,0244	0,149714
PM2,5 AA	0,0782349	0,714746	0,174668	-0,128887
NO2	0,383983	0,424818	0,760336	0,172826
O3OCTNOR	-0,704236	-0,157771	0,455243	-0,23757
SO2	0,326536	0,72773	-0,200042	-0,13505
FOR	0,324599	-0,325889	0,0236463	0,75947

The values of the variables in the equation are standardized by subtracting their means and dividing by their standard deviations. The other three equations for the factors F2, F3 and F4, are obtained in an analogous way, simply by substituting the coefficients of the corresponding column for each factor.

Next, three of the factors obtained will be plotted in 3D to see the contributions of the different variables to each factor.

Figure 6. Load graph of the different factors

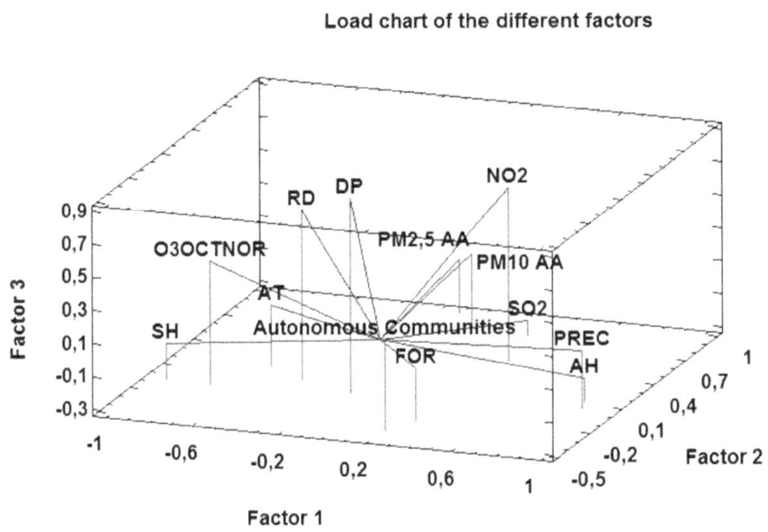

It can be seen that the most important loads for the factor 1 are provided by the variables PREC, AH, FOR and NO2, and contrary to the same factor appear AT, RD, SH and O3OCTNOR. For the factor 2, the variables PM10AA, PM2.5AA and SO2 appear as very important contributors, and the variable of the AACC. For factor 3, variables DP, RD and NO2 appear as favourable, and no candidate variables appear as contrary. For factor four, a parallel study was carried out, with the result that the variables that most favoured this factor were FOR and AT, and not RD and AACC.

Multivariate analysis provides a series of Pearson moment correlations between each pair of variables as indicated above. The relationships found were the following:

The table above shows the pairs of variables found after analysis that show P-values below 0.05, i.e. with 95% confidence level. For a better interpretation of the analysis, a bubble diagram has been made to see the connection between each variable. Two groups have been obtained. In the first one, the connection of the variables AH, AT, O3OCTNOR and SH to the variable PREC stands out, that is to say, there is a good correlation between precipitation and humidity, temperature, ozone and sunshine hours. To this first group and through the ozone variable (O3OCTNOR), RD, DP and NO2 are connected. If we consider the main link in the chain, the number of elderly people who have died in homes (RD), this depends on population density (PD) and ozone (O3OCTNOR), and this in turn depends on the whole chain formed by the other variables analysed. The second group consists of particulate matter and sulphur dioxide. It can be seen that the PM10AA particulate matter acts as a link between the other type of particulate matter (PM2.5AA) and sulphur dioxide (SO2). It is important to note that the formaldehyde variable (FOR) does not appear in any combination of the above, which may indicate that its contribution to indoor air quality pollution in nursing homes is less significant.

Table 6. Significant relationship between variables and bubble diagram

Combination of factors	Bubble diagram
RD+DP RD+O3OCTNOR AH+PREC AH+SH AH+O3OCTNOR AT+PREC AT+SH PREC+SH PREC+O3OCTNOR PREC+NO2	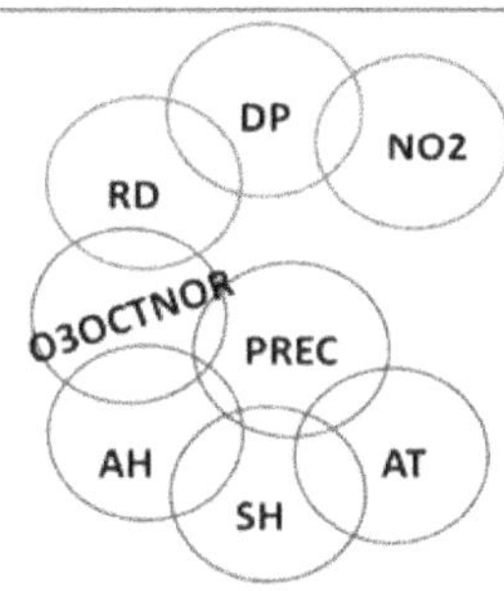
PM10AA+PM2,5AA PM10AA+SO2	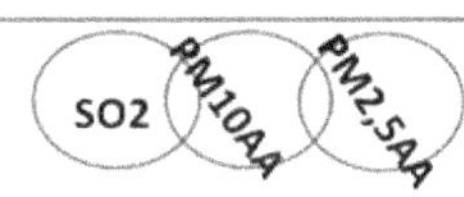

Carrying out a multivariate analysis by cluster or conglomerate methodology, the best combination obtained for the AACC separated into four groups with all the possible combinations with the other 12 variables under study, is the combination obtained with respect to the precipitation variable (PREC), and with it, total significance is not achieved as can be seen in the following graph, due to the fact that there is a very clear intersection of two groups.

Figure 7. Best significance of all parameters against AACC

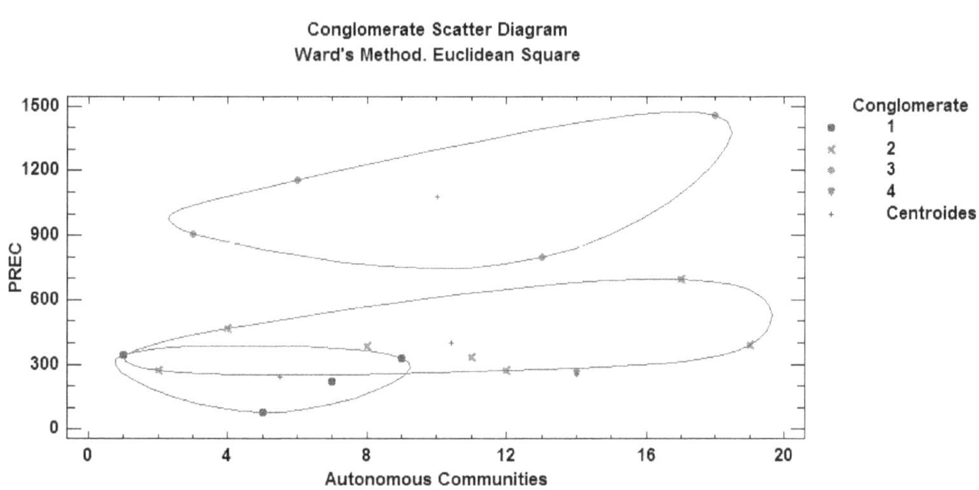

All the possible combinations between all the variables have been carried out, in total 156 combinations studied and the most significant cases of dependence are achieved between the variables humidity (AH), nitrogen dioxide (NO2), precipitation (PREC) and particulate matter (PM10AA), as can be seen in the following graphs (Figure 8). The graph on the upper left shows the best significance between the Autonomous Communities and all the variables, and even so, the significance between the AACC and the humidity (HA) is not very good as there are intersections between the groups.

In the centre of figure 8, the form in which these four variables of influence to be considered in the number of deaths in nursing homes are related has been interwoven by means of red circles.

Figure 8. Best meanings among all the parameters studied

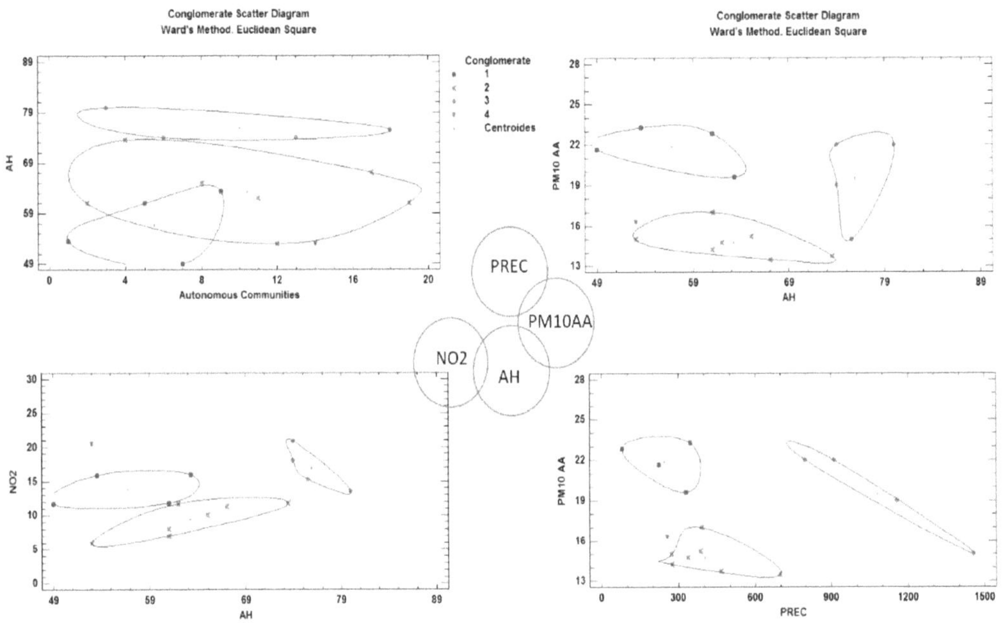

DISCUSSION

After the study carried out in this work, it was found that the highest number of deaths in geriatric homes or senior centers corresponded to the residences with a capacity of over 100 people. The number of elderly persons who have died has been greater in private homes than in public homes, probably due to a greater shortage of health resources and personnel due to the cost of these.

The residences, both private and public, together with the number of deaths in them, have been able to be separated into four groups by common or similar characteristics. The first group is formed by the communities of Cataluña, Andalucía and Castilla y León. A second group formed by Castilla-La Mancha, C. Valenciana, Galicia, Extremadura and País Vasco. A third group formed only by the Community of Madrid, and finally, a fourth group formed by the rest of the Communities.

Establishing a link between atmospheric factors and harmful effects on human health poses some difficulties. The effects on health are very varied, and each factor can generate more than one effect on human beings. The physical agents studied, in addition to population density, have a clear connection with the number of deaths, through ozone (Table 6). The main meteorological factors that have been shown to have a clear effect on human health are: air temperature, humidity, wind speed, hours of solar radiation, pressure and precipitation (Sanglier et al., 2020a, Fernandez-Garcia, 2014), many of which have already been analyzed in this work and their dependency relationship demonstrated.

If the relationship between the number of elderly deaths and the total number of deaths in each community is taken into account, the communities of Castilla-La Mancha (14.40%) and Extremadura (14.13%) present the most significant results. If we study the environmental conditions (physical agents) of these communities compared to the average values of these agents in Spain (AT=16.6ºC, AH=60.4%, SH=2739.8 h and PREC=496.7 mm), what makes these regions present these higher rates of elderly deaths could be having a relative humidity higher than the average, temperatures and hours of sunshine approximately equal to the average values obtained for the whole of Spain, and finally, precipitation lower than the average. If there is an excess of humidity, the proliferation of moulds and mites may be favoured, which increases the risk of infectious pathologies of the respiratory system (Monsalve, 2011). Health problems related to the presence of heat waves (lower rainfall) have worsened, affecting the most vulnerable population groups, i.e., the elderly and children, to a greater extent. The communities of Extremadura and Castilla-La Mancha have a type of Mediterranean climate characterized by hot, dry summers with little rainfall in the summer period, and long, milder winters.

Six pollutants have been studied, including ozone, particulate matter PM2.5 and PM10, sulphur dioxide, nitrogen dioxide and formaldehyde. Formaldehyde has not been significantly presented in the study, the contribution of this agent to indoor air quality pollution due to construction materials and furniture in the residence can be estimated at 3-4% (Guardino, 2012).

The communities of Extremadura and Castilla-La Mancha present ozone values 32% higher than the average for Spain (60.4 µg/m3). Some studies have shown that there is a direct relationship between daily mortality and tropospheric ozone levels (WHO, 2005). As tropospheric ozone concentrations increase above the guide value (100 µg/m3), health effects are aggravated so that prolonged exposure can have chronic effects on the respiratory system. These effects may increase during episodes of very hot weather.

Good significance has been found between particulate matter in its two tested varieties PM10 and PM2.5 and sulphur dioxide (Table 6). Anthropogenic activity is the main cause influencing air pollution levels on hourly and daily time scales. However, a clear relationship has also been found between African desert dust intrusions (Saharan air layer) and the number of exceedances of the daily PM10 limit value. The contribution of these long-range particulate transport processes to the annual average levels of PM has been especially relevant in the southern and central regions of the peninsula, Melilla and in the Canary and Balearic archipelagos. The average annual contribution of African episodes to PM10 and PM2.5 levels has also been quantified. There is also a large variation, from the north of the peninsula, with annual contributions of 1-2 µgPM10/m^3 and < 1 µgPM2.5/m^3, to the Canary Islands, Melilla, the Balearic Islands and the south of the peninsula with 4-6 µgPM10/m^3 and 1-2 µgPM2.5/m^3 (Querol, 2012).

The factors with the greatest influence on the temporal variability of PM concentration levels include locally recorded episodes of anthropogenic pollution (mainly traffic emissions, with some industrial exceptions, demolition-construction, domestic and residential emissions), followed by episodes of natural and anthropogenic pollution produced on a regional scale and by African episodes. The highest rates of incident solar radiation recorded in the summer months favour the formation of secondary particles, as well as the re-suspension of mineral dust by convective processes in semi-arid environments on the peninsula, such as the autonomous communities of Andalucia, Extremadura and Castilla-La Mancha, which have average values of particulate matter PM10/ PM2.5 in µg/m3 of (23.27/13.27), (21.66/10.66) and (15/8) respectively. As can be seen from the data, the community closest to the storm of particulate matter from the Sahara desert has the highest values of particulate matter, compared to the community of Castilla-La Mancha which is the furthest away, but all of them strongly affected by this material.

As for sulphur dioxide, in indoor environments the levels present are much lower than those outside, provided that there are no sources indoors that might cause it, such as kerosene stoves, boilers or chimneys. For this reason, it is not a pollutant that can cause major problems indoors. However, the existence of particles produces a synergistic effect in the presence of sulphur dioxide, since the combination of these two substances produces a greater effect than each substance alone. Sulphur oxides also contribute to the formation of acid rain as do NOx. The communities of Murcia and Asturias are the ones with the highest rates of this pollutant. For the communities with the highest ratio of elderly deaths in residence/community, this parameter affects very little, the highest incidence occurring in the community of Extremadura.

The statistical combination of the population density (PD) variable with all of the variables analysed shows the worst correlation or significance, so that it could be ruled out as having an influence on the number of elderly persons who died in the homes.

One more relation that has been obtained, is the one that relates the parameters of precipitation (PREC), Humidity (AH), nitrogen dioxide (NO2) and particulate matter (PM10AA). The concentration of PM10 is strongly related to relative humidity and precipitation (Figure 8, upper right and lower right respectively). The higher the precipitation, the higher the humidity and the higher the particulate matter concentration (Sanglier et al, 2020c; Villalba, Fajardo & Romero, 2018). On the other hand, relative humidity (RH) is closely related to nitrogen oxide (Figure 8, bottom left). There are several internal sources of nitrogen oxide emissions. On the one hand, those that occur in outdoor environments can infiltrate into indoor environments through air change processes. On the other hand, combustion processes, in indoor environments, responding to the main internal source of nitrogen oxide emissions. The decreases in industrial, transport and commercial activity since the beginning of the coronavirus outbreak have considerably reduced the levels of atmospheric nitrogen dioxide (NO2) in the AACC, so it is considered a minor factor in the analysis. However, the increase in humidity also produces an increase in the concentration of nitrogen dioxide. For the communities of Extremadura and Castilla-La Mancha, NO2 concentrations show values of 10.66 $\mu g/m^3$ and 8 $\mu g/m^3$ respectively, below the national average of 13.89 $\mu g/m^3$ and within the regulations.

Based on the above discussion, it is difficult to establish precisely to what extent indoor air quality may affect the health of elderly people in residential homes as the information available, in terms of data related to exposure and effect of concentrations, is not sufficient. On the other hand, although the effects of acute exposure to many air pollutants are well known, there are important gaps in the data concerning long-term exposures to low concentrations and mixtures of different pollutants. Only experience on the one hand and rational design of ventilation, occupancy and residential compartmentalisation will guarantee better indoor air quality.

In the last 20 years, studies on the presence of pollutants in many indoor environments have shown that it is higher than expected, and in addition, pollutants different from those present in outdoor air have been identified. The presence of microorganisms in indoor air can lead to problems of an infectious and allergic nature. Viruses as it is in this case, by means of Covid-19, can cause acute respiratory diseases, especially in the most vulnerable people, as they are the elderly, and mainly in the residences, where the occupation density is usually high and there is an important air recirculation.

The ventilation of a building, and consequently of a nursing home, is measured in renovations per hour. This means that every hour a volume of air equal to the volume of the nursing home enters from the outside; similarly, every hour a similar volume of air is expelled from the inside. If there is no forced ventilation, this value usually varies between 0.2 and 2 renewals per hour. The concentration of pollutants produced inside the residence will be lower with high values of renewals, although this is not a guarantee of air quality. The dilemma between air quality and energy efficiency is of great importance and especially when the residences are private, where the cost factor is very important.

Efforts to reduce the air velocity inside the building, as well as trying to increase the insulation and waterproofing of residences require the installation of materials that can be sources of indoor air pollution.

In recent decades, the trend towards designing buildings, including nursing homes, with minimal energy use has led to the development of buildings with very low air infiltration and exfiltration, which has generally allowed a higher concentration of airborne microorganisms and other types of pollutants, such as HVAC systems. The placement of air intakes near cooling towers or other sources of microorganisms, in addition to the difficulty of accessing the HVAC system for maintenance and cleaning, are also design, operation and maintenance defects that can affect the health of its occupants, and especially the elderly, who live there..

DISCLOSURE

None of the authors have conflicts of interest associated with this study to report. All authors contributed equally in the preparation of this manuscript.

FUNDING INFORMATION

The authors received no specific funding for this work.

REFERENCES

AEMet. (2020). *Agencia Estatal de Meteorología*. Author.

Berenguer, M. J., Guardino, X., Hernández, A., Martí, M. C., Nogareda, C., & Solé, M. D. (1994). *El síndrome del edificio enfermo. Guía para su evaluación*. Instituto Nacional de Seguridad e Higiene en el Trabajo.

Borsboom, W., De Gids, W., & Logue, J. (2016). *Technical note AIVC 68 residential ventilation and health*. INIVE.

Cao, G., Awbi, H., Yao, R., Fan, Y., Sirén, K., Kosonen, R., & Zhang, J. (2014, March). A review of the performance of different ventilation and airflow distribution systems in buildings. *Building and Environment*, *73*, 171–186. Advance online publication. doi:10.1016/j.buildenv.2013.12.009

CCAyES. (2020). *Centro de Coordinación de Alertas y Emergencias Sanitarias del Ministerio de Sanidad*. https://www.mscbs.gob.es/profesionales/saludPublica/ccayes/home.htm

Ceballos, M. A., Segura, P., Gutiérrez, E., Gracia, J. C., Ramos, P., Reaño, M., & García, B. (2018). *La calidad del aire en el Estado español durante 2018*. Ecologistas en Acción.

Código Técnico de la edificación (CTE). (2018). *Documento Básico HS3: Calidad del aire interior*. Ministerio de Fomento del Gobierno de España.

Daisey, J. M., Angell, W. J., & Apte, M. G. (2003, March). Indoor air quality, ventilation and health symptoms in schools: An analysis of existing information. *Indoor Air*, *13*(1), 53–64. Advance online publication. doi:10.1034/j.1600-0668.2003.00153.x PMID:12608926

Davidson, L., & Olsson, E. (1987). Calculation of age and local purging flow rate in rooms. *Building and Environment*, *22*(2), 111–127. doi:10.1016/0360-1323(87)90031-X

Dimitroulopoulou, C. (2012, January). Ventilation in European dwellings: A review. *Building and Environment*, *47*, 109–125. Advance online publication. doi:10.1016/j.buildenv.2011.07.016

DIRECTIVA. 2008/50/CE del Parlamento Europeo y del Consejo. (2008). *Relativa a la calidad del aire ambiente y a una atmósfera más limpia en Europa*. https://www.boe.es/doue/2008/152/L00001-00044.pdf

ENR. (2019). *Estadísticas sobre residencias. Distribución de centros y plazas residenciales por provincia*. ENR.

European Collaborative Action. (1992). *Indoor Air Quality and its Impact on Man. Report No.11: Guidelines for Ventilation Requirements in Buildings*. Office for Publications of EU.

Fernández, M. (2008). Sistema integral de ventilación de viviendas de acuerdo con el Código Técnico de la Edificación (HS3). *Instalaciones y técnicas del confort*, (197), 10-21.

Fernández-García, J., Dosil Díaz, O., Taboada Hidalgo, J., Fernández, J., & Sánchez-Santos, L. (2014). Influencia del clima en el infarto de miocardio en Galicia. *Medicina Clínica*, *145*(3), 97–101.

González-Lezcano, R. A., & Hormigos-Jiménez, S. (2016). Energy saving due to natural ventilation in housing blocks in Madrid. In *IOP Conference Series: Materials Science and Engineering* (Vol. 138). Institute of Physics Publishing. doi:10.1088/1757-899X/138/1/012002

Guardino, X. (2012). *Calidad del Aire Interior. Instituto Nacional de Seguridad e Higiene en el Trabajo*. INSHT.

Guo, H. (2011). Source apportionment of volatile organic compounds in Hong Kong homes. *Building and Environment*, *46*(11), 2280–2286. https://doi.org/10.1016/j.buildenv.2011.05.008

Health Canadá. (2005). https://www.canada.ca/en/health-canada/services/health -care-system/reports-publications/canada-health-act-annual-reports/annual-report-2004- 2005.html

Hormigos-Jimenez, S., Padilla-Marcos, M. A., Meiss, A., Gonzalez-Lezcano, R. A., & Feijó-MuÑoz, J. (2018a). Experimental validation of the age-of-the-air CFD analysis: A case study. *Science and Technology for the Built Environment, 24*(9), 994–1003. doi:10.1080/23744731.2018.1444885

Hormigos-Jimenez, S., Padilla-Marcos, M. Á., Meiss, A., Gonzalez-Lezcano, R. A., & Feijó-Muñoz, J. (2018b). Computational fluid dynamics evaluation of the furniture arrangement for ventilation efficiency. *Building Services Engineering Research and Technology, 39*(5), 557–571. https://doi.org/10.1177/0143624418759783

INE. (2020). *Instituto Nacional de Estadística*. https://www.ine.es/covid/covid_inicio.htm

Instituto Nacional de Seguridad e Higiene en el Trabajo (INSHT). (2010). *Prevención de la exposición a formaldehído* (Vol. 830). Nota Técnica de Prevención.

Kim, S., Kim, J. A., Kim, H. J., & Do Kim, S. (2006). Determination of formaldehyde and TVOC emission factor from wood-based composites by small chamber method. *Polymer Testing, 25*(5), 605–614. https://doi.org/10.1016/j.polymertesting.2006.04.008

Klepeis, N. E., Nelson, W. C., & Ott, W. R. (2001). The national human activity pattern survey (NHAPS): A resource for assessing exposure to environmental pollutants. *Journal of Exposure Science & Environmental Epidemiology, 11*, 231–252.

Knoppel, H., & Wolkoff, P. (1992). *Chemical, Microbiological, Health and Confort Aspects of Indoor Air Quality-State of the Art in SBS*. Kluwert Academic.

Kostiainen, R. (1995). Volatile organic compounds in the indoor air of normal and sick houses. *Atmospheric Environment, 29*(6), 693–702. https://doi.org/10.1016/1352-2310(94)00309-9

Kwon, K. S., Lee, I. B., Han, H. T., Shin, C. Y., Hwang, H. S., Hong, S. W., … Han, C. P. (2011). Analysing ventilation efficiency in a test chamber using age-of-air concept and CFD technology. *Biosystems Engineering, 110*(4), 421–433. doi:10.1016/j.biosystemseng.2011.08.013

Li, Y., Leung, G. M., Tang, J. W., Yang, X., Chao, C. Y. H., Lin, J. Z., … Yuen, P. L. (2007, February). Role of ventilation in airborne transmission of infectious agents in the built environment - A multidisciplinary systematic review. *Indoor Air*. doi:10.1111/j.1600-0668.2006.00445.x

Liu, M., & Tilton, J. N. (2010). Spatial distributions of mean age and higher moments in steady continuous flows. *AIChE Journal. American Institute of Chemical Engineers, 56*(10), 2561–2572. https://doi.org/10.1002/aic.12151

Monsalve, F. (2011). *Influencia del tiempo y de la contaminación atmosférica sobre enfermedades de los sistemas circulatorio y respiratorio en Castilla-La Mancha*. Academic Press.

Namiesnik, J., Gorecki, T., Krosdon-Zabiegala, B., & Lusiak, J. (1992). Indoor air quality pollutants, their sources and concentration levels. *Building and Environment, 27*(3), 339–356.

Nathanson, T. (1993). *Indoor Air Quality in Office Buildings: A Technical Guide*. Health Canadá.

OMS. (1987). *Air Quality Guidelines for Europe*. European Series, no. 23. Copenhague: Publicaciones Regionales de la OMS.

OMS. (2005). *Guías de calidad del aire de la OMS relativas al material particulado, el ozono, el dióxido de nitrógeno y el dióxido de azufre*. OMS.

OMS. (2012). *Atlas de la salud y del clima*. OMS.

Querol, X., Viana, M., Moreno, T., & Alastuy, A. (2012). *Bases científicos-técnicas para un Plan Nacional de mejora de la Calidad del Aire. Informes CSIC*. Consejo Superior de Investigaciones Científicas.

Rasmussen, M., Foldbjerg, P., Christoffersen, J., Daniell, J., Bang, U., Galiatto, N., & Bjerre, K. (2017). *Healthy Homes Barometer 2017 - Buildings and Their Impact on the Health of Europeans. VELUX Group*.

Real Decreto 102/2011. (2011). *Relativo a la mejora de la calidad del aire*. https://www.boe.es/buscar/act.php?id=BOE-A-2011-1645

RITE. (2020). *Reglamento de Instalaciones Térmicas de los Edificios*. Academic Press.

Saéz, E. (2017). Análisis de la calidad del aire en función de la tipología de ventilación. Aplicación al frototipo. Universidad Politécnica de Valencia.

Sanglier, G., González, R. A., Cesteros, S., & López, E. J. (2020c). Validation of the Mathematical Model Applied to Four Autonomous Communities in Spain to Determine the Number of People Infected by Covid-19. *Modern Applied Science*, *14*(6). https://doi.org/10.5539/mas.v14n6p110

Sanglier, G., Robas, M., & Jiménez, P. A. (2020a). Application of Forecasting to a Mathematical Models for adjusting the Determination of the Number of People Infected by Covid-19 in the Comunity of Madrid (Spain). *Contemporary Engineering Sciences*, *13*(1), 89–101. https://doi.org/10.12988/ces.2020.91453

Sanglier, G., Robas, M., & Jiménez, P. A. (2020b). Gamma Radiation in Aid of the Population in Covid-19 Type Pandemics. *Contemporary Engineering Sciences, 13*(1), 113-129. doi:10.12988/ces.2020.91456

Spengler, J. D., & Chen, Q. (2000). Indoor air quality factors in designing a healthy building. *Annual Review of Energy and the Environment*, *25*, 567–600. https://doi.org/10.1146/annurev.energy.25.1.567

Stymne, H., Axel Boman, C., & Kronvall, J. (1994). Measuring ventilation rates in the Swedish housing stock. *Building and Environment*, *29*(3), 373–379. https://doi.org/10.1016/0360-1323(94)90037-X

Tham, K. W. (2016). Indoor air quality and its effects on humans—A review of challenges and developments in the last 30 years. *Energy and Buildings, 130*, 637–650. doi:10.1016/j.enbuild.2016.08.071

Turiel, I. (1986). *Indoor Air Quality and Human Health*. Stanford Univ. Press.

UNE EN ISO 7730:2006. Determinación analítica e interpretación del bienestar térmico mediante el cálculo de los índices PMV y PPD y los criterios de bienestar térmico local.

Van Buggenhout, S., Zerihun Desta, T., Van Brecht, A., Vranken, E., Quanten, S., Van Malcot, W., & Berckmans, D. (2006). Data-based mechanistic modelling approach to determine the age of air in a ventilated space. *Building and Environment*, *41*(5), 557–567. https://doi.org/10.1016/j.buildenv.2005.02.029

Villalba, D., Fajardo, E. J., & Romero, H. (2018). *Relación entre el material particulado PM10 y variables meteorológicas en la ciudad de Bucaramanga – Colombia: Una aplicación del análisis de datos longitudinal*. Simposio Internacional de Estadística.

Wadden, R. A., & Schelf, P. A. (1983). *Indoor Air Pollution. Characterization, Prediction and Control*. Wiley.

Wagner, H.U., Verhoeff, A.P., Colombi, A., Flannigan, B., Gravesen, S., Moulliseux, A., Nevelainen, A., Papadakis, J., & Sebel, K. (1993). *Biological Particles in In-door Environments. Indoor Air Quality and Its Impact*. Comisión Europea.

Wargocki, P. (2000). The effects of outdoor air supply rate in an office on perceived air quality, sick building syndrome (SBS) symptoms and productivity. *Indoor Air*, *10*(4), 222–236. https://doi.org/10.1034/j.1600-0668.2000.010004222.x

Wargocki, P. (2013). The Effects of Ventilation in Homes on Health. *International Journal of Ventilation*, *12*(2), 101–118. https://doi.org/10.1080/14733315.2013.11684005

Wargocki, P. (2015, August 25). *What are indoor air quality priorities for energy-efficient buildings? Indoor and Built Environment*. SAGE Publications Ltd. doi:10.1177/1420326X15587824

Wargocki, P. (2016). Ventilation, indoor air quality, health, and productivity. In P. Wargocki (Ed.), *Ergonomic Workplace Design for Health, Wellness, and Productivity* (pp. 39–72). CRC Press.

World Health Organization. (2010). *Guidelines for Indoor Air Quality: Selected Pollutants*. WHO Regional Office for Europe.

Xu, C., Luo, X., Yu, C., & Cao, S. J. (2020). The 2019-nCoV epidemic control strategies and future challenges of building healthy smart cities. In *Indoor and Built Environment*. SAGE Publications Ltd. doi:10.1177/1420326X20910408

Yocom, J. E., & McCarthy, S. M. (1991). *Measuring Indoor Air Quality: A Practical Guide*. Wiley.

Chapter 7
Occupants' Habits and Natural Ventilation in a Hot Arid Climate

Hanan Al-Khatri
Sultan Qaboos University, Oman

Farah Al-Atrash
German Jordanian University, Jordan

ABSTRACT

This chapter presents the findings of a questionnaire distributed in the Arabian Gulf region to explore the potential of utilising natural ventilation to moderate thermal conditions in residential buildings. It, additionally, explores the occupants' acceptance of the idea of applying natural ventilation when outdoor thermal environments are acceptable. Natural ventilation is a key sustainable solution to improve thermal conditions in the region considering its extreme climate, huge consumption of cooling energy, and the cultural attitude to depend on mechanical ventilation. The chapter discusses the thermal adaptation including physical or behavioural adaptation, and it sheds light on selected studies discussing similar issues. A detailed climatic analysis is presented with reference to Muscat city. Discussing the questionnaire's findings revealed high acceptance to depend on natural ventilation of around 93.7% of the participants. In addition, the possibility to depend on natural ventilation in the period from November to March, especially during daytime, was revealed.

INTRODUCTION

The American Society of Heating, Refrigerating, and Air-Conditioning Engineers (ASHRAE) defines thermal comfort as: "that condition of mind which expresses satisfaction with the thermal environment and is assessed by subjective evaluation" (ASHRAE, 2017, p. 9.1). This definition emphasises that thermal comfort is a psychological state. Indeed, thermal comfort is greatly affected by psychological adaptation opportunities (Brager & de Dear, 1998). Thermal adaptation processes involve all actions

DOI: 10.4018/978-1-7998-7023-4.ch007

taken by people to acclimatise themselves to the ambient thermal conditions (Brager & de Dear, 1998; Nicol et al., 2012; Nikolopoulou & Steemers, 2003). Defined by the available adaptive opportunities, these processes affect thermal comfort level (Nikolopoulou et al., 2001). Choosing a certain adaptive action is limited by effectiveness, ease, and related economic consequences (Mishra & Ramgopal, 2013).

Psychological adaptation opportunities include the natural appearance of internal spaces, known as naturalness, that increases the probability of accepting wide ranges of thermal conditions. Naturalness is widely achieved by vegetation presence and landscape viewing (Nikolopoulou & Steemers, 2003) and it is related to the expectation of designers' responsibility in creating comfortable indoor environments. Expectation is another psychological adaptive action. It was found that people in mechanically ventilated buildings have less tolerance of temperature variations compared with those in naturally ventilated buildings (Brager & de Dear, 1998). People's expectations are affected by their thermal experiences and usually assist them in accepting ambient conditions via decisions such as consuming cool or hot drinks, changing postures, or changing clothing. In this regard, a Chinese questionnaire explored the effect of thermal history on acceptability and comfort and revealed that adaptation to improved thermal conditions is relatively easy compared with lowering expectation of thermal ambience (Luo et al., 2016). Another study investigated the influence of participants' thermal history through unifying clothing levels, which encourages psychological and physiological adaptation. Exposure to cool conditions resulted in higher sensitivity as participants maintained a narrow comfort range. The study concluded that the thermal history of cool experiences weakens thermal adaptation (Ning et al., 2016).

Furthermore, exposure time affects thermal perception as people tend to accept uncomfortable environments if they spend short periods in such spaces. This is obvious in outdoor spaces as people generally decide their stay time freely. This decision and other similar decisions such as switching on or off fans or air conditioning units, opening or closing windows, changing posture, or clothing level, are part of what is known as the perceived control that positively correlates to accepting thermal conditions (Nikolopoulou et al., 2001; Nikolopoulou & Steemers, 2003; Singh et al., 2011). As noticed, most psychological adaptation mechanisms affect people's perception and, thus, their acceptance of thermal environment through adaptive actions or influencing tolerance.

Thermal adaptation is not limited to psychological adaptation. It besides involves physiological and physical processes. Physical or behavioural adaptation includes consuming food, closing or opening windows besides changing activity, posture, or clothing (Brager & de Dear, 1998; Nikolopoulou & Steemers, 2003; Singh et al., 2011). This type of adaptation is classified as interactive and reactive. The former involves altering the surrounding environment, whereas the latter deals with personal status alterations (Nikolopoulou & Steemers, 2003).

The availability and utilisation of the adaptive opportunities are partially restricted by the buildings' architectural design in addition to the climatic, social, economic, and organisational conventions (Mishra & Ramgopal, 2013; Peeters et al., 2009). Therefore, integrating energy-efficient measures in buildings alone is not enough. It is crucial to understand the interaction between the occupants and these measures to achieve the aim of integrating them. Hence, the dominant approach of evaluating architectural designs based on energy simulations should extend to include the relationships between the occupants and their buildings (Buso et al., 2015; Gianfrate et al., 2017). In connection to energy consumption inside buildings, occupants' habits can affect the thermal performance of the building directly and indirectly. Actions like opening and closing windows as well as switching on or off fans or air conditioning units can directly influence the cooling or heating energy inside buildings. On the other hand, actions like changing clothing level and consuming hot or cold drinks influence the thermal comfort state of the occupants, which, in turn, may act as a trigger for changes that affect the cooling or heating energy.

This chapter aims to investigate the potential of utilising natural ventilation in residential buildings in the Arabian Gulf region through defining the months and days in which people can use natural ventilation. It, additionally, explores their acceptance of the idea of applying natural ventilation when outdoor thermal conditions are acceptable and explores the variables that determine the use of air conditioning units and the set point temperatures. Such explorations are important considering the potential of natural ventilation to moderate indoor thermal conditions on one hand, and the extreme climates, huge cooling energy levels, and cultural attitude that tends to consider mechanical ventilation as the favourable way of ventilation throughout the year on the other hand.

Background

Several studies reported the influence of occupants' habits on energy consumption levels in buildings as presented in the following paragraphs. The lack of research from the hot arid climate is noticed despite the extreme conditions found in this climate. Moreover, it is noted that the majority of available research was conducted in offices or residential buildings.

The influence of behavioural actions taken individually in offices occupied by more than one person on thermal satisfaction and perceived control was investigated in an experimental study. The behavioural actions included tilting windows, closing blinds, using a ceiling fan, and adjusting clothing level. It was observed that increasing the number of people in the one office was positively correlated to the number of actions related to windows and negatively correlated to the number of actions related to blinds and fans. Applying the mixed effect model analysis provides statistical evidence on the association between the investigated behavioural actions and the number of people in the office. Regarding the clothing level, it was affected by the running mean outdoor air temperature and by the operative temperature. Moreover, increasing the number of people in the one office reduces the perceived control, which, in turn, strongly affect the neutral temperature (Schweiker & Wagner, 2016).

With the aim of establishing a model that integrates actions that are rarely integrated into occupants' behavioural models, a study was conducted in offices in the subtropical climate of Brazil. The integrated actions include changing clothing and ingestion of hot or cold beverages in addition to the operation actions of windows, fans, and air conditioning systems. Outdoor thermal conditions affected the operation of windows and air conditioning units. The researchers found that when the outdoor air temperature is 21.5 °C, the probability of either opening the window or turning on the air conditioning unit was estimated as 50%. Moreover, increasing the indoor operative temperature by 1 °C resulted in reducing the likely to increase the clothing level by around 49%. Drinking hot or cold beverages was more complicated; yet, consuming hot drinks was generally associated with changing in occupants' thermal sensations and indoor thermal conditions, whereas consuming cold drinks was generally associated with changes in both indoor and outdoor thermal conditions (Rupp et al., 2021).

Monitoring the occupants' behaviour in an office building in Germany linked non-optimal behaviour regarding windows' opening to an increased level of heating energy in winter and a potential increase in consumed energy in summer. The pattern of windows operation was strongly influenced by indoor and outdoor air temperatures. However, it was not possible to specify if higher indoor air temperatures were the reason for or the result of the lower opening of windows. Besides, it was consistent with the sun movement from east to west with eastern and western windows being opened more during the afternoon and the morning respectively (Schakib-Ekbatan et al., 2015)

Another study attempts to find out the reason behind the occupants' behaviour regarding windows opening and artificial lighting switching on. The possible link between these actions and environmental variables, namely indoor and outdoor air temperatures, was investigated as well as the link with time. The study was conducted in Italy under Mediterranean climate in offices' buildings during summer. The findings confirmed the investigated links. In particular, it was found that opening and closing windows were mainly related to the indoor and outdoor air temperatures, whereas switching lights on and off was mainly related to time. It was, additionally, found that the interaction with windows may be affected by the geographical location, whereas the interaction with lights is not (Naspi et al., 2018).

Moreover, a study conducted to validate the adaptive thermal comfort model examined the adaptive principle by exploring the relationship between the model and some behavioural actions applying Fisher's exact test. The investigated behaviours were clustered in two groups, namely actions related to thermal comfort and metabolic activity. The former group includes actions like opening or closing windows, drinking a hot or cold drink, put on or off clothing, taking a shower, and changing the thermostat setpoint. The latter group includes lying or sleeping, sitting relaxed, walking, jogging, and running. In addition, the relationship between clothing level and thermal sensation votes was explored. The results revealed no significant correlation between the thermal sensation and some behavioural actions like opening or closing windows, taking off clothing, having a shower, or turning down the thermostat. The study concluded that opening windows was mostly for ventilation purposes rather than to improve the indoor thermal conditions (Ioannou et al., 2018).

A relatively recent study in Denmark found a link between the habits of the occupants and the characteristics of single-family detached houses. The study examined the assumption that energy-efficient buildings have a positive impact on occupants' behaviour in relation to energy consumption. Compared with adjusting thermostats and clothing levels, operating windows showed less correlation with the energy efficiency measures integrated into the buildings. Moreover, the changes in the clothing level were related to the properties of buildings' envelopes, whereas adjusting thermostats actions were related to the integrated technologies in the building. Overall, the study found a decrease in the occupants' thermal tolerance in the presence of technological energy efficiency measures (Hansen et al., 2018).

The electricity consumption in residential buildings was examined recently based on the occupants' characteristics. Collected in 2009, the analysed data represented 12083 houses and around 113.6 million people in the USA. It was revealed that some characteristics have a positive effect, whereas some have a negative effect on energy consumption. The former group of characteristics includes the type of occupancy and the length of residency. The latter group includes the age of the householders, their income, and educational level. It may worth mentioning that gender does not correlate with electricity consumption. Moreover, the collective effect of householders' age, household size, income, educational level, type of occupancy, and length of residency was responsible for explaining 10.7% approximately of the variability in the consumed energy levels (Xu et al., 2020).

Additionally, the importance of considering occupants' habits was highlighted in a study conducted to explore the habits' effect on energy refurbishment of residential buildings in the Mediterranean climate. The study considered opening and closing windows, increasing and decreasing heating setpoint, and adjusting shading among other actions. It was revealed that refurbishing the building resulted in reducing the consumed primary energy by 21%. However, wrong actions related to windows opening and closing in the refurbished model increased the primary energy demand by 31%, which wasted the refurbishment benefits (Ascione et al., 2020).

Applying an experimental approach, a study was conducted to explore the effect of the dynamic interaction between the occupants and the buildings with reference to the consumed levels of energy. The "random walk" approach was followed, which is mainly a series of random variables. The findings revealed that no correlation could be proven between the occupants' presence and energy consumption; yet, their active actions of operating the heating or cooling systems were correlated to the consumed levels of energy. It is worth mentioning that not all occupants participated in changing the indoor thermal conditions of the investigated environment (Ahn & Park, 2016).

One of the few studies concerned with the occupants' behaviour in the Arabian region came from Amman, Jordan that is characterised by a hot summer Mediterranean climate. The study was conducted in three office buildings of which two were mixed mode buildings and the third was naturally ventilated. The occupants' perceived control affected their thermal perception and was negatively correlated with the offices' areas. The study suggested designing buildings with operable buildings as the results revealed that occupants depended mainly on opening windows and adjusting thermostats to modify their thermal environments (Al-Atrash et al., 2018).

MAIN FOCUS OF THE CHAPTER

The Arabian Gulf region includes six countries which are Oman, United Arab Emirates (UAE), Saudi Arabia, Bahrain, Qatar, and Kuwait. In most buildings in this region, the balance between modernity and thermal comfort on one hand, and energy efficiency on the other hand, is unfortunately partially achieved (Aboulnaga, 2006; Giusti & Almoosawi, 2017). This is mainly owing to importing international architectural styles without modifying them to consider the local climatic conditions (Katodrytis, 2015; Said & Abdeirahman, 1989 in Zurigat et al., 2003). The extreme climatic conditions of the Arabian Gulf region make it no surprise that buildings rely heavily on air conditioning systems to provide the required levels of thermal comfort. Indeed, these systems are integrated into all buildings despite their different types (Al-Gharibi, 2016; Majid et al., 2014; Qader, 2009b). Consequently, the ecological footprint per person in the Gulf countries exceeds the global level (Global Footprint Network, 2018). Considering Oman for instance, surveying a random sample of residential buildings revealed that air conditioning units were used in all bedrooms besides more than 60% of kitchens, living, and guest rooms (Majid et al., 2014). In another survey, most participants reported daily operation of air conditioning units of around 20-24 hours during summer in Muscat, the capital city of Oman (Al-Gharibi, 2016), which illustrates the consumption of huge energy amounts for cooling in Oman (Al-Hinai et al., 1993). Indeed, air conditioning systems are estimated to consume more than 70% of the electrical energy in most Omani buildings. The majority of the used systems depend on the vapour compression technique (Zurigat et al., 2003) that releases chlorofluorocarbon (CFC) gases. These gases deplete the ozone layer and, hence, are related to the climate change phenomenon. Adding to this, the main source of domestic electrical energy in Oman is fossil fuels like natural gas and oil (Al-Badi et al., 2011).

In addition, it is worth mentioning that the Arabian Gulf countries are considerable contributors of the greenhouse gases (GHG) emissions globally taking into account their relatively small population and large production of oil and gas (Abdul-Wahab, Charabi, Al-Maamari, et al., 2015; Qader, 2009a). The processes of fossil fuels extraction and electricity production are the main sources for these emissions

(Qader, 2009a). Moreover, a relatively recent study indicated that there was an increase in the GHG emissions in Oman (Abdul-Wahab, Charabi, Al-Rawas, et al., 2015), which was possibly reflected in the climatic events and changes that occurred in the country in the last few years. Examples include the increase in heat waves, droughts, diurnal temperatures, hot days' number, and cyclones' damage power (Abdul-Wahab, Charabi, Al-Rawas, et al., 2015).

Among several reasons, the heavy reliance on air conditioning systems in the Arabian Gulf region may be attributed to the lack of adequate natural ventilation inside buildings. This lack, in turn, is positively related to poor indoor air quality (IAQ) conditions due to the concentrations of indoor air pollutants. Indoor air pollution is a major global environmental threat to health according to the World Health Organization (WHO), especially in Gulf countries where ventilation is heavily dependent on air conditioning. Poor indoor quality increases the risk of respiratory infections, heart disease, and lung cancer. Reasons contributing to this lack of adequate ventilation include poor architectural design, privacy and security requirements, increased noise levels, occupants' perception regarding outdoor environments, and unfavourable outdoor conditions.

Natural ventilation is a key sustainable solution for improving thermal comfort, maintaining a healthy indoor environment, and reducing energy consumption when the outdoor conditions are suitable; yet, it is highly dependent on regional climate. The climatic diversity of the Arabian Gulf region allows for utilising natural ventilation in certain climatic zones as well as in certain periods in the year. Such utilisation, if supported by architectural climatic responsive design, contributes positively towards achieving indoor thermal comfort passively and reducing the cooling energy consumption. In this regard, it is crucial to understand the occupants' behaviour before proposing any design measures that utilise natural ventilation. This importance stems from the frequent failure of buildings in achieving their designed targets, especially those related to energy consumption. Such failure is usually known as the energy performance gap and it is mainly attributed to the occupants' behaviour (Gianfrate et al., 2017; Salvia et al., 2020).

This chapter reports the findings of a questionnaire distributed in the Arabian Gulf region with the aim of exploring the relations between the occupants' habits and the utilisation of both natural ventilation and air conditioning. In specific, the questionnaire investigates the periods in which people in the Gulf countries depend on natural ventilation and it explores their perception regarding utilising natural ventilation when the outdoor conditions are thermally acceptable. It, additionally, explores the variables that determine air conditioning use and setpoint temperatures. On a larger scale, the findings of this questionnaire may assist in setting the Government electricity subsidy and define an electricity tariff that changes based on seasons.

The Questionnaire

Consisting of five sections, a questionnaire was prepared to explore the influence of the occupants' habits on the potential of utilising natural ventilation in the Gulf region. The first section gathered personal data about the participants like the country, city, gender, and age. The second section explored the type of residential property, the duration of residence, and the daily stay duration. It, additionally, explored the number of family members and the number of rooms. In the third section, the occupant's perception regarding the role of natural ventilation in achieving thermal comfort and the potential of utilising it throughout the year was explored. This was achieved by identifying the months and the times during the day in which occupants depend on natural ventilation. The occupants' degree of personal control was investigated in the fourth section and their thermal state, preference, and perception about the surrounding environment were explored in the fifth section.

Climatic Analysis

In general, exploring the local context in thermal comfort studies is useful as it assists in understanding the participants' thermal behaviour. This is owing to the influence of the climatic, cultural, and social background on thermal preferences as widely reported (Al-Khatri et al., 2020; Al-Khatri & Gadi, 2019; Humphreys et al., 2016). The following paragraphs give an overview of the climatic conditions dominant in the Arabian Gulf region in general and in Muscat, the capital city of Oman, in specific as the majority of the participants were Omanis who live in Muscat.

According to Koppen-Geiger climate classification, the Arabian Gulf region is located within a hot arid desert climate that is characterised by high annual air temperature and low precipitation (Almazroui et al., 2020). It is possible to divide the region into two climatic zones. One covers the northern part of the region with annual mean air temperatures that extend between 8.6 °C and 28.3 °C. The other covers the southern part with the corresponding temperatures between 26.7 °C and 34.0 °C (Almazroui, 2013). In some locations, the maximum air temperature during summer can be above 50 °C (Almazroui, 2012; Pal & Eltahir, 2016).

Muscat Climatic Conditions

Muscat is the capital city of the Sultanate of Oman and it is located at 23.61 °N latitude and 58.54 °E longitude. Based on Koppen-Geiger climate classification, the city is located in a desert hot arid region. Yet, its mountainous nature and proximity to the Indian Ocean changed its climatic conditions to hot and humid (Al-Azri et al., 2013; Konya & Vandenberg, 2011; Ragette, 2012). In general, there are two distinctive climatic periods; the first starts in April and extends to October and it forms the hot humid period. The rest of the year forms the relatively cooler period.

The climatic data of Muscat city were officially requested from the General Directorate of Meteorology and the obtained data were from Muscat International Airport weather station. Additionally, the average number of wet days and the average daily sunshine hours were obtained from Weather Base website. Details follow below.

Air Temperature

The monthly mean, minimum, and maximum air temperatures of 14 years are presented in Figure 1. In the hot humid period, the mean temperature extended from 29.5 °C roughly in April and October to 34.7 °C approximately in June. The maximum air temperature extended from around 39.0 °C to almost 45.3 °C in October and May, respectively. The range of minimum air temperature was almost 20.9-27.7 °C. Considering the cooler period, the mean temperature roughly extended from 20.8 °C to 25.8 °C in January and November, respectively. The maximum and minimum temperatures were around 28.2-35.8 °C and around 13.2-17.8 °C, respectively.

Figure 1. Monthly mean, maximum, and minimum air temperature for 2001-2014 [Reproduced from the Omani General Directorate of Meteorology data]
Source: Authors

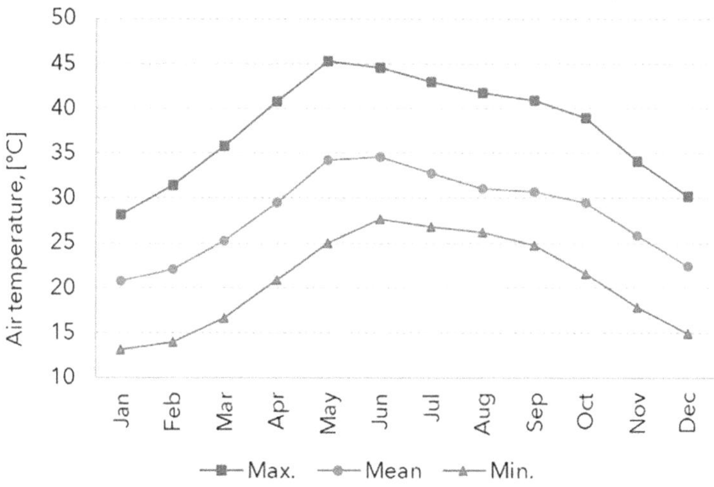

Relative Humidity

Figure 2 presents the monthly mean, maximum, and minimum relative humidity for 14 years. Considering the hot humid period, the mean relative humidity roughly fluctuated between 41.5% and 74.3% in May and August, respectively. The maximum records roughly extended from 93.9% in October to 96.1% in August, whereas the minimum records approximately extended from 6.6% in May to 19% in August. The ranges were smaller in the cooler period. For instance, the mean relative humidity ranged from almost 57.5% to around 65.6% in March and December, respectively. The maximum and minimum ranges were around 91.8%-96.0% and 10.9%-27.1% respectively.

Figure 2. Monthly mean, maximum, and minimum relative humidity for 2001-2014 [Reproduced from the Omani General Directorate of Meteorology data]
Source: Authors

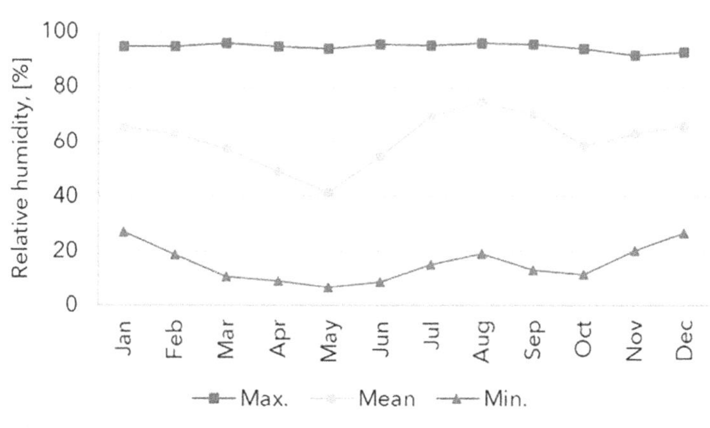

Wind

The hourly wind data obtained from the General Directorate of Meteorology covered a period of four years. The data were measured at a height of 10 m and were used to plot the wind rose of mean velocity as demonstrated in Figure 3. The wind rose was generated using a template excel file from Enviroware web page.

Figure 3. Wind rose of mean air velocity [m/s] for 2006-2010 [Reproduced from the Omani General Directorate of Meteorology data]
Source: Authors

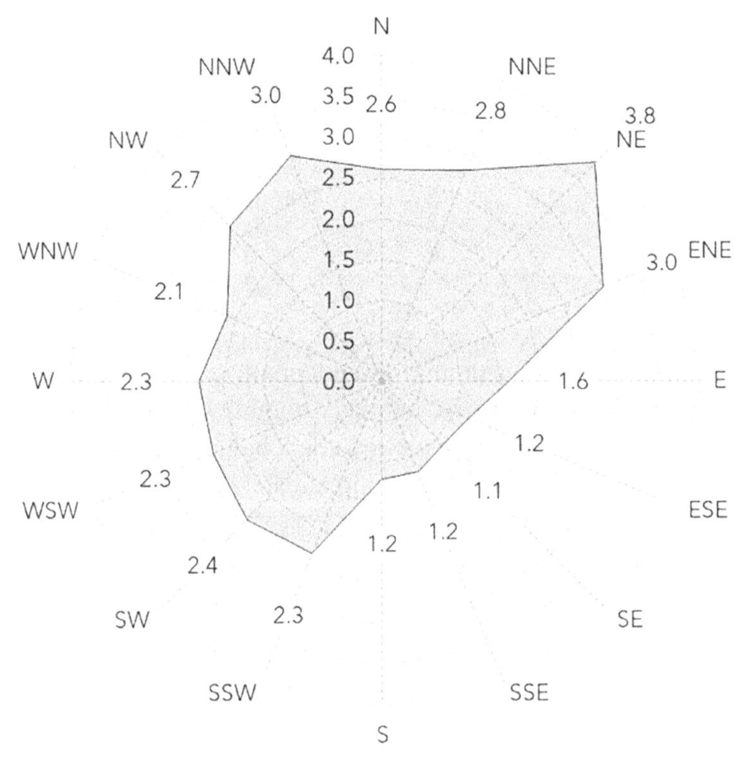

As noted from the figure, the mean wind velocity ranged between 1.1 m/s and 3.8 m/s. The corresponding directions were south-east and north-east. In general, it is possible to classify the mean wind velocity into three categories, namely:

- $0 \leq$ mean wind velocity < 2 from east, east-south-east, south-east, south-south-east, and south directions;
- $2 \leq$ mean wind velocity < 3 from north, north-north-east, south-south-west, south-west, west-south-west, west, west-north-west, and north-west directions; and
- $3 \leq$ mean wind velocity from north-east, east-north-east, and north-north-west directions.

Moreover, the distribution of the wind direction is plotted in Figure 4. The dominant direction for the wind is north-east as it blew from this direction for around 15% of the measured period. The frequencies of the other directions were below 10% and can be categorised into:

- Relative frequency < 5% for the directions between east and south including both besides to west-north-west and north-west directions; and
- 5% ≤ relative frequency < 10% for the remaining directions

Figure 4. Distribution of wind direction for 2006-2010 [Reproduced from the Omani General Directorate of Meteorology data]
Source: Authors

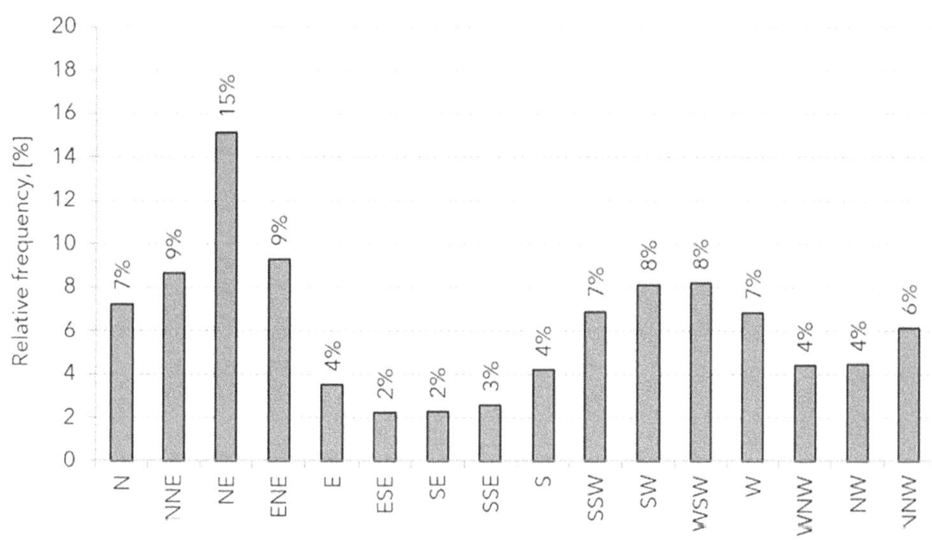

Precipitation

The monthly mean and maximum precipitation were computed for 14 years starting from 2001 as presented in Figure 5. It is possible to distinguish a rainy period from November to April with monthly means of around 10 mm or slightly higher. January was an exception with a mean of 6.6 mm roughly. In the remaining period, thc monthly means were below 3 mm except June that had a mean of almost 6 mm. In addition, September was the driest month with a mean of around 0 mm.

It is noted that the pattern of the maximum amount of precipitation was similar to the of the monthly means. The maximum precipitation in June was noticeably higher because of Gonu and Phet cyclones' effect. Muscat city was subjected to the former cyclone in June 2007 and to the latter in June 2010. Lasted for two days, Gonu cyclone caused 49 deaths, left 20,000 people homeless, damaged the infrastructure services severely, and cost around US $4 billion according to the estimations of the Omani authorities. In Muscat alone, the first day of the cyclone resulted in damaging almost 35% of telecommunications networks, 30% of mobiles networks, and 90% of roads, Around 27% and 23% of the city areas lost electric power and water supply, respectively. Considering Phet cyclone, it caused the death of 24 people (Al-Shaqsi, 2010). It is worth mentioning that the recorded amounts of precipitation in Muscat International Airport weather station during these cyclones were around 257 mm and 81.6 mm, respectively.

Figure 5. Monthly mean and maximum precipitation for 2001-2014 [Reproduced from the Omani General Directorate of Meteorology data]
Source: Authors

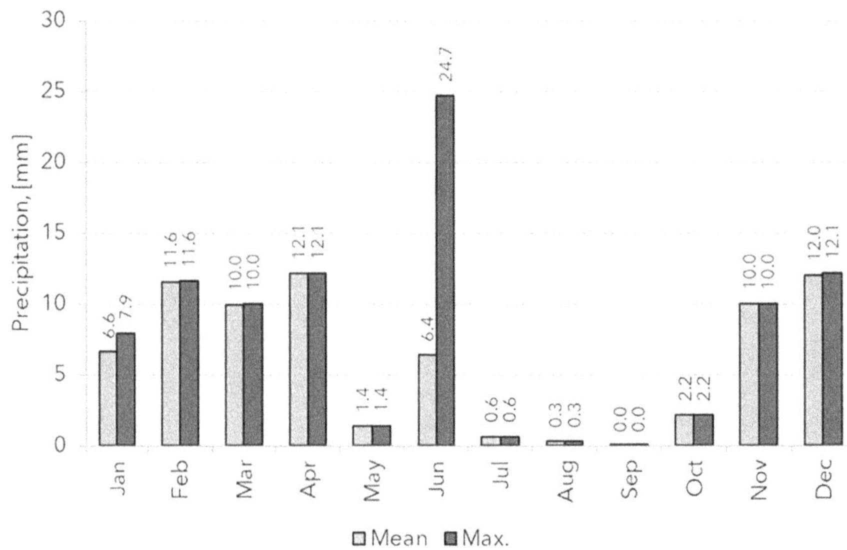

RESULTS AND DISCUSSION

In total, 235 questionnaires were returned. These questionnaires were filtered to exclude those with empty answers to some questions. The remaining questionnaires were analysed using the Statistical Package for the Social Sciences (SPSS) software. The findings are presented and discussed in the following paragraphs.

From the 144 occupants who participated in the questionnaire, 38.2% were females and 61.8% were males. The majority of the occupants were from Oman as they formed almost 64.6% followed by Saudi Arabia and UAE forming 21.5% and 11.1%, respectively. Considering the ages' groups presented in Figure 6, the highest category refers to 31 to 40 years old with 43.1% almost equally distributed between females and males. Followed is the 18 to 30 years old category with 23.6%, which was similarly almost equally distributed between females and males. It is noticed that the highest difference in the gender percentage came from the 41 to 50 years old category with 89.2% for males and 10.8% for females. Considering the age categories to which the majority of the participants belong, it is possible to infer that the questionnaire was answered by the targeted category.

Figure 6. Groups of age with respect to gender (Labels refer to number of occupants)
Source: Authors

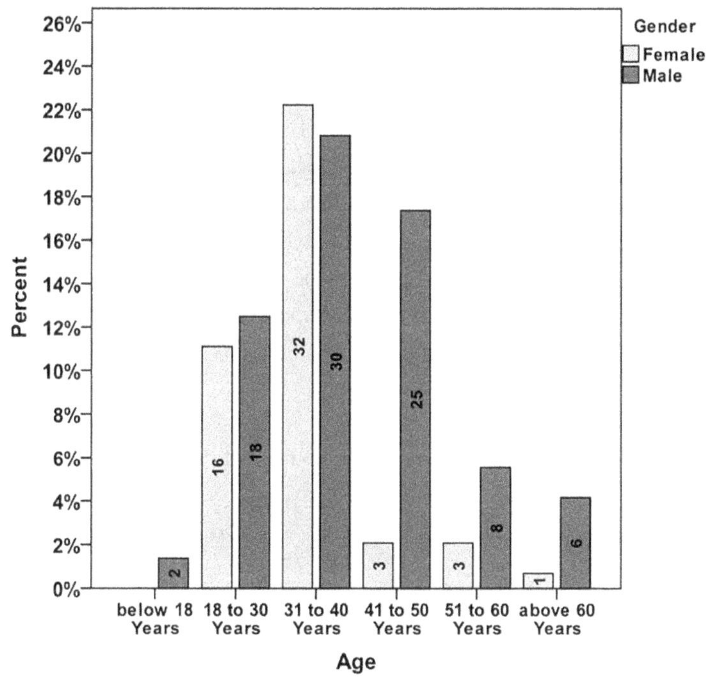

Figure 7. Type of the residential property vs. residency period (Labels refer to number of occupants)
Source: Authors

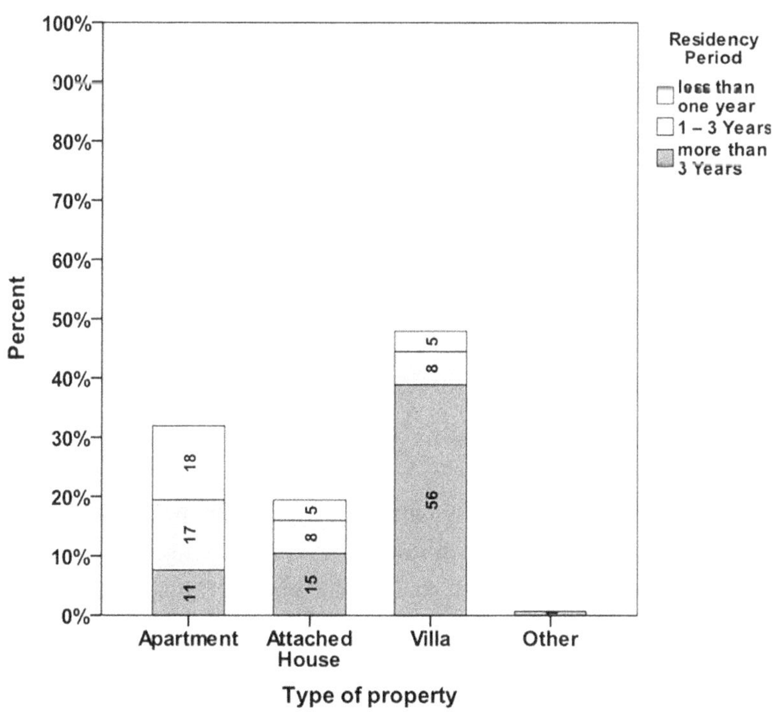

Figure 7 presents the type of the residential property. It can be observed that around 47.9% of the participants live in villas, 31.9% in apartments, and 19.4% in attached houses. The majority of those live in villas and attached houses spent more than three years in their properties with percentages of 81.2% and 53.6%, respectively, whereas the highest percentage for the apartment properties indicates that occupant spent less than one year in their residences with 39.0%.

With reference to Figure 8, the surveyed sample of the occupants exhibited a high acceptance to utilise natural ventilation to satisfy their thermal comfort demands as around 93.7% accepted to depend on natural ventilation provided that outdoor temperatures are suitable. As indicated by the figure, males reported higher acceptability to natural ventilation by almost one third compared with females. The respective percentages were 57.3% and 36.4%. This may be attributed to the cultural background influence as males spend relatively longer periods outdoors compared with females in the Arabian Gulf region. Besides, males are usually responsible for paying the electrical bills of their families.

Figure 8. Natural ventilation acceptability among gender. Numbers in boxes refer to number of occupants
Source: Authors

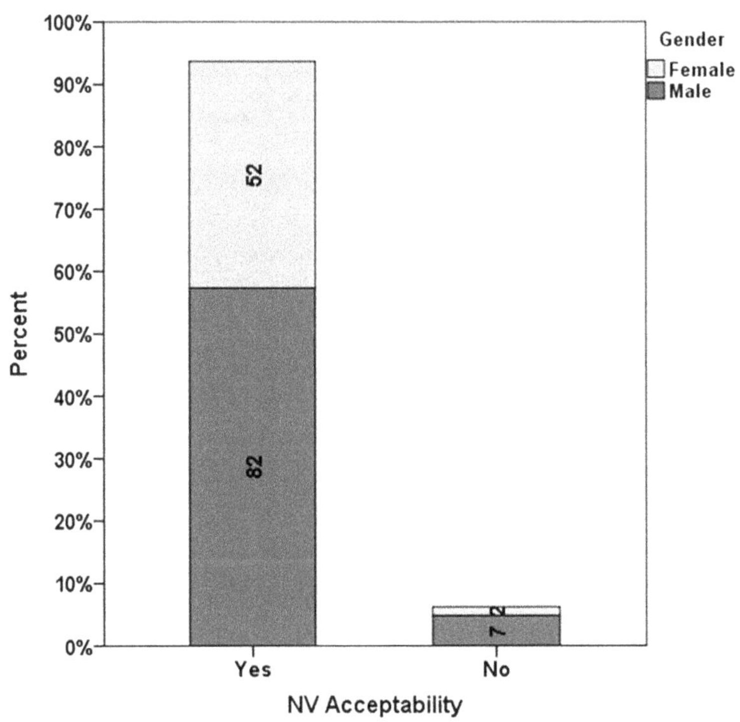

Additionally, Figure 9 displays the monthly pattern of utilising natural ventilation based on four categories, namely "Day only", "Night only", "Day and night", and "No use". The corresponding pattern of utilising air conditioning is displayed in Figure 10. It can be observed that the two patterns are consistent. For instance, the high percentages of "No use" of natural ventilation reported for the period from April to October including both is consistent with the heavy depends on air conditioning units during "Day and night" reported in the same period. Considering natural ventilation, the percentages for these months were 55%, 67%, 70%, 71%, 70%, 64%, and 46%, respectively. The corresponding percentages for air conditioning utilisation were 77%, 92%, 93%, 94%, 93%, 88%, and 75%, consequently. These findings are in agreement with the findings of (Al-Gharibi, 2016; Majid et al., 2014). Further explorations of the utilisation patterns during the period between November and March revealed the dependence of 36%, 37%, 37%, 36%, and 35% of the occupants on natural ventilation during "Day Only". A very similar pattern was reported for utilising natural ventilation during "Day and Night" with percentages of 31%, 37%, 39%, 33%, and 21%.

Figure 9. Natural ventilation use over the year with respect to the usage categories: (a) Day Only, (b) Night Only, (c) Day and Night, and (d) No Use (Labels refer to number of occupants)
Source: Authors

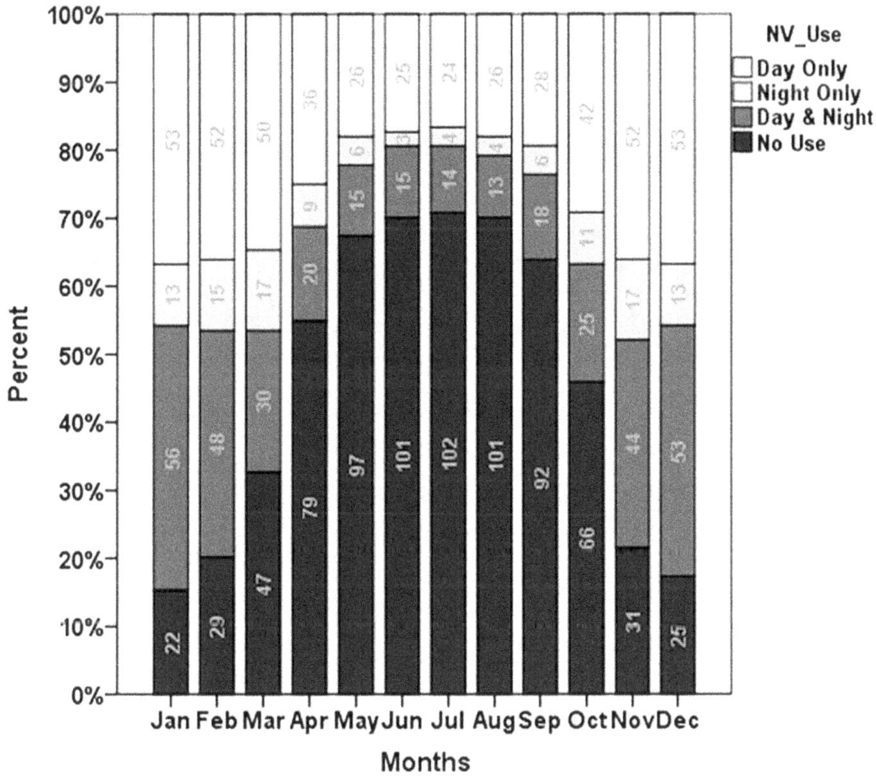

It is worth mentioning that distribution patterns of utilising natural ventilation and air conditioning discussed above follow the seasonal pattern of the Arabian Gulf region. With minor variations, it is possible to distinguish two seasonal periods in the region; one extends from April to October and it represents the hot season. The other period extends between November and March representing the relatively cooler season. Figure 11 and Figure 12 provide a collective reading for the utilisation patterns of natural ventilation and air conditioning in the hot and relatively cooler seasons, respectively. During the hot season, 63.5% roughly of the surveyed occupants revealed "No use" of natural ventilation compared with 20.4% approximately reported utilising natural ventilation during "Day only". As expected, most of the participants, forming around 87%, use air conditioning systems during "Day and night". Considering the relatively cooler season, around 35.9% and 31.9% of the occupants depend on natural ventilation during "Day only" and during "Day and night", respectively. Moreover, around 30.6%, 18.2%, and 16.6% of the occupants revealed their reliance on air conditioning systems during "Day and night", "Night only", and "Day only", respectively. On the other hand, 34.6% approximately of the participants revealed "No use" of air conditioning systems for the same period of the year.

Figure 10. Air conditioning use over the year with respect to the usage categories: (a) Day Only, (b) Night Only, (c) Day and Night, and (d) No Use (Labels refer to number of occupants)
Source: Authors

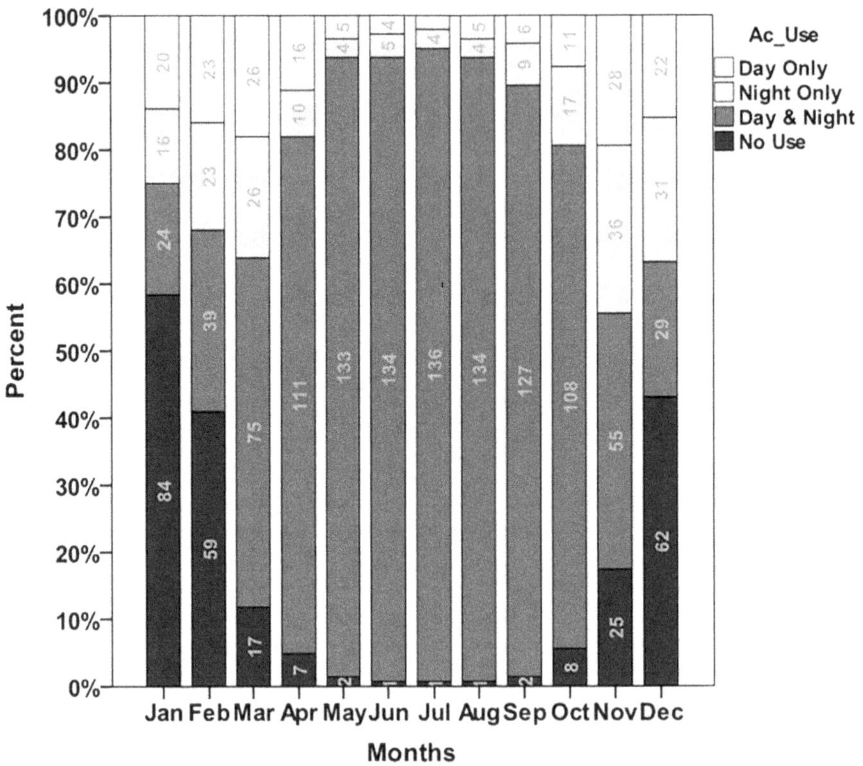

The distribution of the set point temperatures based on gender is presented in Figure 13. The maximum and minimum set point temperatures reported by the occupants were 35 °C and 10 °C, respectively. The median temperatures extended in a narrow range throughout the year, which was between 20 °C and 23 °C for the hot season (April to October) and between 23 °C and 24 °C for the relatively cooler season (November to March). This may indicate narrow thermal tolerance and addiction to cool conditions.

Considering the hot season, the upper limit of most reported temperatures ranged between 25 °C to 27 °C, whereas the lower limit ranged between 11 °C and 18 °C. The mean of the set point temperature was 21.4 °C and the median temperatures were 21°C for June, July and August, 22°C for May and September, and 23°C for April and October.

Figure 11. Periods in the hot season in which occupants depend on: (a) natural ventilation, and (b) air conditioning
Source: Authors

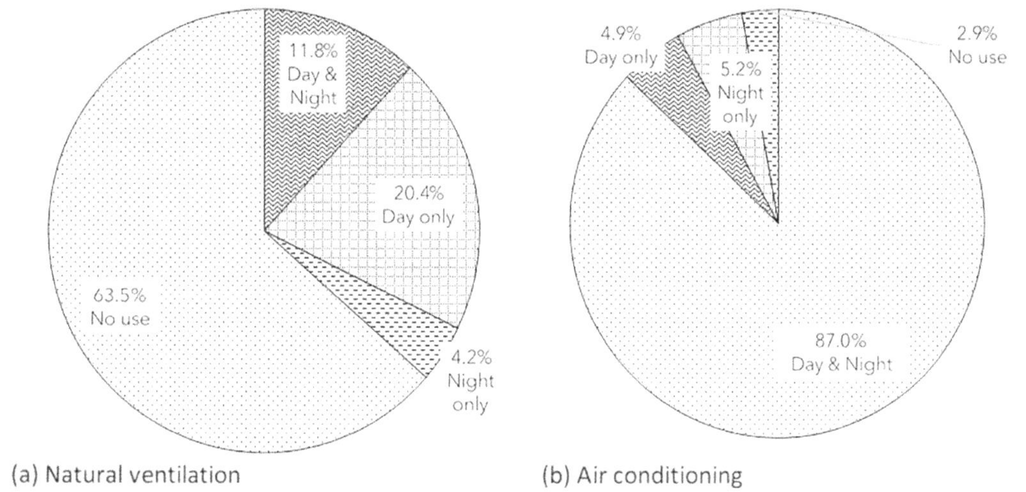

Figure 12. Periods in the relatively cool season in which occupants depend on: (a) natural ventilation, and (b) air conditioning
Source: Authors

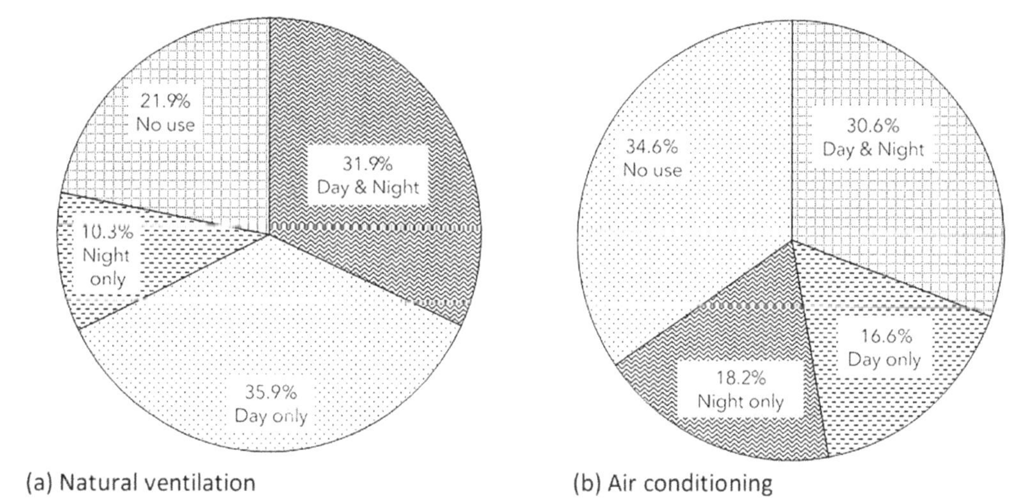

In the relatively cooler season, the corresponding temperatures were from 26 °C to 28 °C for the upper limit and from 18 °C to 20 °C for the lower limit. It is worth mentioning that around 20% of the occupants did not specify the set point temperatures for this season, which may indicate no use of air conditioning systems. Besides, the mean temperature of the set point temperature was 23.4 °C. As expected, the set point temperatures were lower during the hot season compared with those of the relatively cooler season, which is in agreement with (Al-Atrash et al., 2020) and with a familiar trend in the Arabian Gulf region.

It may worth mentioning that both males and females exhibited a similar behaviour regarding specifying the set point temperatures. In specific, the behaviour of choosing the set point temperatures was similar for both males and females during April, May, September, and October causing no change of the median temperatures within each month. On the other hand, the median temperatures for the females in June, July, and August were lower than that of the males. This consists with the neutral temperatures computed, applying Griffiths' method, of 24.3 ± 1.09 °C and 26.1 ± 0.92 °C for Omani females and Saudi males high school students, respectively (Al-Khatri et al., 2020).

Figure 13. The distribution of set point temperatures based on gender (The ends of the whiskers indicate minimum and maximum values. The bottom and the top of the boxes represent the 25th and 75th percentiles, and the horizontal lines within the boxes denote the median values)
Source: Authors

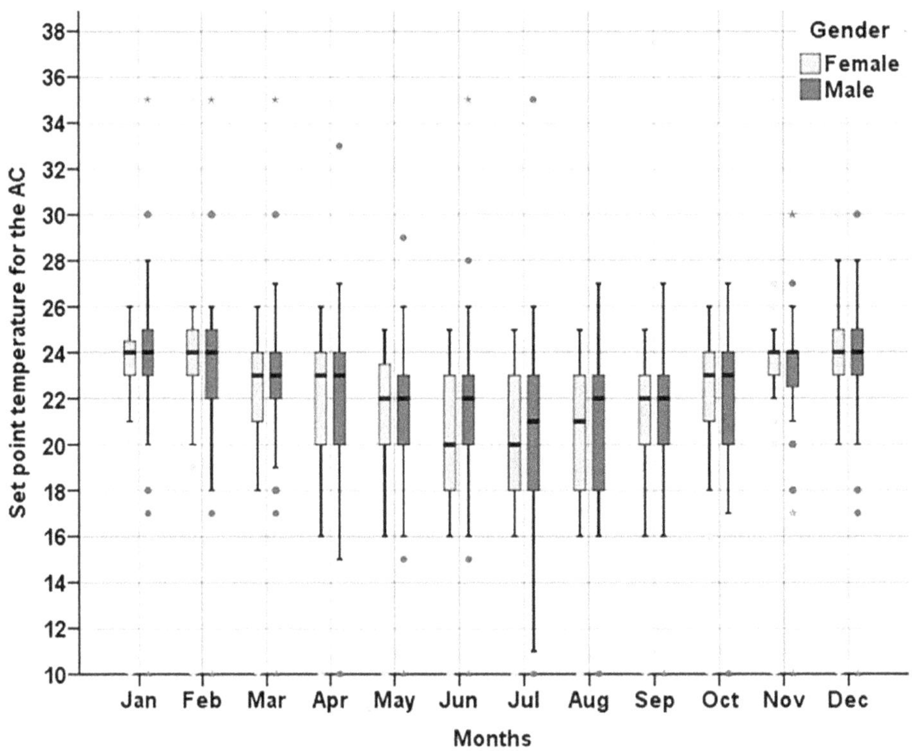

FUTURE RESEARCH DIRECTIONS

The reported study in this chapter sheds light into the dynamic interaction between the occupants and their buildings through exploring the potential of utilising natural ventilation to moderate indoor thermal comfort conditions in the Arabian Gulf region instead of the heavy dependence on mechanical systems. To further understand this interaction, future research may explore the degree of personal control in the hot arid climate via investigating the limitation of desert climate on personal control, how personal control affect the type of ventilation in residential buildings, and which variables affect the use of air conditioning and the set point temperatures. Additionally, it is recommended to view the findings of the current questionnaire in light of the hourly climatic data on one hand and considering the boundaries of thermal comfort zone. Another direction of future research may explore certain architectural systems that aim to facilitate the use of natural ventilation as a main design strategy to improve indoor thermal conditions. The environmental and economic impact of such systems should be explored as well.

CONCLUSION

Achieving a good indoor thermal comfort with minimising the cooling energy use is a big challenge in the Arabian Gulf region owing to the effect of the environmental and cultural backgrounds. The extreme outdoor environment conditions besides the conventional ventilation practices played an important role to set the current acceptable modern opportunities in order to achieve indoor thermal comfort. However, local people nowadays are more open for applying adaptive design and practice strategies due to increased awareness on sustainability as well as the deterioration of the economic environment. Natural ventilation is a key to reduce the energy consumption and improving indoor thermal comfort in certain zones and periods in the Arabian Gulf region. This chapter investigates the occupants' practices and desires for the utilisation of both natural ventilation and air conditioning systems. It addresses the periods in which natural ventilation is utilised and the occupants' perception in this regard. Additionally, it presents the used set point temperatures for the air conditioning systems. The different parts of the questionnaire aim to collect reliable and effective database to better understand the potential of integrating natural ventilation in the Arabian Gulf region.

The surveyed sample of 144 occupants exhibited a high degree acceptability of almost 93.7% to rely on natural ventilation in order to moderate the indoor thermal conditions as long as the outdoor temperatures are acceptable. It was found that males' acceptability was higher by almost one third compared with females' acceptability perhaps because of the cultural background effect. Consistent with the seasonal pattern of the Arabian Gulf region, this acceptability was highly witnessed from November to March, including both, with around 35.9% and 31.9% of the occupants depending on natural ventilation during "Day only" and "Day and night" respectively. In the remaining months, around 63.5% reported "No use" of natural ventilation and almost 87.0% reported depending on air conditioning during "Day and night". Exploring the preferred setpoint temperatures indicated narrow thermal tolerance and addiction to cool conditions as the medians were 20-23 °C in the hot season and 23-24 °C in the relatively cooler season. Overall, the setpoint temperatures ranged between 10 °C and 35 °C and the female occupants preferred cooler temperatures in June, July, and August.

ACKNOWLEDGMENT

This research received no specific grant from any funding agency in the public, commercial, or not-for-profit sectors.

REFERENCES

Abdul-Wahab, S. A., Charabi, Y., Al-Maamari, R., Al-Rawas, G. A., Gastli, A., & Chan, K. (2015). CO2 greenhouse emissions in Oman over the last forty-two years [Review]. *Renewable & Sustainable Energy Reviews*, *52*, 1702–1712. doi:10.1016/j.rser.2015.07.193

Abdul-Wahab, S. A., Charabi, Y., Al-Rawas, G. A., Al-Maamari, R., Gastli, A., & Chan, K. (2015). Greenhouse gas (GHG) emissions in the Sultanate of Oman. Greenhouse Gases. *Science and Technology*, *5*(3), 339–346. doi:10.1002/ghg.1501

Aboulnaga, M. M. (2006). Towards green buildings: Glass as a building element—the use and misuse in the gulf region. *Renewable Energy*, *31*(5), 631–653. doi:10.1016/j.renene.2005.08.017

Ahn, K. U., & Park, C. S. (2016). Correlation between occupants and energy consumption. *Energy and Building*, *116*, 420–433. doi:10.1016/j.enbuild.2016.01.010

Al-Atrash, F., Hellwig, R. T., & Wagner, A. (2018). Personal control over indoor climate in office buildings in a Mediterranean climate - Amman, Jordan. Academic Press.

Al-Atrash, F., Hellwig, R. T., & Wagner, A. (2020). The degree of adaptive thermal comfort in office workers in a hot-summer Mediterranean climate. *Energy and Building*, *223*, 110147. doi:10.1016/j.enbuild.2020.110147

Al-Azri, N. A., Zurigat, Y. H., & Al-Rawahi, N. Z. (2013). Development of bioclimatic chart for passive building design. *International Journal of Sustainable Energy*, *32*(6), 713–723. doi:10.1080/14786451.2013.813026

Al-Badi, A. H., Malik, A., & Gastli, A. (2011). Sustainable energy usage in Oman—Opportunities and barriers. *Renewable & Sustainable Energy Reviews*, *15*(8), 3780–3788. doi:10.1016/j.rser.2011.06.007

Al-Gharibi, H. (2016). Urbanisation and changing lifestyle patterns in Muscat. In S. Nebel & A. von Richthofen (Eds.), *Urban Oman: Trends and perspective of urbanisation in Muscat capital area* (pp. 73–79). LIT Verlag.

Al-Hinai, H., Batty, W. J., & Probert, S. D. (1993). Vernacular architecture of Oman: Features that enhance thermal comfort achieved within buildings. *Applied Energy*, *44*(3), 233–258. doi:10.1016/0306-2619(93)90019-L

Al-Khatri, H., Alwetaishi, M., & Gadi, M. B. (2020). Exploring thermal comfort experience and adaptive opportunities of female and male high school students. *Journal of Building Engineering*, *31*, 101365. doi:10.1016/j.jobe.2020.101365

Al-Khatri, H., & Gadi, M. B. (2019). Investigating the behaviour of ASHRAE, Bedford, and Nicol thermal scales when translated into the Arabic language. *Building and Environment*, *151*, 348–355. doi:10.1016/j.buildenv.2019.01.051

Al-Shaqsi, S. (2010). Care or Cry: Three years from Cyclone Gonu. What have we learnt? *Oman Medical Journal*, *25*(3), 162–167. doi:10.5001/omj.2010.50 PMID:22043331

Almazroui, M. (2012). Temperature Variability over Saudi Arabia and its Association with Global Climate Indices. Journal of King Abdulaziz University-Meteorology. *Environment and Arid Land Agriculture Sciences*, *23*(1), 85–108. doi:10.4197/Met.23-1.6

Almazroui, M. (2013). Simulation of present and future climate of Saudi Arabia using a regional climate model (PRECIS). *International Journal of Climatology*, *33*(9), 2247–2259. doi:10.1002/joc.3721

Almazroui, M., Islam, M. N., Saeed, S., Saeed, F., & Ismail, M. (2020). Future Changes in Climate over the Arabian Peninsula based on CMIP6 Multimodel Simulations. *Earth Systems and Environment*, *4*(4), 611–630. doi:10.100741748-020-00183-5

Ascione, F., Bianco, N., De Masi, R. F., Mastellone, M., Mauro, G. M., & Vanoli, G. P. (2020). The role of the occupant behavior in affecting the feasibility of energy refurbishment of residential buildings: Typical effective retrofits compromised by typical wrong habits. *Energy and Building*, *223*, 110217. doi:10.1016/j.enbuild.2020.110217

ASHRAE. (2017). *2017 ASHRAE Handbook Fundamentals (SI)*. ASHRAE (American Society of Heating, Refrigerating and Air-Conditioning Engineers).

Brager, G. S., & de Dear, R. J. (1998). Thermal adaptation in the built environment: A literature review. *Energy and Building*, *27*(1), 83–96. doi:10.1016/S0378-7788(97)00053-4

Buso, T., Fabi, V., Andersen, R. K., & Corgnati, S. P. (2015). Occupant behaviour and robustness of building design. *Building and Environment*, *94*, 694–703. doi:10.1016/j.buildenv.2015.11.003

Gianfrate, V., Piccardo, C., Longo, D., & Giachetta, A. (2017). Rethinking social housing: Behavioural patterns and technological innovations. *Sustainable Cities and Society*, *33*(May), 102–112. doi:10.1016/j.scs.2017.05.015

Giusti, L., & Almoosawi, M. (2017). Impact of building characteristics and occupants' behaviour on the electricity consumption of households in Abu Dhabi (UAE). *Energy and Building*, *151*, 534–547. doi:10.1016/j.enbuild.2017.07.019

Global Footprint Network. (2018). *Ecological footprint per person*. http://data.footprintnetwork.org/#/

Hansen, A. R., Gram-Hanssen, K., & Knudsen, H. N. (2018). How building design and technologies influence heat-related habits. *Building Research and Information*, *46*(1), 83–98. doi:10.1080/09613218.2017.1335477

Humphreys, M., Nicol, F., & Roaf, S. (2016). *Adaptive Thermal Comfort: Foundations and Analysis* (1st ed.). Routledge.

Ioannou, A., Itard, L., & Agarwal, T. (2018). In-situ real time measurements of thermal comfort and comparison with the adaptive comfort theory in Dutch residential dwellings. *Energy and Building*, *170*, 229–241. doi:10.1016/j.enbuild.2018.04.006

Katodrytis, G. (2015). Performative Urbanism: An Emerging Model of the Gulf. *Architectural Design: UAE and The Gulf Architecture and Urbanism Now,* 121–127.

Konya, A., & Vandenberg, M. (2011). *Design primer for hot climates*. Archimedia Press Limited.

Luo, M., de Dear, R., Ji, W., Bin, C., Lin, B., Ouyang, Q., & Zhu, Y. (2016). The dynamics of thermal comfort expectations: The problem, challenge and impication. *Building and Environment*, *95*, 322–329. doi:10.1016/j.buildenv.2015.07.015

Majid, N. H. A., Takagi, N., Hokoi, S., Ekasiwi, S. N. N., & Uno, T. (2014). Field survey of air conditioner temperature settings in a hot, dry climate (Oman). *HVAC & R Research*, *20*(7), 751–759. doi:10.1080/10789669.2014.953845

Mishra, A. K., & Ramgopal, M. (2013). Field studies on human thermal comfort— An overview. *Building and Environment*, *64*, 94–106. doi:10.1016/j.buildenv.2013.02.015

Naspi, F., Arnesano, M., Zampetti, L., Stazi, F., Revel, G. M., & D'Orazio, M. (2018). Experimental study on occupants' interaction with windows and lights in Mediterranean offices during the non-heating season. *Building and Environment, 127*(September), 221–238. doi:10.1016/j.buildenv.2017.11.009

Nicol, F., Humphreys, M., & Roaf, S. (2012). *Adaptive thermal comfort: Principles and practice* (1st ed.). Routledge. doi:10.4324/9780203123010

Nikolopoulou, M., Baker, N., & Steemers, K. (2001). Thermal comfort in outdoor urban spaces: Understanding the human parameter. *Solar Energy*, *70*(3), 227–235. doi:10.1016/S0038-092X(00)00093-1

Nikolopoulou, M., & Steemers, K. (2003). Thermal comfort and psychological adaptation as a guide for designing urban spaces. *Energy and Building*, *35*(1), 95–101. doi:10.1016/S0378-7788(02)00084-1

Ning, H., Wang, Z., & Ji, Y. (2016). Thermal history and adaptation: Does a long-term indoor thermal exposure impact human thermal adaptability? *Applied Energy*, *183*, 22–30. doi:10.1016/j.apenergy.2016.08.157

Pal, J. S., & Eltahir, E. A. B. (2016). Future temperature in southwest Asia projected to exceed a threshold for human adaptability. *Nature Climate Change*, *6*(2), 197–200. doi:10.1038/nclimate2833

Peeters, L., de Dear, R., Hensen, J., & D'haeseleer, W. (2009). Thermal comfort in residential buildings: Comfort values and scales for building energy simulation. *Applied Energy*, *86*(5), 772–780. doi:10.1016/j.apenergy.2008.07.011

Qader, M. R. (2009). Electricity Consumption and GHG Emissions in GCC Countries. *Energies*, *2*(4), 1201–1213. doi:10.3390/en20401201

Ragette, F. (2012). *Traditional Domestic Architecture of the Arab Region* (1st ed.). American Univeristy of Sharjah.

Rupp, R. F., Andersen, R. K., Toftum, J., & Ghisi, E. (2021). Occupant behaviour in mixed-mode office buildings in a subtropical climate: Beyond typical models of adaptive actions. *Building and Environment, 190*(December). doi:10.1016/j.buildenv.2020.107541

Salvia, G., Morello, E., Rotondo, F., Sangalli, A., Causone, F., Erba, S., & Pagliano, L. (2020). Performance gap and occupant behavior in building retrofit: Focus on dynamics of change and continuity in the practice of indoor heating. *Sustainability (Switzerland)*, *12*(14), 1–25. doi:10.3390u12145820

Schakib-Ekbatan, K., Çakici, F. Z., Schweiker, M., & Wagner, A. (2015). Does the occupant behavior match the energy concept of the building? - Analysis of a German naturally ventilated office building. *Building and Environment*, *84*, 142–150. doi:10.1016/j.buildenv.2014.10.018

Schweiker, M., & Wagner, A. (2016). The effect of occupancy on perceived control, neutral temperature, and behavioral patterns. *Energy and Building*, *117*, 246–259. doi:10.1016/j.enbuild.2015.10.051

Singh, M. K., Mahapatra, S., & Atreya, S. K. (2011). Adaptive thermal comfort model for different climatic zones of North-East India. *Applied Energy*, *88*(7), 2420–2428. doi:10.1016/j.apenergy.2011.01.019

Weatherbase. (2018). *Muscat, Oman: Monthly - Weather averages summary*. Author.

Xu, X., Xiao, B., & Li, C. Z. (2020). Critical factors of electricity consumption in residential buildings: An analysis from the point of occupant characteristics view. *Journal of Cleaner Production*, *256*, 120423. doi:10.1016/j.jclepro.2020.120423

Zurigat, Y. H., Al-Hinai, H., Jubran, B. A., & Al-Masoudi, Y. S. (2003). Energy efficient building strategies for school buildings in Oman. *International Journal of Energy Research*, *27*(3), 241–253. doi:10.1002/er.871

ADDITIONAL READING

Brien, W. O., & Gunay, H. B. (2014). The contextual factors contributing to occupants' adaptive comfort behaviors in of fi ces e A review and proposed modeling framework. *Building and Environment*, *77*, 77–87. doi:10.1016/j.buildenv.2014.03.024

Buso, T., Fabi, V., Andersen, R. K., & Corgnati, S. P. (2015). Occupant behaviour and robustness of building design. *Building and Environment*, *94*, 694–703. doi:10.1016/j.buildenv.2015.11.003

Domínguez-amarillo, S., Fernández-agüera, J., Peacock, A., & Acosta, I. (2020). Energy related practices in Mediterranean low- income housing. *Building Research and Information*, *48*(1), 34–52. doi:10.1080/09613218.2019.1661764

Fabi, V., Vinther, R., Corgnati, S., & Olesen, B. W. (2012). Occupants' window opening behaviour: A literature review of factors in fl uencing occupant behaviour and models. *Building and Environment*, *58*, 188–198. doi:10.1016/j.buildenv.2012.07.009

Laura, A., Lucia, V., Piselli, C., & Fabiani, C. (2020). How peers' personal attitudes affect indoor microclimate and energy need in an institutional building: Results from a continuous monitoring campaign in summer and winter conditions. *Energy & Buildings, 126*(2016), 485–497. doi:10.1016/j.enbuild.2016.05.053

Luo, M., Cao, B., Ji, W., Ouyang, Q., Lin, B., & Zhu, Y. (2016). The underlying linkage between personal control and thermal comfort: Psychological or physical effects? *Energy and Building*, *111*, 56–63. doi:10.1016/j.enbuild.2015.11.004

Salvia, G., Morello, E., Rotondo, F., Sangalli, A., Causone, F., Erba, S., & Pagliano, L. (2020). Performance gap and occupant behavior in building retrofit: Focus on dynamics of change and continuity in the practice of indoor heating. *Sustainability (Switzerland), 12*(14), 1–25. doi:10.3390u12145820

Stazi, F., Naspi, F., & Orazio, M. D. (2017). A literature review on driving factors and contextual events in fl uencing occupants' behaviours in buildings. *Building and Environment, 118*, 40–66. doi:10.1016/j.buildenv.2017.03.021

Yao, M., & Zhao, B. (2017). Window opening behavior of occupants in residential buildings in Beijing. *Building and Environment, 124*, 441–449. doi:10.1016/j.buildenv.2017.08.035

Zhao, B. (2017). ScienceDirect ScienceDirect Factors affecting occupants' interactions with windows in residential buildings in Beijing, China. *Procedia Engineering, 205*, 3428–3434. doi:10.1016/j.proeng.2017.09.857

KEY TERMS AND DEFINITIONS

Arabian Gulf: Six Arab countries namely Bahrain, Kuwait, Oman, Qatar, Saudi Arabia and the United Arab Emirates.

Behaviour: The way an individual reacts to a specific change or stimulus.

Behavioural Adaptation: Conscious or unconscious modifications a person might make to change and control body's thermal balance by adjusting to the surroundings or modifying the surroundings.

Cooling Energy: The energy required to cool the surrounding indoor environment in the building.

Cultural Attitude: A set of familiar values and believes along specific group or society that influence the individuals' understanding towards their surroundings.

Extreme Climate: A climate which differs significantly from average and needs special consideration within the framework of construction design of planning and design.

Questionnaire: A set of questions with a choice of answers, structured and designed for the purposes of a survey or study.

Residential: Comfortable accommodation to live in.

Setpoint Temperature: A set point is the desired temperature value at which a controller endeavour to keep the process variable.

Thermal Comfort: A zone of thermal parameters, which expresses comfort within the surrounding environment. It depends on the individual's experience and expectations, as well as the thermal climate.

Chapter 8
Smart Power Microgrid Impact on Sustainable Building

Hussain Attia
American University of Ras Al Khaimah, UAE

ABSTRACT

This chapter explains the positive impacts of the smart microgrid with respect to the sustainable building performance. The role of the renewable energy sources within the microgrid is also demonstrated. The solar photovoltaic energy source is explained in terms of the working principle and positive effects on reducing CO2 emission through reducing the needs of the traditional power stations' electricity. This chapter explains the advantages of the smart microgrid power system and how reflects on the performance of a sustainable building. The differences between the components of a DC smart grid system and of an AC smart grid system are shown. Different algorithms for maximum power point tracking (MPPT) to improve the PV systems have reviewed in this chapter. A step-by-step overall design of any desired single-phase or three-phase alternating power of any capacity for a PV matrix-based microgrid system, in addition to the role and the importance of inserting DC-DC converter in the photovoltaic systems, is discussed.

INTRODUCTION

The economical level of any country is affected by the level of energy technology and how this technology depends on the best choices of the fuels. For many decades, the fossil fuel plays a main role in this regard to meet the required electrical power to the consumers as discussed by Richard and Lucy (2014), Hussain et al. (2014), and Bierwirth (2020). Using any type of fossil fuels will negatively affect the environmental pollution due to the CO_2 emissions and participate in global warming. At the same time, as mentioned by Min-Joong and Huei (2007), the fossil fuels dwindling leads to search for effective and efficient alternatives considering the human life quality, and reducing the level of air pollution as well. The authors Chapman (2003), Blaabjerg et al. (2006), and Grainger and Stevenson (2008) explained that any traditional power system represents a centralized system which includes the power generation station, and grid of transmission and distribution lines with many step-up and step-down transformers to supply the low voltage alternating electricity to the end users.

DOI: 10.4018/978-1-7998-7023-4.ch008

1. The Infrastructure of a Traditional Electrical Power Grid

The existing power system is representing by a central power generating and delivering to the end user loads. The existing electrical power system includes a traditional electrical power generation station, a set of step up/step down power transformers, and a network of transmission and distribution lines as mentioned by Hussain (2020), and Ali (2020). The electrical loads are dividing into two castigates, residential loads, and industrial loads. Different levels of load current may be drawn from the electrical utility by the residential appliances, in a range of 0.1 A to 25 A based on the electrical load type or function, for example a one set of LED lamp of 20 W/220 V can work with rated power 20 W, while an air conditioner can work with rated current 15 A. Whereas most of industrial field electrical loads can be represented by single or three phase induction motors of a different horse power HP range which can be started from 0.5 HP (1 HP = 746 W) horse power to many tens of HP as shown by Hussain (2017), Hussain (2020), Hussain et al. (2018), Hussain and Ali (2016).

Michael et al. (2017) showed a traditional electrical power grid in Figure 1, which includes electrical generators rotating by a mechanical power of a high pressure and temperature steam produced using fossil fuels. Step up power transformers are also included in the traditional power generation system to increase the level of AC voltage for transmitting with low AC current to reduce the transmission line losses. Whereas on the side of end user, there are step down transformers to have low voltage which should be suitable to be delivered to the connected AC loads as demonstrated by Shahzad et al. (2015), Anyaka, and Olawoore (2014), and Lei et al. (2015).

2. The Operations in a Traditional Electrical Power Grid

A traditional power grid delivers electric power generated by station's generators to the end users through a network of transmission and distribution lines. Ali (2020) showed that the first step is a power plant or generators station which generates an alternating electricity of a voltage range of 11 – 20 kV. Hosham et al. (2018), Darwish et al. (2010), and Karoly (2002) discussed the principle of an electrical generator depends on a magnet rotating inside a stator coil, this leads to flow of a current in the stator coil or in other meaning generates electricity. Different types of turbines can be used to rotate the rotor part of the generator by the effect of external mechanical energy such as steam energy, wind energy, or hydro energy. Steam turbine can be work using one of fossil fuels or nuclear plant as well.

Before transmit the generated electrical power through the transmission lines, the produced voltage is transferring to a higher voltage level in a range of 138-765 kV to reduce the losses due to the transmission lines resistance as shown by Ali (2020). This function is done by a step up transformer of a high power rating, the transformer is inserted between the generator and the high tension transmission lines.

Shahzad et al. (2015), Anyaka, and Olawoore (2014), and Lei et al. (2015) demonstrated that on the consumer side, the electrical appliances are working in a low level of rating voltage, so before delivering the electrical power to the end-user appliances, step down transformer inserted to transfer the high voltage of the electrical network lines to a low voltage level.

Figure 1. A traditional electrical power grid (Michael et al. 2017)

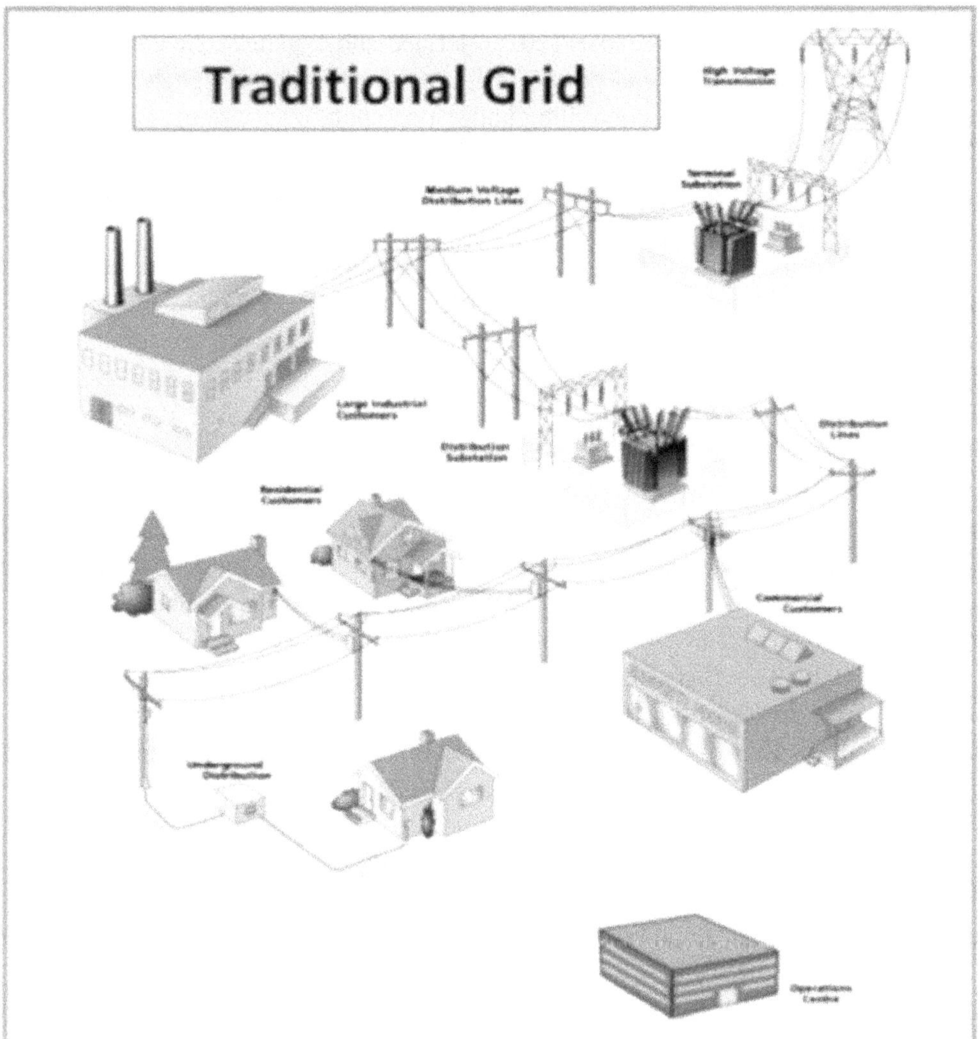

3. The Demerits of a Traditional Electrical Power Grid

Using a centralized one directional power generation of a traditional power electrical grid has many demerits; the electrical power price is charged depending on the energy demand, the instantaneous cost of fossil fuels, operation, and maintenance costs. In traditional grid, there is neither controlling nor manipulating for unbalance load cases, no control to the power losses due to the transmission lines resistance during transmitting the electrical power from the source to the target. In addition to a high cost of generation station and grid expansion due to the difficulty in the expansion of the infrastructure of the traditional power grid as mentioned by Shahzad et al. (2015), Anyaka, and Olawoore (2014), and Lei et al. (2015).

THE SMART POWER MICROGRID: CONCEPT, AND MOTIVATION

Jignesh (2014), Xi et al. (2012), and James (2012) defined and explained the components of a smart grid which is a two-direction electrical power grid, it delivers feedback signals simultaneously with the power suppling, monitoring and controlling under a computerized system supervision. Michael et al. (2017) also showed in Figure 2 a two direction modern smart power grid, smart grid is able to do self-healing, sensing, monitoring, and controlling the system during the power generating, transmitting and delivering to the consumers. Smart grid provides the possibility to benefit from renewable and sustainable energy sources, minimizing the operation and the maintenance costs, and reducing the power transmission losses through the principle of decentralization. The individual renewable energy sources can support the grid by alternating electricity by inserting power electronic devices and equipment in a conversion power system.

Jennifer, and Christopher (2017), Abdullah et al. (2020), and Rahmani et al. (2018) discussed that smart grid effectively contributes in reducing the environmental pollution which yields due to the fossil fuel based technologies. Global warming represents a one of the fossil fuel consumption consequences.

The renewable and sustainable energy sources can be adopted in constricting smart microgrid systems to meet the electrical power demands instead of the traditional power grid, the fossil fuels can be replaced by the sustainable energy sources as discussed by Min-Joong, and Huei (2007).

Solar energy is strongly contributing in delivering electrical power through photovoltaic PV cells. Hussain and Khaled (2020), Wilfried van Sark. (2019), and Hussain and Fernando (2019) demonstrated that a solar PV cell converts the solar incident light into direct current electricity. Direct current electrical power quantity produced by PV panel is proportional to the number of cells connected in the PV panel, and proportional to the number of PV panel in a one PV system.

The electrical power cost regardless drawn from grid or a renewable energy source over a certain period can be expressed as.

$$\text{The load delivered energy} = \text{load power} \times \text{time} = P \times t (W.h) \quad (1)$$

The consumed energy in the connected load is proportional with the consumed electrical power P (W) and the period of the load connected time t, the load consumed power is

$$\text{The consumed power in Watt, } P = V \times I \quad (2)$$

where V is the load voltage (V), and I is the load current (A).

The total cost of the delivered electrical power can be determined by calculating the total number of kW·h multiplying by the cost of 1 kW.h as shown below:

$$\text{Electrical energy cost} = \text{no of } kW{\bullet}h \times \text{cost of } 1kW{\bullet}h \quad (3)$$

Figure 2. A two-way power flow electrical power smart grid (Michael et al. 2017)

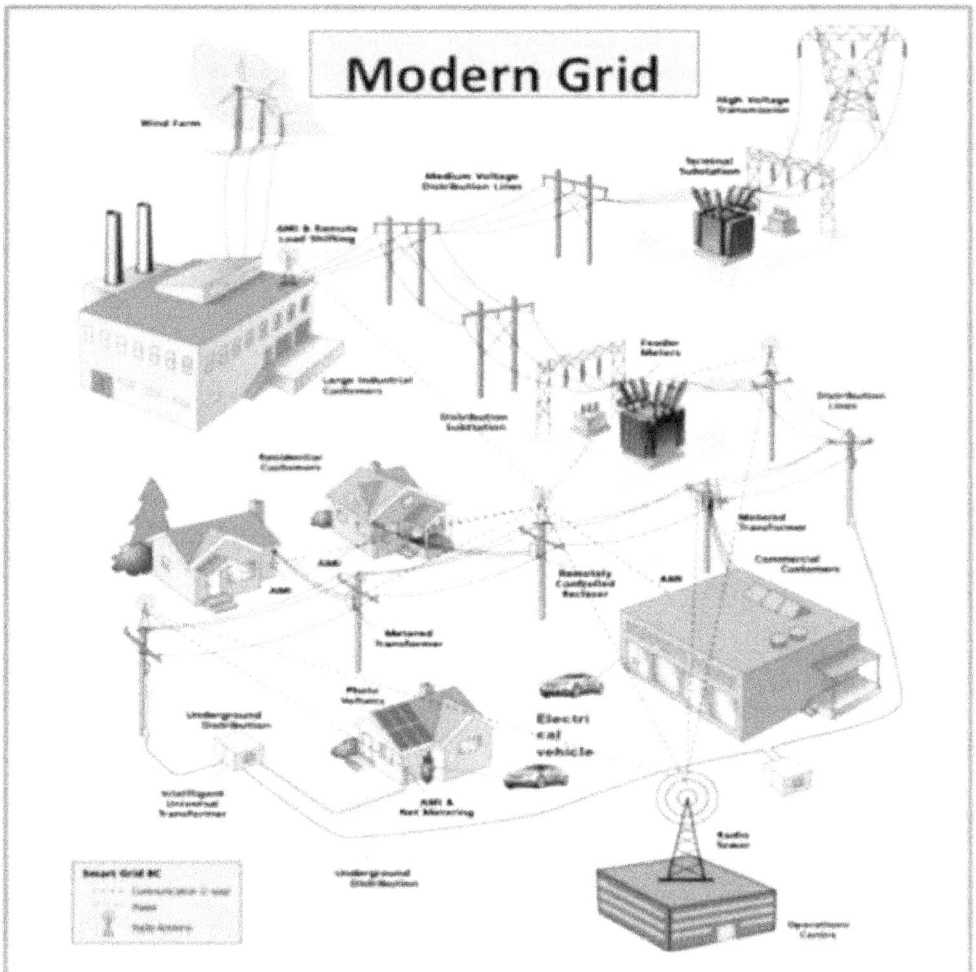

THE SMART POWER MICROGRID: STRUCTURE AND TYPES

Farhad et al. (2016), Galus et al. (2010), Guille and Gross (2009), and Arellano et al. (2013) explained that the fundamental structure of a microgrid usually includes two way feeders for power and information flowing. In addition to different electrical power sources and plants like renewable energy sources (solar photovoltaic arrays, and wind farms), and fossil fuels power plant, nuclear power plant, and storage units, and different of electrical loads such as Electric Vehicles (EV), and home appliances.

Figure 3 shows a schematic of microgrid system, in which, many of power electronic voltage source converters VSCs have been inserted for the power conversion purposes.

Depending on the operating frequency, there are alternating current AC microgrid working with 50 Hz or 60 Hz frequency loads, direct current DC microgrid working with 0 Hz frequency loads, and hybrid microgrid which includes AC and DC systems as shown by Guille and Gross (2009). Figure 4 demonstrates the three types of microgrids, in which, AC loads are connected directly to AC bus of the utility grid. DC loads are connected to DC bus, whereas, DC microgrid is connected to the utility grid through AC/DC power flow coordinator using a suitable rated power bidirectional AC/DC converter. Different types of DC/DC converter, such as, boost, buck, boost buck converters have been used in DC microgrid between the DC bus and the DC power items. Using AC microgrid and DC microgrid in a one system constructs a hybrid microgrid.

Figure 3. Schematic of microgrid system (Arellano et al. 2013)

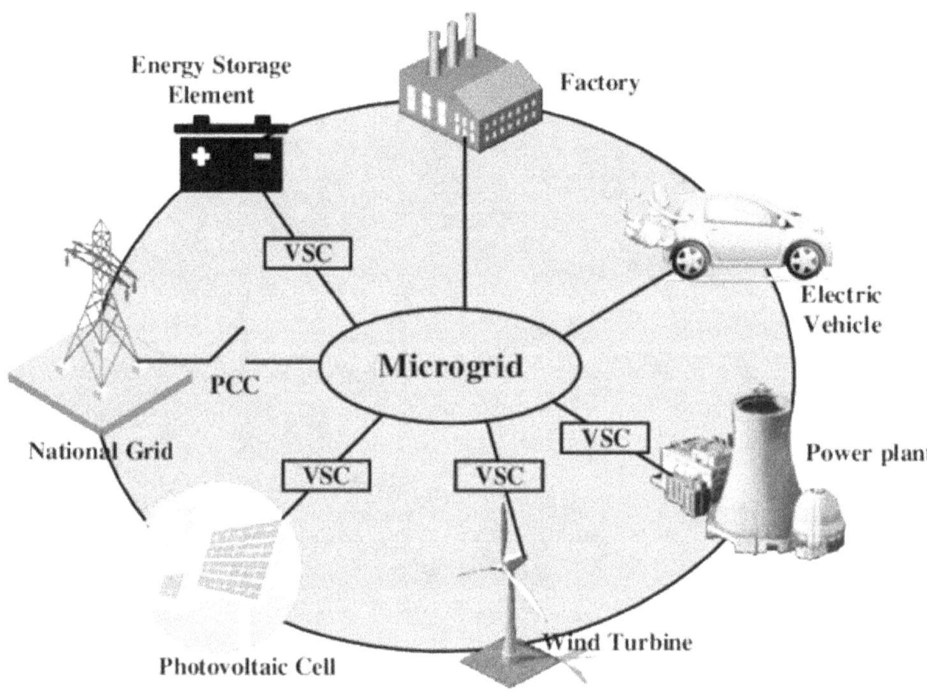

Figure 4. Schematic diagram of hybrid AC/DC microgrid (Guille and Gross 2009)

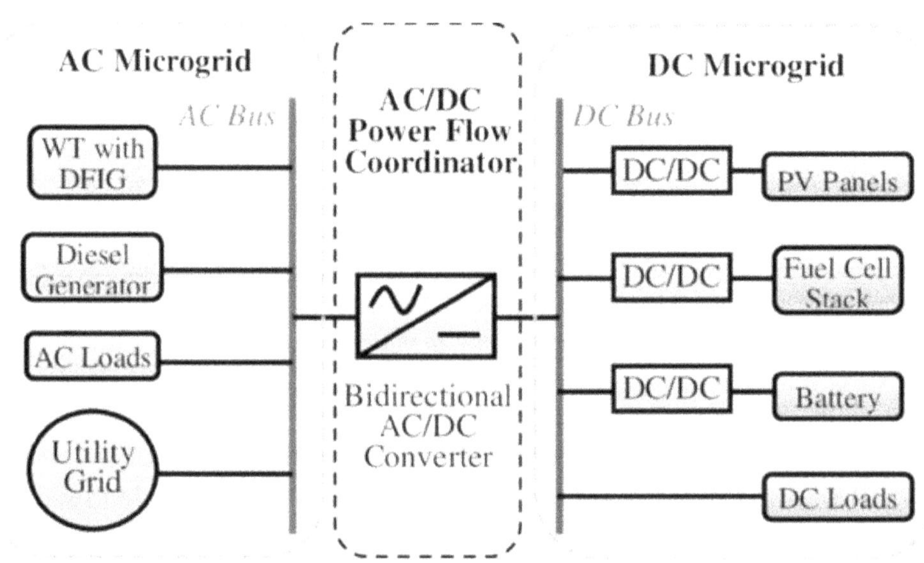

A DC/DC converter converts the direct current (DC) power from a certain voltage level to a higher or lower level based on the system requirements to deliver power to direct current loads. A DC/AC inverter inverts the direct current (DC) power generated by a DC type of a renewable energy source such as a photovoltaic PV panels array to a sinusoidal alternating voltage to deliver it to the alternating current loads as explained by Hussain and Khaled (2020), Wilfried van Sark. (2019), and Hussain and Fernando (2019).

SMART MICROGRID INTEGRATION WITH RENEWABLE SOLAR ENERGY

The sustainable and renewable energy sources such as solar photovoltaic energy source represent desirable alternatives instead of the fossil fuels energy sources. Farhad et al. (2016), Galus et al. (2010), Guille and Gross (2009), and Arellano et al. (2013) discussed that the traditional fossil fuel based vehicles, trains, cranes, tractors, and others all are need to be based on the green energy sources to have better future life quality, to reduce the environmental pollution, and to positively impact the global warming.

The function of a photovoltaic PV cell have explained by Laamami et al. (2017), Hussain (2019), Nguyen and Lehman (2006), and Kawamura et al. (2003). A PV cell generates a fluctuated DC power which is positively proportional with the intensity of falling sunlight on the cell surface. The instantaneous level of the output current is affected nonlinearly by light intensity and ambient temperature. Figure 5 shows the electrical equivalent circuit of a photovoltaic cell. The output levels of the cell current and voltage are demonstrated in (4)–(8), in which, I_D represents a diode current, I_{SC} represents a short circuit current, I_{PV} represents an output PV current, I_P represents a parallel resistor current, R_S represents a serial resistor, R_P represents a parallel resistor, V_{PV} represents a PV output voltage, L_n and L_p represent electron and hole diffusion length, G_r represents a generation rate, and A represents a solar cell area.

$$I_D - I_0 \left[e^{\frac{V_D}{V_T}} - 1 \right] \tag{4}$$

$$I_{SC} = qAG_r[I_n + I_p] \tag{5}$$

$$I_P = \frac{V_D}{R_P} \tag{6}$$

$$I_{PV} = I_{SC} - I_D - I_P \tag{7}$$

$$V_{PV} = V_D - R_S I_{PV} \tag{8}$$

For the PV panel module SLP060-12, which used by Hussain and Khaled (2020), of specifications shown in Table 1 below, Figure 6, Figure 7, Figure 8, and Figure 9 demonstrate the output current and power curves with respect to output voltage variation for the PV module at different light intensity (200 W/m2, 400 W/m2, 600 W/m2, 800 W/m2, and 1000 W/m2,) with constant ambient temperature 25 °C and at different ambient temperature (-5 °C, 10 °C, 25 °C, 40 °C, and 55 °C) and constant light intensity 1000 W/m2.

To have a desired output voltage or current from a PV panel, a suitable serial and parallel connection of many of PV cell should be considered, then to have the desired PV matrix based of the required system rating power, a number of PV panels N_S are connected serially to increase the output PV matrix voltage whereas a number of PV panels N_P are connected in parallel to increase the output PV matrix current as shown in (9), and (10):

$$V_{PV\_Matrix} = N_S \times V_{PV} \tag{9}$$

$$I_{PV\_Matrix} = N_P \times I_{PV} \tag{10}$$

Figure 5. Equivalent electrical circuit of photovoltaic cell (Hussain 2019)

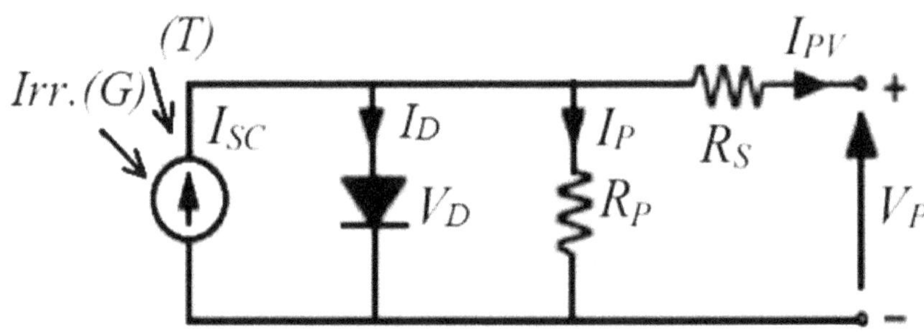

Table 1. Specifications of PV panel module SLP060-12 (Hussain and Khaled 2020)

Parameter	Value
Maximum Power(W)	60 W
Maximum Power Voltage(V_{mpp})	17.8 V
Maximum Operating Current(I_{mpp})	3.37 A
Open Circuit Voltage(V_{oc})	22 V
Short Circuit Current(I_{sc})	3.64 A
Tolerance Wattage (%)	± 5%
No. of Cells and Connections	36 (3×12)
Temperature Coefficient of Power	-(0.5±0.05)% per 1°C
Operating Temperature	-40 °C to +85 °C
Maximum System Voltage	600 V

Figure 6. Solar photovoltaic PV panel module SLP060-12 multi crystalline; Output current at different light intensity, (Hussain and Khaled 2020)

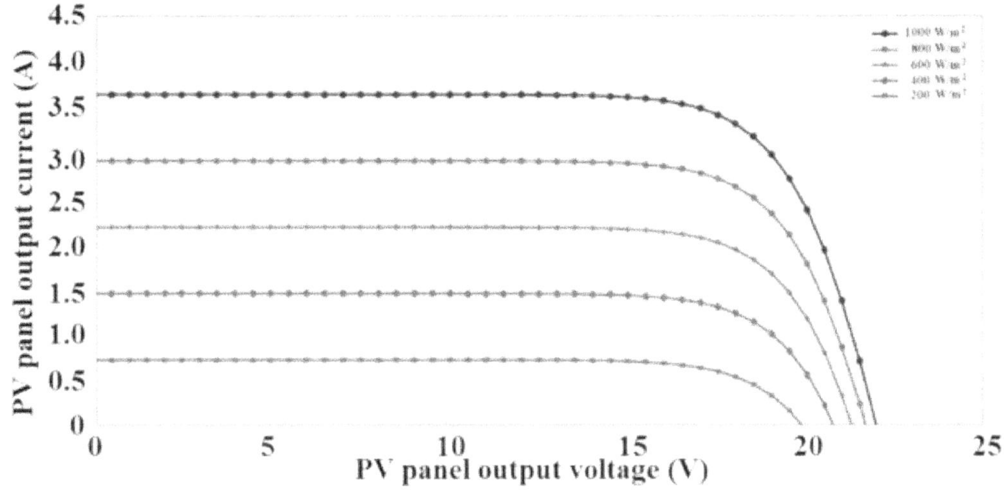

Figure 7. Solar photovoltaic PV panel module SLP060-12 multi crystalline; Output power at different light intensity, (Hussain and Khaled 2020)

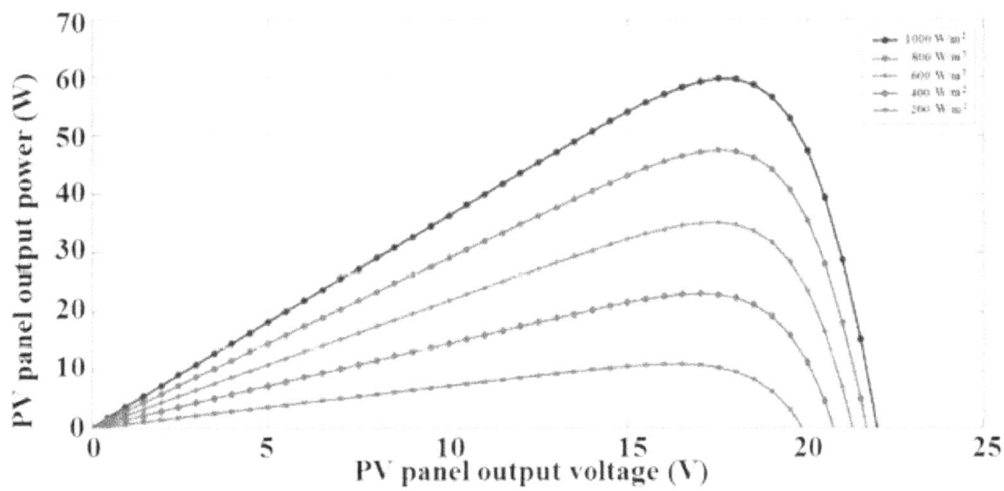

ALGORITHMS OF MAXIMUM POWER POINT TRACKING FOR PV SYSTEMS

The instantaneous power harvesting from a photovoltaic panel is nonlinearly affected by the ambient temperature and incident light intensity. In other meaning, to continuously have a maximum power harvesting it needs to capture and work at the point of maximum power of the power curve of the PV panel. Researchers have been focused on this challenge to guarantee working in the maximum power point tracking MPPT. Hussain et al. (2014), Hussain (2019), Hussain (2018), Yen-Jung et al. (2006), Mohamed et al. (2016), Manakshi et al. (2010), Alajmi et al. (2011), Ahmed et al. (2014), Subiyanto et al. (2009), Samuel et al. (2015), Mathur (2014), Kalaiarasi et al. (2014), and Fathi (2015) have proposed and discussed different algorithms and methods, such as Constant Voltage Method, algorithm of Perturb and Observe algorithm, Incremental Conductance algorithm, short circuit current method, MPPT by Fuzzy Logic Control FLC, and Artificial Neural Networks ANN algorithm as well, have been proposed, tested and evaluated through MPPT algorithm and/or MPPT controller simulation and/or laboratorial experimentation, and results analyzing. The proposed algorithms differ in terms of the method of tracking the maximum power point and also differ in tracking speed and the response during transient and steady state of weather conditions.

Figure 8. Solar photovoltaic PV panel module SLP060-12 multi crystalline; Output current at different ambient temperature, (Hussain and Khaled 2020)

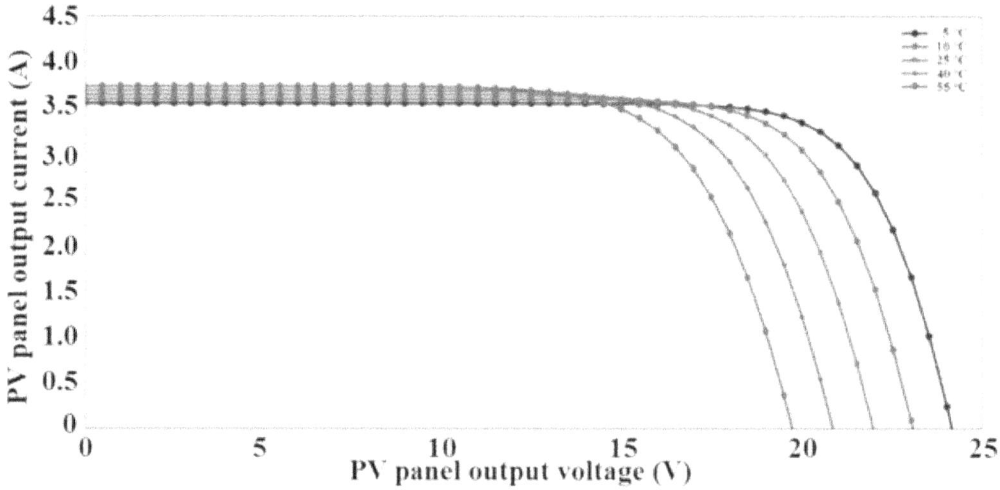

Figure 9. Solar photovoltaic PV panel module SLP060-12 multi crystalline; Output power at different ambient temperature, (Hussain and Khaled 2020)

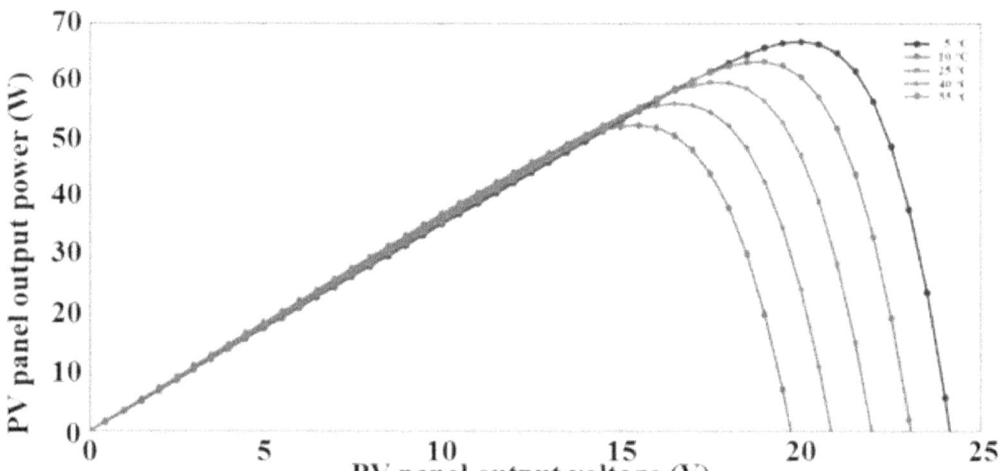

1. Constant Voltage Method

The constant voltage algorithm depends on the fact of the limitation of the voltage range in which the maximum power point is located with respect to the PV panel voltage. Hussain et al. (2014), Yen-Jung et al. (2006), and Mohamed et al. (2016) have explained the way of using this algorithm. Figure 10 shows the limited variation of the MPP voltage during irradiation variation, whereas the agreed voltage for the shown voltage range equals to 76% of the open circuit voltage of the PV panel, in other words $V_{MPP} = 0.76\ V_{OC}$. The fact of approximately constant MPP voltage is highly simplify and accelerate the method response whereas it works with low accuracy due to the panel voltage affecting by the ambient temperature fluctuation.

Figure 10. Constant Voltage Method (Yen-Jung et al. 2006)

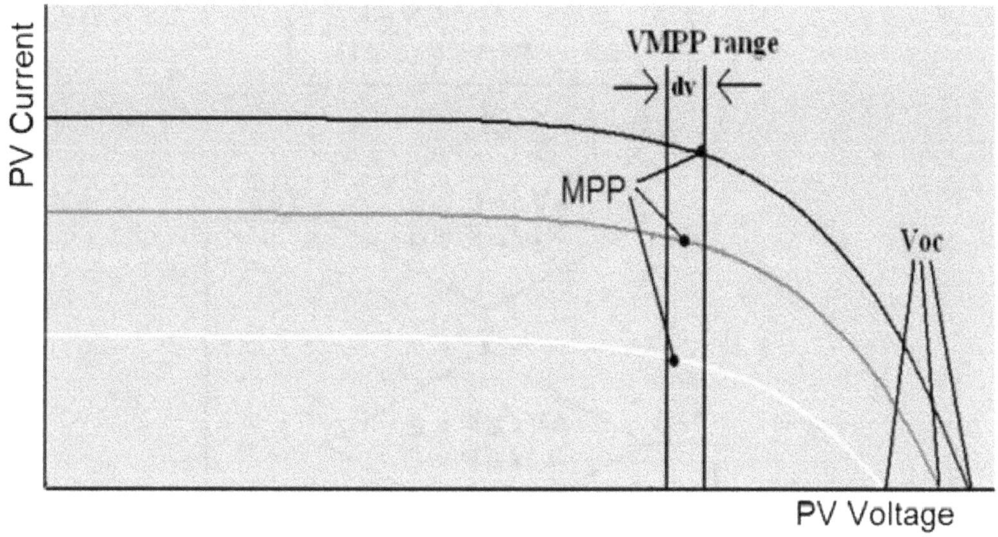

2. Incremental Conductance (IncCond) Algorithm

The process of the incremental conductance algorithm is illustrated in the algorithm flowchart shown in Figure 11. Hussain 2018, and Yen-Jung et al. 2006 have discussed the process this algorithm which considers the curves of the photovoltaic panel output current, and power with respect to the output voltage. Firstly the instantaneous variations in the current, voltage and power will be sensed by current and voltage sensors, and power determined by multiplying the output current and voltage. Then the direction of the perturbation will be recognized and inserted to track and guarantee the maximum power point working condition. The incremental conductance algorithm offered a high accuracy in tracking the maximum power point during fast and slow of weather variation.

Figure 11. The flowchart of the incremental conductance algorithm (Hussain 2018)

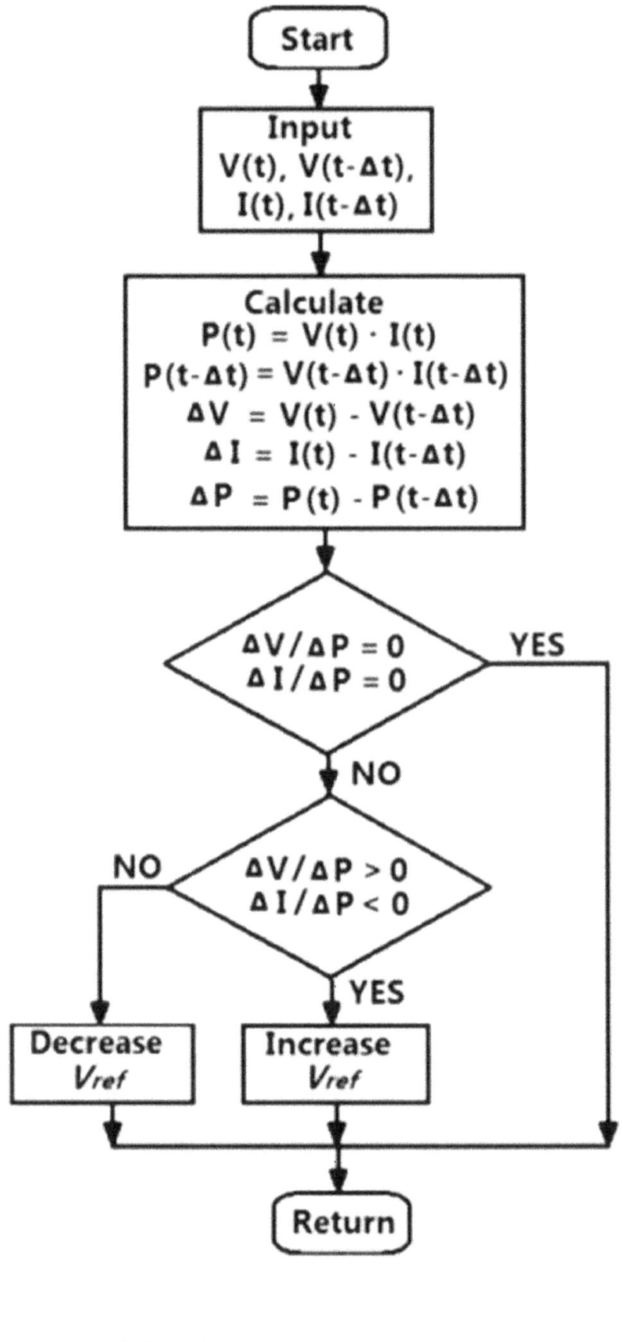

3. Perturb and Observe (P&O) Algorithm

The Perturb and Observe algorithm is different in maximum power point tracking from the incremental conductance algorithm, Figure 12 shows the P&O algorithm flowchart. Manakshi et al. (2010), Hussain (2018), and Yen-Jung et al. (2006) have discussed this algorithm which also called Hill Climbing algorithm. Through this algorithm, the PV panel operating voltage is perturbed by a certain desired increment, and the instantaneous output current of the panel will be observed and then the delivered power will be observed by multiplying the output current and voltage of the panel.

The maximum power point will be tracking by this algorithm, and will be oscillated around the MP point during the steady weather condition or after guaranteeing the MPPT. The algorithm response speed is affected by the duty cycle increment value. During P&O algorithm, there is a tradeoff between the oscillation amplitude and the tracking response time depending on the increment step value, high increment step increase the oscillation level but reduces the response time whereas reducing the increment step improves and reduces the oscillation level at steady state condition but increases the response time.

Due to the process simplicity, this algorithm is a commonly used in implementing different purposes PV systems.

Figure 12. The flowchart of the Perturb and Observe algorithm (Hussain 2018)

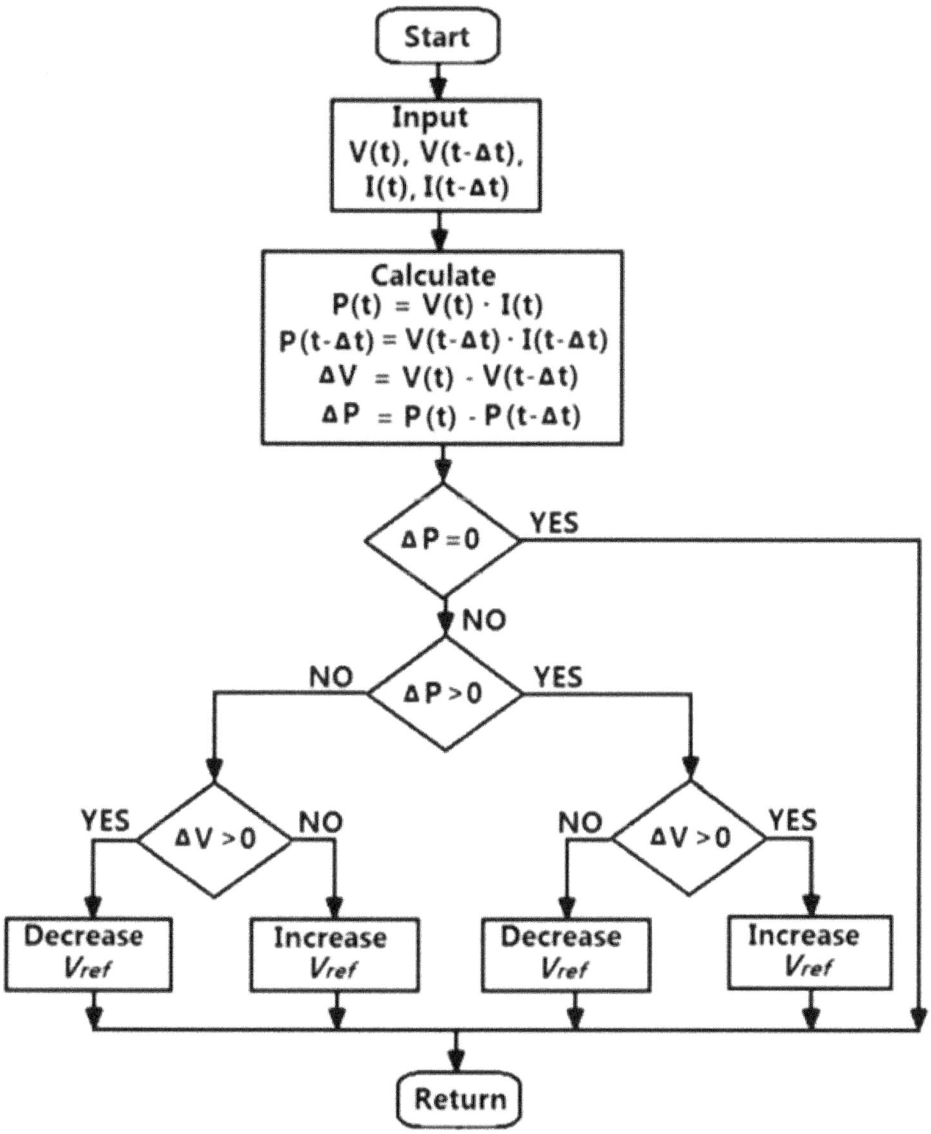

4. Fuzzy Logic Control FLC Method

The method of Fuzzy Logic Controller (FLC) is offering a high robustness with very fast tracking response to the maximum power point. Hussain (2018), Alajmi et al. (2011), Ahmed et al. (2014), Subiyanto et al. (2009), and Samuel et al. (2015) have discussed and used this controlling method.

The controller based on Fuzzy logic method works on capturing the MPP during weather variation. The process is started by measuring the sampled current level and voltage level, then calculating the sampled power. After that the method process calculates the delta or the error of power by subtracting the previous sampled power level from the currently one. Based on the level of error signal, the Fuzzy logic controller is deciding a suitable addition to the initially decided duty cycle to guarantee working at the MPP conditions.

5. Artificial Neural Networks (ANN) Algorithm

Artificial Neural Networks (ANN) has been adopted and discussed in many studies, such as Hussain (2019), Hussain (2018), Mathur (2014), Kalaiarasi et al. (2014), and Fathi (2015). This algorithm is an intelligent machine learning process which works on improving the PV system performance in terms of predicting the maximum power point voltage for any weather condition. ANN algorithm represents a very fast response algorithm in predicting the MPP voltage comparing to different traditional MPPT algorithms or methods.

THE IMPACT OF PV BASED MICROGRID SYSTEM ON SUSTAINABLE BUILDING

Renewable solar photovoltaic system based microgrid effectively supports a sustainable building by the required single phase alternating electricity for both smart grid connected buildings and stand-alone buildings as explained by Ali (2020), and Hussain (2019).

Figure 13 shows the detailed steps for designing a general PV system based microgrid which considers the required rated power to fully support a sustainable building by an alternating power demand.

First step in designing a PV system based microgrid is represented by determining the DC link voltage V_{dc} which should be satisfy system inverter as shown below:

$$V_{dc} = \frac{\sqrt{2}V_{ac}}{M_a} \tag{11}$$

where V_{ac} is the root mean square value of the system AC voltage, M_a is the initial amplitude modulation index of the inverter. To have low harmonics distortion in the output inverter current, the connected inverter is adjusted at a high $M_a \geq 0.9$.

Any connected inverter is operated at the maximum power point MPP voltage V_{MPP} of the system PV matrix. Therefore, the number of serially connected PV panels *NM* in a one string, as discussed by Ali (2020), and Hussain (2019), is

$$NM = \frac{V_{dc}}{V_{MPP}} \quad (12)$$

The string voltage *SV* can be calculated by:

$$SV = NM \times V_{MPP} \quad (13)$$

Figure 13. The steps of designing a PV system based microgrid

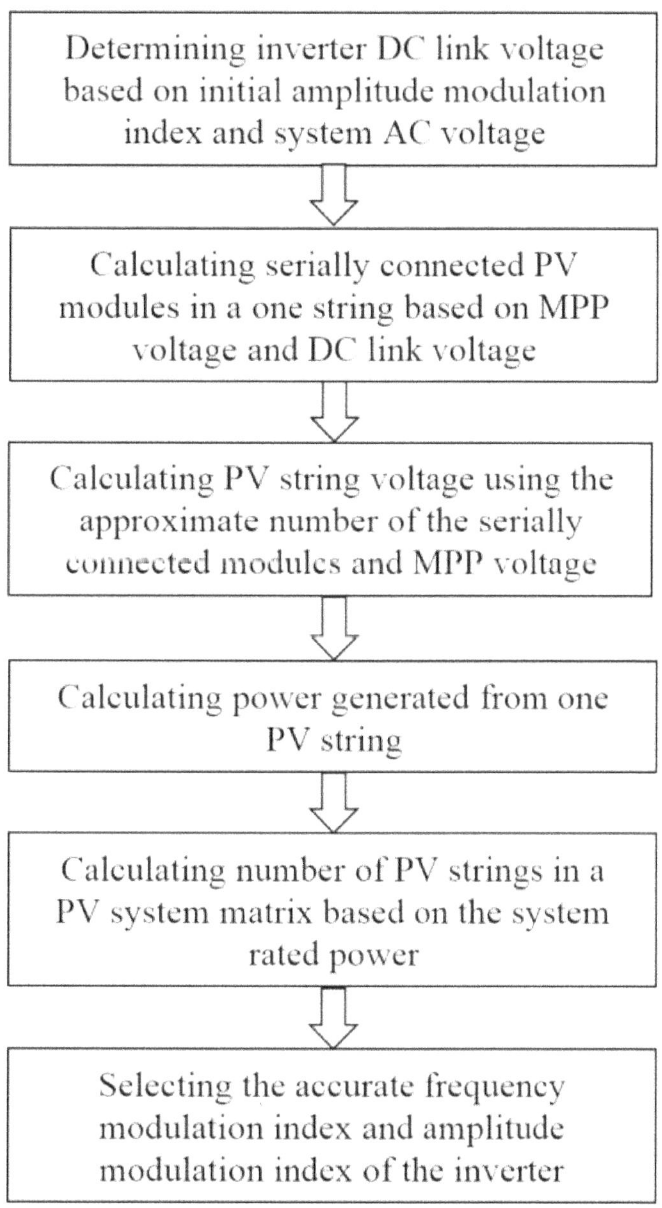

The power generated *SP* by a one PV string can be calculated by considering the PV module power P_{MPP} at MPP condition:

$$SP = NM \times P_{MPP} \tag{14}$$

The number of PV strings *NS* in a system PV matrix can be calculated by considering the required system rated power *AP* and the string rated power *SP*:

$$NS = \frac{AP}{SP} \tag{15}$$

The system inverter can be adjusted at the accurate amplitude modulation index M_a, and frequency modulation index M_f based on the inverter switching frequency f_s and fundamental reference frequency f_o.

$$M_a = \frac{\sqrt{2}V_{ac}}{V_{dc}} \tag{16}$$

$$M_f = \frac{f_s}{f_o} \tag{17}$$

Figure 13 and equations above explain accurately the step-by-step of designing a PV system size and how set a suitable single phase inverter to deliver the required power to a sustainable building with offering the possibility to connect the system to the utility grid as a microgrid system.

SAMPLE OF PV SYSTEM DESIGN

The step-by-step design process of a PV system will focus on a small scale building and will demonstrate how calculate a rate power of the system and how design a suitable PV matrix size and arrangement to meet system requirement.

The design starts by tabulating the building appliances, the quantity of each appliance, the estimated daily working time, and the total kW hours for all appliances per day. In a small scale building, normally the electrical loads can be represented by one air conditioner, desktop computer set, television TV, laptop, electric cooker, two exhaust fans, and four LED lighting units.

Table 2 shows the details of the building electrical appliances, rated power multiplied by the number of appliance of each type, working hours, kilo Watt hours of each load, and the total required kW hours.

Based on considering daily five hours for the solar photovoltaic energy harvesting as discussed in Hussain, A., & Khaled, H. (2020), the rated power of the designed PV system of Table 2 electrical appliances equals to the total required power per day divided by the five hours, i.e. 12000 kW·h / 5 h = 2.4 kW.

So, the rated voltage and power of the desired system are 230 V, and 2.4 kW to deliver the suitable alternating electricity to a small scale building.

Table 2. Power requirement details of selected appliances in a small scale building

No.	Electrical Appliance	No. of same load	Load rated power (W)	Daily working hours	Required kW hours (kW·h)
1	Air conditioner	1	» 1200	6	7200
2	Electric cooker	1	» 1000	2	2000
3	Desktop computer set	1	» 250	4	1000
4	TV	1	» 150	4	600
5	Laptop	1	» 100	6	600
6	Exhaust fans	2	» 20	3	120
7	LED light unit	4	» 20	6	480
The approximate total required power per day equals to					= 12,000

By considering the rated AC voltage 230 V of the system, using an initial value of the amplitude modulation index $M_a = 0.9$, then using eq. (11) yields the DC link voltage V_{dc} as input voltage to the system inverter can be determined:

$$V_{dc} = \frac{\sqrt{2}V_{ac}}{M_a} = \frac{\sqrt{2} \times 230}{0.9} = 361.4 \text{ V}$$

Based on the DC link voltage 361.4 V of the system inverter, and the MPP voltage 17.8 V of the selected PV module SLP060-12, the design needs to connect many PV panels in series to meet the required level of DC link voltage, for this purpose, eq. (12) can be used to calculate number of PV modules *NM* in one PV string:

$$NM = \frac{V_{dc}}{V_{MPP}} = \frac{361.4V}{17.8V} = 20.3 \approx 20 \text{ PV panels per PV string}$$

The accurate string voltage *SV* of the 20 PV panels serially connected can be calculated using eq. (13):

$$SV = NM \times V_{MPP} = 20 \times 17.8V = 256V$$

The total harvested solar energy from one PV string *SP* can be calculated using eq. (14) by considering the MPP working condition of the PV system:

$$SP = NM \times P_{MPP} = 20 \times 60 = 1200W \text{ Power per one PV string}$$

To calculate the number of PV strings *NS* in the PV matrix of the system, the system rated power *SP* and eq. (15) can be used:

$$NS = \frac{AP}{SP} = \frac{2400W}{1200W} = 2\,\text{PV strings in the PV system}$$

For accurate adjusting to the system inverter in terms of amplitude modulation index M_a, eq. (16) can be used:

$$M_a = \frac{\sqrt{2}V_{ac}}{V_{dc}} = \frac{\sqrt{2}\times 230V}{356V} \approx 0.914$$

Adjusting the frequency modulation index M_f, can be known using eq. (17), is depending on the inverter switching frequency f_s (which can be selected by considering the level of total harmonic distortion THD parameter) and reference utility frequency f_o.

Figure 14, and Figure 15 show the arrangement and the power curves, respectively, of the designed PV matrix using the selected module SLP060-12 multi crystalline photovoltaic PV panel.

Figure 14. The arrangement of the designed PV matrix based SLP060-12 multi crystalline photovoltaic PV panel

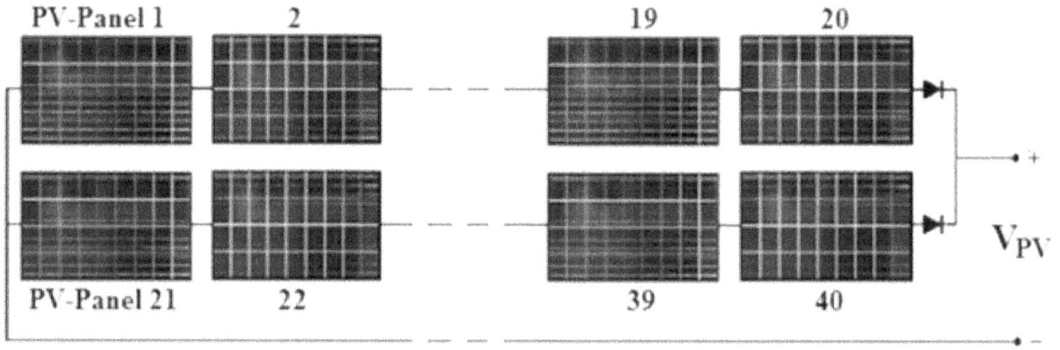

Figure 15. The designed PV matrix based SLP060-12 multi crystalline photovoltaic PV panel; Output power at different light intensity

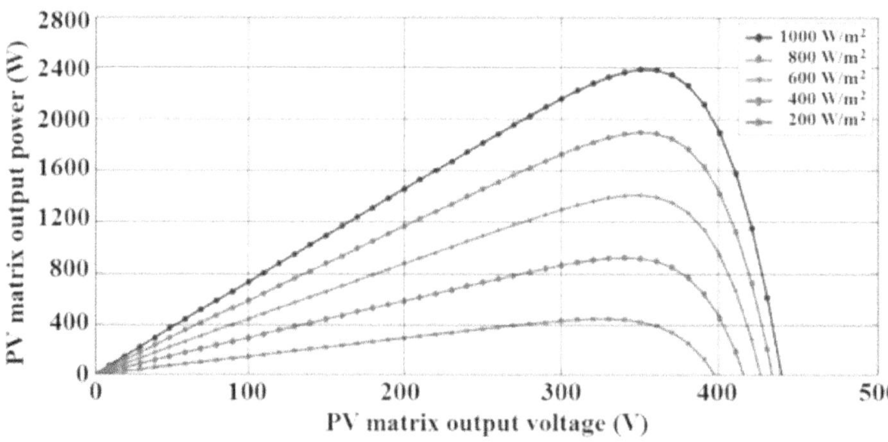

The above mentioned and explained process of PV system design can be followed for mid-scale buildings or large scale buildings, the difference will in the PV system size and arrangement, which will be expanded to meet the power requirement of the additional electrical appliances, and the increment in the number of appliances of the same type.

DC-DC CONVERTERS; ROLE AND IMPORTANCE IN PV SYSTEMS

The DC-DC Converter is a high-frequency power electronic conversion circuit used to convert a fixed-voltage dc source or fluctuated voltage into a regulated voltage. The converter circuit includes switching device (such as Bipolar Junction Transistor BJT, Insulated Gate Bipolar Transistor IGBT, or Metal Oxide Semiconductor Field Effect Transistor MOSFET), diode, inductor, capacitor, and the connected load. This converter function is important in different power electronic systems like the PV systems, and in industrial applications in general. The dc-dc converter is used in energy conversion in the renewable energy based systems. In addition, DC-DC converter is used as driver for traction motor in electric automobiles, forklift trucks, trolley cars, trolley cars, marine hoists as well as mentioned in Daniel (2011). Different types of DC–DC converters have been proposed to convert the input voltage (fixed or varying) to a desired regulated voltage level, a detailed comparison among the DC-DC converter types have shown in Walker & Sernia. (2004).

The circuit of DC-DC converter includes switching component which represented by a type of transistor, this transistor serves as a power electronic switch, which can be controlled (switch fully on/ switch fully off) through periodically driving pulses. Varying the Duty cycle D, which is the rate of the on time t_{on} to the total time (on time t_{on} and off time t_{off}) as shown in (18), leads to control the level of the output DC voltage of the converter.

$$D = \frac{t_{on}}{t_{on} + t_{off}} = \frac{t_{on}}{T} = t_{on} \cdot f_s \tag{18}$$

where D is the duty ratio of the drive pulses of the converter transistor, f_s is the converter switching frequency in Hz, and T is the period of switching time.

There are common three types of DC-DC converters used in PV system, and working either in Continuous Conduction Mode CCM, or working in Discontinuous Conduction Mode DCM. First converter type is a DC-DC boost or step-up converter which receives the input DC voltage of an unregulated DC source and convert it to a regulated higher output voltage level. Second converter type is a DC-DC buck or step-down converter which receives the input DC voltage of an unregulated DC source and convert it to a regulated lower output voltage level. Third converter type is a DC-DC buck-boost or step-down & step-up converter which receives the input DC voltage of an unregulated DC source and convert it to a regulated lower or higher output voltage level based on the value of duty ratio.

1. Boost DC-DC Converter

A boost DC-DC converter or step-up DC-DC converter is a common power electronic type of switch-mode converters. The converter receives an input direct current DC voltage (fixed or fluctuated voltage) and boosts or steps-up it to a desired voltage level. So, the boost converter works as a DC step-up transformer whereas the conventional transformer stepping up or stepping down the alternating current AC voltage level, from low to high or from high to low level respectively with keeping the same power level. In other meaning, for the DC-DC boost converter as a DC transformer or for the AC transformer, stepping-up the voltage decreases the output current at the source power. Figure 16 shows the circuit topology of the boost converter.

Figure 16. Boost DC-DC converter (Daniel (2011)

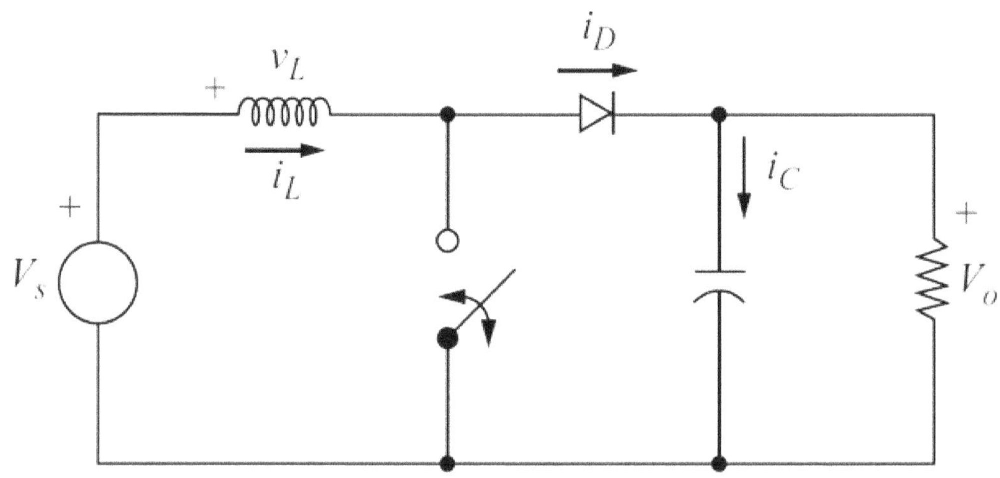

The boost converter is characterized by the simplicity, and the high efficiency can reach up to 99% efficiency. The value of the output voltage can be determined using (19):

$$V_o = \frac{V_{in}}{1-D} \tag{19}$$

where V_o is the converter regulated output voltage, D is the duty ratio, and V_{in} is the unregulated input voltage.

2. Buck DC-DC Converter

A buck DC-DC converter or step-down DC-DC converter is a second common power electronic type of switch-mode converters. This converter receives an input fixed or unregulated DC voltage and bucks or steps-down it to a desired lower voltage level. Comparing to the AC transformer, buck converter serves as a DC step-down transformer to reduce the level of output voltage with respect to the converter input voltage, and at the same time increases the level of the converter output current. Figure 17 shows the circuit topology of the buck converter.

The buck converter is also characterized by the simplicity, and the high efficiency. The output voltage can be determined using (20):

$$V_o = V_{in} \cdot D \tag{20}$$

where V_o is the regulated output voltage of the converter, D is the duty ratio, and V_{in} is the input voltage.

Figure 17. Buck DC-DC converter (Daniel (2011)

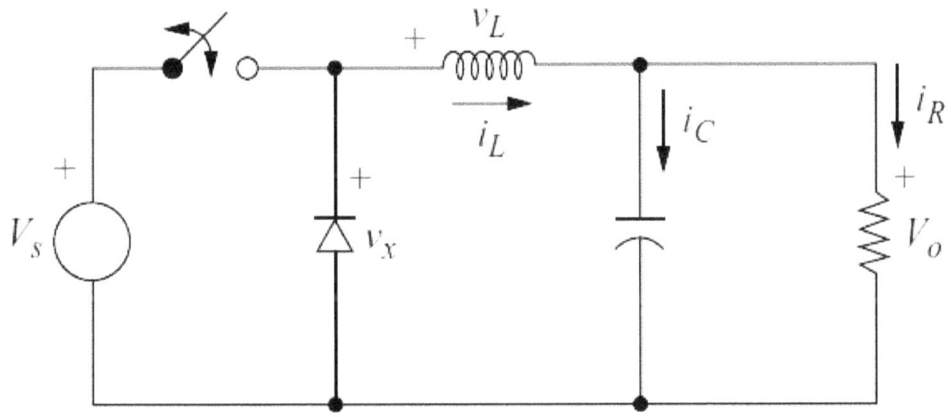

3. Buck-Boost DC-DC Converter

A buck-boost DC-DC converter or step-down & step-up DC-DC converter is a third common power electronic type of switch-mode converters. This converter works in two directions, it receives an input fixed or unregulated DC voltage and it can bucks or boosts it to a desired lower or higher voltage level depending on the value of the duty ratio D. Figure 18 shows the circuit topology of the buck-boost converter.

Figure 18. Buck-Boost DC-DC converter (Daniel (2011)

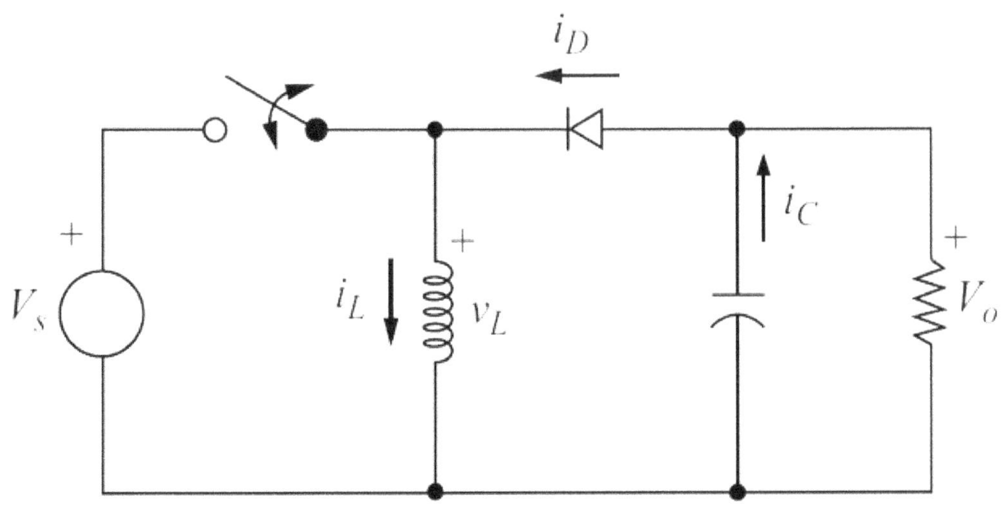

The buck-boost converter is able to decrease or increase the level of the converter output voltage respectively by reverse voltage sign with respect to the input voltage, this can be done by selecting a suitable value of D as shown in (21). For boosting the voltage, duty ratio D should be greater than 0.5, whereas for bucking the voltage, duty ratio should be less than 0.5. When $D = 0.5$, the value of converter output voltage will be equal to the input voltage with minus sign.

$$V_o = -V_{in} \cdot \frac{D}{1-D} \tag{21}$$

where V_o is the regulated output voltage of the converter, D is the duty ratio, and V_{in} is the input voltage.

CONCLUSION

Throughout this chapter; the infrastructure and the demerits of a traditional power grid have demonstrated by explaining the function in details of the traditional grid components. In addition to that it mentioned the negative effects of this grid on the environmental pollution and global warming due to using the fossil fuels in the traditional power station. The chapter explained the reasons for transferring step by step to the smart microgrid. A detailed comparison between the specifications of the traditional power grid and the smart microgrid have shown. Then the merits of a smart power grid as a two-directional electrical power grid and how it monitored and controlled have explained. The smart microgrid structure in terms of power and information feeders as well as the advantages a smart grid, this chapter also explained the direct current DC smart microgrid systems, and the alternating current AC smart microgrid systems. This chapter showed a comparison between the DC smart microgrid system and AC smart microgrid system in terms of the work principle. The major role of solar photovoltaic systems for microgrid implementation have explained and proceed by the demonstrating the principle and behavior of PV panel in terms of panel current and power with respect to panel voltage. The effect of Maximum Power Point Tracking algorithms have revised in this chapter to explain the positive effect of the MPPT algorithms in improving the PV system performance. The function, types, and operating modes of the DC-DC converter with respect to the PV system design and implementation are also discussed and explained in this chapter. Detailed generalized step-by-step of designing a PV system based single phase microgrid have mentioned and explained. Then followed by applying the related relationships in design a small scale PV system in terms of size and arrangement of the PV matrix to meet the power requirement of a small scale buildings. The way of expanding the system size to meet the requirements of mid-scale and large scale buildings have also mentioned.

REFERENCES

Abdullah, H. A., Razzaqul, A., Nasser, H., Reza, G., & Eklas, H. (2020). Survey of Smart Grid Concepts and Technological Demon striations Worldwide Emphasizing on the Oman Perspective. *Applied System Innovation*, *3*(5), 5. Advance online publication. doi:10.3390/asi3010005

Ahmad, E. (2014). Fuzzy Logic Controller Based SEPIC Converter for Maximum Power Point Tracking. *IEEE Transactions on Industry Applications*, *50*(4), 2349–2358. doi:10.1109/TIA.2014.2298558

Alajmi, B. N., Ahmed, K. H., Finney, S. J., & Williams, B. W. (2011). Fuzzy-Logic-Control Approach of a Modified Hill Climbing Method for Maximum Power Point in Microgrid Standalone Photovoltaic System. *IEEE Transactions on Power Electronics*, *26*(4), 1022–1030. doi:10.1109/TPEL.2010.2090903

Ali, K. (2020). *Design of Smart Power Grid Renewable Energy Systems*. Wiley.

Anyaka, B., & Olawoore, O. (2014). Minimization of Power Losses in Transmission Lines. *Journal of Electrical and Electronics Engineering (Oradea)*, *9*(3), 23–26.

Arellano, B., Sena, S., Abdollahy, S., Lavrova, O., Strattonl, S., & Hawkinsl, J. (2013). Analysis of electric vehicle impacts in New Mexico urban utility distribution infrastructure. *IEEE Transportation Electrification Conference and Expo (ITEC)*. 10.1109/ITEC.2013.6573495

Australian National University. (2020). *Carbon dioxide toxicity and climate change: a serious unapprehended risk for human health*. P.N. Bierwirth.

Blaabjerg, F., Teodorescu, T., Liserre, M., & Timbus, A. V. (2006). Overview of control and grid synchronization for distributed power generation systems. *IEEE Transactions on Industrial Electronics*, *53*(5), 1398–1409. doi:10.1109/TIE.2006.881997

Chapman, S. (2003). *Electric machinery and power system fundamentals*. McGraw Hill.

Daniel Hart. (2011). *Power Electronics*. McGraw Hill.

Darwish, Al Awadhi, & Bin Amer. (2010). Combining the nuclear power plant steam cycle with gas turbines. *Energy*, *35*(12), 4562-4571.

Farhad, K., Shamsodin, T., Hamed, T., & Edris, P. (2016). Integration of electric vehicles into a smart power grid: A technical review. *IEEE Electrical Power and Energy Conference (EPEC)*.

Fathi & Aashoor. (2015). *Maximum power point tracking techniques for photovoltaic water pumping system* (Ph.D. dissertation). Dept. Elect. Elect. Eng., Univ. of Bath, Claverton Down, UK.

Galus, M. D., Zima, M., & Andersson, G. (2010). On integration of plug-in hybrid electric vehicles into existing power system structures. *Energy Policy*, *38*(11), 6736–6745. doi:10.1016/j.enpol.2010.06.043

Grainger, J., & Stevenson, W. D. (2008). *Power systems analysis*. McGraw Hill.

Guille, C., & Gross, G. (2009). A conceptual framework for the vehicle-to-grid (V2G) implementation. *Energy Policy*, *37*(11), 4379–4390. doi:10.1016/j.enpol.2009.05.053

Hosham, S., Khalid, F., & Raheel, J. (2018). Enhancement of Performance for Steam Turbine in Thermal Power Plants Using Artificial Neural Network and Electric Circuit Design. *Applied Computational Intelligence and Soft Computing*, 1–9. Advance online publication. doi:10.1155/2018/8042498

Hussain, A. (2017). Novel 9-Steps Automatic Voltage Regulator based on Two Step-Down Transformers. *Iranian Journal of Electrical and Computer Engineering*, *7*(2), 576–583.

Hussain, A. (2019). Supplying DC Electricity to the Isolated Dwellings through MPP Tracked PV System Based on Artificial Neural Network. *6th International Conference of Control, Dynamic Systems, and Robotics (CDSR'19).*

Hussain, A. (2020). A New Intelligent Power Factor Corrector for Converter Applications. *6th International Conference on Control, Modeling and Computing (CMC 2020).*

Hussain, A. (2020). Artificial Neural Network Based Unity Power Factor Corrector for Single Phase DC-DC Converters. *Iranian Journal of Electrical and Computer Engineering*, *10*(4), 4145–4154.

Hussain, A., & Ali, S. (2016). Novel Discrete Components Based Speed Controller for Induction Motor. *International Journal of Power Electronics and Drive Systems*, *7*(4), 1075–1084.

Hussain, A., & Fernando, G. (2019). Stand-alone PV System with MPPT Function Based on Fuzzy Logic Control for Remote Building Applications. *International Journal of Power Electronics and Drive Systems*, *10*(2), 842–851.

Hussain, A., & Khaled, H. (2020). Hybrid Technique for an Efficient PV System through Intelligent MPPT and Water Cooling Process. *International Journal of Power Electronics and Drive Systems, 11*(4), 1835-1843.

Hussain, A., Maissa, F., Mousa, M., Aaesha, A., Abdulrahman, S., Ahmad, C., Moath, A., Somaya, A., & Sumith, S. (2018). Novel Design of AURAK Hydrogen Car: Electrical Aspects. *The International Conference on Energy, Water & Environmental Sciences (ICEWES 2018).*

Hussain, A., Mousa, M., Basil, Q., Mohammed, A., Osman, A., & Zahidur, R. (2018). New Design and Implementation of AURAK Solar Car: Electrical Vision. *International Conference on Energy, Water & Environmental Sciences (ICEWES 2018).*

Hussain, A., Yousif, I., & Beza, N. (2014). *Design and Simulation of a High Performance Standalone Photovoltaic System. In Proceeding ICREGA'14.* Springer.

James, M. (2012). *Smart Grid Fundamentals of Design and Analysis.* IEEE Press.

Jignesh, G. B. (2014). Smart Grid - An Evolutionary Version of Energy Infrastructure. *Smart World*, 32-35.

Kalaiarasi, N., Paramasivam, S., & Sanchari, K. (2014). Control of Z-Source Inverter Based PV System with MPPT Using ANFIS. *International Review on Modelling and Simulations*, *7*(5), 797–806. doi:10.15866/iremos.v7i5.1970

Karoly, V. (2002). Hybrid Combined Cycle Power Plant. *International Conference Nuclear Energy for New Europe 2002.*

Kawamura, H., Naka, K., Yonekura, N., Yamanaka, S., Kawamura, H., Ohno, H., & Naito, K. (2003). Simulation of I-V characteristics of a PV module with shaded PV cells. *Solar Energy Materials and Solar Cells*, *75*(3/4), 613–621. doi:10.1016/S0927-0248(02)00134-4

Laamami, Benhamed, & Sbita. (2017). Analysis of shading effects on a photovoltaic array. *International Conference on Green Energy Conversion Systems (GECS).*

Lei, L., Xingong, C., Xiju, Z., Wen, W., & Chao, W. (2015). Research on Transmission Line Loss and Carrying Current Based on Temperature Power Flow Model. *3rd International Conference on Mechanical Engineering and Intelligent Systems (ICMEIS 2015).*

Manakshi, D., Sumita, D., & Champa, N. (2010). Maximum Photovoltaic Power Tracking Using Perturb & Observe Algorithm in MATLAB/Simulink Environment. *International Journal of Electrical Engineering and Technology, 1*(1), 71–84.

Mathur, D. (2014). Maximum Power Point Tracking with Artificial Neural Network. *International Journal of Emerging Science and Engineering, 2*(3), 38–42.

Michael, H., Damir, N., & Mariesa, W. (2017). Electric Power Grid Modernization Trends, Challenges, and Opportunities. *IEEE Advancing Technology for Humanity.* https://www.ieee.org/content/dam/ieee-org/ieee/web/org/about/corporate/ieee-industry-advisory-board/electric-power-grid-modernization.pdf

Min-Joong, K., & Huei, P. (2007). Power Management and Design Optimization of Fuel Cell/Battery Hybrid Vehicles. *Journal of Power Sources, 165*(2), 819–832. doi:10.1016/j.jpowsour.2006.12.038

Mohamed, L., Ali, K. A., & Mazen, A., & S., Ookawara (2016). Performance Enhancement of Constant Voltage Based MPPT for Photovoltaic Applications Using Genetic Algorithm. *3rd International Conference on Power and Energy Systems Engineering (CPESE 2016).*

Nguyen, D., & Lehman, B. (2006). *Modeling and simulation of solar PV arrays under changing illumination conditions.* IEEE Workshops on Computers in Power Electronics. doi:10.1109/COMPEL.2006.305629

Rehmani, M. H., Reisslein, M., Rachedi, A., Erol-Kantarci, M., & Radenkovic, M. (2018). Integrating Renewable Energy Resources into the Smart Grid: Recent Developments in Information and Communication Technologies. *IEEE Transactions on Industrial Informatics, 14*(7), 2814–2825. doi:10.1109/TII.2018.2819169

Samuel, R. B. (2015). A Comparative Analysis of Integrated Boost Fly back Converter using PID and Fuzzy Controller. *International Journal of Power Electronics and Drive System, 5*(4), 486-501.

Shahzad, S., Umair, U., & Shan, U. (2015). Electric Power Transmission and Distribution Losses Overview and Minimization in Pakistan. *International Journal of Scientific and Engineering Research, 6*(4), 1108–1112.

Smart Buildings © ACEEE. (2017). *Smart Buildings: Using Smart Technology to Save Energy in Existing Buildings.* Author.

Subiyanto & Hannan. (2009), Maximum Power Point Tracking in Grid Connected PV System Using A Novel Fuzzy Logic Controller. *Proceedings of SCOReD.*

The International Institute for Sustainable Development. (2014). *The Impact of Fossil-Fuel Subsidies on Renewable Electricity Generation.* https://www.iisd.org/system/files/publications/impact-fossil-fuel-subsidies-renewable-electricity-generation.pdf

Walker, G. R., & Sernia, P. C. (2004). Cascaded dc-dc converter connection of photovoltaic modules. *IEEE Transactions on Power Electronics, 19*(4), 1130–1139. doi:10.1109/TPEL.2004.830090

Wilfried van Sark. (1826). Photovoltaic System Design and Performance. *Energies*, 12.

Xi, F., Satyajayant, M., Guoliang, X., & Dejun, Y. (2012). Smart Grid – The New and Improved Power Grid: A Survey. *IEEE Communications Surveys and Tutorials*, *14*(2), 944–980.

Yen-Jung, M. T., Aiguo, P. H., & Nirmal, K. N. (2006). Evaluation of Micro Controller Based Maximum Power Point Tracking Methods Using dSPACE Platform. *Australian University Power Engineering Conference*. https://citeseerx.ist.psu.edu/viewdoc/download?doi=10.1.1.526.1472&rep=rep1&type=pdf

Chapter 9
Statistical Understanding and Optimization of Building Energy Consumption and Climate Change Consequences

José Antonio Orosa
University of A Coruña, Spain

Diego Vergara
https://orcid.org/0000-0003-3710-4818
University Catholic of Ávila, Spain

Feliciano Fraguela
University of A Coruña, Spain

Antonio Masdías-Bonome
University of A Coruña, Spain

ABSTRACT

In the present chapter, a new tool was designed to find a better alternative for improving building energy consumption in the next years. In this sense, in the first stage of this calculation procedure, ISO Standard 13790 calculation procedure was developed in accordance with Monte Carlo method and results showed the probability of energy consumption as a Weibull model. Furthermore, a map of different Weibull models in accordance with different input parameters and future climate change effect was developed as a future building design guide. This tool defines the probability of energy consumption of an existing building, or a building that is being designed today and in the near future, preventing the climate change effect. More applications at the time of building retrofitting and healthy indoor ambiences are proposed.

DOI: 10.4018/978-1-7998-7023-4.ch009

INTRODUCTION

In accordance with the International Energy Agency (IEA), one-third of the energy consumption is in indoor ambiences. In this sense, as a consequence of the energy consumption, different humanity problems like climate change, greenhouse gas effect and some types of contaminations are related to the burn of fossil fuels. The problem is related to the fact that these combustibles have higher calorific power respect its weight and, in consequence, it is usually employed in transportation. These same engines employed in maritime transport use to be employed in Power plants due to the higher the thermal focus, the higher the efficiency of the process. In consequence, the electricity generated by these power plants and cogeneration plants is distributed to nearby cities emitting carbon dioxide emissions. As a consequence of burning this fossil fuels, carbon dioxide emissions increase and the previous commented problems will remain.

Despite the new renewable resources, it is possible to complement the energy consumption in buildings with solar energy, wind energy or by improving thermal inertia and constructive characteristics in accordance with the international Energy Agency objectives called "Nearly Zero Energy Buildings" (NZEB).

Energy consumption in indoor ambiences is related with different parameters like, for instance, building characteristics (orientation, thermal inertial, …), air conditioning systems (HVAC) and habits of occupants, between others. All these concepts are grouped into standards that try to define a mathematical model of heat and mass transfer in buildings with the aim to be an optimization tool in the future. The interest of the scientific community is reflected in the growing trend of the number of articles published in recent years which are related to energy consumption in buildings. As an example, in Figure 1 is represented the number of articles indexed in SCOPUS in the last decade through the following searching: (TITLE-ABS-KEY ("energy consumption")) AND (building). Thus, a growing trend is exposed in Figure 1, which suggests the number of articles related to this topic is increasing about 450 per year (indexed in Scopus database).

Figure 1. Growing trend in the number of articles indexed in Scopus related to the energy consumption in buildings

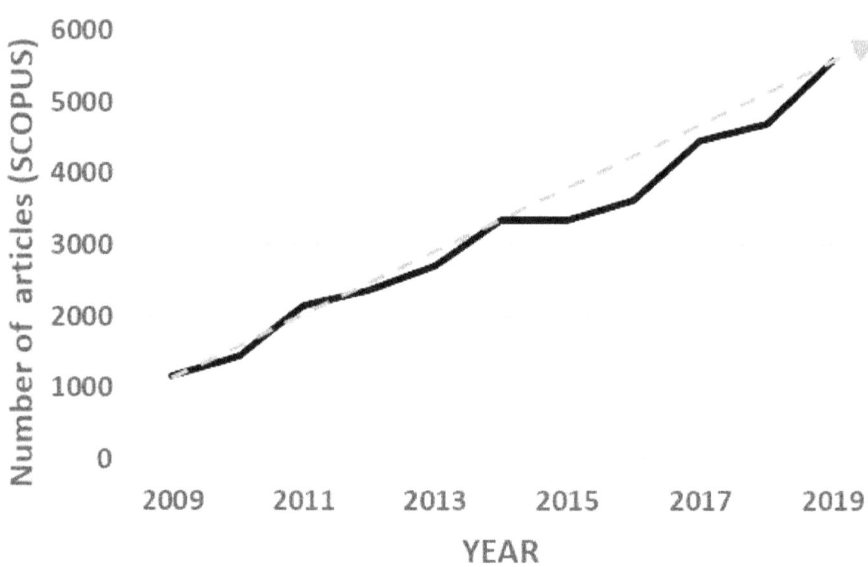

Climate Change is a subsequent problem related with previous commented. Despite the fact it was neglected in the last decades, nowadays there are few doubts about its origin and its consequences now and in a near future. Once again, the combustion of fossil fuels and deforestation are related with greenhouse gases and these are related with a change in natural processes like the Gulf Stream which, as a consequence of the heating, is weaker. In consequence, a lower amount of warm water is sent to the Gulf of Mexico, U.S. and Western Europe. The main translation of this climate change is a change of weathers seasons from four to two extreme; just one hot and one cold weather seasons (Orosa, Costa, Rodríguez-Fernández and Roshan, 2014). This effect was observed in different regions and its initial consequences are an increment of average outdoor temperature. This will imply a modification of rainy seasons and extreme temperatures making wrong most of building designs with an error margin that increases with years. Despite this, few research works investigate this effect due to, in an initial period, an increase of no more than one degree in ten or more years was observed. The problem arrives when this average outdoor temperature increases grows exponentially and non in a linear way. In consequence, few calculation procedures and improvement proposals to prevent or correct this variation in existing buildings are defined.

Recent research works have shown that, in the last years and in the near future, there was and will be an increment in the mean outdoor temperature, and also how to minimize the energy consumption in the next 30 years. Some of the consequences of this increment are related to an increase in the cooling energy consumption and increase of general discomfort in the indoor environments (Orosa and Oliveira, 2010; Roshan, Ghanghermeh and Orosa, 2014; Ye *et al.,* 2021).

On the one hand, different research works showed that climate change will impact on trends in future heating and cooling demands over Europe and its results are proposed for future energy system analysis. In particular, these results highlight in a near period until 2050 highest changes in the north-eastern Europe, but with clear consequences in nearly all Europe (Larsen et al., 2020). It is known that the global annual average temperature in the past hundred years increased significantly (0.6 ± 0.2°C), and the maximum change was noted in the decade of the 1990s. This revealed that Europe suffered a higher increase in overall temperatures (0.95°C) and this heating condition was experienced at a much higher level in some regions such as Spain. In particular, previous research works have shown that in north-western Spain, there will be a clear increment in the value of outdoor temperature during all the seasons. Furthermore, the increment in daily cooling energy consumption was about 3.2 kWh.

Climate change in Spain was analyses by several researchers, most of the time related to their more influence on the economy of the country like agriculture (Vila-Traver et al., 2021) and tourism (Lars Hein et al., 2009), especially in the south of Spain. In particular, its main simulations try to define the causes of water stress and tourism economic consequences. What is more, a Spanish strategy for coastal adaptation to climate change was published in 2019 (Losada et al., 2019).

If we now focus on research works that propose solutions to prevent climate change effects in energy consumption in Spanish building sector, just a few papers can be found. These papers are centered into control the heat island effect in cities due to climate change (Herrera-Gomez et al., 2017) and its main conclusions focus on the definition of the optimal dimensions of this green roof are, which is claimed as a fundamental solution. Other solutions for a reduction in fossil fuels in Spanish building sector was to employ solar energy which experimented a recent weakening due to the elimination of incentives (Fernandez-González et al. 2020). In this sense, just an adequate design and selection of solar energy in buildings with adequate cost-effectiveness may be employed in future buildings.

On the other hand, energy consumption is one of the most important factors to be considered during the designing of new constructions and to be improved in existing buildings based in recent research works. In this sense, there are a lot of software resources that can be used to estimate building energy consumption during the full year and different standards for building energetic qualification (ASHRAE; ISO 13790). Despite this, today, there are very **few options** to prevent changes in energy consumption in building construction related to climate change (Roshan, Orosa and Nasrabadi, 2012). One of the main objectives of Annex 55 of the International Energy Agency (IEA) is to obtain new methodologies for defining the better alternative for preventing energy consumption in buildings.

As a consequence of this situation, the effect of climate change over the present and future buildings energy consumption must be defined and a calculation procedure, in accordance with present standards, must be developed. In the present paper, a new tool was designed for finding a better alternative for improving building energy consumption.

In this sense, just ventilation seems to be the low costs and more effective procedure to fight against this so complex problem. What is more, adequate ventilation is clearly related to energy consumption and the health of occupants due to the numbers of air changes done in a period of time. This basic and classic control of indoor ambiences was employed since the beginning of building construction and still, there is not a detailed understanding of how to employ it and its real effect over building energy consumption.

Despite the fact that standards give output values that show architects and engineers indications about the consequence of each modification of building designs and that it was transformed into simple software resources for energy rating like CE3X in Spain, there is not an in-depth understanding of the real effect of ventilation on indoor ambiences and the physical origin of this effect.

Different research works tried to define the energy consumption in buildings by Artificial Intelligence procedures like Neuronal Networks with genetic algorithms (Divina et al. 2020), or support vector machine (SVM) (Khosravani et al., 2016) time series to replace the traditional statistical analysis due to it needs lower sampled data and simpler models. Despite this, artificial intelligence procedures are based on iterative processes that develop a map of solutions far away from the physical process that happens between input variables and obtained results, being quite difficult to define the origin of each energy consumption variation. In this sense, statistical procedures are centered in real sampled data and can give us the physical relation between variables until reach the particular physical laws of each process.

This is the reason why, in this research work, to look for the origin of the ventilation effect overbuilding energetic behavior, the calculation procedure of the ISO 13790 standard was automated by means of Visual Basic to be, after tested with another software resources, employed as calculation tool of the Monte Carlo Method. The Monte Carlo Method consist of introducing data in a random way in a system and to define the frequency of the output values. This procedure, for the particular case of energy consumption in buildings, will allow to define the type of curve and probabilistic distribution of the energy consumption in buildings as a consequence of modifications. This probabilistic density can be related to the fiscal behavior of this energetic system. In this sense, in previous works (Haarhoff and Mathews, 2006; Keirstead and Shah, 2011; Mavrotas, Florios, and Vlachou, 2010), this calculation procedure showed as main result that the probability of energy consumption takes the distribution of a Weibull model (Gang, Wang, Shan and Gao, 2015; Lu, Huang, and Zhang, 2013; Prabhakar-Murthy, Bulmer and Eccleston, 2004; Rao, 2005; Ren, Yan, and Wang, 2014). Furthermore, a map of different Weibull models in accordance with different input parameters were obtained. It was a really amazing result due to not only ventilation rate and other parameters like thermal inertia, orientation and much other input variables can be identified with building energy consumption as a Weibull model distribution. What is more, each probability density function identified with a Weibull model just experiment a modification of its constants between each building modification. In consequence, to understand the meaning of these constants may allow to travel from actual building conditions to another one considered as optimal with lower energy consumption, healthy indoor conditions or more comfortable indoor ambiences. Research about the origin of this Weibull model and the meaning of its constants may give a retrofitting tool and an in-depth understanding never explained before in scientific research works. What is more, adequate control of this model may give the optimal retrofitting strategy towards an optimal design with the lowest energy consumption (Masdías- Bonome, Orosa and Vergara, 2020).

Finally, it is interesting to highlight that most of the software resources for the energetic understanding of building energy consumption does not have too much feedback options that let researchers to define the optimal building design due to the so complex calculation procedure needed. Just the new versions of CE3X software shows indications of possible input modifications that may help architects to improve their building design. Despite this, this software is based on data bases defined by results from more details software resources like Energy Plus and, in consequence, despite the so easy application and useful improvement proposals, these are not too much concise results and a really limited option is identified. This is another interesting reason to develop a more detailed analysis of the Weibull models obtained.

In summary, the present paper shows a new procedure to be developed to define the effect of different changes in building energy consumption, like climate change, over Weibull model distributions. In particular, this work is centered into air changes due to seeming to be the simpler modification to do in an existing building.

MATERIALS

Weather Stations

Climatic data were collected through a sampling process in 50 weather stations located in central and remarkable points in Galicia (ASHRAE, 2017; MeteoGalicia, 2020) sampling temperature, relative humidity, and wind speed variables with a sampling frequency from about 5 to 10 min. These weather stations were chosen for the purpose of this study because they ignore buildings and other interferences that could modify the sampled data. Finally, an error margin of the sampling devices employed in these weather stations are of 0.1ºC, 0.2%, and 0.1 m/s were considered acceptable.

Building Objective of this Study

A real building was designed and tested in accordance with ISO 13790 calculation procedure and validated with CE3X software in 2.3 version, as it was shown in previous works (Masdías-Bonome, Orosa, and Vergara, 2020). This CE3X software was developed by the Spanish Ministry of Industry as a useful free tool for building energy rating in Spain.

The building objective of the study is a residential building that was analysed and, after this, was retrofitted by the technical offices towards an expected reduction of the energy consumption doing these modifications based on their common sense and working experience. In consequence, both conditions were tested in accordance with the ISO 13790 calculation procedure and the CE3X software. What is more, the initial step was to sample real indoor conditions and real energy consumption to compare the calculation procedures with real sampled data. Despite the fact that these standards and the more complex software resources develop their precise analysis, it was selected the CE3X software due to its reduced time of configuration and calculation. In consequence, the CE3X software has as input parameters:

- Climate zone: C1 (according to D.B. HE1 Spanish construction law)
- Living area: 2343 m^2
- Number of floors: 5
- Year of construction: 1987 (applicable Spanish standard NBE CT 1979)
- Facade wall (double ceramic brick wall with non-ventilated chamber): U=1.69 $W/m^2 \cdot K$
- Building roof (roof with concrete slab and tile): U=2.27 W/m^2K•Soil: U=0.47 $W/m^2 \cdot K$
- Air changes: 0.63 Ach/h•DHW: 1771 litres/day (at 60oC)
- Boiler: standard diesel boiler with 1 burner stage (made in 1980)
- Walls transmittance: 1.69
- Roof cover transmittance: 2.27 $W/m^2 \cdot K$

Software Resources HULC and CE3x

The Spanish Ministry of Industry, Energy and Tourism in collaboration with the Spanish Ministry of Development fixed as a fundamental objective to create a software resource for building energy rating. In consequence, to reach this objective and to define an initial building model, in agreement with the RD 235/2013, these are the homologated software for this function in Spain:

- Lider Calener 2013 unified tool (HULC)
- CYPETHERM HE Plus by CYPE Ingenieros
- SG SAVE
- CE3x SOFTWARE
- The CERMA computer program

The first one software resource is the Lider Calener which is a unified software that joins the Lider and the Calener Vyp developed by the Thermotechnology department of the University of Sevilla (Spain). This tool obeys article 4.1 of the Royal Decree 235/2013 being the procedure for building energy rating in Spain. His calculation engine is derived from DOE 2, which was developed by the Energy department of the United States. What is more, this software considered the last modification of the technical code of edification (HE0) and the last modifications of the regulation of thermal installations in buildings RITE. CE3X is an equivalent software that was developed by Efinovatic and Spanish National Centre for Renewables Energies (CENER) and was validated by the Spanish Ministry of Edification. This software is a simplified tool for energy rating of new or existing buildings and, in recent versions, it attaches the economical investment needed. It can be considered as useful information not defined in most of the previous software resources. Another of its advantages is the reduction the time to introduce the data in the software and to simulate and obtain initial results. In consequence, it is one of the more employed software for energy rating in Spain. As it was explained before, it is interesting to highlight that these advantages are centred into the fact that this software does not solve the physical equations of Calener and just employ the results from a database of previous simulations done in Calener and extrapolate the expected results for the building objective of study.

METHODS

As it was commented before, the methodology employed in the present study aims to define the future probability density function of energy consumption during the winter and summer seasons in accordance with a near climate change.

Climatic Change Predictions

In order to define the climatic change that occurred and will occur in north-western Spain, the models obtained in previous research works were employed to define its effect over buildings energy consumption (Barkhudaryan, Orosa and Roshan, 2014). These research works show an adequate curve fit for each season expressed as linear models of Equations 1–4. Each equation reveals the average rise in the outdoor temperature that occurred in Galicia during the last 30 years with regard to the temperature of 13.63 °C, which corresponds to the average temperature in 1987, as shown in Figure 2.

$$\Delta t_{Winter} = 0.05 \bullet \tau \tag{1}$$

$$\Delta t_{Spring} = 0.07 \bullet \tau \tag{2}$$

$$\Delta t_{Summer} = 0.05 \bullet \tau \tag{3}$$

$$\Delta t_{Autumn} = 0.03 \bullet \tau \tag{4}$$

Where Δt is the mean increment in the seasonal average temperature (°C), and τ is the time (years)

From these models, it was very easy to define the expected climate change in the next 20 years, as shown in Figure 2. Once these relations were obtained, they could be used for the prediction of climate change that will occur in Galicia during the next 20 years. In particular, in accordance with ISO standard calculation procedure, the summer and winter seasons are the more interesting periods. In this sense, the models for these two seasons were defined with the same constant models and, in consequence, present the same increment in the temperature. Finally, the year 2030 presents a special interest because the increment in the outdoor temperature will reach 1°C.

Probability Density Function Models and Monte Carlo Method

In the present research work, a new methodology for selecting a more suitable development for the reduction of building energy consumption due to climate change will be discussed. This methodology is centred in the Monte Carlo method, which is defined as one of the most interesting stochastic approaches to real life problems, such as ventilation strategies between others.

In essence, the proposed procedure considers random input variables to a typical problem defining the probabilistic density distribution of the same problem. Selected random variables, in accordance with ISO Standard 13790 and recent research works (Fei, Zhou, Mai and Li, 2014; Šadauskienė, Paukštys, Šeduikytė and Banionis, 2014), were air changes, thermal inertia, and the number of occupants. Despite this, not all these parameters will exert an effect over building energy consumption in yearly studies. During this calculation procedure, in accordance with previous works (Karlsson and Fahlén, 2008; Mazarrón,

Cid-Falceto and Cañas, 2012; Nematchoua, Tchinda, Orosa and Andreasi, 2015; Orosa, 2012; Orosa and Oliveira, 2012), it was obtained that thermal inertia must only be considered in the instantaneous energetic analysis and not in a yearly study. It is the thermal inertia that resulted constant under the same weather conditions. On the other hand, the other input parameter in this ISO standard is the number of occupants, which can be considered as another heat source. In consequence, the present work will be centered in the number of air changes and its relationship with climate change and energy consumption. In consequence, the study will be centered in the effect of climate change over air consumption.

Once the probabilistic density distribution of each particular case was identified under the same weather conditions, an energetic map can be obtained is the most suitable tool to define the effect of each different input parameter over the energy consumption.

In the present paper, a new methodology will be employed in real buildings, in accordance with ISO 13790 Standard, for defining the most suitable actuation step at the time of improving the energy consumption in buildings with respect to a near climate change. In particular, this research work is centered in just one variable to be optimized; the ventilation rate. In consequence, the aim of this case study is to obtain charts that show probabilistic curves of energy consumption due to ventilation rate remaining constant all the other input parameters of ISO 13790 Standard. In this sense, the selected building is located in the north-west of Spain and was previously analyzed in depth showing real sampled data for long periods of time. The values of outdoor and indoor air temperature were proposed as random within the real sampled data limits. Finally, its corresponding histograms were developed, as shown in Figure 3.

Based on this initial information, in the base case, the limits of the random Monte Carlo process was defined between 20–27° C during the summer and between 14 to 24.4ºC for the winter season. The number of air changes was fixed at 0.8, thermal inertia was 150 kJ/m^{2}K, and the number of occupants was fixed as 15 with a reduced intensity of use, consumption of domestic hot water of 1771 l/day (at 60ºC) and 74910 kWh (m^2 year) of heating energy consumption. More detailed information about the building objective of this study showed in the research work (Masdias-Bonome et al., 2020).

Subsequently, the probability density function of heating and cooling energy consumption for different air change values was defined, as shown in Figures 4–13.

This random process was developed in Visual Basic for Applications with more than 60,000 iterations for each different curve to obtain an adequate correlation factor during the curve fitting process. However, after a curve fitting in accordance with the most common models, a Weibull density distribution was selected. In consequence, this adequate curve fitting could be identified by a correlation factor (over 0.9) between the simulated data and the Weibull model of Equation 5. Results obtained are listed in Tables 1 and 2.

$$y = a\left(\frac{d-1}{d}\right)^{\left(\frac{1-d}{d}\right)}\left[\frac{x-b}{c}+\left(\frac{d-1}{d}\right)^{1/d}\right]^{d-1}\exp\left[-\left(\frac{x-b}{c}+\left(\frac{d-1}{d}\right)^{1/d}\right)^{d}+\frac{d-1}{d}\right] \quad (5)$$

where x is the input variable and a, b, c, and d are the model constants.

RESULTS

As was explained before, the building objective of this study is placed in the city of A Coruña. The information obtained from the weather stations showed that the city, placed few meters above sea level, experiments a warm and temperate climate. The winter months are much rainier than the summer months. In this sense, most of the precipitation here falls in November and the driest month is July. At the same time, the average temperature is 13.5°C being August the warmest month of the year with an average temperature of 18.7°C and February the coldest month, with temperatures averaging 9.0°C.

From Figure 2, it can understand that, in the next 20 years, there will be an increase in the outdoor air temperature of 1ºC for both the summer and winter season. This is because they present the same model constants of Equations 1 and 3.

The sampled building is a residential building built in 1987 with a living area of 2343 m^2 in a total of five floors. At the time to identify the main variables related with building energy consumption, special attention was paid to the façade wall and roof, as it was described in materials and methods section. At the same time, for each probability density function, air changes were defined with different values, but remaining constant the other variables of the iterative process.

Once Monte Carlo Method was applied over building energy consumption Standard 13790, with more than 60,000 runs, the frequency of each different value was defined, as shown in Figure 3. This histogram can be curve fitted in accordance with Weibull distribution model of Equation 5 with an adequate correlation factor; these equations were represented in Figures 4 and 5.

Figure 2. Temperature increment in next years due to climate change over winter and summer seasons in northwest of Spain

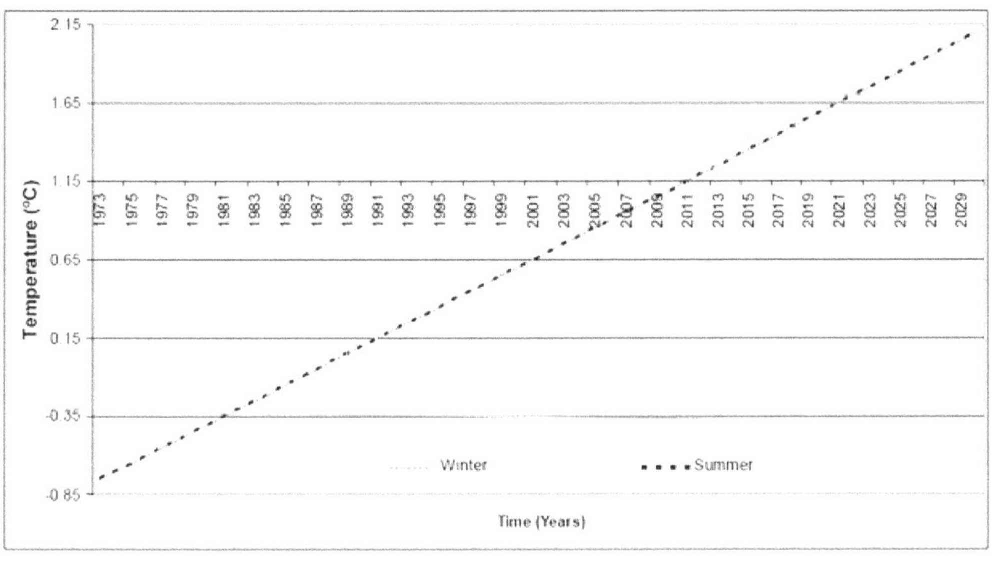

Figure 3. Histogram of energy consumption during the heating season

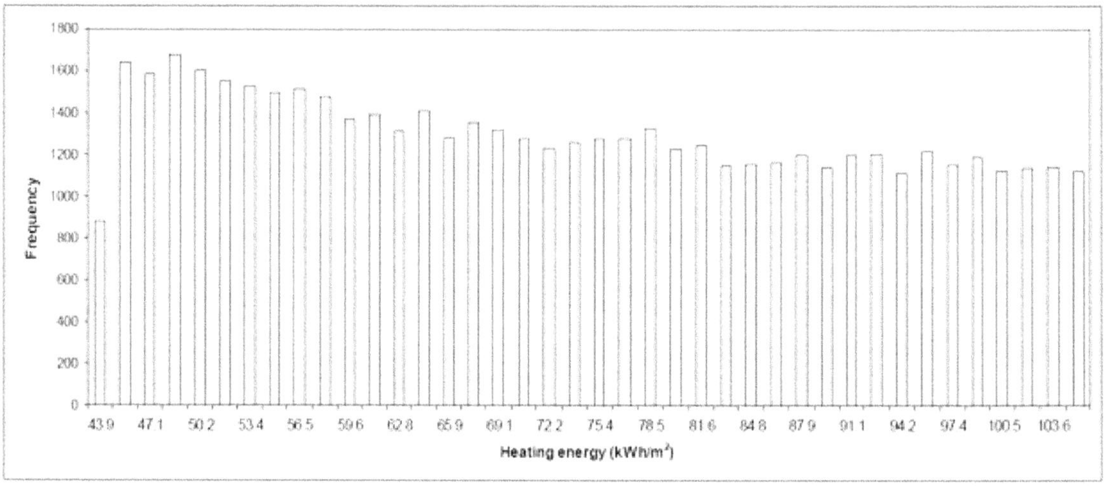

Figure 4. Probability density functions of cooling energy consumption for different air changes (today)

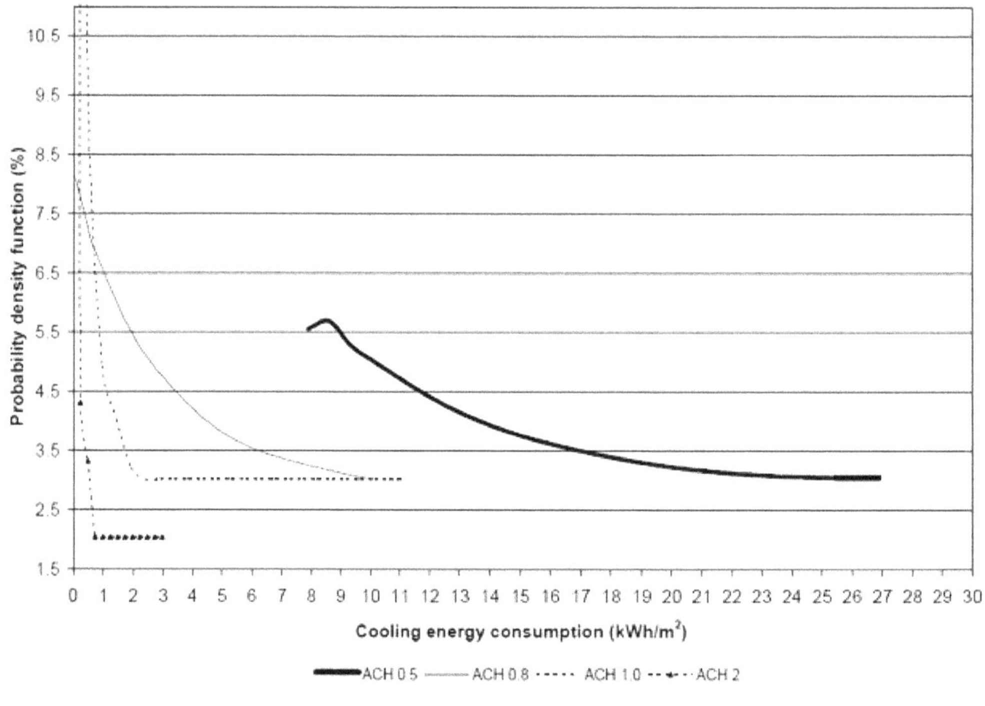

After simulating the actual conditions, in accordance with the values obtained from climate change models, Monte Carlo method was applied over the winter and summer seasons with an increase in the mean outdoor temperature of 0.5°C and 1°C, respectively. It must be remembered that this temperature increment corresponds to the result of a climate change from today to 10 and 20 years later, respectively. As a result, and after a curve fitting process for each different input values and each different weather conditions, Figures 6–13 have been obtained.

Figure 5. Probability density functions of heating energy consumption for different air changes (today)

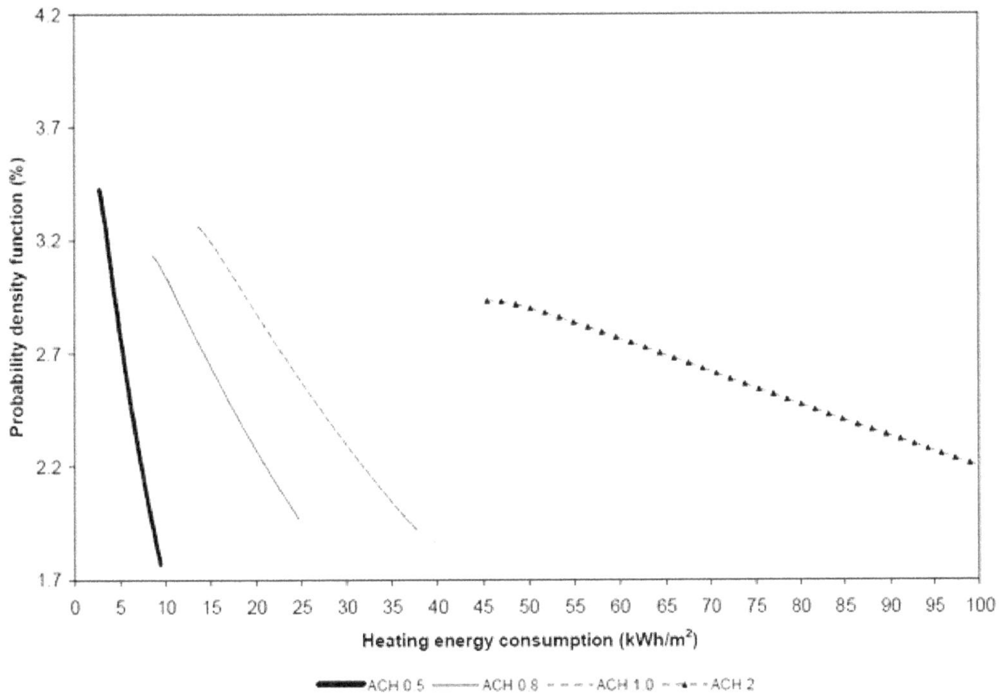

Figure 6. Probability density functions of cooling energy consumption for different air changes in the next 10 years (+0.5ºC)

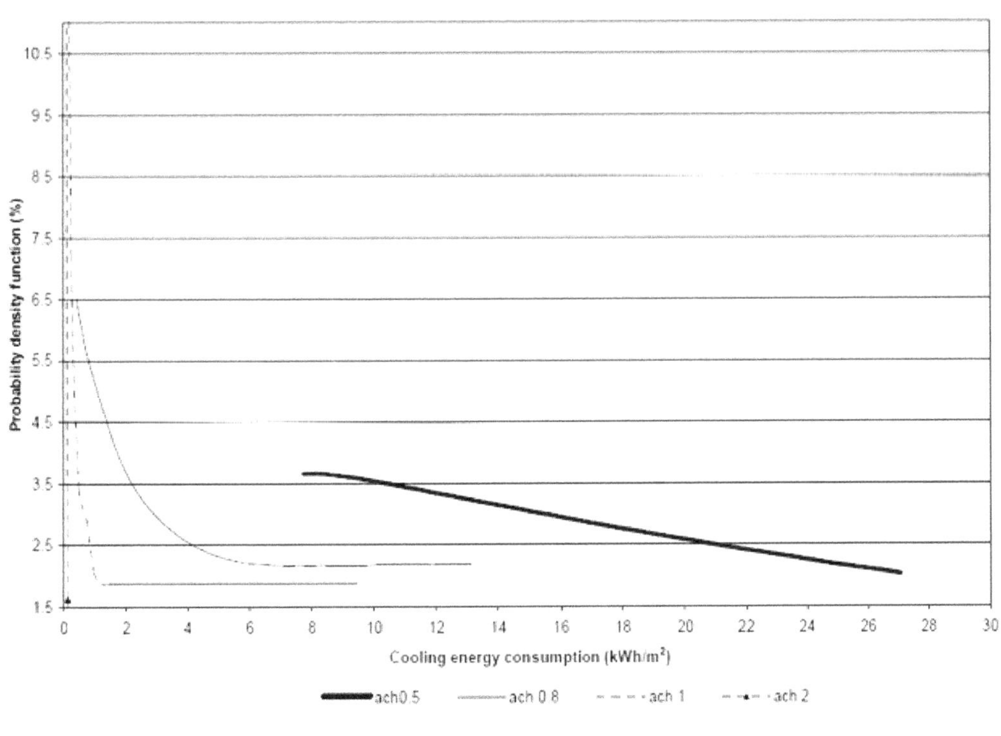

Figure 7. Probability density functions of heating energy consumption for different air changes in the next 10 years (+0.5°C)

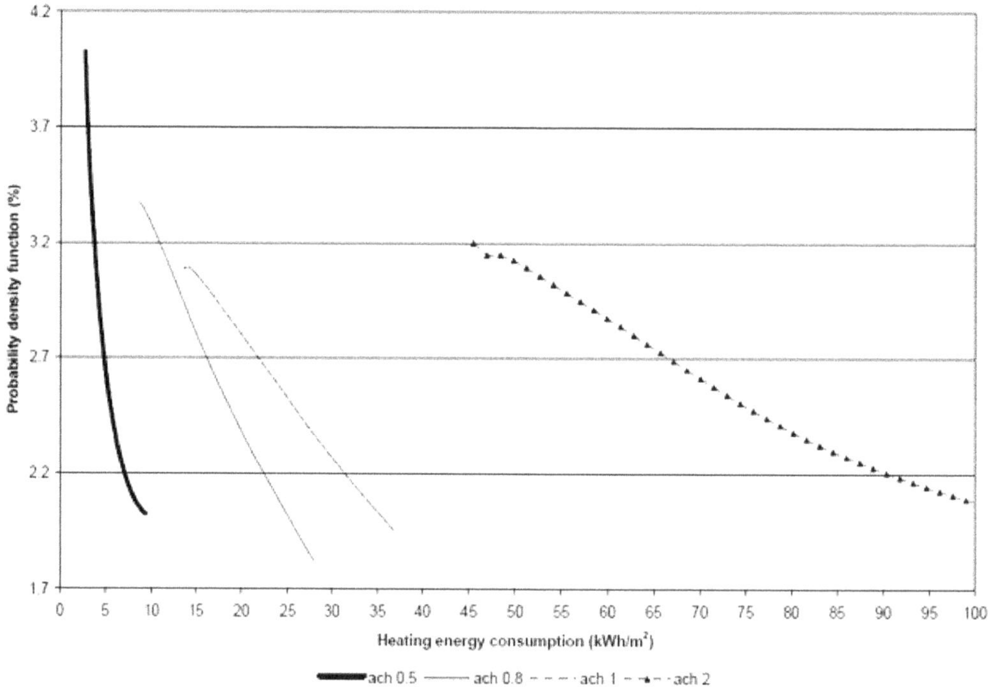

Figure 8. Probability density functions of cooling energy consumption for different air changes in the next 20 years (+1°C)

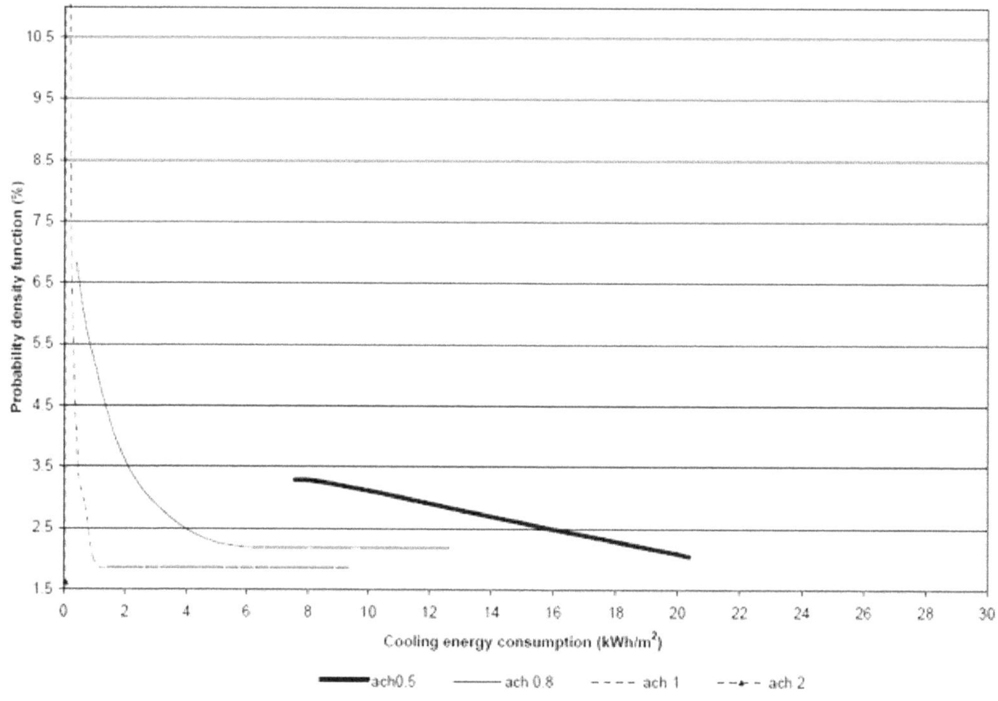

Figure 9. Probability density functions of heating energy consumption for different air changes in the next 20 years (+1ºC)

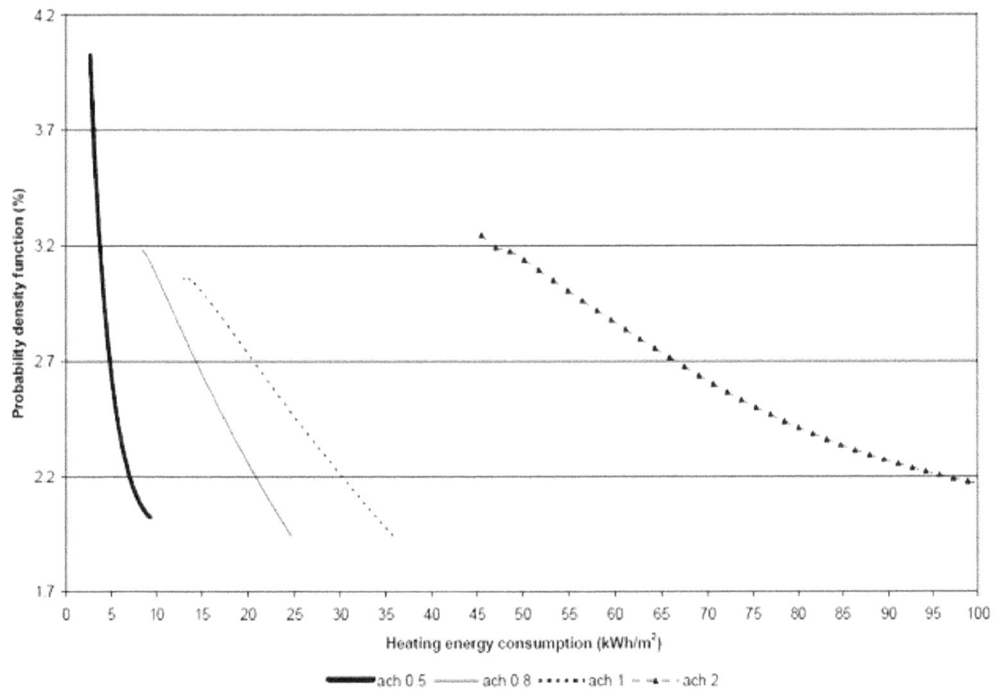

Figure 10. Effect of increment in the outdoor temperature over the probability of cooling energy consumption for low air changes

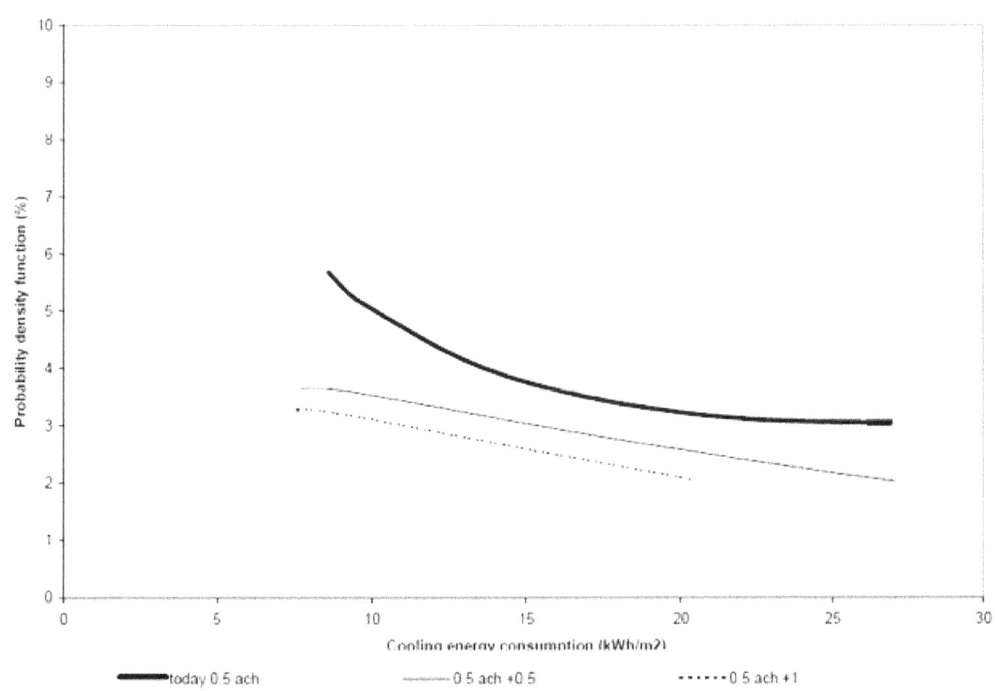

Figure 11. Effect of increment in the outdoor temperature over the probability of heating energy consumption for low air changes

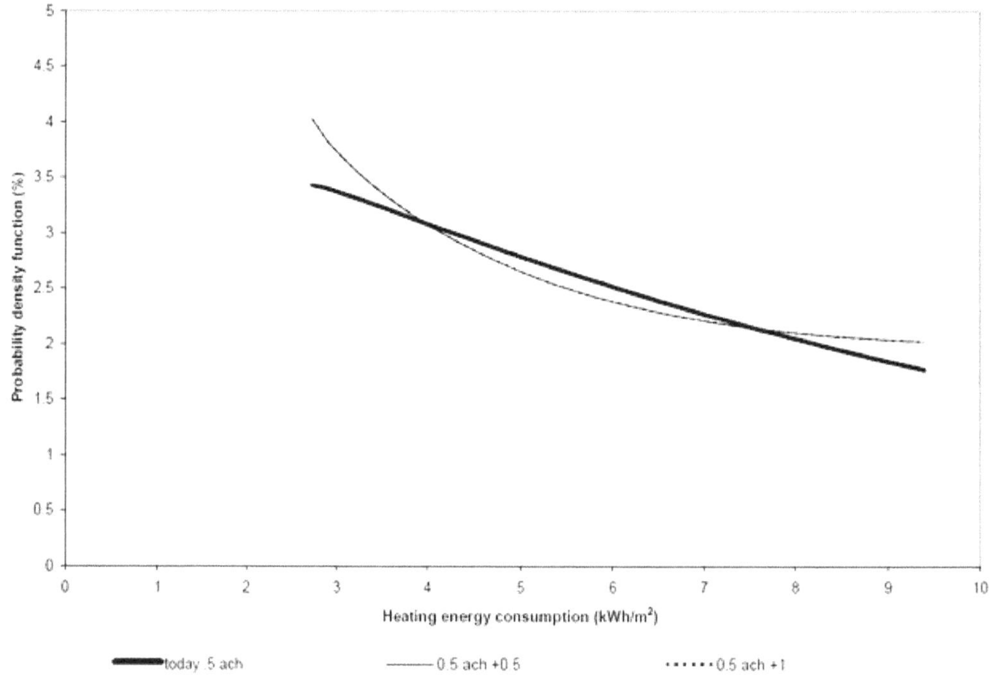

Figure 12. Effect of increment in the outdoor temperature over the probability of cooling energy consumption for high air changes

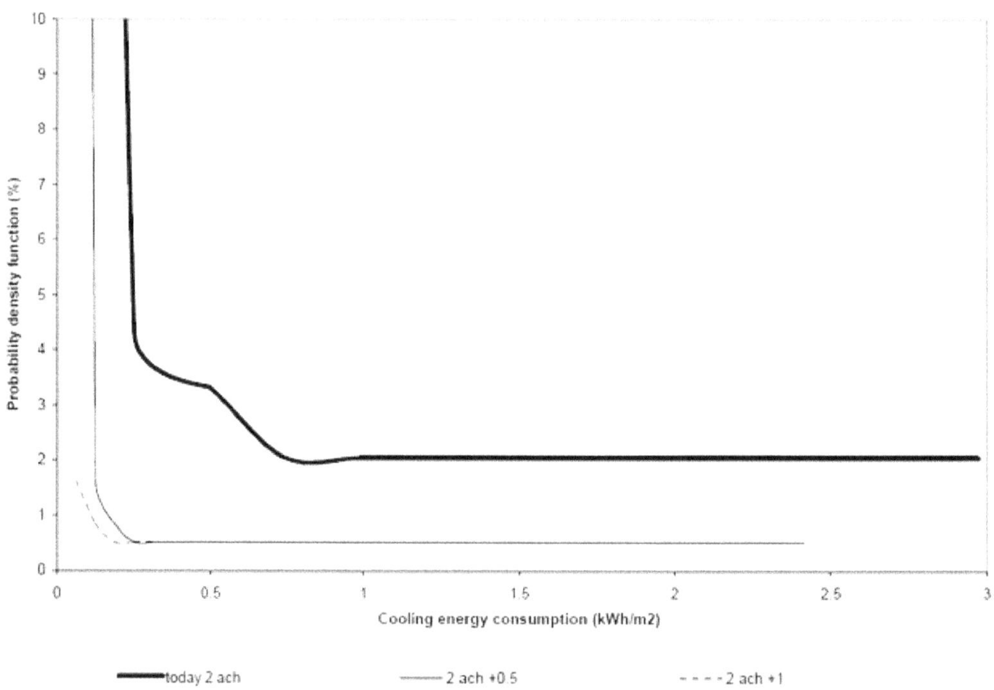

As shown in previous works, the effect of thermal inertia does not exist in this kind of yearly studies about energy consumption. In consequence, instantaneous analyses must be developed to consider this effect and present study will be centered in the effect of the number of air changes and climate change. Other parameters like the number of occupants are considered as another heat sources, and therefore, curve fitting was developed.

As comparative parameter between Weibull models, the value of the constants "*a*" and "*b*" of each model can be used, as shown in Tables 1 and 2, as well as the point of maximum probability value of each curve. For example, in Figure 4, the maximum probability of cooling energy consumption when the air temperature changes by 0.5 ºC is about 5.5% of 8 kWh/m^2. At the same time, the most probable energy consumption under 0.8 ºC air changes is 6.5% of 0.25 kWh/m^2. It is in accordance with the common sense notion because as higher are the air changes, lower are the energy consumption to cool a hot ambience with temperatures under outdoor conditions. In consequence, from Figure 4, it can be concluded that, for the cooling season, as higher are the air changes as lower is the probability of cooling energy consumption.

If the effect of climate change over this present situation is analyzed, the conclusion obtained from Figures 6 and 8 is that the maximum value for the lowest air changes of 0.5 ºC is about 8 kWh/m^2, like in previous cases. Despite this, its probability is reduced as outdoor temperature increases. For example, it is about 3.6% when outdoor air increments 0.5 ºC and 3.3% when outdoor air increases 1 ºC. This effect over time is being diffused when the number of air changes increases.

When the effect of air changes over heating energy consumption is analyzed, the conclusion obtained from Figure 5 is that the number of air changes increases, the heating energy consumption increase maintaining nearly constant the probability of the most probable value of each curve. For example, it can be seen in Figure 5 that the most probable energy consumption for 0.5 ºC air changes is about 3 kWh/m^2 for a probability of 3.5%. At the same time, for two air changes, a probability of 3% for a heating energy consumption of 47 kWh/m^2 is found.

Now, if the effect of climate change over energy consumption during the heating season is analyzed, it can be observed from Figures 7 and 9 that the probability of the most probable value of each curve increments with outdoor temperature. For example, in Figure 7, the maximum probability value for 0.5 ºC air changes is about 4.1% with respect to 3.5% for the same number of air changes today, as shown in Figure 5. Finally, it must be noted that this effect is nearly the same for an increase in the outdoor weather temperature of 1 ºC with respect to 0.5 ºC.

Now, after obtaining the effect of climate change over building probability density function distribution of energy consumption, it is interesting to develop a comparison of the same curves at different time moments, as shown in Figures 10–13.

From Figure 11, it can be observed the effect of climate change over heating energy consumption under low air changes. Furthermore, it can be detected in the same figure that when outdoor temperature increases, heating energy tends to be at a higher probability under lower energy consumption values when compared with present conditions. For example, in Figure 11, an energy consumption of 3 kWh/m^2 presents a probability of 4.2% in the next 10 or 20 years and 2.5% in today's time. A clearer effect can be observed when the number of air changes increases, as shown in Figure 13.

When cooling energy consumption is analyzed, as shown in Figure 10, under low air changes, the following conclusion can be drawn: there is a higher probability of the lower energy consumption (7 kWh/m^2) than in future years. This effect remains for high air changes, as shown in Figure 12.

Figure 13. Effect of increment in the outdoor temperature over the probability of heating energy consumption for high air changes

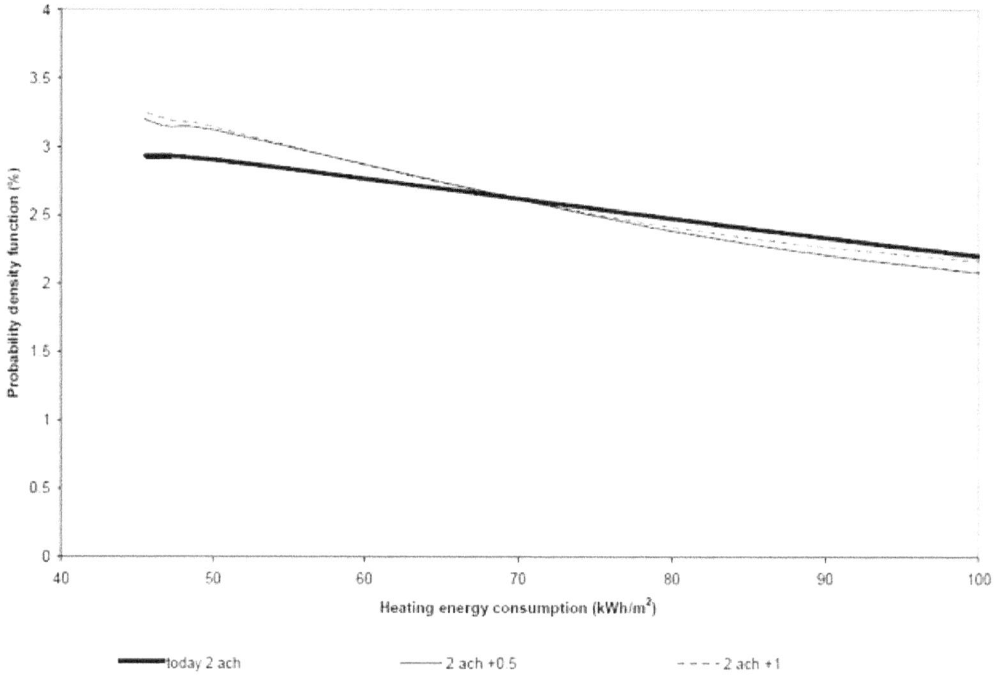

After analyzing the effect of climate change over the probability of building energy consumption, it is necessary to analyses each of Weibull models given in Tables 1 and 2.

Table 1. Curve fitting of probability density function during the heating season

Heating		*a*	*b*	*c*	*d*	*e*	r^2
ACH 0.5	Today	3.363271	2.661263	9.437849	1.011474		0.923202
	+0.5 ºC	1.854899	2.080979	2.565297	2.295291	1.007102	0.971939
	+1 ºC	1.854899	2.080979	2.565297	2.295291	1.007102	0.971939
ACH 0.8	Today	3.090611	8.386542	32.028872	1.010309		0.930000
	+0.5 ºC	3.304539	8.415014	29.33913	1.012122		0.905596
	+1 ºC	3.124542	8.369421	30.40935	1.010302		0.908827
ACH 1.0	Today	3.212779	13.38395	42.19215	1.010308		0.918687
	+0.5 ºC	3.077777	13.48641	45.60571	1.011666		0.925038
	+1 ºC	3.050013	13.48202	45.16667	1.011673		0.928822
ACH 2	Today	2.882833	46.04790	162.5114	1.012475		0.850571
	+0.5 ºC	1.512973	1.519573	48.13923	39.73227	1.093759	0.909674
	+1 ºC	1.862788	1.223835	47.25478	30.64512	1.095933	0.939802

Table 2. Curve fitting of probability density function during the cooling season

Cooling		*a*	*b*	*c*	*d*	*e*	r^2
ACH 0.5	Today	2.794215	2.479324	8.207154	5.324309	1.061680	0.912479
	+0.5 ºC	3.618063	7.566651	29.62327	1.013869		0.829317
	+1 ºC	3.273713	7.463340	24.62232	1.0127		0.84182
ACH 0.8	Today	3.668553	9.123290	-2.01957	3.30911	1.058540	0.95
	+0.5 ºC	3.273713	7.463340	24.62232	1.0127		0.84182
	+1 ºC	1.973009	4.136657	0.203551	1.313985	1.085427	0.811232
ACH 1.0	Today	2.734177	3321.353	-3.61981	3.85073	3.236254	0.99092
	+0.5 ºC	3.077777	13.48641	45.60571	1.011666		0.925038
	+1 ºC	3.050013	13.48202	45.16667	1.011673		0.928822
ACH 2	Today	1.449848	347.4151	-0.27093	0.463775	3.049235	0.999109
	+0.5 ºC	1.512973	1.519573	48.13923	39.73227	1.093759	0.909674
	+1 ºC	0.447502	149105.2	-0.3577	1.293282	10.01984	0.999786

In this sense, it is worth remembering that constants *a* and *b* can be related with each curve and its corresponding energy consumption. It must be observed that most of correlation factors present a high value showing a good agreement between Weibull model and the probability of energy consumption that can be found today and in the next 20 years.

After analyzing these constants, a three-dimensional model that interrelates the most probable energy consumption under each different situation (constants *a* and *b* and the number of air changes for today and in the next 20 years could be of interest. This model was defined for the heating and cooling seasons and was represented in Figure 14 and Equations 6 and 7 with an adequate correlation factor of 0.916104565 and 0.8839247728, respectively:

$$Air\_changes = c + \frac{d \cdot a}{\ln(a)} + \frac{e \cdot \ln(b)}{b} \tag{6}$$

where *a* and *b* are Weibull model constants, $c = -1.6892256$, $d = 1.473976685$, and $e = -6.05859561$.

For the cooling season, new constants for the 3D equation were obtained, as shown in Equation 7. This equation is similar to the previous case but with a lower correlation factor, as it happened with Weibull models.

$$Air\_changes = c + \frac{d \cdot a}{\ln(a)} + \frac{e \cdot \ln(b)}{b} \tag{7}$$

where *a* and *b* are Weibull model constants for cooling season, $c = -2.55911881$, $d = 1.193927337$, and $e = 0.000417658$.

This result is a useful means for defining the most probable energy consumption of an existing building or a building that is being designed today and in the near future. In consequence, it is a tool for building design for preventing the higher energy consumption due to the climate change effect. This amazing result must be analyzed in future research works.

Figure 14. 3D relationship between Weibull model constants for heating energy consumption for the next 20 years

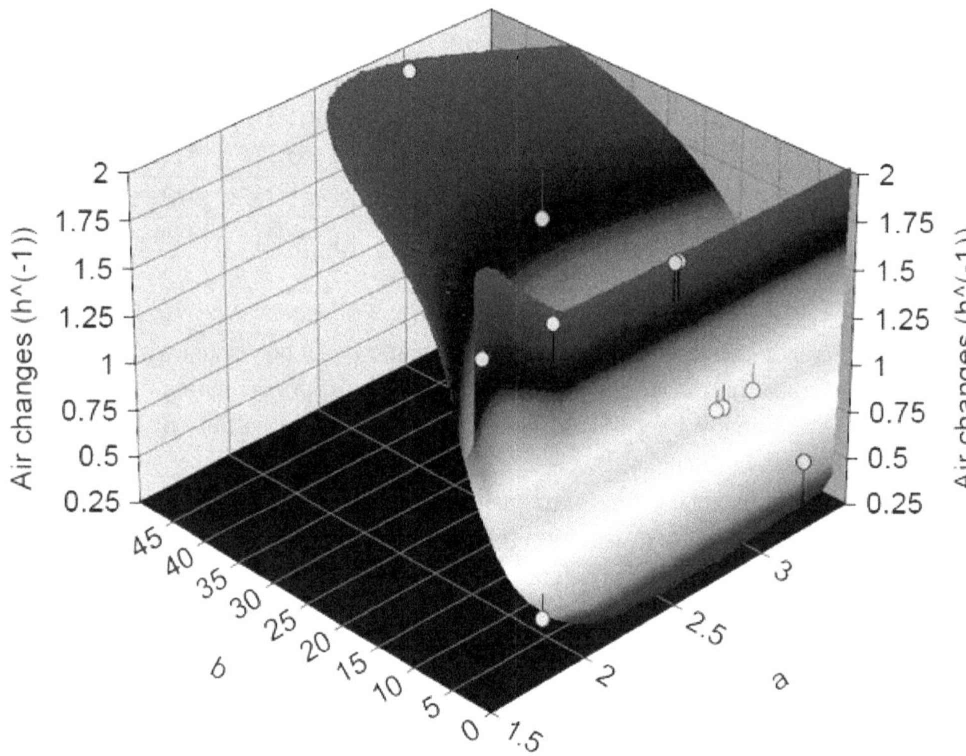

DISCUSSION

The present paper shows an interesting idea about the knowledge in heat and mass transfer and laws of thermodynamics when they are applied in a building system. The identification of these principles at the particular case of buildings was identified by different national standards and international research works. These standards give a response about the energy system (building) behavior under different conditions but there is not a well-known understanding of this response. In consequence, an inverse engineering procedure is needed to understand the meaning and the physical implications of any change. This inverse engineering process will analyses the output from the building model, defined by the standards, and will give probabilistic distributions that must be compared with well-known physical models usually employed in most of engineering and physic processes. Once the related physical models were identified with each building modification or input variable change, it can be developed a posterior analysis to identify how to improve this process towards the design objectives like energy saving, thermal comfort or indoor air quality, for instance. Despite the so difficult process that its inverse engineering analysis implies, it will give the scientific knowledge of the system objective of analysis and will let researches to improve his/her knowledge about this. Similar engineering applications, like artificial neural networks, may give a particular better perdition of the evolution of the objective variable (energy saving, thermal comfort or indoor air quality) and this is the reason why they are proposed as control algorithm in the control systems of Heating Ventilation and Air Conditioning Systems (HVAC). Despite this, it is an engineering solution to a particular case study that is centered in a high mathematical resolution capacity based on iterative processes but it has some control limitation and works without any understanding of the origin of the process and its physical implications.

The objective of the present paper was to show an initial step towards building energy understanding based on one of the traditional well-known mathematical procedures: the Monte Carlo Method. This mathematical procedure is usually employed to understand the response of a system after any input variable modification. In the particular case study of the present research work, this is the ISO13790 standard which, despite the fact it is more complex than nowadays energy rating software resources, it is a well-known standard. What is more, due to the so wide possibility of input variable modifications of building energy systems, it was selected the air changes due to, it has some advantages respect the constructive parameters like, for instance, to correct thermal inertia in walls or size and orientation of windows. In this sense, air changes is easy to be implemented variable in existing buildings and are related to most of the common objectives in building improvements. It is related with the energy saving in air conditioning systems, related with the indoor air quality in accordance with the air changes and related with the general and local thermal comfort conditions based on the way this outdoor air is placed in each indoor ambience.

As it was explained before, another so interesting parameter is considered at the time of doing these simulations; a change in the outdoor average temperature. It may be identified with a change in the placement of the building or a future climate change due to both considerations will imply a change in this average outdoor temperature. It is true that other related changes, like a modification of the average outdoor relative humidity, may imply a real modification of the indoor thermal conditions. Despite this, average outdoor relative humidity is not usually defined as a representative value for a region and, in consequence, complex conclusions can be derived from this analysis.

Once the ISO 13790 standard was calculated in an automatic way, it was undergone to an automatic calculation process in accordance with Visual Basic programming of more than 60,000 iteration per each building system proposed modification. A convergence algorithm was employed to predict the adequacy of this number of iterations. This is related to the real approach of the probability density function obtained.

At the same time, a Matlab Curve fitting process was done with more than 60,000 mathematical models to look for the better model that can explain the behavior of each modification of air changes in the building energy consumption. After this calculation process a Weibull distribution was obtained as the better one with an adequate determination factor and, in consequence, the better that explains the modification of building energy consumption after any modification of air changes or average weather temperature.

It is of real interest to observe a determination factor most of the time over 0.9, as it was showed in the results section, Tables 1 and 2. Few other models may reach a similar accuracy than the Weibull model defined by equation 5. In particular, the other models proposed are again Weibull models with a minor number of parameters and a determination factor that is near 0.9 but always lower than that offered by the proposed model.

Weibull model is a so interesting mathematical model that is employed in different research areas like, for instance, in equipment's maintenance due to it has a great approach to the frequency of failures in accordance with time and can be employed to predict the planned obsolescence and, in consequence, the moment in which corrective action must be done. To reach this objective, the Weibull model experiments a modification of its distribution in accordance with the values of its constants changing form a logarithmic to a normal distribution and to an exponential one depending on the time period in which the process is analyzed.

In the particular case study of the present research work, once identified the Weibull model as the one that can identify the energy consumption as a modification of air changes and outdoor air temperature, in a similar way than with maintenance techniques, it was analyzed the effect of model constants over the curve behavior. These constants were showed in Tables 1 and 2 and can be defined as so interesting parameters due to, in accordance with the value of constant *e*, it can be related with the type of curve obtained. It is the so called "shape parameter" and can help to define its better value of the control variables (air changes and weather temperature) to reach the proposed objective (energy consumption, indoor air quality). Based on a simple extrapolation of the mathematical procedure employed in maintenance tasks of most of the industrial equipment, when this value is similar to 1, the curve reaches a normal distribution and is the moment in which the area below the curve is minimum being an interesting objective to reduce, for instance, the energy consumption.

What is more, it is possible to define a mathematical model that relates all the constants as a function of the air changes and the model constants employed with an adequate determination factor. It must be a really useful model, for future applications of building energy optimization.

The importance of this new methodology can be summarized in a new way to define the better ventilation level for a healthy indoor ambience with the lower energy consumption. What is more, this same procedure can be employed with other input variables of the energy certification procedure like walls transmittance or roof cover transmittance, for instance, and to obtain the probabilistic distribution of the energy consumption due to each different value of each analyzed variables.

There are a lot of applications of this methodology. For instance, recent papers showed that this process can be employed to predict the better yearly period in which an increase in ventilation rate due to pandemic conditions needs. Results showed that, as it is common sense, it is in the summer period when the number of air changes, in a mild climate of the Galicia region, the energy consumption will be the lowest and the risk of aerial infections will be easily prevented.

Despite this, a more in-depth analysis of these results show a new optimization possibility; it is possible to define some modifications that imply a change in the Weibull model constants of a determined building and, in consequence, to change the period in which the normal distribution appears (associated with the maximum energy consumption). In consequence, it is possible to modify some building characteristics to reduce the energy consumption under a high ventilation rate in a cold period, which may be a useful application to improve indoor air quality and prevent aerial contagions, with lower energy consumption.

Finally, these are just examples of energetic understanding and optimization when ventilation rate is considered and remaining constant the other implied variables. It is a so complex process, but it can be done the same calculation for each different input variable, like wall transmittance, and new curves will appear. The final objective is to jump from one initial condition to another optimal condition acting over one or more input variables.

FUTURE RESEARCH DIRECTIONS

As a consequence of the results of these previous research works, there are many engineering applications in which this mathematical tool –once validated with further case studies– can help researchers to optimize building designs. In this sense, these mathematical models may be of interest to design buildings with lower energy consumption at really high ventilation rates, as explained in previous works to prevent contagion in pandemic conditions.

Other works can be developed based on these previous models. For instance, it can be employed to design buildings, considering the real climatic conditions and those expected in the near future (next 50 years). What is more, acting on some design parameters, when necessary, it may help to do buildings more suitable for multiple types of climates, not only for today's climate for which they are usually designed. In this sense, with an adequate understanding of these probabilistic density functions, it is possible to modify some easy to modify implied variables, like air changes and type of glasses employed in windows to make a building more adequate for different climatic regions.

Finally, there are many pending studies to be carried out analyzing the effect of other variables such as new materials, phase change materials, permeability, orientation, glazed areas, etc., that can help to improve the trends of previous designs.

CONCLUSION

In the present research work, a new method to define building design improvements for reducing energy consumption due to climate change will be discussed. In particular, results showed that Monte Carlo method applied over ISO 13790 standards are a useful tool to define the expected effect of building energy consumption. Further, it was obtained that the probability of building energy consumption can be defined by a Weibull model.

Results showed that, for the cooling season, the higher the air changes the lower the probability of cooling energy consumption. Furthermore, results also showed that if the effect of climate change over this present situation is analyzed, it could be concluded that this probability is reducing as outdoor temperature increases and is diffused when the number of air changes increases.

On the other hand, when analyzing the effect of air changes over heating energy consumption, it can be found that when the number of air changes increases, the heating energy consumption increases maintaining nearly constant the probability of the most probable value of each curve.

Now, if the effect of climate change over energy consumption during the heating season is analyzed, the conclusion is that the probability of the most probable value of each curve increments with the outdoor temperature and this effect is nearly the same for an increment in the outdoor weather temperature of 1ºC with respect to 0.5ºC.

After obtaining the effect of climate change over building probability density function distribution of energy consumption, it is interesting to develop a comparison of the same curves at different time moments. Results showed that when outdoor temperature increases, heating energy tends to present a higher probability under lower energy consumption, when it is compared with the present conditions. A clearer effect can be observed when the number of air changes increases.

When analyzing the cooling energy consumption under low air changes, it can be concluded that there is a higher probability of the lower energy consumption than in future years and this effect will remain for high air changes.

After analyzing these constants, it can be defined a three-dimensional model that interrelates the energy consumption of each different situation (constants *a* and *b*) and the number of air changes for today and in next 20 years. This model with an adequate correlation factor was obtained as a useful tool for future building design. This tool defines the most probable energy consumption of an existing building, or a building that is being designed today and in the near future. In consequence, it is a new tool for preventing the climate change effect. This is a tool for building design for preventing the higher energy consumption due to the climate change effect and must be analyzed in the future research works.

ACKNOWLEDGMENT

This research was funded by CYPE Ingenieros S.A. in their research project to reduce energy consumption in buildings and its certification, in collaboration with the University of A Coruña (Spain) and the University of Porto (Portugal). (Grant No. 64900). The authors wish to express their deepest gratitude to the Sustainability Specialization Campus of the University of A Coruña for the administrative and technical support.

REFERENCES

ASHRAE. (2017). *Handbook. HVAC Fundamentals* (Spanish Edition). Editorial Index.

Barkhudaryan, N., Orosa, J. A., & Roshan, G. (2014). A new procedure to analyze the effect of air changes in building energy consumption. *Journal of Environmental Health Science & Engineering*, *12*(37), 1–9. doi:10.1186/2052-336X-12-37 PMID:24456655

CE3X Software. (n.d.). https://certific.es/descargar-ce3x-version-2.1.html

Divina, F., Torres, J. F., García-Torres, M., Martínez-Álvarez, F., & Troncoso, A. (2020). Hybridizing Deep Learning and Neuroevolution: Application to the Spanish Short-Term Electric Energy Consumption Forecasting. *Applied Sciences (Basel, Switzerland)*, *10*(16), 5487. doi:10.3390/app10165487

Fei, F., Zhou, S., Mai, J. D., & Li, W. J. (2014). Development of an indoor airflow energy harvesting system for building environment monitoring. *Energies*, *7*(5), 2985–3003. doi:10.3390/en7052985

Fernández-González, R., Suárez-García, A., Álvarez-Feijoo, M. A., Arce, E., & Díez-Mediavilla, M. (2020). Spanish Photovoltaic Solar Energy: Institutional Change, Financial Effects, and the Business Sector. *Sustainability*, *12*(5), 1892. doi:10.3390u12051892

Gang, W., Wang, S., Shan, K., & Gao, D. (2015). Impacts of cooling load calculation uncertainties on the design optimization of building cooling systems. *Energy and Building*, *94*, 1–9. doi:10.1016/j.enbuild.2015.02.032

Haarhoff, J., & Mathews, E. H. (2006). A Monte Carlo method for thermal building simulation. *Energy and Building*, *38*(12), 1395–1399. doi:10.1016/j.enbuild.2006.01.009

Hein, L., Metzger, M. J., & Moreno, A. (2009). Potential impacts of climate change on tourism; a case study for Spain. *Current Opinion in Environmental Sustainability*, *1*(2), 170–178. doi:10.1016/j.cosust.2009.10.011

Herrera-Gomez, S.S., Quevedo-Nolasco, A., & Pérez-Urrestarazu, L. (2017). The role of green roofs in climate change mitigation. A case study in Seville (Spain). *Building and Environment, 123*, 575-584.

IEA. (n.d.). http://www.ecbcs.org/

ISO 13790. (2011). Thermal performance of buildings-calculation of energy use for space heating and cooling. ISO/DIS 13790:2008, International Organization for Standardization, Geneva.

Jim, C. Y. (2015). Thermal performance of climber greenwalls: Effects of solar irradiance and orientation. *Applied Energy*, *154*, 631–643. doi:10.1016/j.apenergy.2015.05.077

Karlsson, F., & Fahlén, P. (2008). Impact of design and thermal inertia on the energy saving potential of capacity controlled heat pump heating systems. *International Journal of Refrigeration*, *31*(6), 1094–1103. doi:10.1016/j.ijrefrig.2007.12.002

Keirstead, J., & Shah, N. (2011). Calculating minimum energy urban layouts with mathematical programming and Monte Carlo analysis techniques. *Computers, Environment and Urban Systems*, *35*(5), 368–377. doi:10.1016/j.compenvurbsys.2010.12.005

Khosravani, H. R., Castilla, M. M., Berenguel, M., Ruano, A. E., & Ferreira, P. M. (2016). A Comparison of Energy Consumption Prediction Models Based on Neural Networks of a Bioclimatic Building. *Energies*, *9*(1), 57. doi:10.3390/en9010057

Larsen, M. A. D., Petrovic´, S., Radoszynski, A. M., McKenna, R., & Balyk, O. (2020). Climate change impacts on trends and extremes in future heating and cooling demands over Europe. *Energy and Building*, *226*, 110397. doi:10.1016/j.enbuild.2020.110397

Losada, I. J., Toimil, A., Muñoz, A., Garcia-Fletcher, A. P., & Diaz-Simal, P. (2019). A planning strategy for the adaptation of coastal areas to climate change: The Spanish case. *Ocean and Coastal Management*, *182*, 104983. doi:10.1016/j.ocecoaman.2019.104983

Lu, Y., Huang, Z., & Zhang, T. (2013). Method and case study of quantitative uncertainty analysis in building energy consumption inventories. *Energy and Building*, *57*, 193–198. doi:10.1016/j.enbuild.2012.10.046

Masdías-Bonome, A. E., Orosa, J. A., & Vergara, D. (2020). A new methodology for decision-making in buildings energy optimization. *Applied Sciences (Basel, Switzerland)*, *10*(13), 4558. doi:10.3390/app10134558

Mavrotas, G., Florios, K., & Vlachou, D. (2010). Energy planning of a hospital using Mathematical Programming and Monte Carlo simulation for dealing with uncertainty in the economic parameters. *Energy Conversion and Management*, *51*(4), 722–731. doi:10.1016/j.enconman.2009.10.029

Mazarrón, F. R., Cid-Falceto, J., & Cañas, I. (2012). Ground thermal inertia for energy efficient building design: A case study on food industry. *Energies*, *5*(2), 227–242. doi:10.3390/en5020227

MeteoGalicia. (2020). *Anuario climatolóxico de Galicia. Consellería de Medio Ambiente*. Xunta de Galicia. https://www.meteogalicia.gal/observacion/informesclima/informesIndex.action

Modeste, K. N., Mempouo, B., René, T., Costa, A. M., Orosa, J. A., Raminosoa, C. R. R., & Mamiharijaona, R. (2015). Resource potential and energy efficiency in the buildings of Cameroon: A review. *Renewable & Sustainable Energy Reviews*, *50*, 835–846. doi:10.1016/j.rser.2015.05.052

Nematchoua, M. K., Ricciardi, P., Orosa, J. A., & Buratti, C. (2018, August). A detailed study of climate change and some vulnerabilities in Indian Ocean: A case of Madagascar island. *Sustainable Cities and Society*, *41*, 886–898. doi:10.1016/j.scs.2018.05.040

Nematchoua, M. K., Tchinda, R., & Orosa, J. A. (2014). Thermal comfort and energy consumption in modern versus traditional buildings in Cameroon: A questionnaire-based statistical study. *Applied Energy*, *114*, 687–699. doi:10.1016/j.apenergy.2013.10.036

Nematchoua, M. K., Tchinda, R., Orosa, J. A., & Andreasi, W. A. (2015). Effect of wall construction materials over indoor air quality in humid and hot climate. *Journal of Building Engineering*, *3*, 16–23. doi:10.1016/j.jobe.2015.05.002

Orosa, J. A. (2012). Thermal inertia and ISO 13790 - Have its effects on future energy consumption been fully considered? *Energy Education Science and Technology*, *29*(1), 339–352.

Orosa, J. A., Costa, A. M., Rodríguez-Fernández, A., & Roshan, G. (2014). Effect of climate change on outdoor thermal comfort in humid climates. *Journal of Environmental Health Science & Engineering*, *12*(1), 46. doi:10.1186/2052-336X-12-46 PMID:24517127

Orosa, J. A., Nematchoua, M. K., & Reiter, S. (2020). Air changes for healthy indoor ambiences under pandemic conditions and its energetic implications: A Galician case study. *Applied Sciences (Basel, Switzerland)*, *10*(20), 7169. doi:10.3390/app10207169

Orosa, J. A., & Oliveira, A. C. (2010). Impact of climate change on cooling energy consumption. *Journal of the Energy Institute*, *83*(3), 171–177. doi:10.1179/014426010X12759937396876

Orosa, J. A., & Oliveira, A. C. (2012). A field study on building inertia and its effects on indoor thermal environment. *Renewable Energy*, *37*(1), 89–96. doi:10.1016/j.renene.2011.06.009

Parra, J., Guardo, A., Egusquiza, E., & Alavedra, P. (2015). Thermal performance of ventilated double skin façades with venetian blinds. *Energies*, *8*(6), 4882–4898. doi:10.3390/en8064882

Peng, L. L. H., & Jim, C. Y. (2013). Green-roof effects on neighborhood microclimate and human thermal sensation. *Energies*, *6*(2), 598–618. doi:10.3390/en6020598

Prabhakar-Murthy, D. N., Bulmer, M., & Eccleston, J. A. (2004). Weibull model selection for reliability modelling. *Reliability Engineering & System Safety*, *86*(3), 257–267. doi:10.1016/j.ress.2004.01.014

Rao, A. S. R. S. (2005). On joint Weibull probability density functions. *Applied Mathematics Letters*, *18*(11), 1224–1227. doi:10.1016/j.aml.2005.02.011

Ren, X., Yan, D., & Wang, C. (2014). Air-conditioning usage conditional probability model for residential buildings. *Building and Environment*, *81*, 172–182. doi:10.1016/j.buildenv.2014.06.022

Roshan, G., Ghanghermeh, A., & Orosa, J. A. (2014). Thermal comfort and forecast of energy consumption in northwest Iran. *Arabian Journal of Geosciences*, *7*(9), 3657–3674. doi:10.100712517-013-0973-7

Roshan, G., Orosa, J. A., & Nasrabadi, T. (2012). Simulation of climate change impact on energy consumption in buildings, case study of Iran. *Energy Policy*, *49*, 731–739. doi:10.1016/j.enpol.2012.07.020

Šadauskienė, J., Paukštys, V., Šeduikytė, L., & Banionis, K. (2014). Impact of air tightness on the evaluation of building energy performance in Lithuania. *Energies*, *7*(8), 4972–4987. doi:10.3390/en7084972

Setiawan, A. F., Tzu-Ling, H., Chun-Ta, T., & Chi-Ming, L. (2015). The effects of envelope design alternatives on the energy consumption of residential houses in Indonesia. *Energies*, *8*(4), 2788–2802. doi:10.3390/en8042788

Vila-Traver, J., Aguilera, E., Infante-Amate, J., & González de Molina, M. (2021). Climate change and industrialization as the main drivers of Spanish agriculture water stress. *The Science of the Total Environment*, *760*, 143399. doi:10.1016/j.scitotenv.2020.143399 PMID:33293093

Ye, B., Jiang, J., Liu, J., Zheng, Y., & Zhou, N. (2021). Research on quantitative assessment of climate change risk at an urban scale: Review of recent progress and outlook of future direction. *Renewable & Sustainable Energy Reviews*, *135*, 110415. doi:10.1016/j.rser.2020.110415

ADDITIONAL READING

Nematchoua, M. K., Orosa, J. A., & Reiter, S. (2019). Climate change: Variabilities, vulnerabilities and adaptation analysis - A case of seven cities located in seven countries of Central Africa. *Urban Climate*, *29*, 100486. doi:10.1016/j.uclim.2019.100486

Nematchoua, M. K., Orosa, J. A., & Reiter, S. (2019). Energy consumption assessment due to the mobility of inhabitants and multiannual prospective on the horizon 2030–2050 in one Belgium city. *Energy*, *171*, 523–534. doi:10.1016/j.energy.2019.01.032

Nematchoua, M. K., Orosa, J. A., & Reiter, S. (2019). Life cycle assessment of two sustainable and old neighbourhoods affected by climate change in one city in Belgium: A review. *Environmental Impact Assessment Review*, *78*, 106282. doi:10.1016/j.eiar.2019.106282

Nematchoua, M. K., Vanona, J. Ch., & Orosa, J. A. (2020). Energy efficiency and thermal performance of office buildings integrated with passive strategies in coastal regions of humid and hot tropical climates in Madagascar. *Applied Sciences (Basel, Switzerland)*, *10*(7), 2438. doi:10.3390/app10072438

Orosa, J. A. (2009). Research on the origins of thermal comfort. *European Journal of Scientific Research*, *34*(4), 561–567.

Orosa, J. A. (2012). A new validation tool of weather forecast for engineering applications. *Energy Education Science and Technology, Part A. Energy Science and Research*, *29*(1), 381–394.

Orosa, J. A., & Carpente, T. (2009). Thermal inertia effect in old buildings. *European Journal of Scientific Research*, *27*(2), 228–233.

Orosa, J. A., & Oliveira, A. C. (2011). Software tools for HVAC research. *Advances in Engineering Software*, *42*(10), 846–851. doi:10.1016/j.advengsoft.2011.05.028

Roshan, G., Farrokhzad, M., & Orosa, J. A. (2019). A climate-architecture modeling to explore the effect of land use change scenario on building bioclimatic design pattern in order to provide human thermal comfort. *Journal of Environmental Health Science & Engineering*, *17*(2), 1131–1151. doi:10.100740201-019-00427-1 PMID:32030180

Roshan, G., Najafei, M. S., Costa, A. M., & Orosa, J. A. (2014). Effect of climate change on wind energy production in Iran. *Arabian Journal of Geosciences*, *8*(4), 2359–2370. doi:10.100712517-014-1374-2

KEY TERMS AND DEFINITIONS

Building Energy Consumption: The amount of energy consumed in Heating Ventilation and air conditioning of indoor ambiences.

Climate Change: A modification in climate patterns that may affect a region or the whole Earth planet.

Energy Rating Software: Computer application employed to define the level of efficiency of a building energy consumption process.

Indoor Ambience: It is the state of the atmosphere in closed places.

Optimization: Process to define the optimal state of a system in accordance with previously fixed objectives.

Ventilation: Process to introduce outdoor air into an indoor ambience.

Weibull Model: Continuous statistical distribution usually employed to define real life data.

Chapter 10

Studying Physical and Chemical Properties of Graphene Oxide and Reduced Graphene Oxide and Their Applications in Sustainable Building Materials

Omar Ibrahim
Istanbul University-Cerrahpasa, Turkey

Savaş Erdem
https://orcid.org/0000-0003-0906-5233
Istanbul University-Cerrahpasa, Turkey

Ezgi Gurbuz
Istanbul University-Cerrahpasa, Turkey

ABSTRACT

In this chapter, the authors identified graphene oxide (OG)/reduced graphene oxide (RGO) as nano composites by studying its nanoparticles' properties physically, chemically, and mechanically. In this study, they mentioned the photo catalyst materials PC regarding carbon nanostructures such as GO and RGO, which have excellent oxygen functionalities, efficient adsorption areas, and considerable surface area. The compositions of GO and RGO exceed electron-holes pair reinstallation time and minimize energy hiatus by adjusting valence band level (VBL) with conducting band level (CBL) bringing high suction of the exist radiance, which improves photo degeneration achievement of material oxides and composites made from polymers. They also studied the main applications of GO and RGO in engineering fields and summarized the usefulness of intercalation of GO and RGO in construction sectors. Moreover, many synthesis techniques lead to many types of GO. Therefore, in this chapter, the authors tried to collect most GO and RGO properties, structures, and applications.

DOI: 10.4018/978-1-7998-7023-4.ch010

INTRODUCTION

Different governments around the world have a liability to include that there is safe equipping of energy to secure economic improvements. In many sophisticated communities there is usually very few edges between existing power provider and electrical-energy request. With improving electricity employ from current consumers and new links, new offspring needs to be brought on line to apply the increasing amounts in request. Moreover, due to variable climate types and the increasing hazard of dryness, countries that are strongly dependent on electricity from hydropower as their essential source of electricity are forfeiting much of their offspring capability showed in powerful power portioning.

Nowadays, Graphene Oxide (GO) and Reduced Graphene Oxide (RGO) became one an imperative subject in scientific field and engineering realm. Their accepted porosity value, low density, and relatively high surface area, in addition to its ability to conduct electric stream makes the concerns of wide range of researchers and scientists. The efforts and works on analysing Graphene Oxide and Reduced Graphene Oxide is not new. It got material science scholars' interests from 30 years ago. Several materials have been developed regards time and methods to their manufacturing form an actual prospect for the material uses as super capacitors, electric batteries, gas sensors, and actuators. Recently, the interests in Graphene Oxide as materials forerunner significantly increased. This could be referred to its ability to form steady colloidal dispersions in polar solvent (like water) and consequentially transfer into a minimized electrically conductive Graphene like organize under thermal and physiochemical treatment. Moreover, the capability of such composite materials to multi-cyclic distortions, elasticity, and flexibility pulls interests of material researchers.

During building construction and operation process, there is formidable effect on our environments and communities using materials, water resources and different kinds of energy. To reduce the environmental perversions and save the natural resources, demands for applying co-environmental 'Green' building is increasing gradually. This article and its sub-topic address a diversity of surveys to improve construction energy efficiency. The activity of different resources and methods to manage environmental pollution are directed in other topics inside this encyclopaedia. The major energy-using frameworks related to constructions are building construction members (which is, the building cover), ventilation, heating, energy-consuming devices, air conditioning equipment (HVAC), and apparatus, including lighting. The output of all these energy frameworks could be enhanced by applying different standards and by turning to energy-efficient devices. Another way to enhance energy outputs is by applying burden administration technologies.

The swift increasing of publication activities in this topic recently is considered to some extent to the need for the enhancement of light electronic devices, especially energy storage devices which became one of life needs of anyone. Reduced Graphene Oxide also used in expansion of aerogels. Authors in this direction are endeavouring to develop the electrical conductivity, the mechanical properties, surface area, durability, and elasticity of these materials to get such a considerable composite.

Many polymers easily mix with Graphene Oxide to compose nanocomposites, which extremely improve the main polymer's properties such as elastic modulus, tensile strength, electrical conductivity, and thermal stability. When GO chips are firm, they appear to stick together, forming small, extremely stable paper-like structures. Moreover, many researchers and materials scientists developed GO and reform its internal features that have beater and more enhanced properties. These features allow us to use GO in the construction field as well. It has been used directly in concrete or added as an admixture to other cementitious materials which notably improved the physical, chemical, and mechanical properties of binders and different structural members.

Studies on graphene reinforced cementitious composites are very important nowadays when most of the energy consumption is made for heating and cooling purposes in buildings. Graphene can effectively be used energy harvesting in buildings and also provide comfort temperature inside the building. This study, which refers to the types and properties of graphene and reduced graphene oxides and summarizes the numerical and experimental studies on these materials, has shown that graphene can play an important role in reducing energy consumption in buildings.

WHAT IS GRAPHENE?

Graphene is a composite material that consists of carbon atoms which are combining in a cyclic style of hexagons. Graphene is considered as a so fluffy material that represented in two dimensions. Graphene usually fits as flat honeycomb pattern which gives it many exceptional features like being one of the strongest materials around the world, lightest, have huge amount of conductivity and transparency.

The name of Portmanteau of "graphite" with the suffix "ene", reflected the reality that the components of graphene allotrope of carbon takes place in stacked graphene layers (Boehm et al., 1994). In a graphene plate, all atoms are bonded to separate three adjacent by sigma bond (σ-bond) and conduces single electron to a conducting band that broadens of the whole plate or sheet. These transmission tapes make graphene a semimetal that have remarkable features which are the best characterized by hypothesis for massless relativistic particles. The charge power transporters that exist as graphene offer linear, rather more than quadratic, reliance of power on momentum. As further, bipolar conduction could be shown when field-effect transistors are bonded together with graphene. Charge transfer is ballistic on a very big space; the substance shows huge share pulses, and as well as large and non-linear diamagnetism. Graphene behaves heat capacity and electricity properties are highly efficiently along its sheet. The one-only layers of carbon atoms give the basis for many other composite materials. Like the compound in pencil leads, graphite is produced from heaped graphene. Carbon nanotube is made of swiped graphene and it is used in many new applications from kids' accessories to military industries. Figure 1 shows that the graphene has a closely packed honeycomb-like structure.

WHAT IS GRAPHENE OXIDE?

As known, graphene is very expensive material and so, it's very hard to produce. Therefore, lots of efforts and hard works are done to find other ways to use graphene materials or its derivatives in terms of making the production process easier and less expensive. Graphene Oxide (GO) is considered as single-atomic layered substance, produced by powerful oxidation of extracted from graphite, which is relatively cheap and plentiful. GO is graphene oxidized material, tightened with oxygen-containing groups. Since it's dispersible in water (and many other solutions); it is very easy to process and even be used to manufacture graphene materials. Graphene Oxide is a week conductor, but manufactural processes exist augment its physical and chemical properties. Generally, GO is sold in in powder shape, dispersed or as rendering on substrates.

Figure 1. a) Chemical structure b) Crystal structure of graphene compound (Priyadarsini et al., 2018)

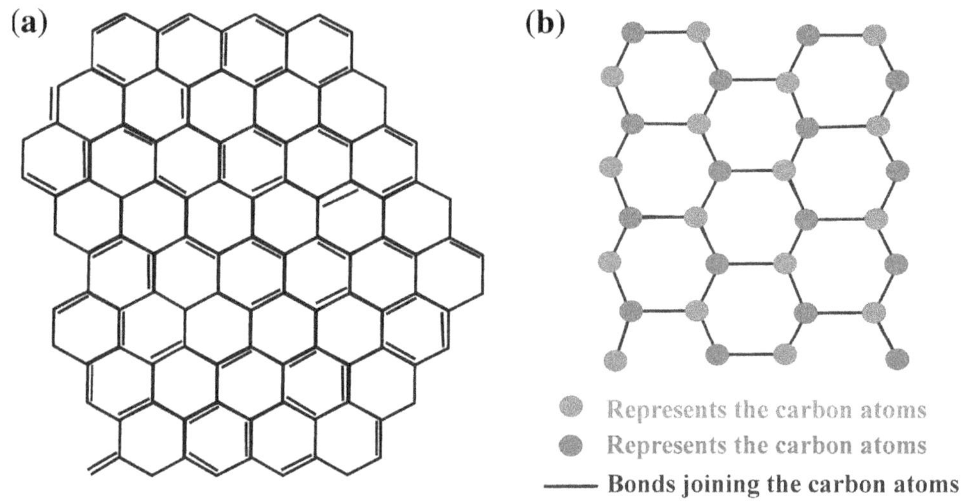

GO is a synthesized material using four main styles: Staudenmaier, Hoffman, Brodie, and Hummers. There are many variations concerns with each type of these methods. Researches and developers set the improvement development priorities to have more quality with less industry costs. However, Hammers method is the most famous to make Graphene Oxide synthesized. Hammers method energetic stirring of graphite strength with conc. Here, H_2SO_4 shatters the epitaxial stratums of graphene. Sodium nitrate and Potassium permanganate are consumed mainly as an oxidizing factor that fertilizes the oxygen performance and eventually forming GO (Hummers & Offeman, 1958). This style could be further modified to achieve different oxygen operations (Chen et al., 2016). The authorship of reduced graphene oxide (RGO) could be fulfilled by lowering of Graphene Oxide using some methods. Figure 2 shows the transformation stages for graphene oxide (GO) to reduced graphene oxide (RGO) and so reduce graphene.

Usually, there are four main ways to minimize the amount of Graphene Oxide (GO). These methods are known as: thermal treatment, chemical reduction, thermal and chemical method, and lots of special techniques as: ultra-sonication, plasma radiation /electron (Chua & Pumera, 2014). The chemical way concerns with reducing graphene oxide (GO) using chemical compounds $NaBH_4$ which own high reducing ability under normal conditions. Reduction of graphene oxide (OG) by thermic handing could be applied by warming graphene oxide molecules in vacuum with an unmoving atmosphere, like nitrogen. This is done generally to collapse the oxygen-containing functionalities while maintaining the hexagonal latticework structure of carbon atoms that are approximately complete (Jung et al., 2008; Li et al., 2016; Lipatov et al., 2013; Lu et al., 2009; Tarjomannejad et al., 2016). Surprisingly, hybrid (combination) method contains of two main methods: chemical method and thermal reduction in terms of obtaining higher amount of chemical reduction. Consequently, these results shown in the conversion to graphene aims to achieve greater amount of carbon to oxygen (Chua & Pumera, 2014). Those ways act for propitious plans for RGO process that aims to get more increased C/O percentage so decreasing the band gap which eventually enhancing the electrical accessibility, physical and chemical features which is important for producing photo catalyst which could be performed under visual area of solar shade (Choi et al., 2015).

Figure 2. Conversion of graphene into GO and RGO (Priyadarsini et al., 2018)

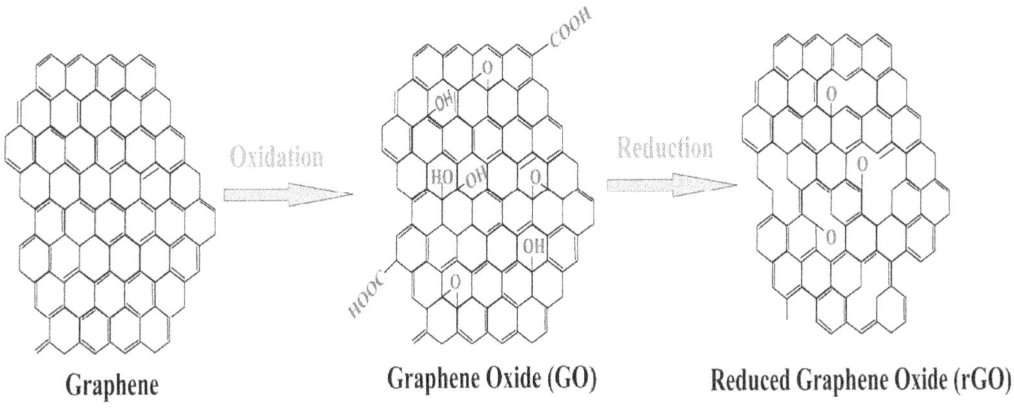

Properties of Graphene Oxide (GO) and Reduced Graphene Oxide (RGO)

Graphene Oxide has different oxygen employments which includes carbonyl, epoxy, hydroxyl, and carboxyl functional groups. Therefore, TiO2 and ZnO that shaped in photo catalyst could binned firmly with these oxygen molecules implicating functional sets and voluntarily widespread on the Graphene Oxide face, that enhances the photocatalytic reduction of the Graphene Oxide (Qiu et al., 2012). All band gaps that extract from Graphene Oxide can be set by altering oxygen functionalities where some of Graphene Oxides particles got the ability to appear as a semiconductor while fully oxidized Graphene Oxide can act as an electrical non-conductor (Loh et al., 2010). As so, the band gap power of Graphene Oxide assigned by the UV-vis absorption spectra is 3.26 eV (Shukla & Saxena, 2011).

Graphene Oxide has very good mechanical properties, excellent electronic features, great catalytic activity, and poor density, beg surface area and high electron transporting characteristics, thus participating in various reactions, and obtaining excellent absorptivity and locative charge in the hybrid structure. Graphene Oxide also has a very good electron acceptor ability to raise the picture induced electron-transfer to operate better and highly efficient photo catalyst (Hu et al., 2013).

Reduced Graphene Oxide has robust features and auspicious functionalities as compared to Graphene Oxide like more efficient optical properties, better electron mobility, physiochemical stability, excellent bond, good surface area (Shetti et al., 2019), high thermal-conductivity, and great resilience (Mishra et al., 2019). Reduced Graphene Oxide have the ability to be as co catalyst when reacted with TiO2 that aims to improve photo catalyst redox capabilities under visual in addition to UV radiation source. Regarding its outstanding electronic conductivity, it has the ability to separate photo generated electrons to override the electron hole reinstallation rate. Nevertheless, the results show that π-π interactivity with organic dyestuff stuffs and formulization of hydrogen connecting all organic pollutants, improves the photo catalyst degeneration procedures (Dai et al., 2017).

The photo conversion performance of photo catalyst related to Graphene Oxide/Reduced Graphene Oxide could be enhanced in avoiding electrons-holes recombination. This step could be applied by doping with noble elements or materials such as Silver, Gold, and Platinum, that has the ability to create and edit the band gap of the given material. For example, integration of Reduced Graphene Oxide in silver particles is used officially to improve photo degeneration action regarding more efficient charge-separation and high operative places or active areas, hence quell all recombination rates of electron and hole (Kumar et al., 2019).

Binary Compositions of Graphene Oxide and Reduced Graphene Oxide

Graphene takes possession of noteworthy electron transmit features and acts as an excellent acceptor (Huang et al., 2001). Graphene Oxide and Reduced Graphene Oxide shapes composite materials with various kinds of material/metal oxide, like Cu_2O, CdSe-TiO2, Nd/TiO2, ZnO, TiO2, $W_{18}O_{49}$, $SnWO_4$, $CuFe_2$, Mn_2O, $COFe_2$, $La_2Ti_2O_7$, and Fe_2O_3/ZnO. These materials are used as photo catalyst for demeaning noxious organic pollutants which they use it in fuel cells generally (Upadhyay et al., 2014).

Integration of Nano-filler especially Carbon Nano-filler like Graphene Oxide and Reduced Graphene Oxide in material oxides such as TiO2 minimize the recombination rate efficiently of electron and hole pairs and modify the band gap of TiO2 (Hosseinzadeh & Ramin, 2018). The technique behind the photo catalyst regression process of TiO2 with Graphene Oxide and TiO2 related with Reduced Graphene Oxide might be epitomized as: when the reflected light touches the surface of Graphene Oxide-TiO2 electrons and holes are manufactured, electrons resulted in the CB of TiO2 get transmitted to the Graphene Oxide or Reduced Graphene Oxide surface. It responds vigorously with power a little bit greater than conduction band related to TiO2 (-4.2eV) (Loryuenyong et al., 2013). The functionally active groups which contain Oxygen located on the outer surface of Graphene Oxide hold p-electron on that surface, letting them catch those traveling electrons coming from TiO2. The p-electrons left unpaired found in Oxygen on the surface of Graphene Oxide have the ability to fasten with the Ti atoms coming from TiO2 to shape Ti-O-C formation which results in increasing the range of light absorption of TiO2. (Zhang et al., 2009). The Graphene Oxide-TiO2 composites show a perfect photochemical reaction when exposed to visible light irradiation (1 > 400 mm). The electrons accepted by the O2 groups of the Graphene Oxide create •O2- superoxide radical however the holes stick on the outer surface of TiO2 forming Hydroxyl radical •OH by the result of a reaction that happens in the aqueous solution with the ions of Hydroxyl however in case of Reduced Graphene Oxide-TiO2, the electrons reduced in the conduction tie of TiO2 are transmitted to the surface of Reduced Graphene Oxide with Fermi level (-4.4 eV) further down than in TiO2. Therefore, the chemical reaction occurs between the π orbital of Reduced Graphene Oxide and d orbital of TiO2 form d–π electron orbital interferes. In the two cases, the traveling electrons are relocated to the Graphene Oxide & Reduced Graphene Oxide surface forming the superoxide radical •O2-. This substantial also interact with H2O molecules forming Hydroxyl radicals (•OH), which are considered a strong oxidant that demeans biological pollutants (Bagherzadeh & Kaveh, 2018). Those interactive kinds demean the dyestuff stuffs into Carbon dioxide & water which are in-noxious as known. In general, the photo catalyst activity of Reduced Graphene Oxide-TiO2 is considered to be bigger than the analogical Graphene Oxide-TiO2 system. This could be imputed to the truth that Reduced Graphene Oxide-TiO2 has better conductivity, therefore exhibiting perfect transmission of electrons. Reduced Carbon Oxide plays an important role in photo catalyst activity, through corresponding and interfering of π orbital of Reduced Carbon Oxide and d orbitals in the conduction tie of TiO2 thus composing a frail chemical tie due to d-π electronic reaction (Bagherzadeh & Kaveh, 2018).

Choi et al. produced the CuI-Reduced Graphene Oxide micro-composites with the assisting of the ultra-sonification alchemical way (Huang et al., 2001). CuI Nanostructures show a does it became crystallite and photosensitive (Sharma & Rabinal, 2013). It shows clearly noticed that CuI catalysts has a frail capability of surface-adsorption and a bad response to visible-light (Wang et al., 2011). Thus, photoexcitation is impossible to occur under visible light range to stimulate photocatalytic activity since the blast space of CuI is so much bigger. Moreover, it demands shortened radiation wavelength and takes possession of high ability to trigger electrons to the conduction band from the valence band. Whereby, the doping of Reduced Graphene Oxide reduces the bandgap energy to form moderate levels of energy which trigger photocatalytic activity under the visible region. Though, Reduced Graphene Oxide Nano sheets play a substantial role in converting CuI-Reduced Graphene Oxide nanocomposites to visible-light photocatalytic. On the other hand, the solar power hits on the Reduced Graphene Oxide nanostructure, electrons are produced which were then elevated to the CuI nanostructures conduction band. The separated holes and electrons and the conduction band electrons get transmitted to the O_2 molecules to product super oxide radicals that returns in the minimizing reaction that way reducing water components to • OH radicals however the holes generated in the valence band starts a reaction producing hydroxyl ion until producing • OH, a vigorous oxidant. These radicals are boosted and extended by the more steady carbon free radical obtained from Reduced Graphene Oxide Nano sheets, occurring in photocatalytic activity under the visible light region (Huang et al., 2001). In the following figure, we can take a look at the basic mechanism of binary composites for the dyestuff stuffs photo degeneration. The photocatalytic activity of Graphene Oxide/ Reduced Graphene Oxide extracted with metal oxide (MOx) or semiconductor (SC) could be explained by chemical reactions (1), (2), (3) and (4) with the existence of hydrogen per oxide.

$$\text{MOx} + h\nu \rightarrow \text{MOx}\ (h+ + e-) \tag{1}$$

$$\text{MOx}\ (h+ + e-) + \text{GO/RGO} \rightarrow \text{MOx}\ (h+) + \text{GO / RGO}\ (e-) \tag{2}$$

$$\text{GO /RGO}\ (-) + \text{H2O2} \rightarrow \bullet\ \text{OH} + \text{OH}- + \text{GO / RGO} \tag{3}$$

$$\text{MOx}\ (h+) + \bullet\ \text{OH} + \text{dyestuff stuff} \rightarrow \text{degeneration produces} \tag{4}$$

Graphene Reinforced Cement Composites

In recent years, there has been a growing interest in incorporating graphene into cementitious composites. The graphene modified cementitious composites have the following features and advantages compared to conventional cementitious composites (Lin & Du, 2020): (a) requiring less cement for achieving identical mechanical properties of the cement composites even with at low dosage of graphene addition, (b) improving the durability performance of the cementitious composites (c) achieving high electrical conductivity and good piezo resistivity of the composites for the structural health monitoring of smart civil infrastructures, and static and dynamic inductive wireless charging of electric vehicles on flexible and rigid cementitious composite pavements, (d) preventing or minimizing of the evolution of thermal cracking in mass concrete structures during construction process and improving fire resistance of the building because of the superior thermal diffusivity, (e) demonstrating the remarkable electromagnetic interference shielding (EMI) effect of the composites which reduces the risk of electromagnetic emission problems on human health, (f) mass production with low manufacturing cost, but at the same time, continue its pristine properties. Figure 3 schematically shows the above-mentioned multifunctionality of graphene reinforced cement composites.

Figure 3. Multifunctionality of graphene reinforced cement composites (Lin & Du, 2020)

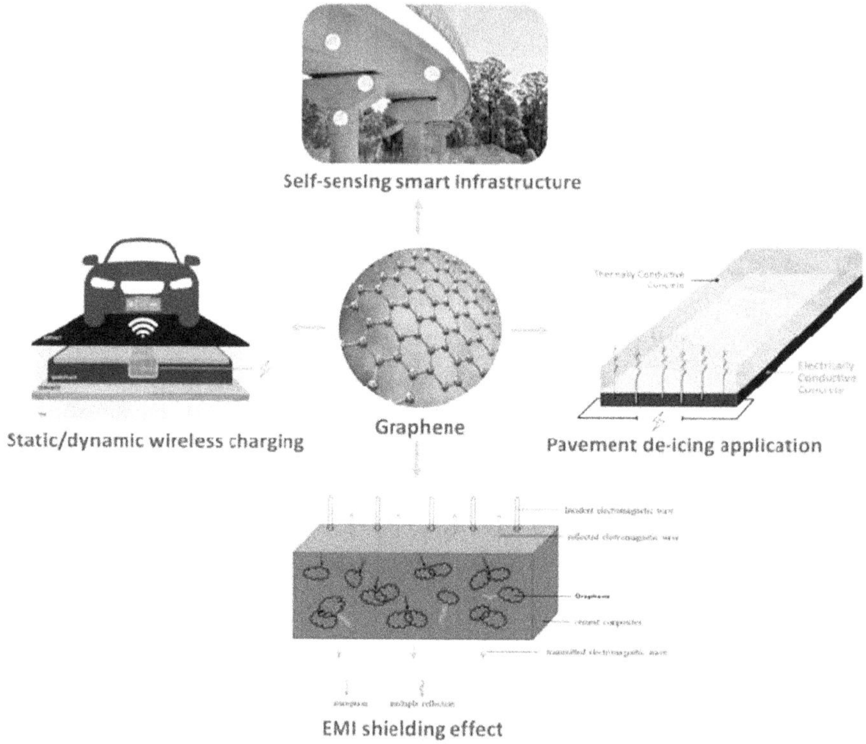

Several studies showed that the addition of GO have adversely affected the workability of cementitious composites. This could be attributed to the fact that the large surface area of GO makes it prone to absorb more water molecule to get wet and a bulky lateral size created by cluster of GO nanosheets have high capacity for water retention (Du & Dai Pang, 2018; Wang et al., 2016). Regardless of the above mention disadvantage of GO, the addition of even small amount of GO (i.e. 1% by weight of cement) have considerably enhanced the compressive strength of the composite (Du et al., 2016). Similarly, Shang et al. have reported an improvement of 15.1% on the compressive strength of the GO (0.04% by weight of cement) modified cement paste compared to that of the plain cement paste (Shang et al., 2015). In other study, an increase over 40% with 0.03% GO inclusion by weight of cement was reported for the compressive strength and flexural tensile strength to the reference sample paste at 28 days curing age (Meng & Khayat, 2016). As shown in Figure 4, uniformly distributed graphene provide crack toughening mechanisms, improves the bridging stress of the cementitious matrix, and makes the micro-structure much denser which, in turn, increases the load carrying capacity of the composite under mechanical stresses.

It is very well-known that one of the degradations mechanisms of the cementitious composites is sourced from the aggressive media such as sulphate and chloride riches environments in which the chemicals penetrates the concrete matrix through the voids. One experimental study conducted by Du et al. (Du et al., 2016) focused on the effect of GO (up to 2.5 wt% of graphene with an increment of 0.05 wt%) on the transport properties of concrete. They found that a refinement of pore structure with the addition of the GO would provide the stronger the interfacial transition zone which makes the difficult the movement of the chemical substances or water across its length, and significantly reduced the water penetration depth, chloride diffusion coefficient and chloride migration coefficient. In addition, they highlighted that when the dosage of the GO in the composite exceeds 1.5% by weight of cement, the influence of GO for the resistance of the composite against moisture and chloride penetration starts to decrease due to the agglomeration of GO in the matrix.

Figure 4. GO reinforced composites under mechanical loading (Wang et al., 2016)

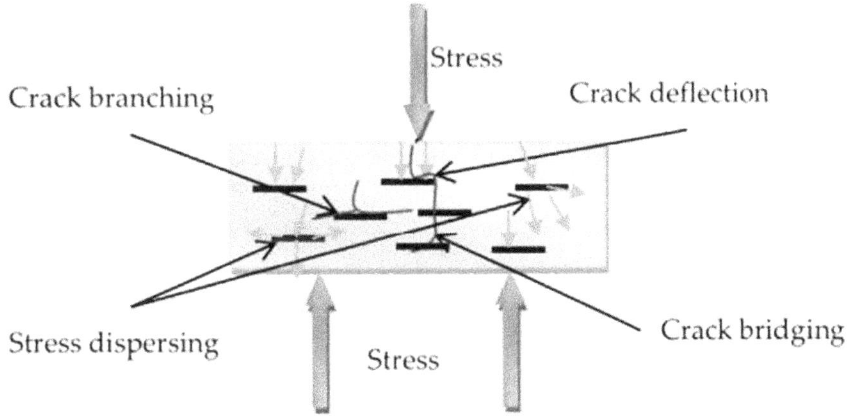

Graphene Reinforced Cement Composites for Energy Harvesting and Thermal Performance

Graphene cement composite have been used by researches for energy harvesting in buildings by improving the thermoelectrical characteristics of cement composite. Thermoelectrical characteristics are gauged in the way of dimensionless figure of merit, ZT, which is equal to (S2σT/κ), where S is Seebeck coefficient in μVK−1, σ is electrical conductivity in Scm−1, T temperature in K and thermal conductivity κ in Wm−1K −1 respectively. Besides, ZT ³ 1 is proposed for energy harvesting application purposes in buildings. In addition, graphene can transform the non-conductive material into conductive material (Rehman et al., 2020). The basic scheme is shown in Figure 5 (Ghosh et al., 2019). According to Balandin (Balandin, 2011), graphene increased the electrical conductive properties of cement composite and decreased the thermal conductive properties.

Ghosh et al. (Ghosh et al., 2019) practiced experimental study using graphene nanoplatelets in cement composite for specify the thermoelectrical characteristics and energy harvesting capability of composite material. Graphene nanoplatelet (GNP) was used as 5%, 10%, 15%, and 20% by mass of cement for preparing graphene cement composite. Four-probe electrical conductivity and Seebeck coefficient measurement system was used to define the Seebeck coefficient and electrical conductivity contemporaneously. Highest specific heat capacity value as 0.88 Jg−1K −1 was found with 10% GNP cement composite. In addition, 20% graphene cement composite showed highest thermal diffusivity as compared to others. It was inferred that graphene-based cement composite considerably contributes to energy harvesting application together with enhancing the quality of indoor environment of buildings. Graphene cement composite can collect the energy, decrease electric consumption, and ensure a substantial financial advantage with this application.

Figure 5. Building energy harvesting using thermoelectric generator of graphene cement composite (Ghosh et al., 2019)

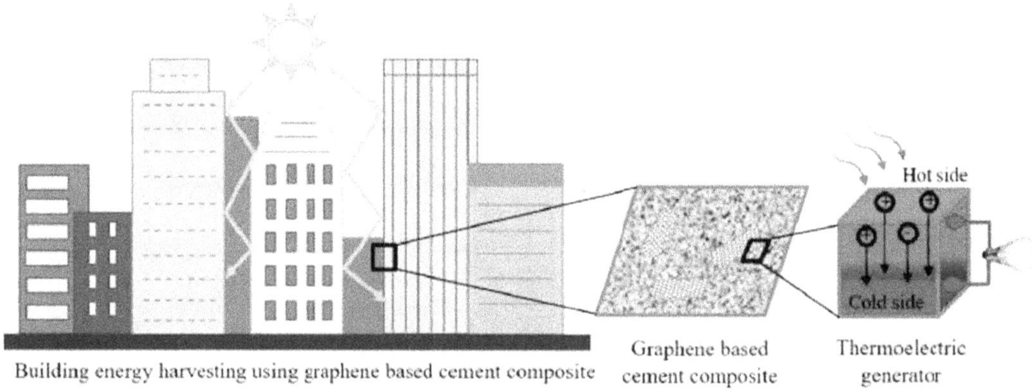

Sedaghat et al. (Sedaghat et al., 2014) tested the hydrated graphene-cement composites under a similar range of temperatures as shown in Figure 6. The general trend is that there is a decline in thermal diffusivity with a rise in temperature, from 25þC to 400þC. It seems that the decline in thermal diffusivity is about 35% for all the mixes, irrespective of the graphene content. The data show that inclusion of 1% graphene did not have any remarkable effect on thermal diffusivity of the mix. On the other hand, inclusion of 5% graphene enhanced the thermal diffusivity by 25% at 25þC and about 30% at 400þC as compared with the simple cement paste or the 1% graphene composite. The mix containing 10% graphene indicates notable improvement in thermal diffusivity of about 75% at 25þC and 60% at 400þC. Generally, it seems that inclusion of graphene in cement paste could considerably enhance thermal diffusivity of the composite. Enhancement of thermal diffusivity of cementitious pastes can decrease the temperature gradient (30þC - 90þC) effect by the reason of cement hydration in mass concrete structures. This can eventually decrease the potential of massive concrete elements to experience thermal cracking so enhancing thermal integrity and durability of concrete structures.

Graphene Oxide Contributions in Solar Power

In heterojunction solar cells, graphene and graphene oxide might be used as photoactive layers in the form of an effective interfacial layer, an electron-hole isolation layer, a hole-transport layer, or an electron-transport layer (Feng et al., 2011). Graphene photoactive layers as a kind of composite materials might have a PCE ranging from 0.4 to 10.3 percent, based on the graphene derivative and the type of photoactive layer being generated. Many of graphene photoactive layers are currently being used in heterojunction solar cells.

The PCE of electroactive sheets composed of few more sheets of graphene oxide shows produced by CVD, flame pyrolysis, or other processes varies between 1.01-2.88 percent (Miao et al., 2012). By merging n-type type silicon with graphene layers, they could be used as n-type heterojunctions. Doping these pure graphene heterojunctions with nitric acid increases the PCE to 4.35 percent, with approximately to 4.18 percent of the PCE staying upon 10 days (Miao et al., 2012). Graphene could also be painted into n-type silicon nanowire arrays, which inhibit and capture radiation somewhat more rapidly

Figure 6. Conversion of graphene into GO and RGO (Balandin, 2011)

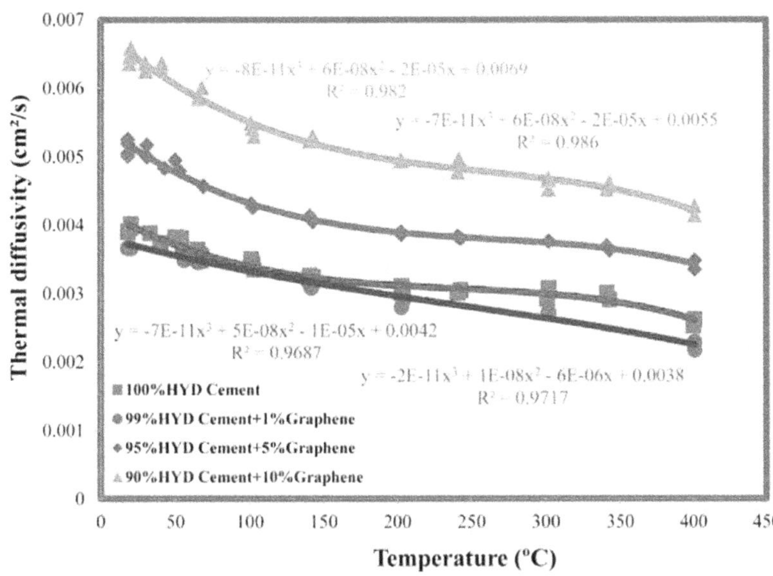

than planar equivalents. And then after doping with thionyl chloride, they have a lower PCE value than planar graphene-silicon heterojunctions. Planar graphene-silicon solar cells could also be drugged with thionyl chloride, that has a PCE of 3.93 percent, which again is lower than nitric acid doping but greater than pure graphene-silicon heterojunctions (Jie et al., 2013). The poorer doping effects of thionyl chloride relative to nitric acid are due to its highly volatile composition. The most crucial element of the heterojunction is the layout, and a flat sheet of CVD-grown graphene on silicon (97 percent transparency, 350 m-2) may well have a PCE of 5.38-7.85 percent, which is much better than multi - layer graphene heterojunctions. By adding a silicon dioxide antireflection surface, this value could be enhanced to 8.94 percent. To avoid efficient charge conjugation at solar cell anodes, graphene quantum dots and crystals silicon could be used as electron trapping layers. Surface passivation could happen in such cells leads to distinct terminal groups, along with oxide, hydrogen, and methyl moieties. The PCE of cells with the methyl marginal group is up to 6.63 percent, relative to 2.24 and 2.92 for hydrogen and oxide terminal groups, respectively (Li et al., 2015). Even then, with time, decay may happen, only with short circuit value dropping by over 5 mAcm-2 and the PCE dropping by up to 1.2 percent. Anodic buffer layers could be produced by graphene oxide and gold nanoparticles. Capping additives, such as glycine or sodium citrate, have been used in these hybrids, and the PCE may range from 2.82 to 3.34 percent. The addition of P3HT and IBCA to the solar cell, but at the other hand, will improve the PCE by up to 5.10 percent (Jie et al., 2013).

Most solar cells can use graphene oxide nanoribbons (GORs) as hole extraction sheets. These materials are developed to support existing ITO-based materials and also have raised the PCE of a solar device from 2.20 percent to 4.19 percent till now (Singh & Nalwa, 2015). In parallel to the PCE, GORs also have low sheet resistance and a stronger shunt resistance than their ITO counterparts.

A modified electrode including graphene oxide, PEDOT: PSS, and n-type silicon nanowires creates among the most efficient graphene photoactive surfaces. The wt percent of graphene oxide has a major effect on the device's PCE, with the highest conditions being 30%, which results in a PCE of up to 9.57

percent (Huang et al., 2012). In contrast, changing from axial silicon to silicon nanowires decreases the PCE by such a significant 4.30 percent. These surfaces have not only a high quantum transparency than non-graphene photoactive layers of comparable structure, but they often have a smaller blazeman decay rate.

IMPACT OF VARIOUS PARAMETERS IN THE PHOTOCATALYTIC ACTIVITY

Impact of Catalyst Amount on Photo Degeneration

Banu et al. produced the Graphene Oxide-CuO nanocomposite (Banu et al., 2019) and noticed that when raising the quantity of photo catalyst, the speed of degeneration increases in order to the increasing in exposed outer surface area of the catalyst. Although, after an additional increase in the photocatalytic dosage (Banu et al., 2019), the degeneration speed starts to decrease which could be resulted from the raising in the number of copper ions. There is a likelihood of a short circuit occurring of Cuprous (I) and Cupric (III) ions (Klauson et al., 2005) that results in smaller hydroxyl radical production therefore causing the reaction speed to decelerate. The photo catalytic action of Graphene Oxide-CuO compound gets raised by raising up, taking advantage of the quantity of photo catalyst reaches 0.07g, and however, raising the photo catalyst dosage causes the speed of degeneration to be decreased (Banu et al., 2019). Maruthamani et al. produced Reduced Graphene Oxide-TiO2 nanocomposites (Maruthamani et al., 2015) and improved the photocatalytic effectiveness depending on the quantity of photo catalyst utilized for the degeneration process. The ratio of degeneration of Rhodamine B dyestuff stuff gets increased when the amount of photo catalyst increased (up to1.5 g L-1) regarding the rise in active site count which formed hydroxyl radicals, resulting in improving the dyestuff stuff adsorption on the catalyst's outer surface. If the quantity of photo catalyst is more improved like (from 1.5 to 2 g L-1) the discoloration ratio will be reduced that could be imputed the catalyst particles bunch due to a rise in the cloudiness of the vine, that depends on the strewing of the light. It results in reducing UV light penetration during the reaction course (Klauson et al., 2005).

Impact of pH

Point of zero charge (pHPZC) is the point reached when a total burden of the absorbent flat comes to 0 point (Li et al., 2015) and that is known to be part of the wide specifications for photo catalyst. The pHPZC of Graphene Oxide-ZnO and Graphene Oxide was specified to reach 8.2 and 7.5 respectively (Klauson et al., 2005). When the pH value of the solvent considered being bigger than pHPZC, then all photo catalyst surface gets a negative charge and carries the suction of the positive charged surfaces. In the same way, the photo catalyst considered to elevate the suction of the negative charged bodies exist the solvent which have a pH number smaller than the pHPZC (Banerjee & Chattopadhyaya, 2017).

Kumaran et al. observed the structure of Graphene Oxide-TiO2 compound by extreme-sonification technique (Kumaran et al., 2020) and defined the charge on its flat. It is notices that plus surface charge of Graphene Oxide-TiO2 nanocomposite presented under pH 6.8, evaluates higher anionic dyestuff stuff solution attraction whereas, at pH values above 6.8, the composite surface charge becomes negative which decreases the negatively charged dyestuff stuff solution adsorption (Kumaran et al., 2020). A pH number lower than pHPZC value is acclamatory for greater suction that increases owing to the rising of positively charged areas amount which elevating the suction of negative charged dyestuff stuff owing to the electrostatic attraction force between positively and negatively charged bodies (Maruthamani et al., 2015). Although, Graphene Oxide-TiO2 micro composite was utilized to the degeneration of orange (ME2RL) dyestuff stuff and was concluded that the greatest color elimination for (60 mg/l) dyestuff stuff concentricity was reached at 6 pH with the usage of 25 mg catalyst in such 24 minutes for touching time. The effectiveness of color elimination of the catalyst was reached to reach 99.65% that reduced only to 1% also after the 5th round on reutilizing the catalyst (Sposito, 1998).

Considering a cationic dyestuff stuff like Methylene Blue, the Methylene Blue adsorption expands in basic medium comparing to acidic medium (Kumaran et al., 2020). Within an acidic substance, a big number of ions of hydrogen found in the mixture, and the dyestuff stuff molecules react with this surplus number of ions of hydrogen for the vacant adsorption areas located on the surface of the catalyst (Huang et al., 2019). Furthermore, the existence of surplus ions of hydroxyl as well elevates the production of OH, which in sequence rises the photocatalytic degeneration speed (Huang et al., 2019). For instance, Reduced Graphene Oxide-160 represents the ultimate proportion degeneration for Methylene Blue at pH 11 (Chiang & Lin, 2013). Degeneration of Methylene Blue escalates from the acidic to the basic medium. In the opposite way, the rising in the pH of the median from 7.0 to 11.0, caused the degeneration of photo catalyst also to rise due to the growing negative charge on the photo catalyst outer face in the alkaline atmosphere which evaluates improved adsorption of Methylene Blue on photo catalyst (Chiang & Lin, 2013).

The amalgam of Graphene Oxide-CuO was explained by Banu et al. by hydrothermal technique. A 7.5 pH value was reached to be perfect for ultimate photocatalytic degeneration due to the good accessibility of the ions of hydroxyl causing the Graphene Oxide-CuO catalyst surface to obtain a negative charge. Therefore, an important amount of molecules of cationic dyestuff stuff takes the position in the Graphene Oxide-CuO catalyst surface because of the attraction electrostatic force. Also, when the pH exceeds 7.5, the photocatalytic degeneration speed is reduced. Come to a conclusion that the dyestuff stuff loses its cationic shape because of the great concentricity of •OH- ions, accordingly electrostatic attraction force amid Graphene Oxide-CuO and BG surface gets reduced. Respectively, the speed of the process is brought to lower levels (Banu et al., 2019).

Impact of Light Concentration

The impact of light concentration is considered to be a highly important factor that affects the speed of dyestuff degeneration. As a result of the studies when the concentration of light rises, the speed of the process rises too. This can result from the rising in the numbers of photons per unit area hitting the outer surface area, this suggests that a bigger number of electrons were immediately eagerly producing extra active oxygen bodies (ROS) causing a rising in the collapse of dyestuff molecules (Chiang & Lin, 2013). Even that a slight rising in the light concentration could trigger most of the thermic reaction which could reduce the space of the reaction. For instance, when it comes to Graphene Oxide, when graphene reacts with CuO composite, It is going to change the light concentration from 10.0 to 70.0 mWcm−2, and biggest values is achieved at 50.0 mW/cm2 whereas an extra rising in the concentration causes the photocatalytic activity to decrease, and it could be with regards to the thermal degeneration (Siong et al., 2019).

FUTURE RESEARCH DIRECTIONS

Reduced Graphene Oxide would be used in expansion of aerogels in a future study. Authors in this direction are endeavouring to develop the electrical conductivity, the mechanical properties, surface area, durability, and elasticity of these materials to get such a considerable composite.

Photo catalyst is also a flourishing research topic which attracted researchers across the board, however, focuses should be switched to the evolution of solar power livened photo catalyst that have better suction and bigger surface area. Freshly, photo degradation implementations had earned interests but still so far from make photo catalyst trendily valid.

CONCLUSION

Studies on graphene reinforced cementitious composites are very important nowadays when most of the energy consumption is made for heating and cooling purposes in buildings. Graphene can effectively be used energy harvesting in buildings and also provide comfort temperature inside the building. This study, which refers to the types and properties of graphene and reduced graphene oxides and summarizes the numerical and experimental studies on these materials, has shown that graphene can play an important role in reducing energy consumption in buildings. In addition, it has been emphasized in the studies that graphene can increase the temperature comfort in buildings. It is very important that graphene, which is used in buildings with a wide variety of methods, is suitable for climatic conditions, and the type of application should be selected according to the climate.

It's crucial to form an oxidation and reduction method that can remove individual carbon stratum and isolate them without changing their composition until graphene oxide can be used as an intermediate in the development of monolayer or few-layer graphene sheets. Although chemical reduction of graphene oxide is generally thought to be the best process for mass production of graphene, scientists have struggled to complete the challenge of manufacturing graphene sheets of the same quality as mechanical exfoliation, but on a much larger scale. We should expect graphene to become much more commonly used in consumer and industrial applications until this problem is resolved.

Finally, graphene modified sustainable building materials provide crack toughening mechanisms, improves the bridging stress of the cementitious matrix, and makes the micro-structure much denser which, in turn, increases the load carrying capacity of the composite under mechanical stresses.

ACKNOWLEDGMENT

This research received no specific grant from any funding agency in the public, commercial, or not-for-profit sectors.

REFERENCES

Bagherzadeh, M., & Kaveh, R. A. (2018). A new SnS2-BiFeO3/reduced graphene oxide photocatalyst with superior photocatalytic capability under visible light irradiation. *Journal of Photochemistry and Photobiology A Chemistry*, *359*, 11–22. doi:10.1016/j.jphotochem.2018.03.031

Balandin, A. A. (2011). Thermal properties of graphene and nanostructured carbon materials. *Nature Materials*, *10*(8), 569–581. doi:10.1038/nmat3064 PMID:21778997

Banerjee, S., & Chattopadhyaya, M. (2017). Adsorption characteristics for the removal of a toxic dye, tartrazine from aqueous solutions by a low-cost agricultural by-product. *Arabian Journal of Chemistry*, *10*, 1629–1638. doi:10.1016/j.arabjc.2013.06.005

Banu, R., Salvi, N., Ameta, C., & Ameta, R. (2019). Visible light driven photocatalytic degradation of brilliant green dye using graphene oxide/copper oxide binary composite. *Indian Journal of Chemistry*, *B581*, 157–166.

Boehm, H. P., Setton, R., & Stumpp, E. (1994). Nomenclature and terminology of graphite intercalation compounds. *Pure and Applied Chemistry*, *66*(9), 1893–1901. doi:10.1351/pac199466091893

Chen, J., Zhang, Y., Zhang, M., Yao, B., Li, Y., Huang, L., Li, C., & Shi, G. (2016). Water-enhanced oxidation of graphite to graphene oxide with controlled species of oxygenated groups. *Chemical Science (Cambridge)*, *7*(3), 1874–1881. doi:10.1039/C5SC03828F PMID:29899910

Chiang, Y.-J., & Lin, C.-C. (2013). Photocatalytic decolorization of methylene blue in aqueous solutions using coupled ZnO/SnO2 photocatalysts. *Powder Technology*, *246*, 137–143. doi:10.1016/j.powtec.2013.04.033

Choi, Y. R., Yoon, Y. G., Choi, K. S., Kang, J. H., Shim, Y. S., Kim, Y. H., Chang, H. J., Lee, J. H., Park, C. R., Kim, S. Y., & Jang, H. W. (2015). Role of oxygen functional groups in graphene oxide for reversible room-temperature NO2 sensing. *Carbon*, *91*, 178–187. doi:10.1016/j.carbon.2015.04.082

Chua, C. K., & Pumera, M. (2014). Chemical reduction of graphene oxide: A synthetic chemistry viewpoint. *Chemical Society Reviews*, *43*(1), 291–312. doi:10.1039/C3CS60303B PMID:24121318

Dai, H., Huang, Y., & Huang, H. (2017). Eco-friendly polyvinyl alcohol/carboxymethyl cellulose hydrogels reinforced with graphene oxide and bentonite for enhanced adsorption of methylene blue. *Carbohydrate Polymers*, *185*, 1–11. doi:10.1016/j.carbpol.2017.12.073 PMID:29421044

Du, H., & Dai Pang, S. (2018). Dispersion and stability of graphene nanoplatelet in water and its influence on cement composites. *Construction & Building Materials*, *167*, 403–413. doi:10.1016/j.conbuildmat.2018.02.046

Du, H., Gao, H. J., & Dai Pang, S. (2016). Improvement in concrete resistance against water and chloride ingress by adding graphene nanoplatelet. *Cement and Concrete Research*, *83*, 114–123. doi:10.1016/j.cemconres.2016.02.005

Feng, T., Xie, D., Lin, Y., Zang, Y., Ren, T., Song, R., Zhao, H., Tian, H., Li, X., Zhu, H., & Liu, L. (2011). Graphene based Schottky junction solar cells on patterned silicon-pillar-array substrate. *Applied Physics Letters*, *99*(23), 233505. doi:10.1063/1.3665404

Ghosh, S., Harish, S., Rocky, K. A., Ohtaki, M., & Saha, B. B. (2019). Graphene enhanced thermoelectric properties of cement-based composites for building energy harvesting. *Energy and Building*, *202*, 109419. doi:10.1016/j.enbuild.2019.109419

Hosseinzadeh, H., & Ramin, S. (2018). Effective removal of copper from aqueous solutions by modified magnetic chitosan/graphene oxide nanocomposites. *International Journal of Biological Macromolecules*, *113*, 859–868. doi:10.1016/j.ijbiomac.2018.03.028 PMID:29524485

Hu, C., Lu, T., Chen, F., & Zhang, R. (2013). A brief review of graphene–metal oxide composites synthesis and applications in photocatalysis. *Journal of the Chinese Advanced Materials Society.*, *1*(1), 21–39. doi:10.1080/22243682.2013.771917

Huang, T., Yan, M., He, K., Huang, Z., Zeng, G., Chen, A., Peng, M., Li, H., Yuan, L., & Chen, G. (2019). Removal of methylene blue dye by solvothermally reduced graphene oxide: A metal-free adsorption and photodegradation method. *Royal Society of Chemistry.*, *543*, 43–51.

Huang, X., Xiaoying, Q., Boey, F., & Zhang, H. (2012). Graphene based composites. *Chemical Society Reviews*, *41*(2), 666–686. doi:10.1039/C1CS15078B PMID:21796314

Huang, X., Yin, Z., Wu, S., Qi, X., He, Q., Zhang, Q., Yan, Q., Boey, F., & Zhang, H. (2001). Graphene-Based Materials: Synthesis, Characterization, Properties, and Applications. *Small*, *7*(14), 1876–1902. doi:10.1002mll.201002009 PMID:21630440

Hummers, W. S. Jr, & Offeman, R. E. (1958). Preparation of Graphene Oxide. *Journal of the American Chemical Society*, *80*(6), 1339. doi:10.1021/ja01539a017

Jie, W., Zheng, F., & Hao, J. (2013). Graphene/gallium arsenide-based Schottky junction solar cells. *Applied Physics Letters*, *103*(23), 233111. doi:10.1063/1.4839515

Jung, I., Dikin, D. A., Piner, R. D., & Ruoff, R. S. (2008). Tunable Electrical Conductivity of Individual Graphene Oxide Sheets Reduced at "Low" Temperatures. *Nano Letters*, *8*(12), 4283–4287. doi:10.1021/nl8019938 PMID:19367929

Klauson, D., Preis, S., Portjanskaja, E., Kachina, A., Krichevskayz, M., & Kallas, J. (2005). The influence of ferrous/ferric ions on the efficiency of photocatalytic oxidation of pollutants in groundwater. *Environmental Technology*, *26*(6), 653–661. doi:10.1080/09593330.2001.9619505 PMID:16035658

Kumar, A., Sadanandhan, A. M., & Jain, S. L. (2019). Silver doped reduced graphene oxide as a promising plasmonic photocatalyst for oxidative coupling of benzylamines under visible light irradiation. *New Journal of Chemistry*, *43*(23), 9116–9122. doi:10.1039/C9NJ00852G

Kumaran, V., Sudhagar, P., Konga, A. K., & Ponniah, G. (2020). Photocatalytic Degradation of Synthetic Organic Reactive Dye Wastewater Using GO-TiO2 Nanocomposite. *Polish Journal of Environmental Studies*, *29*(2), 1683–1690. doi:10.15244/pjoes/109027

Li, X., Zhang, S., Wang, P., Zhong, H., Wu, Z., Chen, H., Liu, C., & Lin, S. (2015). High performance solar cells based on graphene-GaAs heterostructures. *Nano Energy*, *16*, 310. doi:10.1016/j.nanoen.2015.07.003

Li, X., Zhao, Y., Wang, X., Wang, J., Gaskov, A. M., & Akbar, S. A. (2016). Reduced graphene oxide (rGO) decorated TiO2 microspheres for selective room-temperature gas sensors. *Sensors and Actuators. B, Chemical*, *230*, 330–336. doi:10.1016/j.snb.2016.02.069

Lin, Y., & Du, H. (2020). Graphene reinforced cement composites: A review. *Construction & Building Materials*, *265*, 1–16. doi:10.1016/j.conbuildmat.2020.120312

Lipatov, A., Varezhnikov, A., Wilson, P., Sysoev, V., Kolmakov, A., & Sinitskii, A. (2013). Highly selective gas sensor arrays based on thermally reduced graphene oxide. *Nanoscale*, *5*(12), 5426–5434. doi:10.1039/c3nr00747b PMID:23661278

Loh, K. P., Bao, Q., Eda, G., & Chhowalla, M. (2010). Graphene oxide as a chemically tunable platform for optical applications. *Nature Chemistry*, *2*(12), 1015–1024. doi:10.1038/nchem.907 PMID:21107364

Loryuenyong, V., Totepvimarn, K., Eimburanapravat, P., Boon-chompoo, W., & Buasri, A. (2013). Preparation and Characterization of Reduced Graphene Oxide Sheets via Water-Based Exfoliation and Reduction Methods. *Advances in Materials Science and Engineering*, *9*, 1–5. doi:10.1155/2013/923403

Lu, G., Ocola, L. E., & Chen, J. (2009). Reduced graphene oxide for room-temperature gas sensors. *Nanotechnology*, *20*(44), 445502. doi:10.1088/0957-4484/20/44/445502 PMID:19809107

Maruthamani, D., Divakar, D., & Kumaravel, M. (2015). Enhanced photocatalytic activity of TiO2 by reduced graphene oxide in mineralization of Rhodamine B dye. *Journal of Industrial and Engineering Chemistry*, *30*, 33–43. doi:10.1016/j.jiec.2015.04.026

Meng, W., & Khayat, K. H. (2016). Mechanical properties of ultra-high-performance concrete enhanced with graphite nanoplatelets and carbon nanofibers. *Composites. Part B, Engineering*, *107*, 113–122. doi:10.1016/j.compositesb.2016.09.069

Miao, X., Tongay, S., Petterson, M., Berke, K., Rinzler, A., Appleton, B., & Hebard, A. (2012). High Efficiency Graphene Solar Cells by Chemical Doping. *Nano Letters*, *12*(6), 2745 2750. doi:10.1021/nl204414u PMID:22554195

Mishra, A., Mehta, A., Basu, S., Shetti, N. P., Reddy, K. R., & Aminabhavi, T. M. (2019). Graphitic carbon nitride (g–C3N4)–based metal-free photocatalysts for water splitting. *Carbon*, *149*, 693–721. doi:10.1016/j.carbon.2019.04.104

Priyadarsini, S., Mohanty, S., Mukherjee, S., Basu, S., & Mishra, M. (2018). Graphene and graphene oxide as nanomaterials for medicine and biology. *Journal of Nanostructure in Chemistry*, *8*(2), 123–137. doi:10.100740097-018-0265-6

Qiu, J., Zhang, P., Ling, M., Li, S., Liu, P., Zhao, H., & Zhang, S. (2012). Photocatalytic Synthesis of TiO2 and Reduced Graphene Oxide Nanocomposite for Lithium Ion Battery. *Interfaces*, *4*, 3636. PMID:22738305

Rehman, S. K. U., Kumarova, S., Memon, S. A., Javed, M. F., & Jameel, M. (2020). A Review of Microscale, Rheological, Mechanical, Thermoelectrical and Piezoresistive Properties of Graphene Based Cement Composite. *Nanomaterials (Basel, Switzerland)*, *10*(2076), 1–42. PMID:33096675

Sedaghat, A., Ram, M. K., Zayed, A., Kamal, R., & Shanahan, N. (2014). Investigation of Physical Properties of Graphene-Cement Composite for Structural Applications. *Journal of Composite Materials*, *4*(01), 12–21. doi:10.4236/ojcm.2014.41002

Shang, Y., Zhang, D., Yang, C., Liu, Y., & Liu, Y. (2015). Effect of graphene oxide on the rheological properties of cement pastes. *Construction & Building Materials*, *96*, 20–28. doi:10.1016/j.conbuildmat.2015.07.181

Sharma, B., & Rabinal, M. K. (2013). Ambient synthesis and optoelectronic properties of copper iodide semiconductor nanoparticles. *Journal of Alloys and Compounds*, *556*, 198–202. doi:10.1016/j.jallcom.2012.12.120

Shetti, N. P., Malode, S. J., Malladi, R. S., Nargund, S. L., & Shukla, S. S. (2019). Electroanalysis of paracetamol at nanoclay modified graphite electrode. *Microchemical.*, *146*, 387–392. doi:10.1016/j.microc.2019.01.033

Shukla, S., & Saxena, S. (2011). Spectroscopic investigation of confinement effects on optical properties of graphene oxide. *Applied Physics Letters*, *98*(7), 073104. doi:10.1063/1.3555438

Singh, E., & Nalwa, H. (2015). Graphene-based bulk-heterojunction solar cells: A Review. *Journal of Nanoscience and Nanotechnology*, *15*(9), 6237–6278. doi:10.1166/jnn.2015.11654 PMID:26716184

Siong, V. L. E., Lee, K. M., Juan, J. C., Lai, C. W., Tai, X. H., & Khe, C. S. (2019). Removal of methylene blue dye by solvothermally reduced graphene oxide: A metal-free adsorption and photodegradation method. *RSC Advances*, *9*(64), 37686–37695. doi:10.1039/C9RA05793E

Sposito, G. (1998). On Points of Zero Charge. *Environmental Science & Technology*, *32*(19), 2815–2819. doi:10.1021/es9802347

Tarjomannejad, A., Farzi, A., Niaei, A., & Salari, D. (2016). An experimental and kinetic study of toluene oxidation over LaMn1−x B x O3 and La0.8A0.2Mn0.3B0.7O3 (A=Sr, Ce and B=Cu, Fe) nano-perovskite catalysts. *Korean Journal of Chemical Engineering*, *33*(9), 2628–2637. doi:10.100711814-016-0108-4

Upadhyay, R. K., Soin, N., & Roy, S. S. (2014). Role of graphene/metal oxide composites as photocatalysts, adsorbents and disinfectants in water treatment. *RSC Advances*, *4*(8), 3823–3851. doi:10.1039/C3RA45013A

Wang, B., Jiang, R., & Wu, Z. (2016, November 04). Investigation of the mechanical properties and microstructure of graphene nanoplatelet-cement composite. *Nanomaterials (Basel, Switzerland)*, *6*(11), 200–208. doi:10.3390/nano6110200 PMID:28335328

Wang, X., Shen, Y., Xie, A., Qiu, L., Li, S., & Wang, Y. (2011). Novel structure CuI/PANI nanocomposites with bifunctions: Superhydrophobicity and photocatalytic activity. *Journal of Materials Chemistry*, *21*(26), 9641–9646. doi:10.1039/c0jm04558f

Zarrabi, M., Haghighi, M., Alizadeh, R., & Mahboob, S. (2019). State of the art: Synthesis and characterization of functionalized graphene nanomaterials. *IOP Publishing Ltd.*, *211*, 738–752.

Zhang, H., Lv, X., Li, Y., Wang, Y., & Li, J. (2009). P25-Graphene Composite as a High Performance Photocatalyst. *ACS Publications.*, *4*, 380–386. PMID:20041631

Chapter 11
The Importance of Light in Our Lives:
Towards New Lighting in Schools

David Baeza Moyano
Universidad CEU San Pablo, Spain

Roberto Alonso González Lezcano
https://orcid.org/0000-0002-6185-4929
Universidad CEU San Pablo, Spain

ABSTRACT

The light that enters through our eyes is not only for vision. The human circadian system responds to light differently than the visual system. The timing of each biological function in mammals is directed by the main clock located in the Supraquiasmic Nucleus, which is regulated by light. However, until now, only the interaction of light with our visual system has been taken into account when choosing the parameters of indoor lighting sources, including those in the classroom. In the publications about school lighting, the first concern was the common parameters of indoor lighting such as horizontal workplane illuminance, illuminance uniformity, and avoiding reflections on different surfaces. In this chapter, the authors show publications about new findings on the effects of light on people, studies carried out in different countries aimed at improving classroom lighting, current regulations on lighting related to classroom lighting, and new parameters that are being considered, along with those already used for new and better lighting.

1. INTRODUCTION

Human beings have genetically evolved to perform their activities during the day and rest during the night. The evolution of the human species has occurred under periodic and relatively stable cycles of light and darkness. (Moyano, Mónica, & Roberto, 2020) Many days throughout the year when children and adolescents come to school in the morning, they do so in total or partial darkness. The intensity of

DOI: 10.4018/978-1-7998-7023-4.ch011

natural light entering classrooms through windows varies throughout the day and during the different seasons. Students work for many hours in classrooms with lighting that does not resemble natural light in both composition and intensity. When they leave the schools, they usually spend very little time receiving natural light. This results in students living virtually every day with insufficient light, not receiving the proportion and amount of visible light for which we humans are genetically designed.

Since the invention of electric light more than 150 years ago, lighting specialists have been concerned with the development and improvement of luminaires by taking into account the image forming (IF) effects of the light. Following the discovery about 20 years ago of intrinsically photosensitive retinal ganglion cells (ipRGCs) and their role in the NIF effects of the light, active research is being conducted to develop new luminaires that will improve the quality of life of people, also taking into account this latest finding.

In this chapter we are going to talk about the importance of redesigning the lighting of classrooms in educational centres based on the new scientific discoveries on the NIF effects of the light. The new LED luminaires in which the spectral power distribution (SPD) can be varied open up the possibility of creating new sources of interior lighting. Architects have to take all these factors into account both for the design of new constructions and for the refurbishment of buildings.

Good lighting will create a visual environment that enables people to see, to move about safely and to perform visual tasks efficiently, accurately and safely without causing undue visual fatigue and discomfort. The illumination may be daylight, electric light or combination of both. (ISO 8995-1, 2002)

2. IMAGE FORMING AND NOT IMAGE FORMING EFFECTS OF THE LIGHT

Light is defined as any electromagnetic radiation that can create a visual sensation by directly stimulating the retinal photoreceptors of the visual system. In addition to enabling vision, these photoreceptors also drive biological effects that powerfully regulate human health, performance and well-being. (CIE, 2019) Light is a regulator of our psychology and behavior that have evolved over millennia in which lighting provided reliable information about the time of the day. The advent of electric light has now altered this relationship with the patterns of light exposure by reflecting on personal tastes and social pressures. (Edwards & Torcellini, 2002)

The duration, timing, spatial distribution, intensity and power of the spectrally distributed light reaching the eyes can influence circadian rhythms and thus health (Bellia et al., 2013) (Figueiro, 2013) along with additional factors not directly related to building design such as age and work schedule requirements.(Konis, 2017) Light affects mood, attention and influences the synchronization of the biological clocks located in the central nervous system (CNS). This clock regulates the self-sustained rhythms that repeat approximately every 24.2 h (sleep-wake cycle, body temperature, cell division, hormone secretion, behavior ...) that have developed in their biological evolution terrestrial species because, by the rotation of the earth on its axis, there is a predictable and regular pattern of daylight and darkness over most of its surface.(Acosta et al., 2015) The ipRGCs provide the biological clocks with the information of the time of day in which we find ourselves. They are the cells responsible for the non-visual responses of light entering through the retina along with the rest of the photoreceptors (cones and rods) (Lucas et al., 2014). Environmental signals can restart and synchronize the SCN daily, ensuring that the behavior and physiological rhythms of humans are in sync daily with the rhythms found within their environment.

(Figueiro, 2013) There is a direct quantitative relationship between the direct circadian impact (percentage of melatonin suppressed) and the characteristics of the incident light (spectral distribution and amount) on the human cornea. Circadian systems respond differently to the same light stimulus depending on the time of day: a light stimulus in the morning can advance the circadian clock (shortening our biological day) and help maintain the cycle of approximately 24 hours while the same light stimulus in the afternoon can delay the clock (extending the biological day). The response also depends on individual differences from light stimuli and the age of the person. (Bellia et al., 2013)

From the point of view of temporal characteristics, the operation of the visual system does not depend significantly on the moment of exposure to light and therefore has a good response to light stimuli at any time of day or night. While the visual system responds to light in less than a second, the duration of light exposure necessary to affect the circadian system can be minutes. For the visual system the spatial distribution of light is critical while the circadian system does not respond to spatial patterns. (Acosta et al., 2015)

Disruptions in the pathway of circadian rhythms with this 24-h light/dark cycle can lead to chronodisruption and have negative effects such as poor performance, depression, insomnia, heart disease, weight gain, and even cancer. (Figueiro, 2013)

The light that enters through the human eye not only has the function of image formation but also influences the health and well-being of human beings by producing non-imaging effects in the long and short term (acute). (Khademagha et al., 2016) From the light spectrum point of view, melanopsin in ipRGCs has the maximum between 460 and 500 nm while the visual system is more sensitive to mid-wavelengths of the visible spectrum around 555 nm. (Bellia et al., 2013) (Stevens, 2009)

3. DAYLIGHTING AND ELECTRIC LIGHTING

Different researchers have shown that natural light increases productivity (Group, 1999), positively affects human performance (Boyce, 2003) and has biological effects on the production of the hormone cortisol, regulating light-cycledarkness and the ability of concentration of the students. (Winterbottoma & Wilkinsb, 2009)

To apply daylighting, sunlight, diffuse sky light and reflected light from external surfaces inside a building must be taken into account to contribute to the indoor lighting requirements and energy savings through the use of electronic light controls. Recommendations for daylighting design depend on latitude, climate, surrounding buildings (height, color) and the orientation of the building (Reinhart et al., 2006) and the structures of the building itself. Taking all these factors into account should be the first step in the design process to determine the interior electrical lighting design.

The lighting should be sufficient but not excessive for the performance of the required activities, minimizing the possibility of glare. Visual comfort is an important aspect to be achieved with an increasing interest in the contribution of day lighting together with artificial lighting. (Winterbottoma & Wilkinsb, 2009) Window design is the key element for allowing daylight inside buildings and a proper design can improve thermal comfort and save electric lighting energy. (Acosta et al., 2015)

The characteristics of the daylight reaching the human eye inside the buildings influence the magnitude of the IF and NIF responses of the building occupants. The architects determine the possibilities of lighting and therefore the luminous characteristics through the design of the envelope of the building, the material properties of the interior walls, floors, ceilings, as well as of the furniture. The modification of most lighting design parameters requires drastic changes and demolition so it is important that building designers consider the relationships between lighting design parameters and NIF light factors. The quality of the light factors reaching the indoor environment depends on the design parameters of the daytime lighting. These factors can be controlled/adjusted through design parameters such as glazing type, daylight opening (vertical, for example, windows or horizontal, for example, skylights) size and position. (Khademagha et al., 2016)

The methods used so far of daylight analysis provide very limited information compared to the new dynamic methods that integrate factors such as sky type and local light conditions. These new methods serve to improve the evaluation of daylight strategies, taking into account the lighting levels for the students' visual tasks. The system called Climate-based daylight modelling (CBDM) integrates the different variations of light in the simulations in relation to the local climate, generating for a specific moment a series of predictions that can become for every hour of a full year. This system has opened a new perspective of natural light studies. (Reinhart et al., 2006)

Useful daylight illuminance (UDI) is used to assess daylight in buildings. It is very useful when what you want to do is an adaptation intervention in existing buildings, especially in heritage buildings where, as a result, more precise and careful assessments are required. (Nocera et al., 2018) UDI is designed to assist in the interpretation of climatic analyses of daylight lighting levels that are based on hourly meteorological data for a period of one full year. The actual daylight illuminances in buildings vary greatly, much more than the variations in the expected daylight factors suggest. Daytime lighting tends to decrease rapidly as the distance from the windows increases. Likewise, daylight illuminations at one point can vary greatly from one moment to the next due to the changing position of the sun and/or sky conditions. The conceptions of uniformity of illuminance arising from the use of the standard cloudy sky approach are inapplicable for realistic conditions where the contribution of direct sunlight shows large differences between maximum and minimum daylight levels. The occupants prefer a space with variation in the pattern of natural light, and where they have a task illuminance slightly higher than the overall enveloping illuminance since in this way their visual perception can be improved (Reinhart et al., 2006). It has been observed that lighting levels which are significantly higher than the typical lighting level of the work plan (for example, 500 lx) are tolerated by the occupants unless there is glare or direct sun, in which case the occupants may choose to use a shading device. (Tanner, 2008)

The exploitation of natural light is recognized as an effective means of reducing the need for electric lighting in non-domestic buildings. Energy saving is a priority worldwide. There are situations in which the time of electric lighting is limited in order to achieve energy savings. (Ho, 2007) The implementation of appropiate daylight access devices can reduce artificial lighting power costs from 50 to 80% by applying self-responding dimming glass and that fail in terms of overall lighting quality. (CIE, 2019) The growing use of tunable LED lighting systems is being driven by energy considerations. There's the potential for significant energy savings relative to fluorescent systems and non-dimmable LED lighting systems. These savings can be attributed to the combination of two factors: the higher efficacy of LED systems compared to fluorescent systems, and the dimming capability provided by tunable lighting. (IES, n.d.)

Recent research suggests the need to enrich the indoor lighting Spectral Power Distribution (SPD) incorporating violet (VL). It is part of the spectrum of visible light that does not enter the classroom because it is absorbed by the window panes and is not yet emitted by LED luminaires. The authors of these investigations claim that this part of the spectrum is important for not only the prevention of myopia progression but also the onset of myopia. (Torii et al., 2017) Excessive ultraviolet (UV) protection and the absence of VL in artificial light may be possible reasons for the global myopia pandemic. (Jiang et al., 2018)

Daylight is extremely important in the educational environment since it can improve alertness and behavior.(Bellia et al., 2013) Research by Bellia and collaborators in a university classroom in Italy evaluates the circadian impact of light sources to calculate nighttime melatonin suppression as a function of the spectral distribution of light irradiance reaching the eyes. It tries to take into account the contribution of all known photoreceptors to illuminate the circadian system: cones, rods, ipRGCs, the blue-yellow spectral opposition channel and also their interactions. The results are compared with the irradiance distributions detected at eye level and correlated color temperature (CCT). To determine the percentage of melatonin suppression at light exposure represented by Circadian Stimulus (CS), the Circadian Light (CL) has to be calculated first. The SPD of daylight reaching the eyes in a classroom is different on a sunny day than on a cloudy day. The CCT values are always compressed between 4000-6000 K while those in the sky are higher. This phenomenon is due to the interior and exterior reflectances. Indoor and outdoor materials absorb low wavelengths while reflecting medium and high wavelengths. Most of the light that reaches the eyes comes from internal surfaces and therefore has a low content of low wavelengths. This means that the choice of materials and colors for interior surfaces is not merely aesthetic but has important implications for the well-being of the users. The highest values of CCT, melatonin suppression and lighting are obtained on the clearest and sunniest days. The melatonin suppression values calculated for electric light are the lowest and therefore the lighting systems do not seem to produce enough stimuli for the circadian system while the horizontal plane illuminance values comply with EN 12564-1 (EN 12464-1, 2019) and therefore the lighting system guarantees the correct operation of the visual system. Similar melatonin suppression values have been found with different levels of illuminance at eye level suggesting the important role of light SPD. (Bellia et al., 2013)

Studies conducted in elementary schools found a significant positive correlation between the presence of daylight and student performance in three different school districts. Daylighting provided through skylights also has a positive effect on students in their classrooms. Other studies that included classrooms with more daylighting compared to classrooms with less daylighting showed an increase in student performance. Designing a classroom with attention to sustainable IEQ criteria, e.g., thermal conditions, acoustic conditions, lighting conditions, furniture, aesthetics, technology, and visual conditions, is associated with positive student outcomes including their overall satisfaction with the classroom IEQ and its perceived effect on their learning, leading to student satisfaction with the courses. (Choi, 2013-14)

In order to achieve a good optimisation of the daylighting, are proposed the use of new tools.

Table 1. Daylighting parameters

Weather Based Daylight Metrics (CBDM)	Dynamic simulation method for the design of classrooms, which provides quantitative information of the proposed strategies, considering the levels of illumination for the visual tasks of students. (Guan & Yan, 2016)
Useful Daylight Illuminances (UDI)	It aims to determine when daylight levels are 'useful' for the occupant. UDI results in three metrics, that is, the percentages of the occupied times of the year when the UDI was achieved (100–2000 lx), fell-short (<100 lx), or was exceeded (>2000 lx). The last bin is meant to detect the likely appearance of glare. (Reinhart et al., 2006)
Daylight autonomy. (DA)	The percentage of occupied hours of the year in which only a minimum threshold of illuminance is reached in daylight. (Reinhart et al., 2006)
Colour Quality Assessment Tool (CQAT)	It is a parameter that utilizes the spectral reflectance factor of interior finishes and the spectral power distribution (SPD) data of luminaires to accurately calculate the luminous exitance radiated by interior finishes and the colour quality components of the SPD, luminance, CAF, colour coordinates, and CCT based on the colours of the interior finishes and light sources. (Kim et al., 2017)

4. EDUCATIONAL BUILDINGS

Different researchers have shown that correct lighting increases productivity, positively affects human performance and has biological effects, regulating the day-night cycles and the student's ability to concentrate.

Places of learning, including elementary schools, middle schools, high schools and higher education facilities must provide for the needs of teachers and students of all ages. Educational facilities applies to projects where dedicated staff are employed for instructional purposes and students can be of any age. (International Well Building Institute, n.d.)

The structural characteristics of building installations have a profound influence on learning. Inadequate light, noise, poor air quality, and poor heating in classrooms are mentioned as factors relevant to poor student progress (Edwards & Torcellini, 2002). Many research studies have highlighted that ensuring good quality lighting in an educational environment is a rather complex task. (Nocera et al., 2018)

School building spaces are the most important non-residential indoor environments that often have, among other things, inadequate lighting. Educational buildings are designed to last for several years, and their state remains constant for a long time without rehabilitation or diagnosis of their influence on users. It implies that some facilities do not meet the minimum quality standards that are a requirement for the new constructions, such as the case of lighting. (López-Chao et al., 2020)

According to the National Center for Education Statistics (Baraa, 2017), 16% of schools with permanent buildings and 28% of schools with temporary (i.e., portable) buildings have natural lighting that is unsatisfactory or very unsatisfactory. Many studies have revealed a significant relationship between the quality of the physical infrastructure and student performance. Most U.S. public schools have building quality problems, with poor lighting, acoustics, temperature regulation, or air quality. (Cheryan et al., 2014)

Empirical research has shown the influence of architectural spatial variables on student performance. Specifically, the performance is higher for students in classrooms with higher lighting. Similarly, young children can differentiate their light needs according to the task to be performed, while visual comfort is a key element for artistic activities, especially for drawing. It is important to study the indoor environmental quality (IEQ) of the classrooms, taking into account that students spend more time in them than in any other indoor environment of the schools. Classroom IEQ can directly influence student outcomes such as satisfaction and learning. Lighting conditions have long been an important IEQ criterion in the built environment, including electrical and daylight sources and environmental and task uses. Exposure to various light sources can be associated with physiological responses in human performance, and daylight entering through windows provides visual illumination along with a view of the outside or natural environment, which has also been found to positively influence human behavior. (Choi, 2013-14)

A study conducted in Jordan to improve visual comfort in classrooms states that the quality of the indoor environment of classrooms has a great effect on the quality of teaching. In classrooms, light levels are directly related to energy consumption, due to the use of artificial lighting. This study aims to improve visual comfort and energy efficiency in existing classrooms by investigating various methods of adaptation for passive daylighting techniques. Energy for buildings is the most important sector of energy demand, providing cooling, heating, water heating, lighting as main uses. In educational facilities and classrooms, it is important to provide a comfortable indoor environment in parallel with the use of appropriate energy which are then suggested cases of passive daylight techniques to improve visual comfort in the classrooms. They perform 9 different cases of classrooms by increasing the reflectance of surfaces, increasing the transmittance of glass, including clerestory window or anidolic ducts. They claim that according to the European Committee for Standardization none of the proposed cases improved the lighting levels to the required levels in the classroom. The case in which a clerestory is put together with the anidolic ceiling directed toward south is where the best results are obtained. (Heschong Mahone Group, 1999)

A study focusing on the analysis of lighting performance in primary school classrooms has been carried out in China considering the importance of daylight in the performance and well-being of students. Their objective was to apply a methodology that integrates the variations of daylight to know the performance of the annual illumination with the CBDM. One of the tools used is the daylight autonomy (DA). It establishes an illuminance value to guarantee autonomy to work only with daylight; however, we can have an exceptional autonomy without guaranteeing visual comfort, by not establishing the upper illuminance limit, we can have too much daylight at certain times of the year. The methodology developed was used to evaluate the horizontal illuminance in the student's work plan. Likewise, this could be applied to estimate the vertical illuminance on the board; however, due to the scope of this study, this was not considered. A base classroom is defined as 78.24 m^2 by 3.5 m high, with capacity for 36 students, oriented to the north, which has direct and indirect lateral lighting from the room. Its main window corresponds to 23% of the total floor area, based on the optimal proposal for the three zones. To control the penetration of sunlight at whiteboard level, the window was kept at a distance of 1.55 m from the perpendicular wall where the whiteboard is placed. This study compared different passive architectural design strategies under the climatic conditions of five representative Chinese cities and selected the most suitable design scheme for each city. In this study, the dynamic annual graphic tool was used for architectural design in China, and we integrated the optimal design with the Chinese academic calendar to achieve improvements within the occupancy time. A detailed calculation could be made to determine the exact size of the architectural device, such as the precise width or length of the light shelves or shading plate according to certain climatic conditions of a given location. (Guan & Yan, 2016)

In Italy and another European countries, many historical buildings are reused as school buildings, changing their original function. This process of adaptive reuse is a method for preserving heritage buildings somctimes sacrificing the quality of daylight and the well-being of students. It is difficult to reconcile the cultural value of historic buildings with comfort standards. Daylight makes a significant contribution to the aesthetics and physical character of a learning space, as well as limiting the possible harmful effects of prolonged exposure to artificial light. A number of different visual activities are performed within the classroom that require specific visual conditions to be successful. This feature is especially emphasized in historical buildings that are reused for purposes other than their original design or that can no longer be used according to their original design (Khademagha et al., 2016). Nocera and collaborators have developed with the objective of improving the lighting for educational use in a historic building in Italy. A daylight model validated in Radiance was built using the Groundhog plug-in for Google Sketchup. The use CBDM such as Daylight Useful Illumination (DIU) and DA has allowed the identification, on an annual basis and under variable sky conditions, of the main problems of a sample classroom. The main problems found relate to excessive lighting levels near windows and too low lighting values when away from windows. In order to reduce the possible risks of glare for students sitting near windows, without negatively affecting the lighting levels in the remaining areas, the installation of double-glazed windows was suggested along with a highly reflective and diffuse false ceiling. The study of the new geometric arrangement given by the breakage of the window frames and the double glazing (Nocera et al., 2018)

5. STUDENTS

Light does not affect children and adolescents in the same way as it does adults. When choosing interior lighting, among other factors, the age for which the light is intended must be taken into account. Teenagers tend to go to sleep and get up several hours later than normal. This is probably associated with the hormonal changes that occur at puberty. Because they don't sleep as many hours as those who go to bed at normal time, they have difficulty waking up in the morning for school. Schools are not prepared to provide adequate light or daylight to stimulate their circadian system, especially in the months of less natural light. Since the circadian system responds to short wavelength visible light, electrical light in classrooms may be insufficient to stimulate the brain's clock, even though it is perfectly adequate for reading. As teenagers spend more time indoors, they may lose the essential morning light needed to stimulate their circadian systems and promote the dragging of the 24-hour solar day. In late spring, if teens spend more time outdoors after school, the brain's clock may be slowed by afternoon exposure to daylight. Late spring mornings can also pose a concern because of very early sunrises. Students who are late in the phase, and therefore reach their minimum core body temperature later in the morning clock, may be at risk of receiving too much daylight in their delayed portion of the response curve phase while traveling to school early in the morning. To help minimize sleep restriction and possibly circadian disruption, adolescents should receive higher levels of morning light (or daylight) in schools and lower levels of night light (or daylight) at home. (Figueiro, 2013)

Research on the health, behaviour and hormone levels of 88 8-year-old students in four classes during the duration of a school year showed that the four classes had very different conditions of natural and artificial lighting: Two had natural light, two did not; two were illuminated with fluorescent emitters of warm white light (3000 K), two had fluorescent emitters of very cold white light (5500 K). A significant correlation was found between the patterns of natural light level, hormone levels and student behavior and they concluded that no windows in the classroom should be abolished. (Kuller & Lindsten, 1992)

Student performance is better in classrooms with higher light intensity. Young children can differentiate their light needs according to the task at hand, while visual comfort is a key element for artistic activities, especially for drawing. (María & Labarca, 2015) In rooms where lighting is not uniform, with more luminosity over the immediate task area than the surrounding area, discomfort effects may be important. (Winterbottoma & Wilkinsb, 2009) There is multiple evidence of damage to children's vision caused by poorly lit classrooms. (Ho, 2007)

A study has been carried out with students who have been fitted with a filter that does not allow the most important part of the light for the elimination of melatonin to pass through. Eleven students at a North Carolina school with unusually high levels of daylight were fitted with orange filter glasses to remove short-wavelength circadian light for five consecutive days. It was observed that the orange filter glasses allowed them to perform their activities but their DLMO (Dim Light Melatonin Onset) was delayed by half an hour compared to the previous week. Another study compared the behaviour of students between two seasons with a significant difference in the amount of natural light received during the day. In this study of 16 students in New York, their DLMO was found to be delayed by 20 minutes and sleep output by 16 minutes in the spring compared to the winter. (Figueiro, 2013)

Researching study of the relationship between the learning space and mathematics and artistic activities was made in 583 primary school students from Galicia (Spain). The Indoor Physical Environment Perception scale has been adapted and validated and conducted in 27 classrooms. The result of this Exploratory Factor Analysis has shown that the learning space has evidenced that students perform better in classrooms with more luminosity. Young children can differentiate their lighting needs according to the task at hand, while visual comfort is a key element for artistic activities, especially for drawing. (López-Chao et al., 2020)

Regarding the risks of the increasing use of electronic devices by students, there are a growing number of publications that doubt the possible risk of blue light emitted by electronic devices on the retina (Moyano, Sola, & Roberto, 2020) (Escofet & Bará, 2015) but this does not mean that their continuous use by students cannot have negative consequences. The Artificial Light at Night (ALAN) is drawing the attention of researchers and environmentalists for its ever-increasing evidence on its capacity of "desynchronization" of organismal physiology. (Khan et al., 2018) Scientific studies state that an illuminance of 2500 lx is necessary to suppress nocturnal melatonin but under certain conditions, such as 1 lx melatonin can be suppressed in humans. (Glickman et al., 2002) Several authors (), have referred for years to the problem of mobile phone addiction, which they define in different ways as mobile phone dependence, the problematic use of mobile phones or cell phones, mobile phone abuse, and nomaphobia (a fear of being without one's phone). Symptoms include preoccupation with the device, excessive use with loss of control, use of the device in inappropriate or dangerous situations, adverse effects on relationships, symptoms from lack of use (feeling angry, tense, depressed when the mobile phone is not accessible, constantly worry that the battery will run out, etc.), and tolerance symptoms (the need for a better mobile phone, more software or more hours of use). Children are more vulnerable to potential harmful agents than adults because their nervous systems are developing. For anatomical reasons such as the smallness of their heads, thinner skulls and the increased nerve conductivity of their brains, children can absorb more energy from electric devices than adults.(Moyano, Moyano, López et al, 2020) These symptoms are similar to those of substance dependence, which is why some researchers consider mobile phone dependence (MPD) to be an independent diagnosis. (Moyano, Mónica, & Roberto, 2020)

Because of all this, it is very important give more information about the correct use of electronic devices by children and adolescents to avoid using them for many hours during the day and not to use them after 20-21:00. Further research is needed to achieve adequate lighting in schools in order to reset the biorhythms in the mornings, not only for better educational performance but also to improve the health of students.

6. INDOOR LIGHTING STANDARS

The technical characteristics of the luminaires and the lighting conditions at the workplaces where they are used are clearly defined in international standards. (ISO 8995-1, 2002) (EN 12464-1, 2019) Currently, the standar ISO 8995:2002/CIE S 008 "Lighting in workplaces" (2201/220) (ISO 8995-1, 2002) and EN 12464-1:2019 "Lighting in workplaces" (EN 12464-1, 2019) are in the process of being reviewed.

Educational Buildings recommended illuminances for classrooms are between 300, 500 and 750 lux, blackboard maintained with an illuminance of 500 lux. It avoids specular reflections. Classrooms, tutorial rooms maintain an illuminance of 300 lux. The illumination must be controllable. Classrooms for evening classes and adult education maintain an illuminance of 500 lux. (ISO 8995-1, 2002)

Lighting requirements are determined by the satisfaction of three basic human needs: visual comfort, where workers have a sense of well-being; indirectly, also contributing to a high level of productivity; visual performance, where workers are able to perform their visual tasks, even in difficult circumstances and for longer periods; and safety. The parameters that determine the ambient brightness are the distribution of luminance, illuminance, dazzle, color rendering and the appearance of light colors, the directionality of light, flicker, daylight. The ranges of useful reflectances for the main interior surfaces are the ceiling: 0.6 to 0.9; walls: 0.3 to 0.8; working planes: 0.2 to 0.6; floor: 0.1 to 0.5. (EN 12464-1, 2019)

The illuminance for educational premises-educational buildings is between 300, 500, 750 lux but European standar make different recommendations for primary schools and secondary schools:

General activities in primary schools have to maintain a minimum illuminance of 300 lux and a higher illuminance of 750 lux. If the activities in a classroom are not completely clear in the design phase, these general activities will be taken. The lighting must be controllable.

General activities in secondary schools must have a minimum maintained illumination of 500 lux and an upper maintained illumination of 750 lux. Specific Requirements: If activities in a classroom are not completely clear in the design phase, these general activities will be taken. The lighting must be controllable. (EN 12464-1, 2019) Computer activities in schools should only be maintained with a minimum illuminance of 300 lux and a maintained upper illuminance of 500 lux. Specific requirements: Illumination must be controllable. Black, green and white boards with a minimum illuminance maintained of 500 lux and a maximum illuminance maintained of 750 lux. Specific requirements: Vertical illuminances. Specular reflections should be avoided. The presenter/teacher must be illuminated with adequate vertical illuminance. (Baraa, 2017) (EN 12464-1, 2019)

The Illuminating Engineering Society of North America (IESNA) (Illuminating Engineering Society of North America (IESNA), n.d.) has indoor lighting proposals in which the intensity varies depending on age that are not found in the ISO and European standards. The Journal of the Illuminating Engineering Society of North America (IESNA) refers to the tenth edition of the Illuminating Engineering Society (IES) Manual of Lighting from the year 2000 which contains a new illuminance recommendation. (David, 2011)

Table 2. Recommendation of target illuminances (lux) based on the visual ages of the observers (years) for some indoor education situations where at least half are

< 25 years old	25 to 65 years old	> 65 years old
200 lux	400 lux	800 lux
250 lux	500 lux	1000 lux
375 lux	750 lux	1500 lux

A minimum limit value of 300 lux (simple task) and a maximum limit value of 2000 lux (accurate visual perception) are established as defined by Illuminating Engineering Society of North America (IESNA). (Illuminating Engineering Society of North America (IESNA), n.d.)

7. TRADITIONAL AND NEW PARAMETERS OF LIGHTING

The architects have taken into account in their future work both traditional and new criteria that are being discovered and that luminaire manufacturers are incorporating into their new LED light models.

7.1 Visual Task Plane

Currently, indoor lighting design generally focuses on visual characteristics, for example, visual task performance, color rendition, reduction of glare, lit appearance (Heschong Mahone Group, 1999) and illumination for task, as to ensure that 500 lux falls on the horizontal work plane. (Andersen et al., 2012) The "Good lighting design" was evaluated mainly by measuring flow density reaching the "visual task planes" (which is almost invariably interpreted as referring to the horizontal work plan (HWP) (Rea et al., 2012), and compare the values with the lighting tables recommended in the interior lighting standards.

A new concept of lighting design is important due to the facts that technology is able to offer better solutions than high lighting where detail is difficult to see. Paper reading in offices has largely been replaced by self-luminous screen reading. Lighting design focused on task plane illuminance can easily lead to a dimmed appearance (or so-called "cave effect") of an interior space. Task lighting can always help when the nature of the work requires very high local illumination. (Dai et al., 2017)

The dominant parameters in indoor lighting design have been the illuminance on workplanes but suggest little about vertical illuminance values. For a correct circadian stimulation and feeling of wellbeing, the characteristics of the light incident on the cornea must be taken into account. The illuminance of the workplane is not relevant. (Dai et al., 2018)

7.2 Color

Researchers from South Korea have studied the influence of the color of interior lighting on the circadian system. To design an appropriate indoor lighting environment, the final color of the space perceived by an observer must be accurately predicted and its effect on an observer must be considered. Such perception is difficult to predict and color cannot be accurately calculated using conventional lighting simulation programs. When designing a human-centered lighting environment that takes into account the effect of the lighting environment on the observer, a tool similar to the CQAT (Color Quality Assessment Tool) proposed in this study should be used that can consider the effect of furniture, daylight, individual occupant characteristics, and other factors. (Alexander & Lewis, 2014)

7.3 Direction of the Light

The direction of light entering the human eye plays an important role in the magnitude of the NIF effects Four different areas have been investigated in the human retina: lower (upper visual field), upper (lower visual field), nasal (visual field on the side of the nose), and temporal (visual field on the side of the ear). Complete retina after 2 h of experimental light exposure the (relative) level of melatonin was significantly suppressed compared to when the lower retina was illuminated. Complete retinal exposure and lower retinal exposure were significantly more effective in suppressing melatonin compared to upper retinal exposure. (Khademagha et al., 2016)

7.3 Lighting Parameters

Some of the important parameters to consider in a lamp taken from the ICE Technical Report and the COMMISSION'S DELEGATE REGULATION (EU) N° 874/2012 of 12 July 2012 supplementing Directive 2010/30/EU of the European Parliament and of the Council with regard to energy labelling of electric lamps and luminaires include the characteristics of light: Light emission or luminous flux emission, luminaire opening angle, perceived color, CCT, color rendering index (CRI), and luminous intensity. (Delegado, 2012)

The most common measurement systems used in lighting related to visual task performance (e.g. horizontal workplane illuminance and illuminance uniformity), visual discomfort (e.g. probability of glare), lighting energy savings (e.g. electrical lighting energy reduction from photocontrols), and Indoor Environmental Quality (IEQ). (Dai et al., 2017)

Table 3 show the most important parameters of light sources:

Table 3. Parameters of light sources. (Moyano, Moyano, López et al, 2020)

Light emission or luminous flux emission (e-ILV 17-738)	Quantity derived from the radiant flux, Φ_e, by evaluating the radiation according to its action upon the CIE standard photometric observer.
Luminaire opening angle	Angle between two imaginary lines cutting a plane through the axis of the optical beam, so that these lines pass through both the center of the front of the lamp and a series of points where the luminous intensity is equivalent to 50% of the intensity of the central beam.
Correlated color temperature (CCT) (e-ILV 17-258) CIE E-ILV http://eilv.cie.co.at/	Temperature of the Planckian radiator having the chromaticity nearest the chromaticity associated with the given spectral distribution on a diagram where the (CIE 1931 standard observer based) coordinates of the Planckian locus and the test stimulus are depicted.
Color rendering index (CRI) (e-ILV 17-154)	Mean of the CIE 1974 special colour rendering indices for a specified set of 8 test colour samples.

Studies on the parameters of an LED luminaire for optimal IF of the light show that luminaires currently on the market meet the requirements of current standards.(Moyano, Moyano, López et al, 2020) These parameters have been designed for the highest optimisation of IF effects of the light.

Table 4 and 5 show the energy consumption indicators and light source service life parameters to be met by luminaires:

Table 4. Energy consumption indicators. (Moyano, Moyano, López et al, 2020)

Wattage or power	Rated power in watts with an accuracy of 0.1 W (eg, 8 W)
Equivalence in watts	Equivalent power in an incandescent lamp in watts with an accuracy of 0.1 W
Energy efficiency (EE) or Energy efficient label	Indicates the EE level of the bulb (ie, the electricity it consumes to produce the light) from the letter A++ for the most efficient ones, to the letter E for the least efficient ones (eg, A+)

Table 5. Light source service life parameters. (Moyano, Moyano, López et al, 2020)

Bulb life	For light emitting diode lamps, the lamp lifetime is the period of operation between the start of use and the time when only 50% of the total lamps survive or the maintenance factor of the average luminous flux of the sample has fallen below 70%, whichever is the sooner. It may be measured in hours or in years (eg, 25 000 hours)
Number of switch-on and switch-off cycles	Number of switching cycles before a fault occurs (eg, 50 000)
Lumen maintenance factor (LMF)	Ratio between the luminous flux emitted by the lamp at a given time in its life and the initial luminous flux (eg, 0.7)

Currently, there are no minimum requirements for access to daylight in buildings to support the circadian track. The International WELL Building Institute has recently developed a building certification system with the stated objective of, "measuring, certifying and monitoring the performance of building features that impact health and well-being". (U.S. Green Building Council, 2019) One of the preconditions for certification, (entitled "Circadian Lighting Design"), is the provision of sufficient melanopic light intensity for work areas. The term "melanopic" refers to a new photometric measure of light intensity weighed by the sensitivity of the melanopsin containing ipRGCs. (Konis, 2017)

In relation to the way in which the NIFs can be evaluated in a quantitative and qualitative way the Lighting Research Center (LRC) at Rensselaer Polytechnic Institute has developed two metrics, circadian light (CL), and circadian stimulus (CS); while, to characterize light as a stimulus to the biological clock. (Rea & Figueiro, 2018; Rea et al., 2012; Rea et al., 2005). Other researchers mention other parameters to take into account in order to quantify the NIF effects of light such as the circadian action factor (CAF) (Choi, 2013-14) (Hunt & Pointer, 2006), circadian efficacy of radiation (CER) and circadian illuminance (CIL) (Hunt & Pointer, 2006) Hankins &Lucas reported an opsin action spectrum for ipRGCs from the electrical response of the cells of the retina. (Hankins & Lucas, 2002) Enezi et al. proposed a single-opsin, melanopic spectral sensitivity function, the parameter is called **Equivalent α-opic illuminance**. (Enezi et al., 2011)

For a circadian lighting design figure 6 show some of the parameters that have been proposed taking into account the NIF.

Table 6. NIF effects of the light parameters

Circadian light (CL)	The irradiance weighted by the spectral sensitivity of the retinal phototransduction mechanisms stimulating the biological clock.
Circadian stimulus (CS)	The percentage of melatonin suppression under light exposure. (Figueiro et al., 2019)
Circadian action factor (CAF)	It represents the impact of light on the hormones and biorhythm of an observer. (Kim et al., 2017)
Circadian efficacy of radiation (CER)	The ratio of the circadian luminous flux to the radian flux. (Ln et al., 2016)
Circadian illuminance (CIL)	The product of VIL (v illuminance level) and CAF, is considered as a threshold for the relative melatonin suppression regardless of the type of optical sources. (Ln et al., 2016)
Equivalent α-opic illuminance	Based upon the now well-established spectral absorption of melanopsin. (Delegado, 2012)
Circadian efficacy of radiation (CER)	The ratio of the circadian luminous flux to the radian flux. (Ln et al., 2016)

There is currently no clear consensus among researchers as to the best method of measuring NIFs. This is why there are different proposals to measure their effects.

8. CONCLUSION

Indoor lighting is in a process of change due to new knowledge about the influence of light on human life together with technological progress in the manufacture of new luminaires.

The discovery of ipRGCs and NIF, has made us reconsider the traditional parameters of lighting to take into account these effects. To achieve a correct lighting of the classrooms, the IF and NIF effects of the light must be taken into account. We propose that new luminaires need to be developed that emit across the entire visible spectrum and the intensity and composition of which should vary throughout the day and according to the seasons.

Numerous investigations confirm the importance of the daylight and that its contribution is the greatest possible in the whole of the light that illuminates the classrooms. For the study of indoor environmental quality (IEQ), we propose the use of new tools such as CBDM, UDI, DA for maximum optimization at every moment of the day.

Children and especially teenagers, due to their hormonal changes are more sensitive to changes in light proportions so special care must be taken in the projection of the facilities where they are going to carry out their education and where they are going to spend most of the time of the day for many years.

The needs for light and indoor lighting have changed radically in recent years due largely to the increasing use of electronic light-emitting devices that are observed almost in front to the detriment of paper texts that reflect light and are observed on a horizontal surface along with the technological revolution of new LED lights whose SPD differs greatly from the lights we have used so far. The way of studying is therefore changing rapidly and this makes the traditional concepts of lighting (horizontal workplane, illuminance and illuminance uniformity) obsolete. Future lighting in educational classrooms will have to take into account that the new learning elements do not reflect light but emit it and that they are not on a horizontal surface but in front of the eyes. Some researchers argue a new perspective on workplanes lighting, color indoor surfaces and the direction of the light. It has a complex application because it would be necessary change the SPD of the luminaires and the reflectance of all indoor surfaces would have to be modified.

The latest versions of international standards (ISO 8995:2002, EN 12464-1:2019) are beginning to take into account these new factors for their update but it is still necessary a greater scientific knowledge of the NIFs to be able to specify better lighting.

People responsible for the construction and design of classrooms, training institutions and public administration must consider the extent to which existing architectural lighting replicates the biological effects of natural light and how artificial lighting combined with natural lighting could be used, to improve the quality of life of students. The lighting industry and scientists have begun to investigate in that direction. They argue that the first thing you need to know properly is how light impacts human behavior and psychology. It is recommended that all this new knowledge be taken into account by architects both in the design of educational buildings in general and in the design of classrooms in particular.

REFERENCES

Acosta, Leslie, & Figueiro. (2015). Analysis of circadian stimulus allowed by daylighting in hospital rooms. *Lighting Research & Technology*, 1-13.

Alexander, D., & Lewis, L. (2014). Condition of America's public school facilities: 2012-13 (NCES 2014-022). Washington, DC: U.S. Department of Education, National Center for Education Statistics.

Andersen, M., Mardaljevic, J., & Lockley, S. W. (2012). A framework for predicting the non-visual effects of daylight - Part I: Photobiology-based model. *Lighting Research & Technology*, *2012*(44), 37–53. doi:10.1177/1477153511435961

Baraa, J. (2017). Improving visual comfort and energy efficiency in existing classrooms using passive daylighting techniques. *Energy Procedia*, *136*, 102–108. doi:10.1016/j.egypro.2017.10.294

Bellia, Pedace, & Barbato. (2013). Lighting in educational environments: An example of a complete analysis of the effects of daylight and electric light on occupants. *Building and Environment*, *68*, 50-65.

Boyce, P. (2003). *Human Factors in Lighting* (2nd ed.). CRC Press. doi:10.1201/9780203426340

Bragazzi, N. L., & Del Puente, G. (2014). A proposal for including nomophobia in the new DSM-V. *Psychology Research and Behavior Management*, *7*, 155–160. doi:10.2147/PRBM.S41386 PMID:24876797

Cheryan, S., Ziegler, S. A., Plaut, V. C., & Meltzoff, A. N. (2014). Designing Classrooms to Maximize Student Achievement. *Policy Insights from the Behavioral and Brain Sciences*, *1*(1), 4–12. doi:10.1177/2372732214548677

Choi. (2013-14). Indoor Environmental Quality of Classrooms and Student Outcomes: A Path Analysis Approach. *Journal of Learning Spaces*, *2*(2).

CIE. (2019). *CIE Position Statement on Non-Visual Effects of Light. Recommending proper light at the proper time.* http://cie.co.at/publications/position-statement-non-visual-effects-light-recommending-proper-light-proper-time-2nd

Cuttle, C. (2013). A New Direction for General Lighting Practice. *Lighting Research & Technology*, *45*(1), 22–39. doi:10.1177/1477153512469201

Dai, Q., Cai, W., Shi, W., Hao, L., & Wei, M. (2017). A proposed lighting-design space: Circadian effect versus visual Illuminance. *Building and Environment*, *122*, 287–293. doi:10.1016/j.buildenv.2017.06.025

Dai, Q., Huang, Y., Hao, L., Lin, Y., & Chen, K. (2018). Luoxi Haoa, Yi Lin, Kaixuan Chen. Spatial and spectral illumination design for energy-efficient circadian lighting. *Building and Environment*, *146*, 216–225. doi:10.1016/j.buildenv.2018.10.004

David, L. (2011). A Procedure for Determining Target Illuminances, LEUKOS. *The Journal of the Illuminating Engineering Society of North America*, *7*(3), 145–158. doi:10.1582/LEUKOS.2011.07.03001

Delegado, R. (UE) n ° 874/2012 de la Comisión, de 12 de julio de 2012, por el que se complementa la Directiva 2010/30/UE del Parlamento Europeo y del Consejo en lo relativo al etiquetado energético de las lámparas eléctricas y las luminarias Texto pertinente a efectos del EEE (accessed on 15 April 2020) https://www.boe.es/doue/2012/258/L00001-00020.pdf

Edwards, L., & Torcellini, P. (2002). Literature Review of the Effects of Natural Light on Building Occupants. *Lit. Rev. Eff. Nat. Light Build. Occup.*, *55*. Advance online publication. doi:10.2172/15000841

EN 12464-1 2019: Lighting of workplaces – Indoor workplaces. https://www.valosto.com/tiedostot/prEN%2012464-1.pdf

Enezi, J., Revell, V., Brown, T., Wynne, J., Schlangen, L., & Lucas, R. A. (2011). "Melanopic" spectral efficiency function predicts the sensitivity of melanopsin photoreceptors to polychromatic lights. *Journal of Biological Rhythms*, *26*(4), 314–323. doi:10.1177/0748730411409719 PMID:21775290

Escofet, J., & Bará, S. (2015). Reducing the circadian input from self-luminous devices using hardware filters and software applications. *Lighting Research & Technology*, *49*(4), 481–496. doi:10.1177/1477153515621946

Figueiro. (2013). An Overview of the Effects of Light on Human Circadian Rhythms: Implications for New Light Sources and Lighting Systems Design. *Journal of Light & Vision Environment, 37*(2-3).

Figueiro, Steverson Heerwagen, Yucel, Roohan, Sahin, Kampschroer, & Rea. (2019). Light, entrainment and alertness: A case study in offices. *Lighting Research and Technology*.

Glickman, G., Levin, R., & Brainard, G. C. (2002). Ocular input for human melatonin regulation: Relevance to breast cancer. *Neuroendocrinology Letters*, *23*(SUPPL. 2), 17–22. PMID:12163843

Group, H. M. (1999). *Daylighting in Schools: An Investigation into Relationship between Daylighting and Human Performance; Detailed Report; Pacific Gas and Electric Company on behalf of the California Board for Energy Efficiency Third Party Program*. Fair Oaks.

Guan & Yan. (2016). Daylighting Design in Classroom Based on Yearly-Graphic Analysis. *Sustainability, 8*, 604. doi:10.3390u8070604

Hankins & Lucas. (2002). The primary visual pathway in humans is regulated according to long-term light exposure through the action of a nonclassical photopigment. *Curr. Biol., 12*, 191-198.

Heschong Mahone Group. (1999). *Daylighting in schools: An investigation into the relationship between daylighting and human performance*. Author.

Ho. (2007). Optimal sun-shading design for enhanced daylight illumination of subtropical classrooms. *Energy and Buildings*.

Hunt, R. W. G., & Pointer, M. R. (2006). *Measuring Colour* (4th ed.). Wiley.

IES. (n.d.). *Tunable light could potentially improve student learning, increase teacher satisfaction and trim energy use. A recent pilot project explored these benefits.* https://www.ies.org/

Illuminating Engineering Society of North America (IESNA). (n.d.). https://www.ies.org/standards/ies-standards-cross-reference/

International Well Building Institute. (n.d.). *WELL Building Standard v1*. Author.

ISO 8995-1 (2002): Lighting of indoor workplaces (CIE S008/2001)

Jiang, Kurihara, Torii, & Tsubota. (2018). Progress and Control of Myopia by Light Environments. *Eye Contact Lens, 44*(5), 273-278. . doi:10.1097/ICL.0000000000000548

Khademagha, P., Aries, M. B. C., Rosemann, A. L. P., & van Loenen, E. J. (2016). Implementing non-image-forming effects of light in the built environment: A review on what we need. *Building and Environment*, *108*, 263–272. doi:10.1016/j.buildenv.2016.08.035

Khan, Z. A., Labala, R. K., Yumnamcha, T., Devi, S. D., Mondal, G., Sanjita Devi, H., Rajiv, C., Bharali, R., & Chattoraj, A. (2018). Artificial Light at Night (ALAN), an alarm to ovarian physiology: A study of possible chronodisruption on zebrafish (Danio rerio). *The Science of the Total Environment, 628–629*, 1407–1421. doi:10.1016/j.scitotenv.2018.02.101 PMID:30045561

Kim, I. T., Choi, A. S., & Sung, M. K. (2017). Development of a Colour Quality Assessment Tool for indoor luminous environments affecting the circadian rhythm of occupants. *Building and Environment, 126*, 252–265. doi:10.1016/j.buildenv.2017.10.009

King, A. L. S., Valença, A. M., Silva, A. C., Sancassiani, F., Machado, S., & Nardi, A. E. (2014). "Nomophobia": Impact of Cell Phone Use Interfering with Symptoms and Emotions of Individuals with Panic Disorder Compared with a Control Group. *Clinical Practice and Epidemiology in Mental Health, 10*(1), 28–35. doi:10.2174/1745017901410010028 PMID:24669231

Konis. (2017). A novel circadian daylight metric for building design and evaluation. *Building and Environment, 113*, 22-38.

Kuller, R., & Lindsten, C. (1992). Health and behavior of children in classrooms with and without windows. *Journal of Environmental Psychology, 12*(4), 305–317. doi:10.1016/S0272-4944(05)80079-9

Ln, Zhu, Guo, Zheng, Lu, Shih, & Wu. (2016). Multi-function indoor light sources based on light-emitting diodes–a solution for healthy lighting. *Optic Express*. doi:10.1364/OE.24.024401

López-Chao, V., Lorenzo, A. A., Saorín, J. L., De La Torre-Cantero, J., & Melián-Díaz, D. (2020). Classroom Indoor Environment Assessment through Architectural Analysis for the Design of Efficient Schools. *Sustainability, 12*(5), 2020. doi:10.3390u12052020

Lopez-Fernandez, O., Honrubia-Serrano, L., Freixa-Blanxart, M., & Gibson, W. (2014). Prevalence of problematic mobile phone use in british adolescents. *Cyberpsychology, Behavior, and Social Networking, 17*(2), 91–98. doi:10.1089/cyber.2012.0260 PMID:23981147

Lucas, Peirson, Berson, Brown, Cooper, Czeisler, Figueiro, Gamlin, Lockley, O'Hagan, Price, Provencio, Skene, & Brainard. (2014). Measuring and using light in the melanopsin age. *Trends in Neurosciences, 37*(1), 1–9. doi:10.1016/j.tins.2013.10.004 PMID:24287308

María, B. P. M., & Labarca, C. Y. (2015). Methodology for Assessing Daylighting Design Strategies in Classroom with a Climate-Based Method. *Sustainability, 7*(1), 880–897. doi:10.3390u7010880

Moyano, Moyano, López, Aznal, & González Lezcano. (2020). Photometric and colorimetric analysis of LED luminaires for interior lighting design. *Optik*. doi:10.1016/j.ijleo.2020.165599

Moyano, D. B., Mónica, S. J. F., & Roberto, A. G. L. (2020). Towards a Sustainable Indoor Lighting Design: Effects of Artificial Light on the Emotional State of Adolescents in the Classroom. *Sustainability, 12*(10), 4623.

Moyano, D. B., Sola, Y., & Roberto, A. G. L. (2020). Blue-light levels emitted from portable electronic devices compared to sunlight. *Energies, 13*(16), 4276. doi:10.3390/en13164276

Nabil, A., & Mardaljevic, J. (2006). Useful daylight illuminances: A replacement for daylight factors. *Energy and Building, 38*(7), 905–913. doi:10.1016/j.enbuild.2006.03.013

Nocera, Faro, Costanzo, & Raciti. (2018). Daylight Performance of Classrooms in a Mediterranean School Heritage Building. *Sustainability*, *10*, 3705. doi:10.3390u10103705

Rea, M. S., & Figueiro, M. G. (2018). Light as a circadian stimulus for architectural lighting. *Lighting Research & Technology*, *50*(4), 497–510. doi:10.1177/1477153516682368

Rea, M. S., Figueiro, M. G., Bierman, A., & Hamner, R. (2012). Modelling the spectral sensitivity of the human circadian system. *Lighting Research & Technology*, *44*(4), 386–396. doi:10.1177/1477153511430474

Rea, M. S., Figueiro, M. G., Bullough, J. D., & Bierman, A. (2005). A model of phototransduction by the human circadian system. *Brain Research. Brain Research Reviews*, *50*(2), 213–228. doi:10.1016/j.brainresrev.2005.07.002 PMID:16216333

Reinhart, C. F., Mardaljevic, J., & Rogers, Z. (2006). Dynamic daylight performance metrics for sustainable building design. *Leukos*, *3*(1), 7–31. doi:10.1582/LEUKOS.2006.03.01.001

Stevens, R. G. (2009). Light-at-night, circadian disruption and breast cancer: Assessment of existing evidence. *International Journal of Epidemiology*, *38*(4), 963–970. doi:10.1093/ije/dyp178 PMID:19380369

Tanner, C. K. (2008). Explaining relationships among student outcomes and the school's physical environment. *Journal of Advanced Academics*, *19*(3), 444–471. doi:10.4219/jaa-2008-812

Torii, H., Kurihara, T., & Yuko, S. K. N. (2017). Violet Light Exposure Can Be a Preventive Strategy Against Myopia Progression. *EBioMedicine*, *15*, 210–219. doi:10.1016/j.ebiom.2016.12.007 PMID:28063778

U.S. Green Building Council. (2019). *LEED v4 for Building Design Construction*. Author.

Winterbottoma, M., & Wilkinsb, A. (2009). Lighting and discomfort in the classroom. *Journal of Environmental Psychology*, *29*(1), 63–75. doi:10.1016/j.jenvp.2008.11.007

Chapter 12
The Uses of Renewable Energy in Buildings:
A Case Study of Australia

Paul C. Okonkwo
Dhofar University, Salalah, Oman

El Manaa Barhoumib
Dhofar University, Salalah, Oman

Wesam H. Beitelmal
https://orcid.org/0000-0001-6386-0639
Dhofar University, Salalah, Oman

Israr I. U. Hassan
Dhofar University, Salalah, Oman

Michael Nnamdi Azubuike
Federal University of Technology, Owerri, Nigeria

Fadhil Khadoum Al Housnia
Dhofar University, Salalah, Oman

Florence Jonathan
The Gateway Polytechnic Saapade, Nigeria

ABSTRACT

Only 5% of Australia's energy utilization comes from renewables, while 86.3% of the electricity is produced from fossil fuels. Nonetheless, this pattern has been disturbed by the ongoing decommissioning and closure of old coal power plants, alongside the Australian policy to reduce fossil fuel emissions. Presently, Australia is at a pivotal phase of its change to renewable energy power generation and utilization specifically in residential and commercial buildings. Sustainability in renewable energy utilization is being achieved through guided government policies, reasonable energy costs, and improved energy technology transfer approaches. To give a refreshed delineation of renewable energy, related government policy, and the route ahead in the Australian setting, this chapter presents a deliberate Australia update with renewable energy generation and utilization in Australian buildings.

DOI: 10.4018/978-1-7998-7023-4.ch012

1. INTRODUCTION

Energy generation and utilizations have been a challenge for an industrial and economical world. During the last two decades, the quick development of the global economy has made a monstrous interest in energy (Samy, Barakat, & Ramadan, 2020). The utilization of renewable energy sources has expanded impressively lately, carrying with it new specialized difficulties for the energy sector, rural and industrial applications (Ellabban, Abu-Rub, & Blaabjerg, 2014). Renewable energy has become one of the major sources of energy (Østergaard, Duic, Noorollahi, Mikulcic, & Kalogirou, 2020). Using sustainable power such as renewable energy sources in building assists with a decrease in the utilization of fossil fuel energy and enhance a low carbon economy. The previous decades have seen a fast advancement of sustainable renewable energy sources in building globally. Some of the renewable energy sources commonly in use in different countries for buildings are; Photovoltaic (PV), wind turbines. Among these renewable energy sources, PV associated systems are encountering quick market development (Liu, Chen, Cao, & Yang, 2019). This is because of the ceaselessly decreasing pattern in PV cost along with the government uphold programs and policies (Xiaojun Li et al., 2019). Renewable energy source assists with relieving the pressure between the demands and supplies for fossil fuel energy sources and assumes a basic function in accomplishing a low carbon economy, particularly when used in buildings. It was reported that energy utilization from renewable energy source assets will represent 9.4% of all the energy utilization in Australia (He, Reynolds, Li, & Boland, 2019). With the fast development of the economy, the building sector of the economy has gotten probably the greatest part of energy utilization and carbon dioxide emission.

Consumers are central participants in this specific circumstance, as adaptability in demand is necessarily required to embrace the discontinuous nature of most renewable energy sources. Active demand-side support is especially essential to guarantee the productive utilization of locally and global renewable energy in the buildings. Ultimately, the building needs to incorporate the utility of coordinated inexhaustible systems for heated water warming, cooling systems, ventilation, solar photovoltaic charge, and so forth (J. Li, 2019). At long last, the improvement plan for a sustainable renewable energy power source in buildings has been embraced globally and continued development has been reported worldwide (Tushar, Bhuiyan, Sandanayake, & Zhang, 2019). Various types of buildings have been evolved and presented in European countries to guarantee normalized low-energy buildings (Reda & Fatima, 2019). While the building segment is growing at a quick pace, the current day buildings have emerged as the third biggest consumer of fossil fuel after industry and farming (Xiaoma Li et al., 2019). Sources place the degree of electricity used in the building department in Europe to around 40–45% of entire energy utilization (Ahmad et al., 2014); around sixty-six percent of this sum is utilized in residential buildings. Different resources guarantee, that in industrialized countries, energy use in buildings is chargeable for roughly 1/2 of carbon dioxide emissions. The fundamental targets of the possible plan were to reduce the exhaustion of basic assets, as an instance, energy, water, and crude materials; prevent natural degradation introduced by way of facilities and device arrangement and duration cycle; and make better conditions which can be the secure, useful and feasible utility of the water and solar energy. Private families' residences in buildings have the enormous unused potential for a decrease in the energy that could be acknowledged through additions in energy productivity, efficiency, social changes, and broadened utilization of low-emission energy (Csoknyai, Legardeur, Abi Akle, & Horváth, 2019). The focal point of energy strategy has been on organizations instead of on private family units, which are just halfway caught by direct arrangement measures.

Recently, the European Union embraced a directive specifying that before the end of 2020, Member States must guarantee that all recently built buildings expend 'almost zero' energy (Annunziata, Frey, & Rizzi, 2013). In Germany, radical decreases of energy interest and demand for space heating have just become an arrangement and policy focus in the course of the most recent decade, both for new and existing buildings. The Australian government has executed arrangements for renewable energy in both new and existing buildings (Roberts, Bruce, & MacGill, 2019). The construction regulations have been reinforced multiple times in recent years and energy demand decreased for space heating and residential heated water. More so, the increase in the family salary globally has resulted in acquiring more of the cooling and heating systems thereby increasing energy consumption. To lessen high energy utilization levels and assess potential situations for renewable energy, it is important to have information on the real energy attributes of the buildings. This information, alongside other related rules, can demonstrate an extremely helpful tool for efficient energy planning, proper usage of renewable energy by building occupants and appliances globally.

To accomplish the objectives of carbon emission reduction approaches, the current political and logical development should progressively consider renewable energy sources as potential energy sources for buildings that not only supply the required energy but also help in reducing carbon emission. It was reported that the potential of renewable energy for buildings and reduction in CO_2 potential can be influenced through different government policies and strategy measures (Nastasi, 2015). Several countries have promulgated various policies, arrangements, guidelines, and key intends to advance the improvement of renewable energy sources for buildings (Foley & Olabi, 2017). Effective government policies and programs will encourage the sustainability of the renewable energy used in the building worldwide (Bürger et al., 2008). This study aims to research the use of renewable energy sources in building applications in Australia by surveying energy generation, utilization, and considering the important strategies, guidelines, and key policies in Australia. This study also offers a far-reaching and precise reference for sustainable renewable energy in building in Australia.

2. METHODOLOGICAL ASSESSMENT OF RENEWABLE ENERGY

2.1 Renewable Energy Potential in Australia

The renewable energy source is characterized as the energy that produces low degrees of greenhouse gas emissions. The energy division is dependent upon equal collaborations among a wide scope of equal and interfacing powers of progress, (for example, strategy, innovation, and system). Appreciation of these powers is fundamental to empowering the sustainable and renewable energy change of a country (Saidi & Omri, 2020). Some of the predominant renewable energy sources are; solar, wind, geothermal, biomass, hydro, tidal, wave, and ocean thermal energy which are all famously accessible and versatile to Australia's geophysical setting (H. X. Li, Edwards, Hosseini, & Costin, 2020). Effective exploitation of these renewable energy energies depends intensely upon the improvement of innovation equipped for gathering such resources (Carbajo & Cabeza, 2018).

2.1.1. Solar Energy

Sunlight is one of our planet's commonly ample and uninhibitedly accessible energy resources. The proportion of solar energy that shows up at the world's surface in one hour is more than the planet's all-out energy requirement for a whole year. Even though it appears to be an ideal renewable energy source, the proportion of solar energy we can use changes according to the hour of the day and the time of the year similar to a topographical region. In Australia, solar energy transformation is a verifiably notable way to deal with upgrade energy utilization. Solar heating is the most prompt and supportive utilization of solar energy applications. Solar energy is a promising decision fit for being one of the fundamental energy hotspots for residential and commercial buildings (Chwieduk, 2003; Tsalikis & Martinopoulos, 2015). In Australia, the elevated levels of solar resources in Queensland give huge chances to solar energy power generation and utilization. The Queensland Solar Bonus Scheme has seen more than 30,000 clients introduce rooftop top photovoltaic (PV) systems, and more than 100 MW introduced limits during 2008 and 2010. The state government improved this with the Queensland Solar Homes Program where a mass acquisition of solar boards allowed 1000 homes to be fitted with PV systems (Martin & Rice, 2012). A state solar program projected at A$115 million is being used to finance different towns and communities solar energy projects in Australia (Martin & Rice, 2012). Solar energy is plentiful and innovative headways in photovoltaic (PV) have fundamentally improved the viability of solar energy generation while lessening establishment cost (H. X. Li et al., 2020). Essentially, enhancements in battery productivity have fundamentally improved the feasibility of this solar renewable energy resource as a quick backup system to the grid amid significant generation failure (Shaw-Williams, Susilawati, Walker, & Varendorff, 2019). However, because of the size of the cluster required and the power transmission system, the thought is restrictive on a cost premise and isn't industrially reasonable until more upgrades in the innovation have been made.

In 2002, 830 GWh of solar energy was generated in Australia and from the generated solar energy, 61%, was utilized simply for residential solar water heating (Bahadori & Nwaoha, 2013). Consequently, starting in 2002 about 36.5 MW of electricity generated from solar energy was taken care of into the public network (Bahadori & Nwaoha, 2013). Solar energy has incredible potential for solar energy generation because of enormous regions of the inland desert with high average temperature all year and low level of overcast spread and downpour. The infrastructure essential requirements for such projects would involve a lot of capital and the expense per unit of electricity generated is commonly higher than other fossil fuel dependent on sustainable power sources. While the areas of most noteworthy solar radiation in Australia are regularly found inland, there are some grid connect territories that have moderately high solar radiation. Australia's unassuming production and utilization of solar energy are based on off-grid and residential buildings. While solar thermal water heating has been the prevalent type of solar energy used to date, the production of electricity from PV and concentrating solar thermal advancements is expanding in Australia.

2.1.2. Wind Energy

The wind is a bountiful source of clean energy. To generate power from wind energy, turbines are used to drive generators that feed electricity into the National Grid. In Australia, the wind farm in north Queensland has a 12 MW of installed limit. AGL Limited has bought the improvement rights to the Coopers Gap wind farm site which has a normal 500 MW age limit (Martin & Rice, 2012). Recently, the

OCE has similarly published a win- map perceiving which areas might be sensible for energy generation ventures. In Australia, tackling wind energy has a long history and is maintained by experienced advancement and strategy inspirations. Again, improved efficiency and capacity of the battery storage system will enhance the advancement of wind energy resources as a reasonable and strong elective energy source for building applications (Loukatou, Johnson, Howell, & Duck, 2021).

2.1.3. Hydro Energy

By building a dam or hindrance, an enormous hydro energy power supply can be utilized to make a controlled progression of water that will drive a turbine, producing electricity. This energy source can frequently be more dependable than solar or wind energy and permits electricity to be stored for use when demand reaches a peak. Like wind energy, in specific circumstances, hydro can be more reasonable as a commercial energy source yet relying especially upon the sort of property, it tends to be utilized for residential building electricity generation. As an antiquated innovation, hydropower is being challenged by climate change and other ecological concerns. The Queensland state in Australia operates three small hydroelectricity stations. The Queensland government has signed a Memorandum of Cooperation with the Papua New Guinea (PNG) government, PNG Energy Developments Limited, and Origin Energy, to encourage the improvement of an enormous hydroelectricity venture on the Purari River in PNG (Martin & Rice, 2012). Power will be provided to the National Electricity Market (NEM) through subsea transmission lines that interface with Townsville, North Queensland, and wanted to upgrade the utilization of renewable energy in buildings to diminish the effect of greenhouse emission.

2.1.4. Tidal Energy

This is another type of hydro energy that utilizes twice-daily tidal currents to drive turbine generators. Albeit tidal flow dissimilar to some other hydro energy sources isn't consistent, it is exceptionally unsurprising and can thusly make up for the periods when the tide momentum is low. In Australia, the Queensland state has given two little innovative work awards to a sum of $0.28 million to help with building a tidal turbine system (Martin & Rice, 2012). Significantly, these projects have seen the renewable energy generation capacities of Queensland develop by 29 percent over 2008 and 2010. While this development is introduced limit has merit, the supply of electricity from renewable energy is still well beneath the 20 percent target in Australia.

2.1.5. Geothermal Energy

By outfitting the normal heat beneath the world's surface, geothermal energy can be utilized to heat buildings legitimately or to generate electricity. Even though it tackles a power legitimately underneath our feet, geothermal energy is of negligible significance in Australia compared with other nations, where geothermal heat is considerably more uninhibitedly accessible (Carr-Cornish, Romanach, & Huddlestone-Holmes, 2019). The Queensland state has given A$15 million to the Queensland Geothermal Energy Center of Excellence that will look at chances to generate geothermal energy utilizing new turbines, cooling towers, and heat exchanger plans. Geothermal energy is one of the underutilized energy sources in Australia, notwithstanding it being the most seasoned of the renewable energy sources accessible.

Although Australia's distinguished geothermal potential is multiple times the countries yearly essential energy utilization, only one electrical power generation plant is at present being used for power generation.

2.1.6. Biomass Energy

Biomass energy is the transformation of solid fuel produced using plant materials into electricity. By converting agricultural, industrial, and domestic waste into solid, liquid, and gas fuel, biomass generates power at a much lower practical and natural expense. In Australia, sugar cane waste generates around 70 percent of the state's renewable energy supplies (Van Holsbeeck & Srivastava, 2020).

3. RESULTS AND ANALYSIS

3.1. Renewable Energy Generation in Australia and Challenges

Hydropower remains the biggest supporter of Australian renewable energy generation, representing 40% in 2016 and 2017 as shown in Fig. 1. In any case, with the ascent in the wind and solar energy generation, there has been a huge drop from 2001 when it was 95%. Wind-determined energy has risen consistently over this equivalent period, contributing 31% of the complete renewable energy source (H. X. Li et al., 2020). Solar energy has likewise expanded drastically as of late, representing 20% of renewable energy in 2016 and 2017 as displayed in Fig.1. Bioenergy, in any case, has remained sensibly static throughout the most recent ten years, demonstrating a minor normal fall of 0.5% for the period.

Figure 1. Renewable energy generation in Australia. Sources; (H. X. Li et al., 2020)

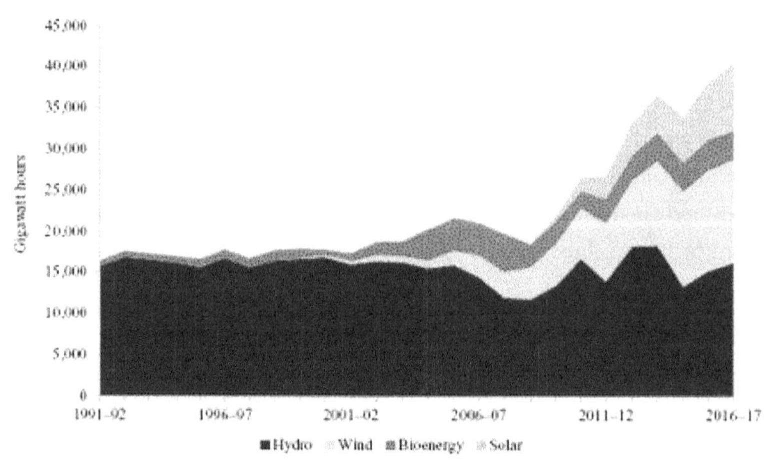

Notwithstanding the Federal Government initiatives in embarking on renewable energy generation and utilization in Australia, various states additionally have set up projects to explore renewable energy resources. Regardless of the apparent multitude of renewable energy activities compared with other developed countries, the extent of renewable energy in electricity generation in Australia is still exceptionally little (H. X. Li et al., 2020). Somewhere in the range of 1990 and 2001, Australia just had a normal development pace of 4% rather than OECD nations' normal pace of 17% (Gan & Smith, 2011). There are four challenges at present facing Australia from deploying renewable energy. The principal challenge among the four factors is Australia's immense geographic region. This needs an outlay in complementary grid and transmission lines. Also, a feebly evolved market for carbon under the 2009 proposed emission trading scheme (ETS) neglected to be executed and the interest for renewables now depends just on the RET and green voluntary sectors (Jotzo & Betz, 2009). The third setback in the Australian power industry structure which is overwhelmed by fossil fuels is the coordination of the renewable energy current electrical systems. The last challenge is the absence of investments from the private sector. Every one of these challenges represents some genuine boundaries to the effective organization and deployment of renewable energy in Australia. Other attributes hinder the development and advancement of renewable energy in Australia. Firstly, the driver for the change to renewable energy originates from outside of the electricity industry. Albeit a few consumers request green electricity, the certified provider originates from outside the industry, this being the need to lessen carbon emissions that cause global warming and climate change. These factors have made the change to renewable energy troublesome as the industry is comfortable with fossil fuel previously giving modest to get to and truly beneficial electricity.

Though renewable energy has the problems of storage and supply, the current renewable energy advancements themselves are low effective in energy conversion (Babatunde, Munda, & Hamam, 2019). For instance, solar innovation can typically change over about 15% of solar energy into electricity while coal and gas with a consolidated cycle can accomplish around 35% efficiency individually (Shoaib, Siddiqui, Rehman, Khan, & Alhems, 2019). While geothermal energy innovation using volcanic heat source has reached some degree of development, geothermal energy that bridles non-volcanic sources, for example, hot stone rock and lower-temperature flowing waters in sedimentary basins are still in the formative stage (Ballesteros, Pujol, Walsh, & Teubner, 2019). A similar case likewise applies to wave and tidal energies since they are as of now in the beginning phase of commercialization (Scherelis, Penesis, Hemer, Cossu, & Wright, 2020). These different diverse renewable energy sources and their distinctive commercialization stages imply that renewable energy improvement despite everything requires constant huge financing and needs different relating advancement arrangements to make them seriously practical versus fossil fuels (Brockway, Owen, Brand-Correa, & Hardt, 2019). Another issue confronting renewable energy advancement is the market rivalry from different sorts of fossil fuel energy sources, for example, gas, clean coal, and nuclear energy. Such advancements are profoundly concentrated and can fit into the current transmission framework without any problem. However, on the public agenda, these elective fossil fuel innovations are seen as suitable options in contrast to renewable energy advancements in alleviating the carbon emission issue (Brockway et al., 2019).

However, to manage the electricity supply efficiently and address the problem of a blackout, solar PV systems are being installed on the rooftops as a measure to venture towards clean energy. Because of this reason, there is an expanded number of solar PV installations in buildings for power generation in Australia. It has been seen that these installations are developing at a yearly increment of 30% (H. X. Li et al., 2020). The PV cells are arranged in series or parallel known as PV modules or PV boards to generate the required electricity. These boards have preferences like simple installation and they don't need a lot of maintenance as compared to other power generation systems (H. X. Li et al., 2020). A few different advantages incorporate noiseless operation, more secure activity, lower operational expenses, and with no air contamination (Baniyounes, Ghadi, Rasul, & Khan, 2013; Morse, Cooper, & Proctor, 1974). It is profoundly projected that by 2050, PV-generated electricity will share a higher than 20%

portion of all global energy demands and will be used to power most buildings in Australia (Wang, Chen, & Ren, 2010).

3.1. Energy Consumption by Sectors

The worldwide commitment from buildings towards energy utilization, both residential and commercial, has consistently expanded arriving at figures somewhere in the range of 20% and 40% in developed nations and has surpassed the other significant sectors: industrial and transportation (Gassar, Yun, & Kim, 2019; Lu & Lai, 2019). Development in the populace, expanding interest for building administrations and comfort levels, along with the increase in time spent inside buildings, guarantee the upward pattern in energy demand will proceed in the future. However, in Australia, each energy consumption tends to converge towards mean energy consumption across most sectors as shown in Fig. 2.

Figure 2. Per capita energy uses in various sectors relative to the mean energy use (Mishra & Smyth, 2017)

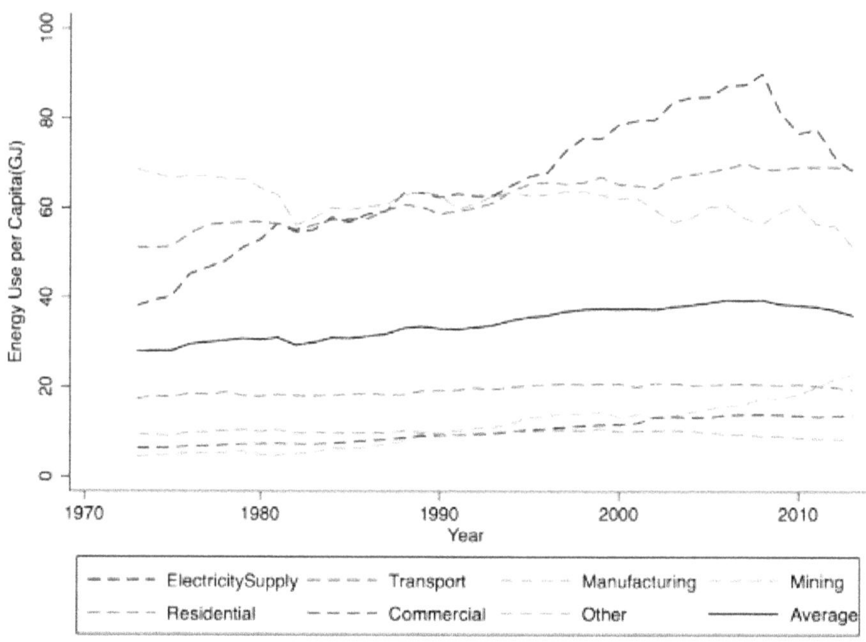

It can be observed from Fig. 2 that the significant energy utilizing sectors such as electricity generation, transport, and manufacturing together accounts for around 76% of Australia's energy consumption. The transport sector represents the biggest portion of Australia's end-user utilization. As indicated by the Bureau of Resources and Energy Economy (BREE) assessment, energy utilization of transport increased by a normal of 2.4 percent annually (Wilson, Trieu, & Bowen, 1994). The other biggest energy expending divisions are the mining, residential, commercial, and services sectors. It can be seen in Fig.2 that there is a gradual increase in the energy consumption across each sector which is in agreement with the energy generation shown in Fig.1 because if relative energy consumption is stationary for a particular sector, it is then assumed that consumption in that sector is converging (Mishra & Smyth, 2017). It was reported that in 2017-2018, own use natural gas consumption in Australia's LNG facilities increased to 286 petajoules accounting for one-fifth of Australian gas consumption ("Australian Energy Update," 2019).

In terms of commercial and residential buildings, energy efficiency and effectiveness in buildings is today a prime goal for energy policy at local, national, and worldwide levels. Among building services, the development in HVAC systems energy use is especially critical. Presently, it is expected to distinguish the sector which requests more energy utilization in Australia. Renewable energy is currently utilized both in the residential and commercial sectors, including heating, ventilation, and air conditioning. When considering the energy consumption in the residential buildings, heating appliances consume more electricity than other electric loads as shown in Table 1.

Table 1. Energy end uses of the residential sector in residential building (Karunathilake, Hewage, Brinkerhoff, & Sadiq, 2019)

Residence type	Single-family detached	Single-family attached	Apartments (multi-unit)
Space heating	57.53%	42.90%	34.13%
Water heating	24.75%	35.89%	40.41%
Space cooling	0.89%	1.06%	0.64%
Lighting	4.77%	4.41%	3.26%
Appliances	12.06%	15.75%	21.56%

During the previous few decades in Australia, the building sector consumes a significant portion of energy utilization worldwide and this offer rises during the summer seasons putting an expanded burden on the grid (Laslett, Carter, Creagh, & Jennings, 2017). The explanation behind this high energy demand throughout the late spring seasons is the utilization of conventional air-conditioning units in the buildings (Laslett et al., 2017). The solar water heater can save between 10 and 15% energy utilization and a solar heating system can make 45% energy saving in buildings (Yang, Yan, & Lam, 2014).

3.2. Greenhouse Gas (GHG) Emission From Energy Utilization

Buildings by and large fall into two parts, for example, residential and commercial. The US Energy Information Administration anticipated that the energy utilization in residential and commercial buildings will increase by 1.1% and 1.5% every year, separately, from 2008 to 2035 (Pérez-Lombard, Ortiz, & Pout, 2008). The pattern of energy utilization demonstrated that, as the expectation for everyday comforts of individuals improved, along with a fast advancement of tertiary industry, the quick development pattern of energy utilization in buildings will also increase and the level of building energy utilization in public buildings will as well increase much further. Buildings are considered as one of the most significant foundation segments in current social orders. The building sector today represents 40% of the world's total energy utilization (Virote & Neves-Silva, 2012). The building sector is likewise liable for about 27% of the nation's total greenhouse emissions and that incorporates commercial buildings that represented 10% of the nation's total greenhouse emissions (Bribián, Usón, & Scarpellini, 2009). Fig. 4 shows the gas house emission of buildings in eleven countries.

Figure 3. The GHG emissions of eleven selected countries (Lu & Lai, 2020)

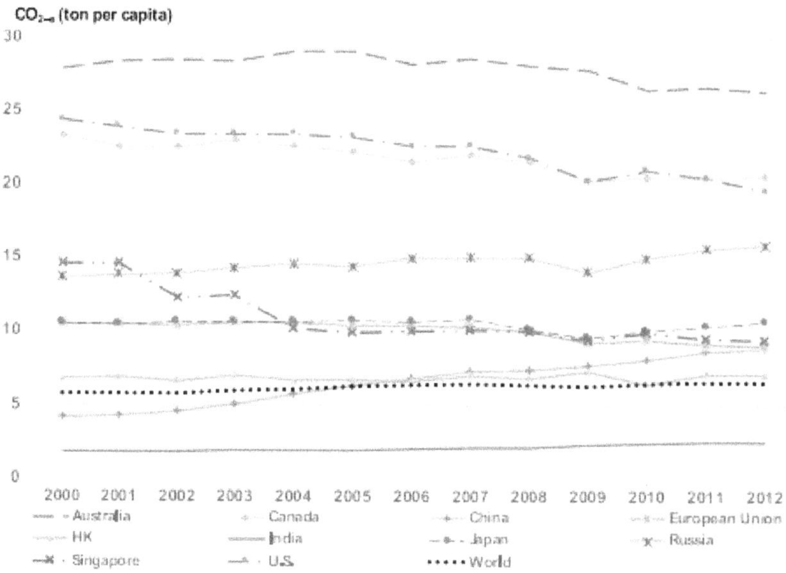

The patterns of both global greenhouse gas emission and carbon dioxide emission have been increased significantly as evident in Fig. 2. This might be because, albeit the emission of the developing nations has continued rising, the developed nations have made carbon decrease moves in light of the response to climate change agreements (Lu & Lai, 2020). The patterns of GHG emanation per capita that appeared in Fig. 3, other than that relating to the entire world, are those of the primarily developed countries. It was revealed that the developed countries (for example Canada, Japan, the U.S.) are more carbon-concentrated than the developing l nations (India and China). The finding showed that the highest GHG emission was found in Australia (26–29 tons of CO2-e per capita), trailed by the U.S and Canada (Lu & Lai, 2020).

For the past two decades in Australia, the commercial buildings have shown poor-performing plans particularly those over the age of 20 years, and is ascribed to atmosphere variety, low protection levels, coating materials, the presence of air holes, and the utilization of costly cooling methods (Leaman, Thomas, & Vandenberg, 2007). The Australian government has invested many funds in establishing renewable energy sources and is expected to forestall the planet against pollution with a boundless renewable energy source, particularly solar photovoltaic systems (Kent & Mercer, 2006; Shafiullah, Amanullah, Ali, Jarvis, & Wolfs, 2012). Solar energy has a significant bit of leeway of zero-emission and along these lines, it is more environmentally friendly than other renewable energy sources (Tsalikis & Martinopoulos, 2015). To fulfill the overall energy needs and limit greenhouse gas emissions, it is imperative to consider the net-zero energy building approach (Wells, Rismanchi, & Aye, 2018). The net-zero energy building being utilized in various states of Australia is pointed toward expanding renewable energy utilization and lessening carbon emission (Wells et al., 2018). The usage of the solar desiccant cooling system in institutional buildings can assist with accomplishing 60% of energy-related cost-saving and a noteworthy decrease of greenhouse gas emission (Yang et al., 2014). Thus, using renewable energy has become a significant aspect of the design and improvement of green buildings. There is extraordinary potential to use solar energy, wind, geothermal energy, and biomass in buildings and the innovation is generally improving. In Australia, the residential buildings represent 13% of Australia's greenhouse emissions (Pérez-Lombard et al., 2008; Wang et al., 2010). Complete residential energy demand in Australia expanded by 88% somewhere in the range of 1973 and 2009 (Pérez-Lombard et al., 2008). Effectively generating, utilizing, and storing the energy will not only improve the quality of life in the buildings but also guarantee that these energy sources noteworthy contributes to global and Australia's electricity grid supply in the future. With the administration policies and strategies set up, Australia will be very much placed to help with meeting its developing clean energy demand.

4. AUSTRALIAN RENEWABLE ENERGY VISION AND FORECAST

The atmospheric temperature has been stable until the 19th century when the industrial revolution started with increasing surface air temperature (Ritchie & Roser, 2020). The rising surface air temperature is attributed to the emission and concentration of greenhouse gas emission in the atmosphere due to increasing human activities dominated by fossil fuel combustion.

Australia, similar to the remainder of the world, is as of now confronting the impacts produced by the development of greenhouse gases in the environment. Most countries on the planet have consented to the Paris Arrangement, which comprises a progression of measures to handle global warming. The principle steps incorporate restricting the increment of temperature by 2050 to 1.5°C, considering a most extreme 2°C increment contrasted with industrial temperatures, and growing new pathways to diminishing greenhouse emission (Lawrence & Schäfer, 2019). As a component of the Paris Agreement, Australia has focused on bringing down its gas emission in any event 26% under 2005 levels by 2030 (Lawrence & Schäfer, 2019). Australia is still a long way from this accomplishment primarily because of the great reliance of the energy sector (and specifically the electricity sector) on fossil fuel. Since the electricity sector addresses the principle commitment of greenhouse gas emission, new methodologies and strategies are important to handle environmental and climate change. In this way, an energy change towards a renewable energy system can significantly diminish the carbon dioxide emanations in the electricity sector in the impending years and replace several fossil fuel power stations

Australia has plentiful and differing renewable sources with a critical potential for future development. Directly, renewable energy sources are utilized for heating and cooling, electricity generation, and transportation fuels like bio-fuel in Australia (H. X. Li et al., 2020). The clean energy resources being used on a business scale incorporate hydro and wind energy for electricity generation, bioenergy, and solar energy for heating, cooling, and electricity generation (El-Bidairi, Nguyen, Mahmoud, Jayasinghe, & Guerrero, 2020). Other renewable resources are generally undeveloped at present and include advancements that are still at the evidence of concept stage or beginning phases of commercialization (H. X. Li et al., 2020; Shafiullah et al., 2012). Various critical boundaries are as yet being developed for enormous scope use of Australia's clean energy resources. Changes in administrative and endorsement processes are influencing the advancement in certain areas. It is frequently expressed that the organization of alternative energy resources will require a lot of new innovative work exertion. The renewable energy advancements are too refined and complex compared to the regular energy transformation method. Despite these difficulties, the deployment of clean energy innovations is gathering pace and is required to assume a basic function in moving to a low emissions future while fulfilling Australia's sustained demand for energy (De Rosa & Castro, 2020).

4.1. Australian Government Renewable Energy Incentive Support Policy and Energy Acts

Australia as a nation has formulated renewable energy policies and regulations in diminishing reliance on fossil fuel and expanding electricity production by solar energy and renewable energy at large. Government policies and strategies have been actualized at a few phases of the solar energy production chain in Australia. Decrease in the photovoltaic and support for solar heating systems and residential home PV installations have decreased the expense of these advancements for consumers and empower their take-up (Pérez-Lombard et al., 2008; Yang et al., 2014). Solar energy credits by the government provide an up-front capital sponsorship towards the establishment of little solar-based PV systems. The Australian Government has likewise allotted financing supports to set up the Australian Solar Institute (ASI), which is situated in Newcastle. It has strong collaborative links with CSIRO and Universities' R&D in solar energy innovations. The establishment is aimed at increasing the solar thermal and PV advancement in Australia, including the territories of proficiency and cost adequacy. Other government approaches, includes; feed-in tariffs, which are proposed or as of now set up in most Australian states and regions, may likewise support the take-up of solar energy. In 2000, the Australian Government gave the Mandatory Renewable Energy Target (MRET) which set 9500 GWh by 2010 to enable renewable energy management through renewable energy certificates (RECs). This regulatory target was met ahead of 2007. In 2009, the Australian Government executed the Renewable Energy Target (RET) project which was stretched out from MRET. This goal was planned to ensure that renewable energy resources will supply 20% of total Australian electricity by 2020 including meeting the 45,000 GWh renewable energy target. In January 2011, the RET was disengaged into two segments: the Large-Scale Renewable Energy Target (LRET) and Small-scale Renewable Energy Target (SRET) (Nelson, Nelson, Ariyaratnam, & Camroux, 2013). This change hopes to make separate motivators for the enormous scope of renewable energy projects and little scope innovation which can diminish the competition with one another in the RET scheme (Cludius, Forrest, & MacGill, 2014). To enhance the execution of the RET scheme, the Australian Government announced public acts to safeguard that the target be achieved in the future.

The obligatory structural Acts place a legitimate commitment onto the at-risk substances which implies that electricity retailers and other huge purchasers need to buy renewable energy certificates (RECs) from certifying renewable energy providers (Holt & Wiser, 2007). If the purchasers can't meet the necessary rules and regulations, they would get fined $65 per MWh for deficits. The RECs can be utilized to demonstrate the liability entities' compliance with the necessity of RET as a 'currency' structure (Holt & Wiser, 2007). Somewhat, it has upgraded the competition of renewable energy with fossil fuel through expanding the expense of non-renewable energy generation and making renewable energy more feasible. This is because the cost is the biggest challenge for renewable energy advancement which requires higher cost compared with fossil fuel generators. Another explanation is that the cost and danger of fossil fuel generation are truly externalized, which prompts the lower private expenses yet higher social cost compared with renewable energy, and further declines the seriousness of renewable energy (Byrnes, Brown, Foster, & Wagner, 2013). Carbon Price in Australia is planned to disguise the natural expenses of non-renewable energy sources because of its emission. Nonetheless, after an adjustment in Government in 2013 the current government, as one of its first authoritative changes, canceled the Carbon Price from 1 July 2014. The RET usage is supervised by the Clean Energy Regulator (CER) which likewise administrates the consistency with the related requirement by enactment. National policy and the lawful system are overseen by the Department of Industry (DOI). The other dependable legislative organization in the Australian Renewable Energy Agency (ARENA) initiated an activity on 1 July 2012 to flexibly subsidizing help for renewable energy projects, innovative work exercises along with information catching and sharing activities. In Australia, the State government assumes a critical part of renewable energy usage. Furthermore, the Commonwealth Scientific and Industrial Research Organization (CSIRO) is a central government-subsidized public science office investigating and developing solar, wind, geothermal, and other renewable technologies for residential and industrial buildings as well as other applications. Renewable energy policy in Australia is dependent upon guidelines and impacted financially by every one of the three circles of governments which are the federal, state, and local governments.

CONCLUSION

The chapter uncovers the abundance of renewable energy in Australia and revealed the emission from energy utilization. It was also discussed that the global carbon emission has proceeded to rise and, in acknowledgment of this issue, most nations have presented plans for carbon mitigation and reduction. Given that buildings, particularly commercial buildings, are the significant producer of carbon emission, a considerable lot of the plans have included incentives or subsidies for the execution of carbon decrease measures in buildings. Our discoveries have significant ramifications for Australia's energy strategy as it looks to understand the emission reductions to which it concurred in Paris. This chapter illustrates that Australia is on the correct way to move towards a clean sustainable renewable energy system as a result of its obligation to the Paris Agreement. This is especially observable given the historical backdrop of Australia's high reliance on fossil fuel resources. However, the target can be achieved by implementing government policies and acts that guide towards carbon emission reduction. To enhance renewable energy, the Australian government has supported the dissemination of rooftop PV in the Australian system which will positively help in providing electricity and reducing carbon emission as we have exhibited in our study. Promoting the production and utilization of renewable energy despite the challenges faced needs consolidated socio-political effects from both government and citizenry. Furthermore, increasing renewable energy generation will require agreement between innovative technological development and social-economic improvements. More so, government policy and energy acts should be a longer-term innovative technological plan to ensure that renewable energy generation encourages future research, technology, and manufacturing advancement. The Australian government should progressively consider renewable energy sources as potential energy sources for buildings that not only supply the required energy but also help in reducing carbon emission.

REFERENCES

Ahmad, A., Hassan, M., Abdullah, M., Rahman, H., Hussin, F., Abdullah, H., & Saidur, R. (2014). A review on applications of ANN and SVM for building electrical energy consumption forecasting. *Renewable & Sustainable Energy Reviews*, *33*, 102–109. doi:10.1016/j.rser.2014.01.069

Annunziata, E., Frey, M., & Rizzi, F. (2013). Towards nearly zero-energy buildings: The state-of-art of national regulations in Europe. *Energy*, *57*, 125–133. doi:10.1016/j.energy.2012.11.049

Babatunde, O. M., Munda, J. L., & Hamam, Y. (2019). A comprehensive state-of-the-art survey on power generation expansion planning with intermittent renewable energy source and energy storage. *International Journal of Energy Research*, *43*(12), 6078–6107. doi:10.1002/er.4388

Bahadori, A., & Nwaoha, C. (2013). A review on solar energy utilisation in Australia. *Renewable & Sustainable Energy Reviews*, *18*, 1–5. doi:10.1016/j.rser.2012.10.003

Ballesteros, M., Pujol, M., Walsh, F., & Teubner, J. (2019). *Geothermal Energy Electricity Generation in Australia: Recent Developments and Future Potential*. Academic Press.

Baniyounes, A. M., Ghadi, Y. Y., Rasul, M., & Khan, M. M. K. (2013). An overview of solar assisted air conditioning in Queensland's subtropical regions, Australia. *Renewable & Sustainable Energy Reviews*, *26*, 781–804.

Bribián, I. Z., Usón, A. A., & Scarpellini, S. (2009). Life cycle assessment in buildings: State-of-the-art and simplified LCA methodology as a complement for building certification. *Building and Environment*, *44*(12), 2510–2520. doi:10.1016/j.buildenv.2009.05.001

Brockway, P. E., Owen, A., Brand-Correa, L. I., & Hardt, L. (2019). Estimation of global final-stage energy-return-on-investment for fossil fuels with comparison to renewable energy sources. *Nature Energy*, *4*(7), 612–621. doi:10.103841560-019-0425-z

Bürger, V., Klinski, S., Lehr, U., Leprich, U., Nast, M., & Ragwitz, M. (2008). Policies to support renewable energies in the heat market. *Energy Policy*, *36*(8), 3150–3159. doi:10.1016/j.enpol.2008.04.018

Byrnes, L., Brown, C., Foster, J., & Wagner, L. D. (2013). Australian renewable energy policy: Barriers and challenges. *Renewable Energy*, *60*, 711–721. doi:10.1016/j.renene.2013.06.024

Carbajo, R., & Cabeza, L. F. (2018). Renewable energy research and technologies through responsible research and innovation looking glass: Reflexions, theoretical approaches and contemporary discourses. *Applied Energy*, *211*, 792–808. doi:10.1016/j.apenergy.2017.11.088

Carr-Cornish, S., Romanach, L., & Huddlestone-Holmes, C. (2019). An Application of Social Science to Inform the Stakeholder Engagement of an Emerging Geothermal Industry in Australia. In *Geothermal Energy and Society* (pp. 71–90). Springer. doi:10.1007/978-3-319-78286-7_6

Chwieduk, D. (2003). Towards sustainable-energy buildings. *Applied Energy*, *76*(1-3), 211–217. doi:10.1016/S0306-2619(03)00059-X

Cludius, J., Forrest, S., & MacGill, I. (2014). Distributional effects of the Australian Renewable Energy Target (RET) through wholesale and retail electricity price impacts. *Energy Policy*, *71*, 40–51. doi:10.1016/j.enpol.2014.04.008

Csoknyai, T., Legardeur, J., Abi Akle, A., & Horváth, M. (2019). Analysis of energy consumption profiles in residential buildings and impact assessment of a serious game on occupants' behavior. *Energy and Building*, *196*, 1–20. doi:10.1016/j.enbuild.2019.05.009

De Rosa, L., & Castro, R. (2020). Forecasting and assessment of the 2030 australian electricity mix paths towards energy transition. *Energy, 205*, 118020. doi:10.1016/j.energy.2020.118020

El-Bidairi, K. S., Nguyen, H. D., Mahmoud, T. S., Jayasinghe, S., & Guerrero, J. M. (2020). Optimal sizing of Battery Energy Storage Systems for dynamic frequency control in an islanded microgrid: A case study of Flinders Island, Australia. *Energy, 195*, 117059. doi:10.1016/j.energy.2020.117059

Ellabban, O., Abu-Rub, H., & Blaabjerg, F. (2014). Renewable energy resources: Current status, future prospects and their enabling technology. *Renewable & Sustainable Energy Reviews, 39*, 748–764. doi:10.1016/j.rser.2014.07.113

Foley, A., & Olabi, A. G. (2017). *Renewable energy technology developments, trends and policy implications that can underpin the drive for global climate change*. Elsevier. doi:10.1016/j.rser.2016.12.065

Gan, J., & Smith, C. (2011). Drivers for renewable energy: A comparison among OECD countries. *Biomass and Bioenergy, 35*(11), 4497–4503. doi:10.1016/j.biombioe.2011.03.022

Gassar, A. A. A., Yun, G. Y., & Kim, S. (2019). Data-driven approach to prediction of residential energy consumption at urban scales in London. *Energy, 187*, 115973. doi:10.1016/j.energy.2019.115973

He, H., Reynolds, C. J., Li, L., & Boland, J. (2019). Assessing net energy consumption of Australian economy from 2004–05 to 2014–15: Environmentally-extended input-output analysis, structural decomposition analysis, and linkage analysis. *Applied Energy, 240*, 766–777. doi:10.1016/j.apenergy.2019.02.081

Holt, E. A., & Wiser, R. H. (2007). *The Treatment of Renewable Energy Certificates, EmissionsAllowances, and Green Power Programs in State Renewables*. PortfolioStandards.

Jotzo, F., & Betz, R. (2009). Australia's emissions trading scheme: Opportunities and obstacles for linking. *Climate Policy, 9*(4), 402–414.

Karunathilake, H., Hewage, K., Brinkerhoff, J., & Sadiq, R. (2019). Optimal renewable energy supply choices for net-zero ready buildings: A life cycle thinking approach under uncertainty. *Energy and Building, 201*, 70–89. doi:10.1016/j.enbuild.2019.07.030

Kent, A., & Mercer, D. (2006). Australia's mandatory renewable energy target (MRET): An assessment. *Energy Policy, 34*(9), 1046–1062. doi:10.1016/j.enpol.2004.10.009

Laslett, D., Carter, C., Creagh, C., & Jennings, P. (2017). A large-scale renewable electricity supply system by 2030: Solar, wind, energy efficiency, storage and inertia for the South West Interconnected System (SWIS) in Western Australia. *Renewable Energy, 113*, 713–731. doi:10.1016/j.renene.2017.06.023

Lawrence, M. G., & Schäfer, S. (2019). Promises and perils of the Paris Agreement. *Science, 364*(6443), 829-830.

Leaman, A., Thomas, L., & Vandenberg, M. (2007). 'Green'buildings: What Australian users are saying. *EcoLibrium*.

Li, H. X., Edwards, D. J., Hosseini, M. R., & Costin, G. P. (2020). A review on renewable energy transition in Australia: An updated depiction. *Journal of Cleaner Production*, *242*, 118475. doi:10.1016/j.jclepro.2019.118475

Li, J. (2019). Optimal sizing of grid-connected photovoltaic battery systems for residential houses in Australia. *Renewable Energy*, *136*, 1245–1254. doi:10.1016/j.renene.2018.09.099

Li, X., Pan, F., Sun, C., Zhang, M., Wang, Z., Du, J., ... Zhang, Z.-G. (2019). Simplified synthetic routes for low cost and high photovoltaic performance n-type organic semiconductor acceptors. *Nature Communications*, *10*(1), 1–11. doi:10.103841467-019-08508-3 PMID:30705277

Li, X., Zhou, Y., Yu, S., Jia, G., Li, H., & Li, W. (2019). Urban heat island impacts on building energy consumption: A review of approaches and findings. *Energy*, *174*, 407–419. doi:10.1016/j.energy.2019.02.183

Liu, J., Chen, X., Cao, S., & Yang, H. (2019). Overview on hybrid solar photovoltaic-electrical energy storage technologies for power supply to buildings. *Energy Conversion and Management*, *187*, 103–121. doi:10.1016/j.enconman.2019.02.080

Loukatou, A., Johnson, P., Howell, S., & Duck, P. (2021). Optimal valuation of wind energy projects co-located with battery storage. *Applied Energy*, *283*, 116247. doi:10.1016/j.apenergy.2020.116247

Lu, M., & Lai, J. (2020). Review on carbon emissions of commercial buildings. *Renewable & Sustainable Energy Reviews*, *119*, 109545. doi:10.1016/j.rser.2019.109545

Lu, M., & Lai, J. H. (2019). Building energy: A review on consumptions, policies, rating schemes and standards. *Energy Procedia*, *158*, 3633–3638. doi:10.1016/j.egypro.2019.01.899

Martin, N. J., & Rice, J. L. (2012). Developing renewable energy supply in Queensland, Australia: A study of the barriers, targets, policies and actions. *Renewable Energy*, *44*, 119–127. doi:10.1016/j.renene.2012.01.006

Mishra, V., & Smyth, R. (2017). Conditional convergence in Australia's energy consumption at the sector level. *Energy Economics*, *62*, 396–403. doi:10.1016/j.eneco.2016.07.009

Morse, R., Cooper, P., & Proctor, D. (1974). *The status of solar energy utilization in Australia for industrial commercial and domestic purposes*. CSIRO Solar Energy Studies Report No. 74/1.

Nastasi, B. (2015). Renewable hydrogen potential for low-carbon retrofit of the building stocks. *Energy Procedia*, *82*, 944–949. doi:10.1016/j.egypro.2015.11.847

Nelson, T., Nelson, J., Ariyaratnam, J., & Camroux, S. (2013). An analysis of Australia's large scale renewable energy target: Restoring market confidence. *Energy Policy*, *62*, 386–400. doi:10.1016/j.enpol.2013.07.096

Østergaard, P. A., Duic, N., Noorollahi, Y., Mikulcic, H., & Kalogirou, S. (2020). *Sustainable development using renewable energy technology*. Elsevier. doi:10.1016/j.renene.2019.08.094

Pérez-Lombard, L., Ortiz, J., & Pout, C. (2008). A review on buildings energy consumption information. *Energy and Building*, *40*(3), 394–398. doi:10.1016/j.enbuild.2007.03.007

Reda, F., & Fatima, Z. (2019). Northern European nearly zero energy building concepts for apartment buildings using integrated solar technologies and dynamic occupancy profile: Focus on Finland and other Northern European countries. *Applied Energy*, *237*, 598–617. doi:10.1016/j.apenergy.2019.01.029

Ritchie, H., & Roser, M. (2020). CO_2 and greenhouse gas emissions. *Our World in Data.*

Roberts, M., Bruce, A., & MacGill, I. (2019). Opportunities and barriers for photovoltaics on multi-unit residential buildings: Reviewing the Australian experience. *Renewable & Sustainable Energy Reviews*, *104*, 95–110. doi:10.1016/j.rser.2018.12.013

Saidi, K., & Omri, A. (2020). The impact of renewable energy on carbon emissions and economic growth in 15 major renewable energy-consuming countries. *Environmental Research*, *186*, 109567. doi:10.1016/j.envres.2020.109567 PMID:32361260

Samy, M., Barakat, S., & Ramadan, H. (2020). Techno-economic analysis for rustic electrification in Egypt using multi-source renewable energy based on PV/wind/FC. *International Journal of Hydrogen Energy*, *45*(20), 11471–11483. doi:10.1016/j.ijhydene.2019.04.038

Scherelis, C., Penesis, I., Hemer, M. A., Cossu, R., & Wright, J. T. (2020). Dataset for concurrent echo-sounder and ADCP measurements at a tidal energy candidate site in Australia. *Data in Brief*, *31*, 105873. doi:10.1016/j.dib.2020.105873 PMID:32642508

Shafiullah, G., Amanullah, M., Ali, A. S., Jarvis, D., & Wolfs, P. (2012). Prospects of renewable energy–a feasibility study in the Australian context. *Renewable Energy*, *39*(1), 183–197. doi:10.1016/j.renene.2011.08.016

Shaw-Williams, D., Susilawati, C., Walker, G., & Varendorff, J. (2019). Valuing the impact of residential photovoltaics and batteries on network electricity losses: An Australian case study. *Utilities Policy*, *60*, 100955. doi:10.1016/j.jup.2019.100955

Shoaib, M., Siddiqui, I., Rehman, S., Khan, S., & Alhems, L. M. (2019). Assessment of wind energy potential using wind energy conversion system. *Journal of Cleaner Production*, *216*, 346–360. doi:10.1016/j.jclepro.2019.01.128

Tsalikis, G., & Martinopoulos, G. (2015). Solar energy systems potential for nearly net zero energy residential buildings. *Solar Energy*, *115*, 743–756. doi:10.1016/j.solener.2015.03.037

Tushar, Q., Bhuiyan, M., Sandanayake, M., & Zhang, G. (2019). Optimizing the energy consumption in a residential building at different climate zones: Towards sustainable decision making. *Journal of Cleaner Production*, *233*, 634–649. doi:10.1016/j.jclepro.2019.06.093

Van Holsbeeck, S., & Srivastava, S. K. (2020). Feasibility of locating biomass-to-bioenergy conversion facilities using spatial information technologies: A case study on forest biomass in Queensland, Australia. *Biomass and Bioenergy*, *139*, 105620. doi:10.1016/j.biombioe.2020.105620

Virote, J., & Neves-Silva, R. (2012). Stochastic models for building energy prediction based on occupant behavior assessment. *Energy and Building*, *53*, 183–193. doi:10.1016/j.enbuild.2012.06.001

Wang, X., Chen, D., & Ren, Z. (2010). Assessment of climate change impact on residential building heating and cooling energy requirement in Australia. *Building and Environment*, *45*(7), 1663–1682. doi:10.1016/j.buildenv.2010.01.022

Wells, L., Rismanchi, B., & Aye, L. (2018). A review of Net Zero Energy Buildings with reflections on the Australian context. *Energy and Building*, *158*, 616–628. doi:10.1016/j.enbuild.2017.10.055

Wilson, B., Trieu, L. H., & Bowen, B. (1994). Energy efficiency trends in Australia. *Energy Policy*, *22*(4), 287–295. doi:10.1016/0301-4215(94)90003-5

Yang, L., Yan, H., & Lam, J. C. (2014). Thermal comfort and building energy consumption implications– a review. *Applied Energy*, *115*, 164–173. doi:10.1016/j.apenergy.2013.10.062

Chapter 13
Water Efficiency Evaluation Analysis Among Environmental Certification Methods:
LEED, BREEAM, DGNB, HQE, EDGE, and BONO VERDE

Estela Karem Samamé Zegarra
https://orcid.org/0000-0003-2317-5076
Navarra University, Spain

ABSTRACT

Developing countries such as Peru are not strangers in promoting sustainability in buildings through local certification programs such as My Green House funded by international organizations; or architectural design competitions for houses such as "Build to Grow"; compliance standards such as the one published in March 2018, "Climate Change Law"; in addition to other optional legal standards such as "EM.110 Thermal and Light Comfort with Energy Efficiency": a technical code of voluntary sustainable construction; and the entry of international environmental certifications such as EDGE and LEED, which are currently very welcomed by real estate developers due to the incentives. One of them is the height bonus, which is promoted by some municipal ordinances, mostly located in the capital city of Lima as a product of a project developed and promoted by the IFC and World Bank. On the other hand, in the retail or office sector, they are promoted by green corporate policies; however, there is a long way to go.

INTRODUCTION

Peru is a country that is divided into three important geographical regions: Coast, Sierra and Selva, where the Peruvian coast or coastal strip is made up of deserts and fertile valleys bathed by the Pacific Ocean (Arid Zones Research Center, National Agrarian University, 1982) and it extends about 3080 km in its length from the Sechura desert to the Nazca pampas and the Atacama desert (Oñativia, 2011).

DOI: 10.4018/978-1-7998-7023-4.ch013

The water resource for this desert region of the Coast of Peru is very important, therefore, this chapter aims to explain the reason for its size in the framework of sustainable construction in addition to knowing the typical residential sector of the region. There is also a general, historical view of the local regulations or environmental certifications that are currently being applied in Peru, for the residential sector, and how they address the evaluation of the water resource. For this, the Peruvian national voluntary regulations of the GREEN BOND and the international certification EDGE ((Excellence in Design for Greater Efficiencies) for housing are taken into account, with the purpose of comparing them with other international environmental certification systems such as LEED ((Leadership in Energy Environmental Design) created by the United States,

The comparison of the water resource is carried out in four packages, which make known in a simple and orderly way the criteria that each certification addresses in actions such as: Water management inside the houses, Water management outside the house, Use of non-traditional water sources, water monitoring system and others. Similarities and differences, degree of importance (scores and / or weightings), mandatory prerequisites, tools and approaches are described.

The conclusions obtained as a result of the analysis of the comparisons, is intended to be a useful reference both for industry professionals, for the development of objectives and goals for urban plans, and for academics interested in green buildings. Likewise, in order to contribute to the existing evaluation systems for residential buildings in Peru or countries with similar characteristics, a focused water efficiency standard is suggested for the desert area of Peru to apply it to the residential sector.

THE IMPORTANCE OF THE WATER RESOURCE FROM AN ENVIRONMENTAL PERSPECTIVE FOR HOUSING ON THE COAST OF PERU.

Developing countries such as Peru join to build sustainable, for this reason it has been promoting local certification programs such as My Green House or architectural design competitions for sustainable homes such as "Build to Grow" (Fondo Mivivienda SA, 2021), compliance standards such as the one published in March 2018 "Climate Change Law", in addition to other optional legal standards such as "EM.110 Thermal and Light Comfort with Energy Efficiency"(Ministry of Housing, Construction and Sanitation, 2014), an updated technical Code of Sustainable Construction (Ministry of Housing, Construction and Housing, 2021) and the entry of international environmental certifications such as EDGE and LEED.

For example, the EDGE certification in Peru is currently very popular with real estate developers. (EDGE, nd)Due to the incentives they receive for certifying any of their residential buildings, one of the incentives, among others, is the height bonus. The green municipal ordinances have been key to this, however they are mostly managed in the capital districts of Lima as part of the project developed and promoted by the IFC World Bank called "Green Building and Municipalities Peru(IFC -International Finance Corporation, 2017-2018). On the other hand, in the Retail or Office sector, green corporate policies promote the use of the internationally known LEED certification.

Speaking precisely of environmental certifications, such as LEED or BREEAM for example, these seek to comply with criteria or efficiency indicators more related to the very serious consumption of energy, however, should energy management be valued, evaluated and / or monitored in the same way in buildings anywhere in the world ?, a question that arises from observing the scores awarded to the Energy and Water Category in the BREEAM ES HOUSING and LEED certification manuals.(BREEAM, 2011), (USG20). Well, in a Peruvian context, experts on the subject of Sustainability and Construction mention

Figure 1. On the left, a satellite image extracted from Google earth that shows the light color that represents the desert area of the coastal strip of Peru, indicating the numbers associated with the climatic zone of: Marine Desert and Desert (source: self made). On the right the map showing the regions of the coast from Tumbes to Tacna (source: http://ualrealidad.blogspot.com/2012/04/realidad-nacional-alcances.html).

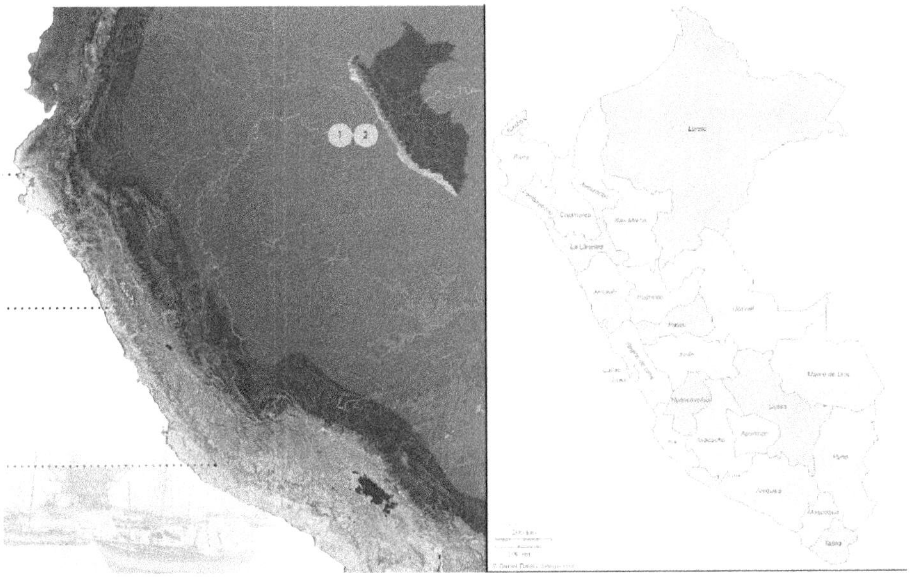

that the water resource is a priority (Biondi, 2012)Likewise, in an interview with Dr. Martín Wieser in April 2018 with the question: If there were an energy certification specifically for social housing in Peru, which of the three Impact categories (energy, water, materials) would consider more important for a weighting? It said that "Apart from that, I think that in the specific case of Lima and the Peruvian coast, the issue of Water / Drainage should have, if not the highest consideration, yes much more than what is assigned to it in the LEED or EDGE certifications."

Peru, a privileged country for its water supply, has an average annual volume of 2 046 287 MMC of water, ranking among the 20 richest countries in the world with 72,510 cubic meters / inhabitant / year (Fernandez). The poor distribution of the population in its three regions Costa, Sierra and Selva with reference to the water availability of its hydrographic units as seen in figure 2, added to the poor management of water by the inhabitants, generates inequalities of opportunity to water drinking water and serious problems of water stress, adding to this, the contamination of this resource. The majority of the Peruvian population (about 70%) lives on the coast and it is precisely this region that receives approximately 1.8% of the country's available water(PUPC, 2018) (OXFAM, nd). To give an example, one of the cities on the coast is the capital, Lima, which is home to 9,674,755 inhabitants as of 2020, representing 29.7% of the total population of Peru (32,625,948 inhabitants).(INEI, 2020)and its Rimac river only has an average annual flow of 42 m3 / s. On the other hand, Lisa Bunclark, specialist doctor in sustainable water management, thinks the following: "If we look at the entire country, we do not have water stress at the national level. If we look at the volume of water and the number of people, it is not a problem. The problem is that the water is not in all the areas we need ".(Bunclark, 2020) According to Oxfam, between 7 and 8 million Peruvians do not have access to drinking water.

Figure 2. Population statistics living on the coast and the availability of water. Fountain:https://es.slideshare.net/iagua/humberto-cruz-ana-per-la-gestin-del-agua-por-cuencas-experiencias-exitosas-para-el-desarrollo-del-per-24292830 and translated into English as own elaboration.

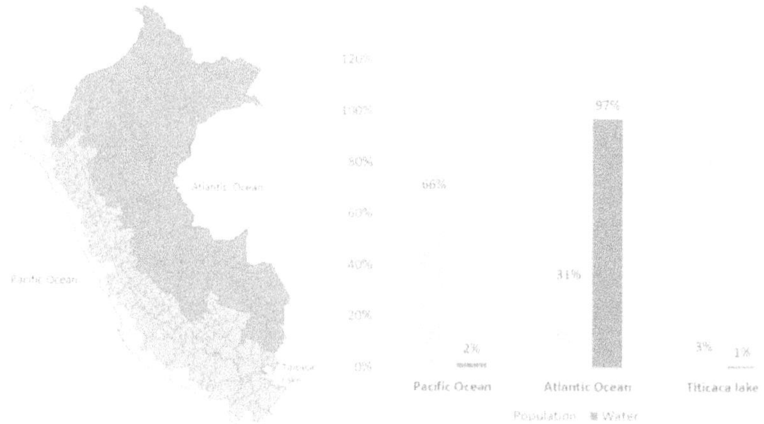

To be precise, the areas of Peru most at risk of running out of water according to a report published by the World Resources Institute (WRI) arethe coastal areas of the regions: La Libertad, Lambayeque, Piura, Tumbes, Ica, Arequipa, Moquegua and Tacna, which are the most affected, and are approaching “day zero”. The report indicates that governments still have time to mitigate the effects of this crisis. “By taking action now and investing in better management.

For Lima, Mirbel Epiquien, researcher of the Territory, Socio-Ecology and Ecosystem Services (TSESE) group of the INTE-PUCP and head of Environmental Management and Ecosystem Services of Sedapal, mentions that climate change can “accelerate processes of prolonged drought and scarcity ” in the capital. “The problem has already been felt for 20 years, and as a consequence other basins, such as the Mantaro, are used to feed Lima.”

Users must understand the importance of conserving water ”, explains the TSESE researcher, who points out that between 25% and 30% of the water produced by Sedapal is lost due to a bad culture of use, problems in distribution, damaged pipes or permanent filtration.

The impact that residential buildings have on the consumption of water resources throughout their life cycle (planning, construction, operation and deconstruction), mainly during the operation phase in residential areas of Lima, has resulted in very high and unsustainable consumption patterns, for example the direct Blue Water Footprint (Blue HH) of Metropolitan Lima for 2016 was 638 MMC and is represented mainly by the residential sector (54%) as it concentrates a third of the national population (AQUAFONDO, 2018)

Professor Liliana Miranda in Peru(Miranda, 2018)Through an interview, he mentions that the water resource is important because it is used to produce electricity through hydroelectric plants. Climate change brings drought consequences and this has already been felt through the El Niño phenomenon (Senamhi, nd) After this statement, it is worth mentioning that hydroelectric production, as shown in figure 3, is 42.7% of total consumption, which is why it is considered the second source of energy production in the country after natural gas and other fuels (53%), a percentage corresponds to solar and wind energy (3.2%).

Figure 3. Percentages of energy sources in Peru (SEIN public network), August 2017
Fountain: https://deltavolt.pe/energia-renovable/renovable-peru translated into English by own elaboration

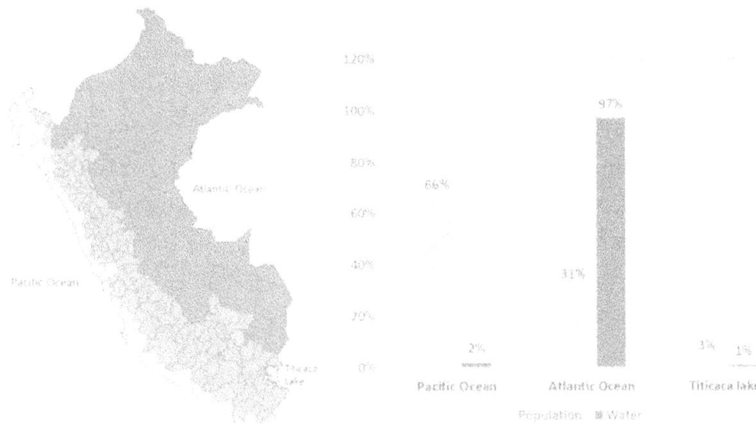

Another current reality up to now facing the housing sector in Peru is informality in construction. Of the 10 million existing homes, 7 million are informal, this means that 70% of homes live in informality, without regulations in a framework of self-construction, according to the Ministry of Housing, Construction and Sanitation and Capeco.

Figure 4. Some types of single-family homes on the Peruvian coast (Source: self made)

In this global and Peruvian context it is required that the preservation of natural resources become a central point of any project strategy and in the territory of the Coast, water is of vital importance for human life and its activities, therefore so much must be conserved to ensure its availability for a long period of time and to avoid major social, economic, environmental and health problems due to its depletion.

Arboleda (2011) says "Water cannot be allowed to flow into the sea without first having been used properly, this is a factor that future governments must take into account"

Of the general characteristics of the certifications BONO VERDE, EDGE, LEED for homes, BREEAM® ES Vivienda, DNGB and HQE TM.

This section presents the comparison of the general aspects taken from the 6 certification systems for the understanding of its origin, the certification levels they have, the schemes available for the type of housing as well as the categories that they take into account for the evaluation.

Figure 5. Comparison of the scores of the BONO VERDE, EDGE, LEED for homes, BREEAM® ES Vivienda, DNGB and HQE TM Certification Categories highlighting the water category in dark color (Source: self made)

BONO VERDE	
Criteria	Requirement
Bioclimatic	R
Urban sustainability	R
Energy	R
Water	R
Materials	R
Waste	R

EDGE	
Criteria	Saving
Energy	20%
Water	20%
Materials	20%

LEED FOR HOMES	
Criteria	points
Integrative processes / education	2
Indoor environment quality	16
Location and Transportation	15
Sustainable sites	7
Regional Priority	4
Innovation	6
Energy and Atmosphere	38
Efficiency in water	12
Materials and Resources	10
	110

BREEAM ES VIVIENDA		
Criteria	points	
Management	10	12%
Health & Wellness	14	14%
Transport	10	8%
Land use and Ecology	10	10%
Contamination	12	10%
Innovation	10	10%
Energy	twenty-one	18%
Water	6	10.5%
Materials	12	12%
Waste		7%
	105	110%

DNGB	
Criteria	
Process quality	12.5%
Quality of sociocultural and functional aspects	22.5%
Site quality	5.0%
Economic objectives	22.5%
Technical quality	15.0%
Environmental objectives	20.1%
Water	2.4%
	100%

HQE	
Criteria	points
Site	2
Components	7
Worksite	15
Energy	7
Waste	9
Maintenance	10
Water	16
Hygrothermal comfort	13
Acoustic comfort	3
Visual comfort	4
Olfactory comfort	6
Spaces quality	17
Air quality	8
Water quality	3
	120

1. My Green Housing Certification (Green Bond- V7)

It is a sustainable housing certification program promoted by the Peruvian state and that is part of the My Home Fund of the FONAPE corporation (National Fund for the Financing of State Business Activity), it is a voluntary certification, whose first procedure manual is published in 2016.

This program encourages the buyer of certified homes or apartments through financial aid and preferential rates. Buildings can be certified in the design stage, that is, in the preliminary or project stage (new construction), being able to achieve three levels of achievement: Grade I +, Grade II +, and Grade III +.

With a record of 331 certified projects by 2020, the MI Vivienda Verde methodology is given through the technical evaluation of documents such as plans, reports, specifications and / or technical product sheets, national or international quality certificates, letters of commitment., licenses and eligibility criteria. Only homes in the range of $ 16,662 to $ 118,744 can qualify for the assessment, as it is a social program.

MY V7 Green House, as can be seen in figure 5, requires compliance with 6 categories, being the following: Water, Energy, Bioclimatic, Materials, Waste and Urban Sustainability.(Mivivienda, 2020)

2. EDGE

EDGE "Excellence in Design for Greater Efficiencies" is a program for developing countries of the International Finance Corporation (IFC), a member of the World Bank Group, financed by the Government of the United Kingdom, launched in July 2014 and by 2020 used in more of 140 countries of the world including Peru.

This voluntary certification can be used to certify buildings at any stage of their useful life cycle, which includes the conceptual idea, design, a new construction, existing buildings and renovations of Residences, Hotels, Hospitals, Shops, Offices and of Education assessing buildings through the use of a free online software application(EDGE, nd)which predicts the savings of three resources: Water, Energy and Materials, comparing the building to be certified with a standard local model case to also estimate the capital cost and return on investment period, all this is possible since it contains a base of climate data, costs, consumption patterns and algorithms of cities to predict performance results.

With the evaluation of the buildings, 3 levels of certification can be achieved: EDGE Certification, EDGE Advance and EDGE Zero Carbon, being the EDGE Certification the easiest to achieve since it requires the building to sustain a minimum saving of 20% in energy operational, 20% savings in water consumption and 20% savings in embedded energy of construction materials as a minimum. The three levels of achievement can be obtained in 2 main stages: preliminary certification (design stage) and final certification (post-construction stage). (IFC, EDGE, 2018)

3. LEED for Homes

The LEED certification system (1998) is under the guidelines of the North American regulation ASHRAE 90.1-2007 and contemplates within its evaluation the Regional Priority category, awarding 4 points for addressing exclusive environmental problems in the region. The system is also used in 135 other countries (USGBC, 2012a)

The LEED toolkit consists of versions for new construction and major renovations, existing building operations and maintenance, commercial interiors, core and exterior development, retail, schools, homes, neighborhood development, and healthcare (USGBC, 2012c)

As for the version for homes, LEED for Homes was created in 2008, where its comprehensive evaluation, that is, in the design stage and later a verification visit is required mid-construction, as well as at the end of construction. LEED for Homes(US Green Building Council, 2017) is available for building design and construction projects for single-family homes and multi-family projects of up to eight stories.

The criteria evaluated in this version are nine: Integrative Process, Location and Transportation, Sustainable Sites, Water efficiency, Energy and Atmosphere, Materials and Resources, Indoor Environmental Quality, Innovation and Regional Priority and the levels that can be reached is by the sum of points where ^{3} 40 points are obtained at Certificate Level, ^{3} 50 points at Silver Level, ^{3} 60 points at Gold Level, and ^{3} 80 points at Platinum Level.

4. BREEAM® ES Housing

It is the environmental certification developed by the BRE Global organization in the United Kingdom, a pioneer at a European level and has shown that it can adapt, with the motto "Think global and act local", to the climate regulations and language of the country as is done in: Countries The Netherlands / BREEAM NL, Spain / BREEAM® ES Housing, Norway / BREEAM NOR, Sweden / BREEAM SE, Germany / BREEAM DE and United Arab Emirates / BREEAM Gulf. It is worth mentioning that it can be certified with BREEAM in Peru with the BREEAM International typology. This seal began to be developed in 1988, however, in the version of BREEAM® ES Vivienda, analyzed in this chapter, it was developed later. This typology values the certification from the sustainability of single-family homes or build-to-rent.

It has a methodology that evaluates characteristics already defined in different phases of development. The categories evaluated are ten(BREEAM, 2011): management, health and well-being, energy, transport, water, materials, waste, ecological use of the soil, pollution and innovation. All these criteria are taken into account at the time of the evaluation to reach the certificate levels: Approved, Good, Very good, Excellent and the highest Exceptional for this, the accompaniment of an accredited professional as a BREEAM advisor is needed, independent of the team of the project according to the type of Certification.

5. DNGB

The DGNB System certification is a system developed by the German Sustainable Building Council DGNB to assess and certify the sustainability of buildings and districts in Germany and internationally, with the ability to adapt to climatic, structural, legal and cultural variations.

It was founded in 2007 and has certified more than 5,900 projects in approximately 30 countries around the world as of 2019.

The certification system is available in different variants: for buildings (new, existing, renovations and in use), districts and interiors based on three key paradigms that correspond to the evaluation of the life cycle, a holistic methodology of sustainability, encompassing factors environmental, economic and sociocultural, and a performance orientation(DGNB System ES, 2020). Instead of individual measures, the DGNB assesses the overall performance of the project emphasizing the circular economy, which is why to achieve a certificate in platinum, gold, silver or bronze a uniform performance weighting is carried out.

The weightings are divided into 6 blocks: Environment, economic objectives, sociocultural-functional aspects and two transversal groups: technical and process quality and site quality. The first three have the same weighting weight in the evaluation, 22.5% each, and the next 15%, 12.5% and 5% respectively adding 100%.

6. HQE TM

Since 2013 the HQE TM (Haute Qualité Environnementale) (High Environmental Quality) building certification is an internationally registered trademark and is the exclusive property of the HQE TM

Association in France covering construction, renovation and operation, as well as urban planning and development. It attaches importance to the analysis of the life cycle on a construction scale and the impacts of a project in 14 criteria distributed from 4 objectives: Energy, Environment, Health and Comfort. Five rankings are possible depending on the number of stars obtained on each topic. HQE Pass 14 PR targets HQE Good 1 to 4 Stars, HQE Very Good 5 to 8 Stars, HQE Excellent 9 to 11 Stars and HQE Exceptional [3] 12 Stars.

The certification requires at least three categories in excellent level, four in superior and seven in good performance level.

The 14 evaluated criteria are: Site, Components, Worksite, Energy, Water, Waste, Maintenance, Hygrothermal comfort, Acoustic comfort, Visual comfort, Olfactory comfort, Spaces quality, Air quality, Water quality (HQE, 2014)

It is important to mention that in South America, in Brazil, the certification is also known as the AQUA process adapted from the original HQE model for the Brazilian version in 2007.

Of the weights, pre-requisites and instruments used for the evaluation of the WATER resource in homes in the BONO VERDE, EDGE, LEED for homes, BREEAM® ES Vivienda, DNGB and HQE TM certifications.

Water efficiency is one of the categories addressed in the six certifications subject to study in this section, for example the certification to obtain the Green Bond - Version 7, considers 9 indicators within the sub-criteria: Rational consumption of water and Reuse of water, of which the first 7 are mandatory for all the first two levels (grade 1+, grade 2+) that correspond to the use of low-consumption hydrosanitary equipment (installation of taps, shower heads, water reserve tanks, technical irrigation in gardens) and the installation of independent meters. However, to reach the higher grade corresponding to 3+, it is mandatory to add the fulfillment of 2 indicators to the 7 indicators mentioned above. These correspond to the Reuse of water: The installation of a gray or black water treatment plant for garden irrigation and another system for use in toilets. Thus, the indicators are not rated by scores or weightings, they are rated by the degree of compliance or non-compliance that corresponds to each level of achievement.

Regarding the instruments or tools used for the evaluation in the design stage, Bono Verde uses a check list that details the technical specifications and the evidentiary documents for approval - compliance with the indicators and does not use or demand the use of a simulation calculator of water consumption of the project, later in use of the building it is necessary to send a consumption report through invoices that evidences the low water consumption, however in the construction stage there is no control of the implementation of the indicators.

As for the EDGE certification, soon, with the new Beta 3.0 version, it will evaluate 8 more indicators in the Water category, compared to its version that is still in force 2.1. Therefore, there would now be 14 indicators, however, it is still unknown which of them will be mandatory to declare, since reference guide 3.0 has not been published to date. The indicators added in this version are: Low consumption bidet evaluation, use of low consumption dishwashers, kitchen rinse spray valves, energy efficient washing machines, pool covers, efficient garden irrigation system, water recovery condensed and smart meters.

On the other hand, in the current version 2.1, the mandatory indicators to declare for the evaluation of the percentage (%) of water savings are 5: The consumption in L / m of the shower heads, bathroom and kitchen taps and the consumption of the L / flush of the toilets of one or two flushes evaluating these with the use of a calculator incorporated in the online software tool where the flows, types and number of devices are entered to obtain an average flow weighted in liters per minute or liters per download, likewise, design drawings, technical sheets of the manufacturers and additional calculations are requested to obtain the evaluation in the preliminary certification stage, then delivery vouchers and photographs of the implementation for the on-site audit for the final certification,which are performed by a certified EDGE auditor.

It is worth mentioning that EDGE allows to be selective with the WATER indicators that appear in the online software for its evaluation, its qualification methodology is not through summative scores but by percentage of savings above a baseline, therefore it is allowed select the indicators that are required until reaching a minimum of 20% savings to achieve certification.

Leed for homes (V4) within the evaluation of the criterion called Water Efficiency, evaluates four indicators, being the use of the water consumption meter the mandatory prerequisite compared to the other three that correspond to: The reduction of the total demand for water, the management of water use inside the building and outdoors, which are evaluated online evidencing their compliance with supporting documents.

This certification gives a total score of 12 points for the evaluation of Water Efficiency of 110 points considered for the evaluation of the other environmental criteria, the 12 points can be compared in a check list and are given in this way:

1 point to 12 points are achieved according to the reduction of% of integral water, for example, if 20% of the total use of water is reduced (this includes the reduction of the use of water inside and outside) above the traditional practices of use, 3 points out of 12 points would be awarded. The tools used to calculate the average flow rate and percentage savings in LEED for water use inside and outside the building are the Water Reduction calculator and the EPA WaterSense Water Budget calculator.

On the other hand, if you want to apply to an evaluation of the indicators that correspond to the management of indoor water use (taps, sanitary ware and selection of washing machines), the points awarded are from 1 point to 6 points which are measurable If they meet the technical conditions, for example, for efficient faucets with Water Sense certificate, 2 points are awarded if 3.7 liters per minute are not exceeded at a pressure not greater than 414kPa.

Likewise, for the management of water use in gardens, from 1 point to 4 points are awarded per% reduction of grass and increase of native plants, providing complete information by other means of the native plants of some states of North America as a guide.

BREEAM® ES Housing, on the other hand, has an evaluation system that gives points and percentage (%) of weighting for the evaluation of the use of Water, where 6 of 110 points with 10.5% (of 110%) of weighting are for this criterion .

The indicator that is assigned the highest score corresponds to: Water consumption (3 points) and the indicators continue: Irrigation system (2 points) and Water treatment at the site (2 points), finally it is awarded 1 point for the Water Meter indicator and 1 point for Water Recycling.

Its indicators are evaluated in groups; To mention an example, if you want to achieve two points in the criterion of water consumption (from the group of taps, toilets, shower heads and electrical appliances such as washing machines and dishwashers) you must incorporate the fulfillment of the criterion AG5 (prerequisite) that corresponds to the criterion of Recycling of water. Therefore, in this way the indicators are interconnected with each other and also connected with the overall certification score. This evaluation system is all or nothing, that is, if it does not comply with a group indicator, despite being efficient, it is not awarded the desired group score.

The tool used by BREEAM® ES Vivienda to evaluate buildings is an Excel format with questions and answers, as well as using another complementary tool such as the water calculator. The tool should be used by qualified consultants - certified who have undergone rigorous evaluations since the data entered is of a technical nature.

As for DNGB, the German certification, assesses water efficiency as a sub-criterion "Consumption of drinking water and volume of wastewater" within the evaluation of the criterion of Environmental Objectives - Use of resources and generation of waste, with a factor of weighting of 2 and a contribution to the total valuation of 2.4% of the 22.5% that has been granted to Environmental Objectives. The total score awarded to this sub-criterion is 100 points. There are three indicators evaluated: Water use index (90 points), Water in outdoor areas (5 points), Integration to the district's infrastructure (5 points). A selective analysis of individual criteria is not allowed, as this would conflict with the holistic evaluation of the certification. The water use index with 90 points goes according to the result in m3 per year as a limit value,

As for the tools used for the calculation, DNGB provides a calculator where it predicts the demand for drinking water and the volume of wastewater + a data on the distribution of rainfall in Germany and other countries. Additionally, mention that by means of another criterion corresponding to LCA (Life Cycle Analysis) the consumption of resources is evaluated throughout the entire life cycle and the indicators are: Water consumption: Net use of fresh water (FW) and Eutrophication of surface waters.

As in DNGB, the HQE TM certification integrates the environmental evaluation of the Water resource as a sub-criterion within the Criterion called Environment, which is one of the 4 criteria.

However, the qualification as it is known is valued by stars, that means that the Environmental Management of Water is valued within a package with other sub-criteria (Energy and Maintenance) the three should obtain from 1 to 4 stars. One star corresponds to P = Performance and 2 stars correspond to P + HP (Performance + High Performance).

In order to obtain P for Performance in the environmental qualification of Water, the 4 indicators that are prerequisites must be met (Plan hot water meters in cases of collective DHW production, Include a hot water reproduction system, Use of hot water flow devices water (plumbing, discharge, etc.) with a distinctive characteristic to report on its quality and include sanitation to handle wastewater treatment when there is no connection to the public service) in addition to achieving a minimum of 4 points. On the other hand, if you want to achieve High Performance HP, you must meet the 4 prerequisites mentioned in addition to achieving a score of 9 points.

HQETM not only evaluates the water from an environmental point of view, but also from the quality where they are prerequisites (mandatory): Plan the washing and disinfection of all pipes after use and before the installation of the plumbing, in charge of the installation company, Each home must be equipped with a retention system (backflow, check valve system, etc.) in the cold water supply and, if necessary, hot water supply in the case of collective DHW, Maintain a feedback system to the point of entry for the supply of hot water to collective buildings when it comes to collective DHW and Limit the risks associated with the development of Legionella in showers or bathtubs.As a tool, HQETM also provides a calculator in Excel format to predict consumption and a very detailed guide for its use.

From the comparison of indicators grouped in Water management inside and outside the home, Use of non-traditional sources and Water monitoring systems in the BONO VERDE, EDGE (Peru), LEED for homes, BREEAM® ES Vivienda, DNGB certifications and HQE™.

This item analyzes the indicators of the water category of the six certifications that are grouped into four packages for a better understanding, as shown in table 1:

Table 1. Comparison of the criteria considered in the water category of the BONO VERDE, EDGE (Peru), LEED for homes, BREEAM® ES Vivienda, DNGB and HQETM certifications

	Water management inside homes	BONO VERDE	EDGE	LEED	BREEAM	DNGB	HQE
1	Faucets in sinks*	X	X	X	X	X	X
2	Shower heads *	X	X	X	X	X	X
3	Faucets inbathtubs				X	X	X
4	Faucets in kitchen*	X	X	X	X		X
5	Faucets in laundry or garages*						X
5	Discharge in toilets *	X	X	X	X	X	X
6	Urinals				X		X
7	Bidet		X				
8	Washing machines*		X	X	X	X	X
9	Dishwasher		X	X	X	X	X
10	Recovery for air conditioning systems for comfort		X				
	Water management to the immediate exterior of homes	**BONO VERDE**	**EDGE**	**LEED**	**BREEAM**	**DNGB**	**HQE**
1	Intelligent or technified systems for garden irrigation *	X	X	X	X		
2	Native plants orxerophylls in gardens*	X		X	X		
3	Water sources (swimming pools or others)		X				X
4	Green roofs or walls						X
5	No irrigation, no gardens					X	
	Use of non-traditional water sources	**BONO VERDE**	**EDGE**	**LEED**	**BREEAM**	**DNGB**	**HQE**
1	Rainwater recovery and use	X	X	X	X	X	X
2	Gray water recovery and use	X	X	X	X	X	X
3	Balck water recovery and use		X	X			X
	Water monitoring system	**BONO VERDE**	**EDGE**	**LEED**	**BREEAM**	**DNGB**	**HQE**
1	Water meters	X	X	X	X		X
2	Communication plan to experts or owners	X		X			
	Others	**BONO VERDE**	**EDGE**	**LEED**	**BREEAM**	**DNGB**	**HQE**
1	Drinking water quality						X
2	Hot water						X
3	Life cycle analysis (ACV)					X	

Package 1: Water Management Inside the Building

1. Water management inside the building (taps, toilets, washing machines)
2. Of water management in heating and cooling systems

This section presents the indicators that take into account the 6 certifications for the management of water inside the home. The implements, devices or systems that are most used for the typical single-family house on the coast have been marked with an asterisk (*) as shown in Table 1.

A typical house on the coast of Peru, according to the sociocultural behaviors and economic activities of a medium to low level, usually has inside their homes; bathrooms equipped with sink, shower, toilet, also a kitchen, laundry and / or garage, a washing machine and the use of hot water for short periods of low temperatures during the year. Not all homes in Peru have a bathtub, bidet, urinal in the bathroom, or dishwasher in kitchens, or air conditioning systems for air conditioning, the latter due to its high cost of installation, for example; however, some more affluent multi-family residential buildings may at times.

Regarding washbasin faucets, the six certifications if they are considered for the evaluation and the minimum requirements for each of them are as described: The Green Bonus at this point only requires that the faucet or aerator have a National seal called SEDAPAL or an international label like Water-Sense. In the case of EDGE in version 2.1 for Peru, it requests to improve the baseline of the flow of 8 liters / minute, this can be achieved with the type of faucet or by adding aerators that achieve a flow of

6 liters / minute at a hydraulic pressure of 43.5 PSI, at a minimum, to demonstrate significant savings; for the new Beta 3.0 version the requirement would be much higher. For LEED for homes (£ 5.6 liters / minute at 60 PSI) additionally the requirement that the taps have the Watersense label, and BREEAM® ES Vivienda (<

The evaluation of the shower heads in Bono Verde have the same technical requirement as the basin taps and do not take into account the evaluation of bathtubs, while EDGE to improve its incorporated baseline requires equal or less flow to 8 liters / minute at 43.5 PSI pressure and also does not rate bathtubs. LEED for homes (£ 6.6 liters / minute at 60 PSI) BREEAM® ES Housing (£ 9 liters / minute at 43 PSI for showers and 160 liters / up to overflow for bathtubs) in terms of DNGB and HQETM, the same happens as described for washbasin taps, in general the same methodology also applies to assess consumption in bathtubs.

Kitchen faucets are evaluated both in Green Bonus, not specifying flow rates, only the requirement: SEDAPAL seal and another international water saving label, as well as in EDGE (£ 6 liters / minute at 43.5 PSI) and it is worth mentioning that in the next new Beta 3.0 version contemplates the indicator of "Pre-rinse spray valves for kitchen use". In both LEED for homes (£5.6 liters / minute at 60PSI) and BREEAM® ES Housing (<6 liters / minute at 43 PSI). In DNGB it does not mention the kitchen taps and for HQETM it is only considered in the calculator, likewise, as part of the indicator "Reduction of distributed water consumption".

The laundry faucet flow rate is not taken into account by the certifications shown, although HQETM could be included in the calculations within the calculator, likewise, as part of the Indicator "Reduction of distributed water consumption" and it is a tap.

Regarding liters per flush for toilets, Bono Verde does not require a minimum flow except for the aforementioned labeling, EDGE has the option to choose between one flush (6 liters / flush) and two flushes (3 liters / 1st flush and 6 liters / 2nd flush) in LEED for homes for one flush (£ 4.1 liters / flush) and BREEAM® ES Housing, dual flush toilets only (£3 liters / 1st flush and 4.5 liters / 2nd flush) is required, otherwise the score, on the other hand, again here, DNGB and HQETM request the flows to enter it in their respective calculators to count it in the total calculations of Reduction which if they are valued globally with the other teams. Mention that the evaluation of the efficiency of the bidet is only considered in EDGE but in its new Beta 3 version.

In relation to the evaluation of electrical appliances that use the water resource, such as the washing machine, EDGE will take it into account in the V3.0 version (35 liters / load) plus the Green Bonus does not consider it in its evaluation despite that many households have the facility to acquire the appliance and use it in Peru. It is also taken into account by LEED for homes for its evaluation (£ 35 liters / use - 7 Kg. + Energy Star labeling) and BREEAM® ES Housing (£ 45 liters / use), if they have the washing machine. In both DNGB and HQE TM they consider entering the liters in their calculators for the result of general consumption.

The dishwasher is also one of the evaluation indicators in EDGE V3.0 (3.75 liters / grid) in case it is used in the building, also BREEAM® ES Housing (£10 liters / use) and Leed for homes (2.4 liters / day). As in the other evaluations of consumption of faucets DNGB and HQETM do not specify the required liters, but if the data must be included in their water calculators, at this point the Green Bond does not consider the evaluation of dishwashers either.

Finally, in case the house or group of houses (multi-family) installs a refrigeration system, it must be demonstrated that the HVAC system is equipped with a device for collecting 100% of the recovered condensed water, as a requirement for EDGE V3 certification, no However, this indicator is not taken into account in the other certifications seen in this chapter.

Package 2: Outdoor Water Management, in Gardens

At this point, we will start with the EDGE certification, for example, which does not contemplate an evaluation for the outdoor use of water in gardens in Version 2.1, more in the Beta 3.0 version if it considers it (4 liters / m2 / day) which is the improved baseline. On the other hand, for the certification of the Green Bonus, it is requested to implement any irrigation system, being able to also opt for a system with mist-trapping micro-sprinklers when there is a green area> 100 m2 and to execute dry gardens and / or Bio-gardens (5% of the total of the green area) as a recommendation, but it does not require it in a mandatory way. BREEAM® ES Vivienda, also evaluates the incorporation of irrigation systems in gardens through: drip irrigation, zoned irrigation control, soil moisture sensors and / or rainfall station to obtain 1 point in the Irrigation system category. On the other hand, in terms of vegetation and plant types adapted to the climate and with low water consumption, it is added to this LEED for Homes evaluation, this being more precise in the calculation using the EPA WaterSense Water Budget Calculator (https://www.epa.gov/watersense/water-budget-tool), granting a higher score when the percentage (%) of grass is decreased (from 5% to 60% of sowing) and a higher score when a higher percentage (%) of native species is planted, for example if only 5% is planted. of area with grass and 75% of native plants is valued with a score greater than 4.

HQETM takes into account the evaluation of garden treatment qualitatively within the indicator "Reduction of distributed water consumption", however, quantitative data can also be incorporated into the calculator in case the building has green walls or roofs. On the other hand, the DNGB within the External Areas Indicator raises two questions to evaluate this indicator: If irrigation is foreseen without drinking water and if the external areas have a rainwater retention device.

Sometimes in gardens, water sources such as swimming pools or pools are incorporated when they are residential complexes in Peru, but it is not very common, however, in the HQE water consumption simulation calculator there is a box for filling the liters that they are incorporated when using a water source. EDGE (V3.0 Beta) will also take it into account as one of the indicators to evaluate, demanding 30% coverage from the pools.

Package 3: Use of Non-Traditional Water Sources such as Rainwater, Gray or Black Water

The Green Bond evaluation system, in this matter, specifies the use of treatment plants for gray or rainwater (only for areas with high rainfall according to SENAMHI) for the reuse of toilets and green areas, these two indicators are evaluated by separate and are mandatory, which must be evidenced with calculations developed by specialists to certify with the major Grade 3+.

Likewise, EDGE AND BREEAM® ES Housing consider detailed specifications in the evaluation of this issue, unlike LEED for Homes, which does not consider it as a specific indicator, but takes it into account tacitly as a recommendation to be effective the criterion of Total Use of the water as a total reduction strategy. Regarding EDGE in Version 2.1, it considers the recycling of rainwater, gray water destined for reuse only in toilets and the reuse of sewage. For the reuse of rainwater, consider 50% of the roof area for collection. On the other hand, BREEAM® ES Vivienda mentions that recycled rainwater must satisfy at least 75% of the total expected demand for: flushing toilets, watering plants,

The objective of DNGB is the sustainable management of this indicator by promoting reducing the volume of wastewater to disturb the natural water cycle as little as possible, therefore it calculates the sum of the determined demand for drinking water and the volume of wastewater (use of rainwater or greywater) which results in the "water use value" to this value is assigned 3 scores, on the other hand, the saved drinking water and the reduced wastewater is incorporated in the life cycle assessment .

As for HQETM, it considers that the reuse of rainwater should include a collection system that provides the necessary information to certify the maintenance of sanitary conditions for use in the project, with considerable emphasis on the leakage rate and the Use of a treatment plant is mandatory to include it in the sanitation for when the connection to the public sewage network is not planned, complying with the sanitation specifications.

Package 4: Water use Monitoring Systems

The Green Bond requires the incorporation of water meters for individual homes or departments with INACAL quality certificates so that they can have true data and then the consumption can be reported to the certification team.

Water meters or meters are also considered in the evaluation by BREEAM® ES Housing and LEED for Homes, however, EDGE in version 2.1 does not take it into account as an indicator, however it has been considered in the current Beta Version 3.0 as an indicator .

For LEED for Homes it is a prerequisite throughout the assessment, although homes that use only well water and are not connected to a municipal water system are exempt from this prerequisite. Landlords or tenants are encouraged to share water usage data with USGBC through a USGBC-approved third party.

Prerequisite and for BREEAM® ES Housing it is part of a requirement to obtain a point where the water meter must have an impulse output and be connected to a Building Management System (SGE) to monitor and control consumption of water. In the case of apartments, the use of a meter in common areas should be considered.

In this topic, HQETM demands to implement the use of a meter not only for cold water but also for hot water in cases of collective production of DHW, on the other hand, DNGB does not consider it in the evaluation.

HQETM not only considers the evaluation of water from the environmental perspective but also from the quality and in that sense it evaluates reducing the risk of legionella in addition to "Carrying out a water analysis in the general meter on the ground floor of buildings or residences and performing an analysis of the water in the plumbing outlet after the works, rinsing and disinfection. In case of discrepancies with local regulations or, where appropriate, ISO 147 Water Quality Standard, the Applicant must take the necessary steps to resolve them, These results should be delivered to future occupants. Tests will be conducted per building, at the residence furthest from the building's water supply point, as well as at a randomly chosen residence. For independent residences,a 5% sampling rate applies to residences with a minimum of one residence. "

CONCLUSION AND RECOMMENDATIONS

Although currently the BONO VERDE and EDGE versions have been updated to June 2021, the water calculators used in certifications such as BREEAM, LEED, HQE and DNGB could serve as a reference to provide more precision in calculating the total flow consumed or saved. Considering the evidence of the importance of the resource on the coast of the Peruvian desert, it should be approached with greater ambition.

The calculator used in LEED for the management of water in gardens is very interesting that it could be a reference for the elaboration of a local starting in the capital since they are implementing the promotion of green roofs in the municipal ordinances.

The local or regional governments of the coastal zone of Peru could implement a program for the promotion and placement of flow reducers, mainly in the shower and bathroom sink, or another strategy that amplifies the impact of reducing water on all in the informal housing sector, in use, which is close to 70% or granting incentives to direct owners.

Considering that the washing machine is a very common appliance in Peru, it should be taken into account in the evaluation to specify consumption and savings.

Homeowner education plans are a good criterion to maintain and replicate the Green Bond to raise awareness of water management for the residential sector.

Although the LEED, BREEAM, HQE and DNGB certifications are more demanding in terms of WATER than EDGE and BONO VERDE, the simplicity of use of the EDGE online tool, as well as the simplicity of the methodology of the GREEN BOND allow them to be easily fulfilled. You need to upload the go, but in simple.

REFERENCES

AQUAFONDO. (2018). *Water Footprint of water users in Metropolitan Lima*. Lime.

Arid Zones Research Center, National Agrarian University. (1982, September). *Arid zones*. Retrieved 2021, from http://www.lamolina.edu.pe/zonasaridas/za11/pdfs/ZONAS%20ARIDAS%20VOL%20I.pdf

Biondi, S. (2012, November 2). *Sustainable architecture*. Retrieved from https://www.youtube.com/watch?v=wxIOK1V1joc&ab_channel=PUCPPUCP

BREEAM. (2011). *Manual BREEAM is housing*. Author.

Bunclark, L. (2020, February 13). *They consider Peru to be at a medium-high level of water stress*. Retrieved from https://rpp.pe/peru/actualidad/consideran-al-peru-en-un-nivel-medio-alto-de-estres-de-agua-noticia-1245508

DGNB System ES. (2020). *New Construction Criteria Version 2020*. Author.

EDGE. (n.d.a). *EDGE at Peru*. Retrieved June 2021, from https://edgebuildings.com/project-studies/?lang=es&_sft_project_countries=peru-es

EDGE. (n.d.b). *EDGE software*. Retrieved from https://edgebuildings.com/software/?lang=es

Fernández, A. (n.d.). *Wastewater in Peru, Problems and use in agriculture*. Retrieved from https://www.ais.unwater.org/ais/pluginfile.php/356/mod_page/content/128/Peru%20INFORME%20DE%20PAIS.pdf

Fondo MiviviendaS. A. (2021, June 16). *Build to Grow*. Retrieved from http://www.construyeparacrecer.com/Inicio.aspx

HQE. (2014). *HQE™ Scheme Environmental Performance Residential Building*. HQE.

IFC, EDGE. (2018). *EDGE User Guide _Version 2.1*. Author.

IFC. (2017-2018). *Terms of Reference-IFC. Cities Program Latin America and the Caribbean "Green Building and Municipalities-Peru"*. IFC.

INEI. (2020). *Press Release: The population of Lima exceeds nine and a half million inhabitants*. Retrieved from http://m.inei.gob.pe/media/MenuRecursivo/noticias/notadeprensa006.pdf

Ministry of Housing, Construction, and Sanitation. (2014, May 13). *Standard EM.110 Thermal and Light Comfort with Energy Efficiency*. Retrieved from http://www3.vivienda.gob.pe/dnc/archivos/Estudios_Normalizacion/Normalizacion/normas/DS-006-2014-VIVIENDA.pdf

Ministry of Housing, Construction, and Housing. (2021). *Peru GBC & MVCS - Technical Code of Sustainable Construction Peru 2021*. Retrieved 2021, from https://issuu.com/perugbc/docs/codigo_-_ctcs

Miranda, A. (2018, May 28). *Water or Energy?* Academic Press.

Mivivienda, F. (2020). *Certification Procedure for projects of the Sustainable Housing Program*. Retrieved from https://www.mivivienda.com.pe/PORTALCMS/archivos/documentos/8586096294727711719.PDF

Oñativia, D. A. (2011, May 10). *The Geographical Regions of Peru*. Retrieved from Blog HISTORY 2DO YEAR IEM 2011: http://historiasegundoiemunsa2011.blogspot.com/2011/05/las-regiones-geograficas-del-peru.html

OXFAM. (n.d.). *OXFAM in Peru. Retrieved from Between 7 and 8 million Peruvians do not have access to drinking water*. https://peru.oxfam.org/qu%C3%A9-hacemos-ayuda-humanitaria/entre-7-y-8-millones-de-Peruvians don't have access to drinking water

PUPC. (2018). *Puntoedu report on water stress in Peru*. Retrieved from https://puntoedu.pucp.edu.pe/noticia/informe-puntoedu-sobre-estres-hidrico-en-el-peru/

Senamhi. (n.d.). *Climatic Events - Phenomenon of the Child*. Retrieved from Senamhi: http://senamhi.gob.pe/?p=fenomeno-el-nino

Traditional geographical regions of Peru. (n.d.). Obtained from Wikipedia: https://es.wikipedia.org/wiki/Regiones_geogr%C3%A1ficas_tradicionales_del_Per%C3%BA

US Green Building Council. (2017). *LEED v4 for Homes design and Construction*. LEED.

Compilation of References

Abdullah, H. A., Razzaqul, A., Nasser, H., Reza, G., & Eklas, H. (2020). Survey of Smart Grid Concepts and Technological Demon striations Worldwide Emphasizing on the Oman Perspective. *Applied System Innovation*, *3*(5), 5. Advance online publication. doi:10.3390/asi3010005

Abdulmohsin, H., Aritra, G., Senthilarasu, S., & Tapas, K. M. (2020). Evaluation of thermal performance for a smart switchable adaptive polymer dispersed liquid crystal (PDLC) glazing. *Solar Energy*, *195*, 185–193. doi:10.1016/j.solener.2019.11.024

Abdul-Wahab, S. A., Charabi, Y., Al-Maamari, R., Al-Rawas, G. A., Gastli, A., & Chan, K. (2015). CO2 greenhouse emissions in Oman over the last forty-two years [Review]. *Renewable & Sustainable Energy Reviews*, *52*, 1702–1712. doi:10.1016/j.rser.2015.07.193

Abdul-Wahab, S. A., Charabi, Y., Al-Rawas, G. A., Al-Maamari, R., Gastli, A., & Chan, K. (2015). Greenhouse gas (GHG) emissions in the Sultanate of Oman. Greenhouse Gases. *Science and Technology*, *5*(3), 339–346. doi:10.1002/ghg.1501

Aboulnaga, M. M. (2006). Towards green buildings: Glass as a building element—the use and misuse in the gulf region. *Renewable Energy*, *31*(5), 631–653. doi:10.1016/j.renene.2005.08.017

Acosta, Leslie, & Figueiro. (2015). Analysis of circadian stimulus allowed by daylighting in hospital rooms. *Lighting Research & Technology*, 1-13.

AEMet. (2020). *Agencia Estatal de Meteorología*. Author.

Ahmad, A., Hassan, M., Abdullah, M., Rahman, H., Hussin, F., Abdullah, H., & Saidur, R. (2014). A review on applications of ANN and SVM for building electrical energy consumption forecasting. *Renewable & Sustainable Energy Reviews*, *33*, 102–109. doi:10.1016/j.rser.2014.01.069

Ahmad, E. (2014). Fuzzy Logic Controller Based SEPIC Converter for Maximum Power Point Tracking. *IEEE Transactions on Industry Applications*, *50*(4), 2349–2358. doi:10.1109/TIA.2014.2298558

Ahn, K. U., & Park, C. S. (2016). Correlation between occupants and energy consumption. *Energy and Building*, *116*, 420–433. doi:10.1016/j.enbuild.2016.01.010

Alajmi, B. N., Ahmed, K. H., Finney, S. J., & Williams, B. W. (2011). Fuzzy-Logic-Control Approach of a Modified Hill Climbing Method for Maximum Power Point in Microgrid Standalone Photovoltaic System. *IEEE Transactions on Power Electronics*, *26*(4), 1022–1030. doi:10.1109/TPEL.2010.2090903

Al-Atrash, F., Hellwig, R. T., & Wagner, A. (2018). Personal control over indoor climate in office buildings in a Mediterranean climate - Amman, Jordan. Academic Press.

Al-Atrash, F., Hellwig, R. T., & Wagner, A. (2020). The degree of adaptive thermal comfort in office workers in a hot-summer Mediterranean climate. *Energy and Building*, *223*, 110147. doi:10.1016/j.enbuild.2020.110147

Al-Azri, N. A., Zurigat, Y. H., & Al-Rawahi, N. Z. (2013). Development of bioclimatic chart for passive building design. *International Journal of Sustainable Energy*, *32*(6), 713–723. doi:10.1080/14786451.2013.813026

Al-Badi, A. H., Malik, A., & Gastli, A. (2011). Sustainable energy usage in Oman—Opportunities and barriers. *Renewable & Sustainable Energy Reviews*, *15*(8), 3780–3788. doi:10.1016/j.rser.2011.06.007

Alexander, D., & Lewis, L. (2014). Condition of America's public school facilities: 2012-13 (NCES 2014-022). Washington, DC: U.S. Department of Education, National Center for Education Statistics.

Al-Gharibi, H. (2016). Urbanisation and changing lifestyle patterns in Muscat. In S. Nebel & A. von Richthofen (Eds.), *Urban Oman: Trends and perspective of urbanisation in Muscat capital area* (pp. 73–79). LIT Verlag.

Alguacil, J. (2011). *Experiencia y metodología para un modelo de desarrollo comunitario. El caso de San Cristóbal de Los Ángeles*. Editorial Popular.

Al-Hinai, H., Batty, W. J., & Probert, S. D. (1993). Vernacular architecture of Oman: Features that enhance thermal comfort achieved within buildings. *Applied Energy*, *44*(3), 233–258. doi:10.1016/0306-2619(93)90019-L

Ali, K. (2020). *Design of Smart Power Grid Renewable Energy Systems*. Wiley.

Al-Khatri, H., Alwetaishi, M., & Gadi, M. B. (2020). Exploring thermal comfort experience and adaptive opportunities of female and male high school students. *Journal of Building Engineering*, *31*, 101365. doi:10.1016/j.jobe.2020.101365

Al-Khatri, H., & Gadi, M. B. (2019). Investigating the behaviour of ASHRAE, Bedford, and Nicol thermal scales when translated into the Arabic language. *Building and Environment*, *151*, 348–355. doi:10.1016/j.buildenv.2019.01.051

Allegretti, G., & Torquato Luiz, J. (2007). Quando lo spazio urbano si fa metafora di un cambiamento politico: la riscossa del territorio nel Piano di Urbanizzazione. In Viaggio in Portogallo. Dentro e fuori territori dell'architettura. Roma: ARACNE editrice S.r.l.

Allen, K., Connelly, K., Rutherford, P., & Wu, Y. (2017). Smart windows—Dynamic control of building energy performance. *Energy and Building*, *139*, 535–546. doi:10.1016/j.enbuild.2016.12.093

Almazroui, M. (2012). Temperature Variability over Saudi Arabia and its Association with Global Climate Indices. Journal of King Abdulaziz University-Meteorology. *Environment and Arid Land Agriculture Sciences*, *23*(1), 85–108. doi:10.4197/Met.23-1.6

Almazroui, M. (2013). Simulation of present and future climate of Saudi Arabia using a regional climate model (PRECIS). *International Journal of Climatology*, *33*(9), 2247–2259. doi:10.1002/joc.3721

Almazroui, M., Islam, M. N., Saeed, S., Saeed, F., & Ismail, M. (2020). Future Changes in Climate over the Arabian Peninsula based on CMIP6 Multimodel Simulations. *Earth Systems and Environment*, *4*(4), 611–630. doi:10.100741748-020-00183-5

Al-Shaqsi, S. (2010). Care or Cry: Three years from Cyclone Gonu. What have we learnt? *Oman Medical Journal*, *25*(3), 162–167. doi:10.5001/omj.2010.50 PMID:22043331

Alves, S., Fernández-Agüera, J., & Sendra, J. J. (2014). Infiltration rate performance of buildings in the historic centre of Oporto. *Informes de la Construcción*, *66*(535), e033. Advance online publication. doi:10.3989/ic.13.009

Al-Yasiri, Q., & Szabó, M. (2021). Incorporation of phase change materials into building envelope for thermal comfort and energy saving: A comprehensive analysis. *Journal of Building Engineering*, *36*, 1–23. doi:10.1016/j.jobe.2020.102122

Amann, M., Kiesewetter, G., Schöpp, W., Klimont, Z., Winiwarter, W., Cofala, J., Rafaj, P., Höglund-Isaksson, L., Gomez-Sabriana, A., Heyes, C., Purohit, P., Borken-Kleefeld, J., Wagner, F., Sander, R., Fagerli, H., Nyiri, A., Cozzi, L., & Pavarini, C. (2020). Reducing global air pollution: The scope for further policy interventions: Achieving clean air worldwide. *Philosophical Transactions - Royal Society. Mathematical, Physical, and Engineering Sciences*, *378*(2183), 20190331. Advance online publication. doi:10.1098/rsta.2019.0331 PMID:32981437

Andersen, M., Mardaljevic, J., & Lockley, S. W. (2012). A framework for predicting the non-visual effects of daylight - Part I: Photobiology-based model. *Lighting Research & Technology*, *2012*(44), 37–53. doi:10.1177/1477153511435961

Anderson, J. E., Wulfhorst, G., & Lang, W. (2015). Energy analysis of the built environment - A review and outlook. Renewable and Sustainable Energy Reviews, 44, 149–158. doi:10.1016/j.rser.2014.12.027

Anderson, G. B., Dominici, F., Wang, Y., McCormack, M. C., Bell, M. L., & Peng, R. D. (2013). Heat-related emergency hospitalizations for respiratory diseases in the medicare population. *American Journal of Respiratory and Critical Care Medicine*, *187*(10), 1098–1103. doi:10.1164/rccm.201211-1969OC PMID:23491405

Andresen, I. (2018). Definitions, Targets, and Key Performance Indicators for New and Renovated Zero Emission Buildings. In R. Wang & X. Zhai (Eds.), *Handbook of Energy Systems in Green Buildings* (pp. 35–60). Springer. doi:10.1007/978-3-662-49120-1_47

Annunziata, E., Frey, M., & Rizzi, F. (2013). Towards nearly zero-energy buildings: The state-of-art of national regulations in Europe. *Energy*, *57*, 125–133. doi:10.1016/j.energy.2012.11.049

Anyaka, B., & Olawoore, O. (2014). Minimization of Power Losses in Transmission Lines. *Journal of Electrical and Electronics Engineering (Oradea)*, *9*(3), 23–26.

AQEG. (2012). *Fine Particulate Matter in the United Kingdom*. Available at: https://uk-air.defra.gov.uk/assets/documents/reports/cat11/1212141150_AQEG_Fine_Particulate_Matter_in_the_UK.pdf

AQUAFONDO. (2018). *Water Footprint of water users in Metropolitan Lima*. Lime.

Arellano, B., Sena, S., Abdollahy, S., Lavrova, O., Strattonl, S., & Hawkinsl, J. (2013). Analysis of electric vehicle impacts in New Mexico urban utility distribution infrastructure. *IEEE Transportation Electrification Conference and Expo (ITEC)*. 10.1109/ITEC.2013.6573495

Ares, B. E. (2016). *Zero Carbon Homes*. House of Commons Library.

Arıcı, M., Bilgin, F., Nižetić, S., & Karabay, H. (2020). PCM integrated to external building walls: An optimization study on maximum activation of latent heat. *Applied Thermal Engineering*, *165*, 1–13. doi:10.1016/j.applthermaleng.2019.114560

Arid Zones Research Center, National Agrarian University. (1982, September). *Arid zones*. Retrieved 2021, from http://www.lamolina.edu.pe/zonasaridas/za11/pdfs/ZONAS%20ARIDAS%20VOL%20I.pdf

Arnstein, S. (1969). A Ladder of Citizen Participation. *Journal of the American Planning Association*, *35*(4), 216–224.

Arpino, F., Cortellessa, G., & Frattolillo, A. (2015). Experimental and numerical assessment of photovoltaic collectors performance dependence on frame size and installation technique. *Solar Energy*, *118*, 7–19. doi:10.1016/j.solener.2015.05.006

Ascione, F., Bianco, N., De Masi, R. F., Mastellone, M., Mauro, G. M., & Vanoli, G. P. (2020). The role of the occupant behavior in affecting the feasibility of energy refurbishment of residential buildings: Typical effective retrofits compromised by typical wrong habits. *Energy and Building*, *223*, 110217. doi:10.1016/j.enbuild.2020.110217

Ashrae. (2002). *Ashrae Guide and Data Book (Fundamentals)*. Ashrae. Retrieved from https://www.ashrae.org/File Library/docLib/Public/2003627132256_326.pdf

ASHRAE. (2007). ASHRAE standard 62.1-2007 Ventilation for Acceptable Indoor Air Quality. ASHRAE.

ASHRAE. (2017). *2017 ASHRAE Handbook Fundamentals (SI)*. ASHRAE (American Society of Heating, Refrigerating and Air-Conditioning Engineers).

ASHRAE. (2017). *Handbook. HVAC Fundamentals* (Spanish Edition). Editorial Index.

Attia, H. A., & Del Ama Gonzalo, F. (2019). Stand-alone pv system with mppt function based on fuzzy logic control for remote building applications. *International Journal of Power Electronics and Drive Systems*, *10*(2), 842–851. doi:10.11591/ijpeds.v10.i2.pp842-851

Australian National University. (2020). *Carbon dioxide toxicity and climate change: a serious unapprehended risk for human health*. P.N. Bierwirth.

Babatunde, O. M., Munda, J. L., & Hamam, Y. (2019). A comprehensive state-of-the-art survey on power generation expansion planning with intermittent renewable energy source and energy storage. *International Journal of Energy Research*, *43*(12), 6078–6107. doi:10.1002/er.4388

Baek, J., Huang, K., Conner, L., Tapangan, N., Xu, X., & Carrillo, G. (2019). Effects of the home-based educational intervention on health outcomes among primarily Hispanic children with asthma: A quasi-experimental study. *BMC Public Health*, *19*(1), 912. doi:10.118612889-019-7272-5 PMID:31288792

Baetens, R., Jellea, B. P., & Gustavsend, A. (2010). Phase change materials for building applications: A state-of-the-art review. *Energy and Building*, *42*(9), 1361–1368. doi:10.1016/j.enbuild.2010.03.026

Bagherzadeh, M., & Kaveh, R. A. (2018). A new SnS2-BiFeO3/reduced graphene oxide photocatalyst with superior photocatalytic capability under visible light irradiation. *Journal of Photochemistry and Photobiology A Chemistry*, *359*, 11–22. doi:10.1016/j.jphotochem.2018.03.031

Bahadori, A., & Nwaoha, C. (2013). A review on solar energy utilisation in Australia. *Renewable & Sustainable Energy Reviews*, *18*, 1–5. doi:10.1016/j.rser.2012.10.003

Bahrar, M., Djamai, Z. I., El Mankibi, M., Larbi, A. S., & Salvia, M. (2018). Numerical and experimental study on the use of microencapsulated phase change materials (PCMs) in textile reinforced concrete panels for energy storage. *Sustainable Cities and Society*, *41*, 455–468. doi:10.1016/j.scs.2018.06.014

Balandin, A. A. (2011). Thermal properties of graphene and nanostructured carbon materials. *Nature Materials*, *10*(8), 569–581. doi:10.1038/nmat3064 PMID:21778997

Ballesteros, M., Pujol, M., Walsh, F., & Teubner, J. (2019). *Geothermal Energy Electricity Generation in Australia: Recent Developments and Future Potential*. Academic Press.

Banerjee, S., & Chattopadhyaya, M. (2017). Adsorption characteristics for the removal of a toxic dye, tartrazine from aqueous solutions by a low-cost agricultural by-product. *Arabian Journal of Chemistry*, *10*, 1629–1638. doi:10.1016/j.arabjc.2013.06.005

Baniyounes, A. M., Ghadi, Y. Y., Rasul, M., & Khan, M. M. K. (2013). An overview of solar assisted air conditioning in Queensland's subtropical regions, Australia. *Renewable & Sustainable Energy Reviews*, *26*, 781–804.

Banu, R., Salvi, N., Ameta, C., & Ameta, R. (2019). Visible light driven photocatalytic degradation of brilliant green dye using graphene oxide/copper oxide binary composite. *Indian Journal of Chemistry*, *B581*, 157–166.

Baraa, J. (2017). Improving visual comfort and energy efficiency in existing classrooms using passive daylighting techniques. *Energy Procedia*, *136*, 102–108. doi:10.1016/j.egypro.2017.10.294

Baranzelli, C., Lavalle, C., Sgobbi, A., Aurambout, J., Trombetti, M., Jacobs, C., Cristobal Garcia, J., Kancs, D., & Kavalov, B. (2016). *Regional Patterns of Energy Production and Consumption Factors in Europe—Exploratory Project EREBILAND—European Regional Energy Balance and Innovation Landscape; EUR 27697*. Publications Office of the European Union.

Barkhudaryan, N., Orosa, J. A., & Roshan, G. (2014). A new procedure to analyze the effect of air changes in building energy consumption. *Journal of Environmental Health Science & Engineering*, *12*(37), 1–9. doi:10.1186/2052-336X-12-37 PMID:24456655

BEIS. (2019). *Digest of United Kingdom Energy Statistics 2018*. Available at: https://assets.publishing.service.gov.uk/government/uploads/system/uploads/attachment_data/file/736148/DUKES_2018.pdf

Bellia, Pedace, & Barbato. (2013). Lighting in educational environments: An example of a complete analysis of the effects of daylight and electric light on occupants. *Building and Environment, 68*, 50-65.

Bere, J. (2013) *Introduction to passive house*. RIBA Publishing.

Berenguer, M. J., Guardino, X., Hernández, A., Martí, M. C., Nogareda, C., & Solé, M. D. (1994). *El síndrome del edificio enfermo. Guía para su evaluación*. Instituto Nacional de Seguridad e Higiene en el Trabajo.

Berglund, B., Clausen, G., Kettrup, A., Lindvall, T., & Maroni, M. (1997). *Total Volatile Organic Compounds (TVOC) in Indoor Air Quality Investigation*. European Communities. doi:10.1111/j.1600-0668.1997.00002.x

Biondi, S. (2012, November 2). *Sustainable architecture*. Retrieved from https://www.youtube.com/watch?v=wxIOK1V1joc&ab_channel=PUCPPUCP

Blaabjerg, F., Teodorescu, T., Liserre, M., & Timbus, A. V. (2006). Overview of control and grid synchronization for distributed power generation systems. *IEEE Transactions on Industrial Electronics*, *53*(5), 1398–1409. doi:10.1109/TIE.2006.881997

Boehm, H. P., Setton, R., & Stumpp, E. (1994). Nomenclature and terminology of graphite intercalation compounds. *Pure and Applied Chemistry*, *66*(9), 1893–1901. doi:10.1351/pac199466091893

Boemi, S. N., & Tziogas, C. (2016). Indicators for Buildings' Energy Performance. In S. N. Boemi, O. Irulegi, & M. Santamouris (Eds.), *Energy Performance of Buildings* (pp. 79–93). Springer. doi:10.1007/978-3-319-20831-2_5

BOPV. (2018). Orden de 28 de noviembre de 2018, del Consejero de Medio Ambiente, Planificación Territorial y Vivienda, por la que se regulan y se convocan para el ejercicio 2019, las ayudas para la promoción y difusión del diseño urbanístico y la arquitectura contemporánea de la Comunidad Autónoma del País Vasco. *Boletín oficial del País Vasco, 236*. Accessed April 18, 2020. https://www.euskadi.eus/y22-bopv/es/bopv2/datos/2018/12/1806029a.shtml

Borsboom, W., De Gids, W., & Logue, J. (2016). *Technical note AIVC 68 residential ventilation and health*. INIVE.

Bouchama, A., & Knochel, J. P. (2002). Heat Stroke. *The New England Journal of Medicine*, *346*(25), 1978–1988. doi:10.1056/NEJMra011089 PMID:12075060

Boussaba, L., Lefebvre, G., Makhlouf, S., Grados, A., & Royon, L. (2021). Investigation and properties of a novel composite bio-PCM to reduce summer energy consumptions in buildings of hot and dry climates. *Solar Energy*, *214*, 119–130. doi:10.1016/j.solener.2020.11.060

Boyce, P. (2003). *Human Factors in Lighting* (2nd ed.). CRC Press. doi:10.1201/9780203426340

Bragazzi, N. L., & Del Puente, G. (2014). A proposal for including nomophobia in the new DSM-V. *Psychology Research and Behavior Management*, *7*, 155–160. doi:10.2147/PRBM.S41386 PMID:24876797

Brager, G. S., & de Dear, R. J. (1998). Thermal adaptation in the built environment: A literature review. *Energy and Building*, *27*(1), 83–96. doi:10.1016/S0378-7788(97)00053-4

BREEAM. (2011). *Manual BREEAM is housing*. Author.

Bribián, I. Z., Usón, A. A., & Scarpellini, S. (2009). Life cycle assessment in buildings: State-of-the-art and simplified LCA methodology as a complement for building certification. *Building and Environment*, *44*(12), 2510–2520. doi:10.1016/j.buildenv.2009.05.001

Brockway, P. E., Owen, A., Brand-Correa, L. I., & Hardt, L. (2019). Estimation of global final-stage energy-return-on-investment for fossil fuels with comparison to renewable energy sources. *Nature Energy*, *4*(7), 612–621. doi:10.103841560-019-0425-z

Brown, S. K., Sim, M. R., Abramson, M. J., & Gray, C. N. (1994). Concentrations of volatile organic compounds in indoor air–a review. *Indoor Air*, *4*(2), 123–134. doi:10.1111/j.1600-0668.1994.t01-2-00007.x

Bunclark, L. (2020, February 13). *They consider Peru to be at a medium-high level of water stress*. Retrieved from https://rpp.pe/peru/actualidad/consideran-al-peru-en-un-nivel-medio-alto-de-estres-de-agua-noticia-1245508

Buonomano, A., Forzano, C., Kalogirou, S. A., & Palombo, A. (2019). Building-façade integrated solar thermal collectors: Energy-economic performance and indoor comfort simulation model of a water based prototype for heating, cooling, and DHW production. *Renewable Energy*, *137*, 20–36. doi:10.1016/j.renene.2018.01.059

Bürger, V., Klinski, S., Lehr, U., Leprich, U., Nast, M., & Ragwitz, M. (2008). Policies to support renewable energies in the heat market. *Energy Policy*, *36*(8), 3150–3159. doi:10.1016/j.enpol.2008.04.018

Buso, T., Fabi, V., Andersen, R. K., & Corgnati, S. P. (2015). Occupant behaviour and robustness of building design. *Building and Environment*, *94*, 694–703. doi:10.1016/j.buildenv.2015.11.003

Byrnes, L., Brown, C., Foster, J., & Wagner, L. D. (2013). Australian renewable energy policy: Barriers and challenges. *Renewable Energy*, *60*, 711–721. doi:10.1016/j.renene.2013.06.024

Cabeza, L. F., Castell, A., Barreneche, C., de Gracia, A., & Fernández, A. I. (2011). Materials used as PCM in thermal energy storage in buildings: A review. *Renewable & Sustainable Energy Reviews*, *15*(3), 1675–1695. doi:10.1016/j.rser.2010.11.018

Cabeza, L. F., Castellón, C., Nogués, M., Medrano, M., Leppers, R., & Zubillaga, O. (2007). Use of microencapsulated PCM in concrete walls for energy savings. *Energy and Building*, *39*(2), 113–119. doi:10.1016/j.enbuild.2006.03.030

Cao, G., Awbi, H., Yao, R., Fan, Y., Sirén, K., Kosonen, R., & Zhang, J. (2014, March). A review of the performance of different ventilation and airflow distribution systems in buildings. *Building and Environment*, *73*, 171–186. Advance online publication. doi:10.1016/j.buildenv.2013.12.009

Cao, V. D., Pilehvara, S., Salas-Bringas, C., Szczotok, A. M., Buia, T. Q., Carmona, M., Rodriguez, J. F., & Kjøniksen, A. L. (2019). Thermal analysis of geopolymer concrete walls containing microencapsulated phase change materials for building applications. *Solar Energy*, *178*, 295–307. doi:10.1016/j.solener.2018.12.039

Cao, V. D., Pilehvar, S., Salas-Bringas, C., Szczotok, A. M., Rodriguez, J. F., Carmona, M., Al-Manasir, N., & Kjøniksen, A. L. (2017). Microencapsulated phase change materials for enhancing the thermal performance of Portland cement concrete and geopolymer concrete for passive building applications. *Energy Conversion and Management*, *133*, 56–66. doi:10.1016/j.enconman.2016.11.061

Carbajo, R., & Cabeza, L. F. (2018). Renewable energy research and technologies through responsible research and innovation looking glass: Reflexions, theoretical approaches and contemporary discourses. *Applied Energy*, *211*, 792–808. doi:10.1016/j.apenergy.2017.11.088

Carr-Cornish, S., Romanach, L., & Huddlestone-Holmes, C. (2019). An Application of Social Science to Inform the Stakeholder Engagement of an Emerging Geothermal Industry in Australia. In *Geothermal Energy and Society* (pp. 71–90). Springer. doi:10.1007/978-3-319-78286-7_6

Carrillo, G. S.-A. E., Lucio, R.L, Chong-Menard, B., & Smith, K. (2015, Sep 1). Improving Asthma in Hispanic Families Through a Home-Based Educational Intervention. *Pediatric Allergy Immunology and Pulmonology, 28*(3), 165-171. doi:10.1089/ped.2015.0523

Casini, M. (2015). Smart windows for energy efficiency of buildings. *International Journal of Civil and Structural Engineering*, *2*, 230–238.

Cayre, E., Allibe, B., Laurent, M., & Osso, D. (2011). There are people in the house! How the results of purely technical analysis of residential energy consumption are misleading for energy policies. *Proceedings of the ECEEE 2011 Summer Study on Energy Efficiency First: The Foundation of a Low-Carbon Society*, 1675–1683.

CCAyES. (2020). *Centro de Coordinación de Alertas y Emergencias Sanitarias del Ministerio de Sanidad*. https://www.mscbs.gob.es/profesionales/saludPublica/ccayes/home.htm

CE3X Software. (n.d.). https://certific.es/descargar-ce3x-version-2.1.html

Ceballos, M. A., Segura, P., Gutiérrez, E., Gracia, J. C., Ramos, P., Reaño, M., & García, B. (2018). *La calidad del aire en el Estado español durante 2018*. Ecologistas en Acción.

CEC. (2007). *California Energy Commission*. Energy Policy.

CEC. (2010). *Directive 2010/31/E.U. of the European Parliament and of the Council of 19 May 2010 on the energy performance of buildings, Official Journal of the European Union*. Commission of the European Communities. doi:10.1039/ap9842100196

Chandel, S. S., & Agarwal, T. (2017). Review of current state of research on energy storage, toxicity, health hazards and commercialization of phase changing materials. *Renewable & Sustainable Energy Reviews*, *67*, 581–596. doi:10.1016/j.rser.2016.09.070

Chapman, S. (2003). *Electric machinery and power system fundamentals*. McGraw Hill.

Chen, J., Zhang, Y., Zhang, M., Yao, B., Li, Y., Huang, L., Li, C., & Shi, G. (2016). Water-enhanced oxidation of graphite to graphene oxide with controlled species of oxygenated groups. *Chemical Science (Cambridge)*, *7*(3), 1874–1881. doi:10.1039/C5SC03828F PMID:29899910

Cheryan, S., Ziegler, S. A., Plaut, V. C., & Meltzoff, A. N. (2014). Designing Classrooms to Maximize Student Achievement. *Policy Insights from the Behavioral and Brain Sciences*, *1*(1), 4–12. doi:10.1177/2372732214548677

Chiang, Y.-J., & Lin, C.-C. (2013). Photocatalytic decolorization of methylene blue in aqueous solutions using coupled ZnO/SnO2 photocatalysts. *Powder Technology*, *246*, 137–143. doi:10.1016/j.powtec.2013.04.033

Choi. (2013-14). Indoor Environmental Quality of Classrooms and Student Outcomes: A Path Analysis Approach. *Journal of Learning Spaces*, *2*(2).

Choi, Y. R., Yoon, Y. G., Choi, K. S., Kang, J. H., Shim, Y. S., Kim, Y. H., Chang, H. J., Lee, J. H., Park, C. R., Kim, S. Y., & Jang, H. W. (2015). Role of oxygen functional groups in graphene oxide for reversible room-temperature NO2 sensing. *Carbon, 91*, 178–187. doi:10.1016/j.carbon.2015.04.082

Chow, T. T., & Li, C. (2013). Liquid-filled solar glazing design for buoyant water-flow. *Building and Environment, 60*, 45–55. doi:10.1016/j.buildenv.2012.11.010

Chow, T. T., Li, C., & Lin, Z. (2010). Thermal characteristics of water-flow double-pane window. *International Journal of Thermal Sciences, 50*(2), 140–148. doi:10.1016/j.ijthermalsci.2010.10.006

Chow, T. T., & Lyu, Y. (2017). Numerical analysis on the advantage of using PCM heat exchanger in liquid-flow window. *Applied Thermal Engineering, 125*, 1218–1227. doi:10.1016/j.applthermaleng.2017.07.098

Chua, C. K., & Pumera, M. (2014). Chemical reduction of graphene oxide: A synthetic chemistry viewpoint. *Chemical Society Reviews, 43*(1), 291–312. doi:10.1039/C3CS60303B PMID:24121318

Chwieduk, D. (2003). Towards sustainable-energy buildings. *Applied Energy, 76*(1-3), 211–217. doi:10.1016/S0306-2619(03)00059-X

CIBSE. (2011). *CIBSE KS17 Indoor Air Quality and Ventilation*. Available at: https://www.cibse.org/Knowledge/CIBSE-KS/KS17-Indoor-Air-Quality-Ventilation

CIBSE. (2015). CIBSE Guide A: Environmental Design (8th ed.). Norwich: CIBSE Publications.

CIE. (2019). *CIE Position Statement on Non-Visual Effects of Light. Recommending proper light at the proper time.* http://cie.co.at/publications/position-statement-non-visual-effects-light-recommending-proper-light-proper-time-2nd

Cludius, J., Forrest, S., & MacGill, I. (2014). Distributional effects of the Australian Renewable Energy Target (RET) through wholesale and retail electricity price impacts. *Energy Policy, 71*, 40–51. doi:10.1016/j.enpol.2014.04.008

Código Técnico de la edificación (CTE). (2018). *Documento Básico HS3: Calidad del aire interior*. Ministerio de Fomento del Gobierno de España.

Cohen, J. (1996). Procedure and Substance in Deliberative Democracy. In Democracy and Difference. Princeton University Press.

Cohen, J. (1989). Deliberation and Democratic Legitimacy. In A. Hamlin & P. Pettit (Eds.), *The Good Polity*. Blackwell.

Commission, E. (2015). *Air Quality Standards, Air Quality*. Available at: https://ec.europa.eu/environment/air/quality/standards.htm

Cooperative Energy Futures. (n.d.). *Energy Smart Homes*. Available at: https://www.cooperativeenergyfutures.com/energy-smart-homes

Crump, D., Dengel, A., & Swainson, M. (2009). *Indoor air quality in highly energy-efficient homes—a review, Report NF18*. Milton Keynes: National House.

Crump, D., Raw, G. J., Upton, S., Scivyer, C., Hunter, C., & Hartless, R. (2002). A protocol for the assessment of indoor air quality in homes and office buildings. London: BRE Bookshop.

Csoknyai, T., Legardeur, J., Abi Akle, A., & Horváth, M. (2019). Analysis of energy consumption profiles in residential buildings and impact assessment of a serious game on occupants' behavior. *Energy and Building, 196*, 1–20. doi:10.1016/j.enbuild.2019.05.009

Cuerda, E., Guerra-Santin, O., Sendra, J. J., & Neila, F. J. (2020). Understanding the performance gap in energy retrofitting: Measured input data for adjusting building simulation models. *Energy and Building*, *209*, 109688. doi:10.1016/j.enbuild.2019.109688

Curwell, S. R., March, C., & Venables, R. (1990). *Buildings and Health: The Rosehaugh Guide to the Design, Construction, Use and Management of Buildings*. RIBA Publ.

Cuttle, C. (2013). A New Direction for General Lighting Practice. *Lighting Research & Technology*, *45*(1), 22–39. doi:10.1177/1477153512469201

Da Cunha, J. P., & Eames, P. (2016). Thermal energy storage for low and medium temperature applications using phase change materials – A review. *Applied Energy*, *177*, 227–238. doi:10.1016/j.apenergy.2016.05.097

Dahy, H., & Knippers, J. (2013). Influence of the three Rs on Modern Approaches in Sustainable Architecture. *The International Journal of Environmental Sustainability*, *3*(4), 43–53. doi:10.18848/2325-1077/CGP/v08i04/55063

Dai, H., Huang, Y., & Huang, H. (2017). Eco-friendly polyvinyl alcohol/carboxymethyl cellulose hydrogels reinforced with graphene oxide and bentonite for enhanced adsorption of methylene blue. *Carbohydrate Polymers*, *185*, 1–11. doi:10.1016/j.carbpol.2017.12.073 PMID:29421044

Dai, Q., Cai, W., Shi, W., Hao, L., & Wei, M. (2017). A proposed lighting-design space: Circadian effect versus visual Illuminance. *Building and Environment*, *122*, 287–293. doi:10.1016/j.buildenv.2017.06.025

Dai, Q., Huang, Y., Hao, L., Lin, Y., & Chen, K. (2018). Luoxi Haoa, Yi Lin, Kaixuan Chen. Spatial and spectral illumination design for energy-efficient circadian lighting. *Building and Environment*, *146*, 216–225. doi:10.1016/j.buildenv.2018.10.004

Daisey, J. M., Angell, W. J., & Apte, M. G. (2003, March). Indoor air quality, ventilation and health symptoms in schools: An analysis of existing information. *Indoor Air*, *13*(1), 53–64. Advance online publication. doi:10.1034/j.1600-0668.2003.00153.x PMID:12608926

Daniel Hart. (2011). *Power Electronics*. McGraw Hill.

Darwish, Al Awadhi, & Bin Amer. (2010). Combining the nuclear power plant steam cycle with gas turbines. *Energy*, *35*(12), 4562-4571.

David, L. (2011). A Procedure for Determining Target Illuminances, LEUKOS. *The Journal of the Illuminating Engineering Society of North America*, *7*(3), 145–158. doi:10.1582/LEUKOS.2011.07.03001

Davidson, L., & Olsson, E. (1987). Calculation of age and local purging flow rate in rooms. *Building and Environment*, *22*(2), 111–127. doi:10.1016/0360-1323(87)90031-X

Davies, I., & Harvey, V. (2008). *Zero carbon: what does it mean to homeowners and housebuilders?* IHS BRE Press.

de la Flor, F. J. S., Domínguez, S. Á., Félix, J. L. M., & Falcón, R. G. (2008). Climatic zoning and its application to Spanish building energy performance regulations. *Energy and Building*, *40*(10), 1984–1990. doi:10.1016/j.enbuild.2008.05.006

de Poza-Vilches, M. F., Gutiérrez-Pérez, J., & López-Alcarria, A. (2019). Participation and Sustainable Development. In W. Leal Filho (Ed.), *Encyclopedia of Sustainability in Higher Education*. Springer., doi:10.1007/978-3-319-63951-2_57-1

De Rosa, L., & Castro, R. (2020). Forecasting and assessment of the 2030 australian electricity mix paths towards energy transition. *Energy*, *205*, 118020. doi:10.1016/j.energy.2020.118020

Delegado, R. (UE) n ° 874/2012 de la Comisión, de 12 de julio de 2012, por el que se complementa la Directiva 2010/30/UE del Parlamento Europeo y del Consejo en lo relativo al etiquetado energético de las lámparas eléctricas y las luminarias Texto pertinente a efectos del EEE (accessed on 15 April 2020) https://www.boe.es/doue/2012/258/L00001-00020.pdf

Delia, A. (2012). Total Volatile Organic Compounds (Total VOCs) in the air. *State of Knowledge Report*, (March), 1–4. Available at: http://www.environment.gov.au/atmosphere/airquality/publications/sok/vocs.html

DGNB System ES. (2020). *New Construction Criteria Version 2020*. Author.

Dimitroulopoulou, C. (2012). Ventilation in European dwellings: A review. Building and Environment, 47, 109–125. doi:10.1016/j.buildenv.2011.07.016

DIRECTIVA. 2008/50/CE del Parlamento Europeo y del Consejo. (2008). *Relativa a la calidad del aire ambiente y a una atmósfera más limpia en Europa*. https://www.boe.es/doue/2008/152/L00001-00044.pdf

Divina, F., Torres, J. F., García-Torres, M., Martínez-Álvarez, F., & Troncoso, A. (2020). Hybridizing Deep Learning and Neuroevolution: Application to the Spanish Short-Term Electric Energy Consumption Forecasting. *Applied Sciences (Basel, Switzerland)*, *10*(16), 5487. doi:10.3390/app10165487

Do, D. C., Zhao, Y., & Gao, P. (2016). Cockroach allergen exposure and risk of asthma. *Allergy*, *71*(4), 463–474. doi:10.1111/all.12827 PMID:26706467

Domínguez-Amarillo, S., Fernández-Agüera, J., Campano, M. Á., Acosta, I., Domínguez-Amarillo, S., Fernández-Agüera, J., ... Acosta, I. (2019). Effect of Airtightness on Thermal Loads in Legacy Low-Income Housing. *Energies*, *12*(9), 1677. doi:10.3390/en12091677

Domínguez-Amarillo, S., Fernández-Agüera, J., Peacock, A., & Acosta, I. (2019). Energy related practices in Mediterranean low-income housing. *Building Research and Information*. Advance online publication. doi:10.1080/09613218.2019.1661764

Domínguez-Amarillo, S., Fernández-Agüera, J., Sendra, J. J., & Roaf, S. (2018). Rethinking User Behaviour Comfort Patterns in the South of Spain—What Users Really Do. *Sustainability*, *10*(12), 4448. doi:10.3390u10124448

Domínguez-Amarillo, S., Sendra, J. J., Fernández-Agüera, J., & Escandón, R. (2017). *La construcción de la vivienda social en Sevilla y su catalogación 1939-1979* (U. de Sevilla, Ed.).

Domínguez-Amarillo, S., Sendra, J. J., & Oteiza San José, I. (2016). *La envolvente térmica de la vivienda social: el caso de Sevilla, 1939 a 1979 Title* (1st ed.). Editorial Consejo Superior de Investigaciones Científicas.

Du, H., & Dai Pang, S. (2018). Dispersion and stability of graphene nanoplatelet in water and its influence on cement composites. *Construction & Building Materials*, *167*, 403–413. doi:10.1016/j.conbuildmat.2018.02.046

Du, H., Gao, H. J., & Dai Pang, S. (2016). Improvement in concrete resistance against water and chloride ingress by adding graphene nanoplatelet. *Cement and Concrete Research*, *83*, 114–123. doi:10.1016/j.cemconres.2016.02.005

Du, K., Calautit, J., Wang, Z., Wu, Y., & Liu, H. (2018). A review of the applications of phase change materials in cooling, heating and power generation in different temperature ranges. *Applied Energy*, *220*, 242–273. doi:10.1016/j.apenergy.2018.03.005

ECA. (1992). *European collaborative action Indoor air quality & its impact on man: Report No.11: Guidelines for ventilation requirements in buildings*. Available at: https://www.inive.org/medias/ECA/ECA_Report11.pdf

EDGE. (n.d.a). *EDGE at Peru*. Retrieved June 2021, from https://edgebuildings.com/project-studies/?lang=es&_sft_project_countries=peru-es

EDGE. (n.d.b). *EDGE software.* Retrieved from https://edgebuildings.com/software/?lang=es

Edwards, L., & Torcellini, P. (2002). Literature Review of the Effects of Natural Light on Building Occupants. *Lit. Rev. Eff. Nat. Light Build. Occup., 55.* Advance online publication. doi:10.2172/15000841

Edwards, R., Larivé, J. F., Rickeard, D., & Weindorf, W. (2014). *Well-To-Tank Report Version 4.a. Well-to-Wheels Analysis of Future Automotive Fuels and Powertrains in the European Context.* Publications Office of the European Union.

El-Bidairi, K. S., Nguyen, H. D., Mahmoud, T. S., Jayasinghe, S., & Guerrero, J. M. (2020). Optimal sizing of Battery Energy Storage Systems for dynamic frequency control in an islanded microgrid: A case study of Flinders Island, Australia. *Energy, 195*, 117059. doi:10.1016/j.energy.2020.117059

Ellabban, O., Abu-Rub, H., & Blaabjerg, F. (2014). Renewable energy resources: Current status, future prospects and their enabling technology. *Renewable & Sustainable Energy Reviews, 39*, 748–764. doi:10.1016/j.rser.2014.07.113

EN 12464-1 2019: Lighting of workplaces – Indoor workplaces. https://www.valosto.com/tiedostot/prEN%2012464-1.pdf

EN 410. (2011). *Glass in Building—Determination of Luminous and Solar Characteristics of Glazing.* German Institute for Standardization.

EN 673. (2011). *Glass in Building—Determination of Thermal Transmittance (U Value)—Calculation Method.* German Institute for Standardization.

ENEDI. (2020). *Energética en la Edificación.* Accessed April 18, 2020. https://www.ehu.eus/es/web/enedi/grupo-enedi

Enezi, J., Revell, V., Brown, T., Wynne, J., Schlangen, L., & Lucas, R. A. (2011). "Melanopic" spectral efficiency function predicts the sensitivity of melanopsin photoreceptors to polychromatic lights. *Journal of Biological Rhythms, 26*(4), 314–323. doi:10.1177/0748730411409719 PMID:21775290

ENR. (2019). *Estadísticas sobre residencias. Distribución de centros y plazas residenciales por provincia.* ENR.

EPA. (2018). *U.S. Greenhouse Gas Emissions and Sinks, 1990-2016, Epa 430-R-18-003.* Doi: EPA 430-R-12-001.

Erdem, S., & Gürbüz, E. (2019). Influence of microencapsulated phase change materials on the flexural behavior and micromechanical impact damage of hybrid fibre reinforced engineered cementitious composites. *Composites. Part B, Engineering, 166*, 633–644. doi:10.1016/j.compositesb.2019.02.059

Escofet, J., & Bará, S. (2015). Reducing the circadian input from self-luminous devices using hardware filters and software applications. *Lighting Research & Technology, 49*(4), 481–496. doi:10.1177/1477153515621946

European Collaborative Action. (1992). *Indoor Air Quality and its Impact on Man. Report No.11: Guidelines for Ventilation Requirements in Buildings.* Office for Publications of EU.

European Union. (2018). *Directive (EU) 2018/844 of the European Parliament and of the Council of 30 May 2018. Amending Directive 2010/31/EU on the Energy Performance of Buildings and Directive 2012/27/EU on Energy Efficiency* https://eur-lex.europa.eu/legal-content/EN/TXT/PDF/?uri=CELEX:32018L0844&from=EN

EUSTAT (Instituto Vasco de Estadística). (2020). Accessed April 18, 2020. https://www.eustat.eus/municipal/datos_estadisticos/donostia_san_sebastian_c.html

Fairweather, V., Hertig, E., & Traidl-Hoffmann, C. (2020). A brief introduction to climate change and health. *Allergy. European Journal of Allergy and Clinical Immunology, 75*(9), 2352–2354. doi:10.1111/all.14511 PMID:32779793

Fanger, P. O. (2006). What is IAQ? *Indoor Air, 16*(5), 328–334. doi:10.1111/j.1600-0668.2006.00437.x PMID:16948709

Fang, L., Clausen, G., & Fanger, P. O. (1998). Impact of temperature and humidity on the perception of indoor air quality. *Indoor Air*, *8*(2), 80–90. doi:10.1111/j.1600-0668.1998.t01-2-00003.x

Faraj, K., Khaled, M., Faraj, J., Hachem, F., & Castelain, C. (2020). Phase change material thermal energy storage systems for cooling applications in buildings: A review. *Renewable & Sustainable Energy Reviews*, *119*, 79–97. doi:10.1016/j.rser.2019.109579

Farhad, K., Shamsodin, T., Hamed, T., & Edris, P. (2016). Integration of electric vehicles into a smart power grid: A technical review. *IEEE Electrical Power and Energy Conference (EPEC).*

Fathi & Aashoor. (2015). *Maximum power point tracking techniques for photovoltaic water pumping system* (Ph.D. dissertation). Dept. Elect. Elect. Eng., Univ. of Bath, Claverton Down, UK.

Fei, F., Zhou, S., Mai, J. D., & Li, W. J. (2014). Development of an indoor airflow energy harvesting system for building environment monitoring. *Energies*, *7*(5), 2985–3003. doi:10.3390/en7052985

Feijó-Muñoz, J., Pardal, C., Echarri, V., Fernández-Agüera, J., Assiego de Larriva, R., Montesdeoca Calderín, M., Poza-Casado, I., Padilla-Marcos, M. Á., & Meiss, A. (2019). Energy impact of the air infiltration in residential buildings in the Mediterranean area of Spain and the Canary islands. *Energy and Building*, *188–189*, 226–238. doi:10.1016/j.enbuild.2019.02.023

Feist, W., Schnieders, J., Dorer, V. & Haas, A. (2005). Re-inventing air heating: Convenient and comfortable within the frame of the Passive House concept. *Energy and Buildings*, *37*(11), 1186–1203. . doi:10.1016/j.enbuild.2005.06.020

Feng, T., Xie, D., Lin, Y., Zang, Y., Ren, T., Song, R., Zhao, H., Tian, H., Li, X., Zhu, H., & Liu, L. (2011). Graphene based Schottky junction solar cells on patterned silicon-pillar-array substrate. *Applied Physics Letters*, *99*(23), 233505. doi:10.1063/1.3665404

Ferguson, A. C., & Yates, C. (2016). Federal Enactment of Healthy Homes Legislation in the United States to Improve Public Health. *Frontiers in Public Health*, *4*, 48–48. doi:10.3389/fpubh.2016.00048 PMID:27047913

Fernández, A. (n.d.). *Wastewater in Peru, Problems and use in agriculture*. Retrieved from https://www.ais.unwater.org/ais/pluginfile.php/356/mod_page/content/128/Peru%20INFORME%20DE%20PAIS.pdf

Fernández, M. (2008). Sistema integral de ventilación de viviendas de acuerdo con el Código Técnico de la Edificación (HS3). *Instalaciones y técnicas del confort*, (197), 10-21.

Fernández-Agüera, J., Sendra, J. J., & Domínguez, S. (2011). Protocols for measuring the airtightness of multi-dwelling units in Southern Europe. In Procedia Engineering (Vol. 21). doi:10.1016/j.proeng.2011.11.1992

Fernández-Agüera, J., Domínguez-Amarillo, S., Alonso, C., & Martín-Consuegra, F. (2019). Thermal comfort and indoor air quality in low-income housing in Spain: The influence of airtightness and occupant behaviour. *Energy and Building*, *199*, 102–114. doi:10.1016/j.enbuild.2019.06.052

Fernández-Agüera, J., Dominguez-Amarillo, S., Fornaciari, M., & Orlandi, F. (2019). TVOCs and PM 2.5 in Naturally Ventilated Homes: Three Case Studies in a Mild Climate. *Sustainability*, *11*(22), 6225. doi:10.3390u11226225

Fernández-Agüera, J., Domínguez-Amarillo, S., Sendra, J. J., & Suárez, R. (2016). An approach to modelling envelope airtightness in multi-family social housing in Mediterranean Europe based on the situation in Spain. *Energy and Building*, *128*, 236–253. doi:10.1016/j.enbuild.2016.06.074

Fernández-Agüera, J., Domínguez-Amarillo, S., Sendra, J. J., Suárez, R., & Oteiza, I. (2019). Social housing airtightness in Southern Europe. *Energy and Building*, *183*, 377–391. doi:10.1016/j.enbuild.2018.10.041

Fernández-García, J., Dosil Díaz, O., Taboada Hidalgo, J., Fernández, J., & Sánchez-Santos, L. (2014). Influencia del clima en el infarto de miocardio en Galicia. *Medicina Clínica, 145*(3), 97–101.

Fernández-González, R., Suárez-García, A., Álvarez-Feijoo, M. A., Arce, E., & Díez-Mediavilla, M. (2020). Spanish Photovoltaic Solar Energy: Institutional Change, Financial Effects, and the Business Sector. *Sustainability, 12*(5), 1892. doi:10.3390u12051892

Figueiro, Steverson Heerwagen, Yucel, Roohan, Sahin, Kampschroer, & Rea. (2019). Light, entrainment and alertness: A case study in offices. *Lighting Research and Technology.*

Figueiro. (2013). An Overview of the Effects of Light on Human Circadian Rhythms: Implications for New Light Sources and Lighting Systems Design. *Journal of Light & Vision Environment, 37*(2-3).

Foley, A., & Olabi, A. G. (2017). *Renewable energy technology developments, trends and policy implications that can underpin the drive for global climate change.* Elsevier. doi:10.1016/j.rser.2016.12.065

Fondo MiviviendaS. A. (2021, June 16). *Build to Grow*. Retrieved from http://www.construyeparacrecer.com/Inicio.aspx

Frattolillo, A., Canale, L., Ficco, G., Mastino, C. C., & Dell'Isola, M. (2020). Potential for Building Façade-Integrated Solar Thermal Collectors in a Highly Urbanized Context. *Energies, 13*(21), 5801. doi:10.3390/en13215801

Galus, M. D., Zima, M., & Andersson, G. (2010). On integration of plug-in hybrid electric vehicles into existing power system structures. *Energy Policy, 38*(11), 6736–6745. doi:10.1016/j.enpol.2010.06.043

Gang, W., Wang, S., Shan, K., & Gao, D. (2015). Impacts of cooling load calculation uncertainties on the design optimization of building cooling systems. *Energy and Building, 94*, 1–9. doi:10.1016/j.enbuild.2015.02.032

Gan, J., & Smith, C. (2011). Drivers for renewable energy: A comparison among OECD countries. *Biomass and Bioenergy, 35*(11), 4497–4503. doi:10.1016/j.biombioe.2011.03.022

Ganuza. (2010). *La democracia en acción, una visión desde las metodologías participativas*. Documentos OIDP. Madrid: Antígona, procesos participativos

Gao, Y., He, F., Meng, X., Wang, Z., Zhang, M., Yu, H., & Gao, W. (2020). Thermal behavior analysis of hollow bricks filled with phase-change material (PCM). *Journal of Building Engineering, 31*, 1–11. doi:10.1016/j.jobe.2020.101447

Gassar, A. A. A., Yun, G. Y., & Kim, S. (2019). Data-driven approach to prediction of residential energy consumption at urban scales in London. *Energy, 187*, 115973. doi:10.1016/j.energy.2019.115973

Ghosh, A., Norton, B., & Duffy, A. (2016). Behaviour of a SPD switchable glazing in an outdoor test cell with heat removal under varying weather conditions. *Applied Energy, 180*, 695–706. doi:10.1016/j.apenergy.2016.08.029

Ghosh, S., Harish, S., Rocky, K. A., Ohtaki, M., & Saha, B. B. (2019). Graphene enhanced thermoelectric properties of cement-based composites for building energy harvesting. *Energy and Building, 202*, 109419. doi:10.1016/j.enbuild.2019.109419

Gianfrate, V., Piccardo, C., Longo, D., & Giachetta, A. (2017). Rethinking social housing: Behavioural patterns and technological innovations. *Sustainable Cities and Society, 33*(May), 102–112. doi:10.1016/j.scs.2017.05.015

Gil-Lopez, T., & Gimenez-Molina, C. (2013). Influence of double glazing with a circulating water chamber on the thermal energy savings in buildings. *Energy and Building, 56*, 56–65.

Giusti, L., & Almoosawi, M. (2017). Impact of building characteristics and occupants' behaviour on the electricity consumption of households in Abu Dhabi (UAE). *Energy and Building, 151*, 534–547. doi:10.1016/j.enbuild.2017.07.019

Glickman, G., Levin, R., & Brainard, G. C. (2002). Ocular input for human melatonin regulation: Relevance to breast cancer. *Neuroendocrinology Letters*, *23*(SUPPL. 2), 17–22. PMID:12163843

Global Footprint Network. (2018). *Ecological footprint per person.* http://data.footprintnetwork.org/#/

Gómez, M. A., Feijoo, M. A. A., Comesaña, R., & Eguía, P. (2012). CFD Simulation of a Concrete Cubicle to Analyze the Thermal Effect of Phase Change Materials in Buildings. *Energies*, *5*(7), 2093–2111. doi:10.3390/en5072093

González-Lezcano, R. A., & Hormigos-Jiménez, S. (2016). Energy saving due to natural ventilation in housing blocks in Madrid. In *IOP Conference Series: Materials Science and Engineering* (Vol. 138). Institute of Physics Publishing. doi:10.1088/1757-899X/138/1/012002

Grainger, J., & Stevenson, W. D. (2008). *Power systems analysis*. McGraw Hill.

Gram-hanssen, K. (2010). Residential heat comfort practices: Understanding users. *Building Research and Information*, *38*(5), 175–186. doi:10.1080/09613210903541527

Group, H. M. (1999). *Daylighting in Schools: An Investigation into Relationship between Daylighting and Human Performance; Detailed Report; Pacific Gas and Electric Company on behalf of the California Board for Energy Efficiency Third Party Program.* Fair Oaks.

Guan & Yan. (2016). Daylighting Design in Classroom Based on Yearly-Graphic Analysis. *Sustainability, 8*, 604. doi:10.3390u8070604

Guardino, X. (2012). *Calidad del Aire Interior. Instituto Nacional de Seguridad e Higiene en el Trabajo.* INSHT.

Guille, C., & Gross, G. (2009). A conceptual framework for the vehicle-to-grid (V2G) implementation. *Energy Policy*, *37*(11), 4379–4390. doi:10.1016/j.enpol.2009.05.053

Guo, H. (2011). Source apportionment of volatile organic compounds in Hong Kong homes. *Building and Environment*, *46*(11), 2280–2286. https://doi.org/10.1016/j.buildenv.2011.05.008

Gürbüz, E. (2018). *Investigation of Mechanical and Impact Behavior of Engineered Cement-Based Composites Incorporating MicroPCM* (Master Thesis). Istanbul University, Turkey.

Gürbüz, E., & Erdem, S. (2020). Development and thermo-mechanical analysis of high-performance hybrid fibre engineered cementitious composites with microencapsulated phase change materials. *Construction & Building Materials*, *263*, 139–152. doi:10.1016/j.conbuildmat.2020.120139

Gutai, M., & Kheybari, A. G. (2020). Energy consumption of water-filled glass (WFG) hybrid building envelope. *Energy and Building*, *218*, 110050.

Gutai, M., & Kheybari, A. G. (2021). Energy consumption of hybrid smart water-filled glass (SWFG) building envelope. *Energy and Building*, *230*, 110508.

Haarhoff, J., & Mathews, E. H. (2006). A Monte Carlo method for thermal building simulation. *Energy and Building*, *38*(12), 1395–1399. doi:10.1016/j.enbuild.2006.01.009

Habraken, N. J. & Mignucci, A. (2009). *Soportes: vivienda y ciudad* [Supports: Housing and City]. Barcelona: Máster Laboratorio de la Vivienda del Siglo XXI.

Habraken, N. J. (1979). El diseño de soportes. Barcelona: Gustavo Gili.

Haghighat, F., & De Bellis, L. (1998). Material emission rates: Literature review, and the impact of indoor air temperature and relative humidity. *Building and Environment*, *33*(5), 261–277. doi:10.1016/S0360-1323(97)00060-7

Hankins & Lucas. (2002). The primary visual pathway in humans is regulated according to long-term light exposure through the action of a nonclassical photopigment. *Curr. Biol., 12*, 191-198.

Hansen, A. R., Gram-Hanssen, K., & Knudsen, H. N. (2018). How building design and technologies influence heat-related habits. *Building Research and Information, 46*(1), 83–98. doi:10.1080/09613218.2017.1335477

Harrison, R. M. (2010). Size distribution of airborne particles controls outcome of epidemiological studies. *Science of the Total Environment. Elsevier B., 409*(2), 289–293. doi:10.1016/j.scitotenv.2010.09.04 PMID:21109288

Hattan, H. A., Madhkhan, M., & Marani, A. (2021). Thermal and mechanical properties of building external walls plastered with cement mortar incorporating shape-stabilized phase change materials (SSPCMs). *Construction & Building Materials, 270*, 1–13. doi:10.1016/j.conbuildmat.2020.121385

Hauptman, M., Bruccoleri, R., & Woolf, A. D. (2017). An Update on Childhood Lead Poisoning. *Clinical Pediatric Emergency Medicine, 18*(3), 181–192. doi:10.1016/j.cpem.2017.07.010 PMID:29056870

Health and Safety Executive. (2011). *EH4O / 2005 Workplace exposure limits EH4O / 2005 Workplace exposure limits* (2nd ed.). HSE Books.

Health and Urban Department. (n.d.). *The Healthy Homes Program*. Retrieved March 26, 2020, from https://www.hud.gov/program_offices/healthy_homes/hhi

Health Canadá. (2005). https://www.canada.ca/en/health-canada/services/health-care-system/reports-publications/canada-health-act-annual-reports/annual-report-2004-2005.html

Healy, J. D. (2017). *Housing, Fuel Poverty and Health : A Pan-European Analysis*. Routledge.

Heffernan, E., Pan, W., Liang, X. & Wilde, P. D. (2016). Zero carbon homes : Perceptions from the U.K. construction industry. *Energy Policy, 79*(2015), 23–36. doi:10.1016/j.enpol.2015.01.005

He, H., Reynolds, C. J., Li, L., & Boland, J. (2019). Assessing net energy consumption of Australian economy from 2004–05 to 2014–15: Environmentally-extended input-output analysis, structural decomposition analysis, and linkage analysis. *Applied Energy, 240*, 766–777. doi:10.1016/j.apenergy.2019.02.081

Hein, L., Metzger, M. J., & Moreno, A. (2009). Potential impacts of climate change on tourism; a case study for Spain. *Current Opinion in Environmental Sustainability, 1*(2), 170–178. doi:10.1016/j.cosust.2009.10.011

Hekimoğlu, G., Nas, M., Ouikhalfan, M., Sarı, A., Tyagi, V. V., Sharma, R. K., Kurbetci, Ş., & Saleh, T. A. (2021). Silica fume/capric acid-stearic acid PCM included-cementitious composite for thermal controlling of buildings: Thermal energy storage and mechanical properties. *Energy, 219*, 1–10. doi:10.1016/j.energy.2020.119588

Hennig, C., Meila, M., Murtagh, F., & Rocci, R. (2015). *Handbook of Cluster Analysis*. CRC Press. doi:10.1201/b19706

Hens, H., Parijs, W., & Deurinck, M. (2010). Energy consumption for heating and rebound effects. *Energy and Building, 42*(1), 105–110. doi:10.1016/j.enbuild.2009.07.017

Hernandez Sampieri, R., Fernandez Collado, C., & Baptista Lucio, M. del P. (2014). *Metodología de la investigación. Metodología de la investigación*. Retrieved from https://www.casadellibro.com/libro-metodologia-de-la-investigacion-5-ed-incluye-cd-rom/9786071502919/1960006

Herrera-Gomez, S.S., Quevedo-Nolasco, A., & Pérez-Urrestarazu, L. (2017). The role of green roofs in climate change mitigation. A case study in Seville (Spain). *Building and Environment, 123*, 575-584.

Heschong Mahone Group. (1999). *Daylighting in schools: An investigation into the relationship between daylighting and human performance*. Author.

Ho. (2007). Optimal sun-shading design for enhanced daylight illumination of subtropical classrooms. *Energy and Buildings.*

Holgate, S. (2020). *The inside story: Health effects of indoor air quality on children and young people.* doi:10.1039/9781839160431-00151

Holt, E. A., & Wiser, R. H. (2007). *The Treatment of Renewable Energy Certificates, EmissionsAllowances, and Green Power Programs in State Renewables.* PortfolioStandards.

Hopfe, C. J., & McLeod, R. S. (2015). *The PassivHaus designer's manual: A technical guide to low and zero energy buildings* (C. J. Hopfe & R. S. McLeod, Eds.; 1st ed.). Routledge. doi:10.4324/9781315726434

Hormigos-Jimenez, S., Padilla-Marcos, M. A., Meiss, A., Gonzalez-Lezcano, R. A., & Feijó-MuÑoz, J. (2018a). Experimental validation of the age-of-the-air CFD analysis: A case study. *Science and Technology for the Built Environment, 24*(9), 994–1003. doi:10.1080/23744731.2018.1444885

Hormigos-Jimenez, S., Padilla-Marcos, M. Á., Meiss, A., Gonzalez-Lezcano, R. A., & Feijó-Muñoz, J. (2018b). Computational fluid dynamics evaluation of the furniture arrangement for ventilation efficiency. *Building Services Engineering Research and Technology, 39*(5), 557–571. https://doi.org/10.1177/0143624418759783

Hosham, S., Khalid, F., & Raheel, J. (2018). Enhancement of Performance for Steam Turbine in Thermal Power Plants Using Artificial Neural Network and Electric Circuit Design. *Applied Computational Intelligence and Soft Computing,* 1–9. Advance online publication. doi:10.1155/2018/8042498

Hosseinzadeh, H., & Ramin, S. (2018). Effective removal of copper from aqueous solutions by modified magnetic chitosan/graphene oxide nanocomposites. *International Journal of Biological Macromolecules, 113,* 859–868. doi:10.1016/j.ijbiomac.2018.03.028 PMID:29524485

HQE. (2014). *HQE™ Scheme Environmental Performance Residential Building.* HQE.

Huang, T., Yan, M., He, K., Huang, Z., Zeng, G., Chen, A., Peng, M., Li, H., Yuan, L., & Chen, G. (2019). Removal of methylene blue dye by solvothermally reduced graphene oxide: A metal-free adsorption and photodegradation method. *Royal Society of Chemistry., 543,* 43–51.

Huang, X., Xiaoying, Q., Boey, F., & Zhang, H. (2012). Graphene based composites. *Chemical Society Reviews, 41*(2), 666–686. doi:10.1039/C1CS15078B PMID:21796314

Huang, X., Yin, Z., Wu, S., Qi, X., He, Q., Zhang, Q., Yan, Q., Boey, F., & Zhang, H. (2001). Graphene-Based Materials: Synthesis, Characterization, Properties, and Applications. *Small, 7*(14), 1876–1902. doi:10.1002mll.201002009 PMID:21630440

Hu, C., Lu, T., Chen, F., & Zhang, R. (2013). A brief review of graphene–metal oxide composites synthesis and applications in photocatalysis. *Journal of the Chinese Advanced Materials Society., 1*(1), 21–39. doi:10.1080/22243682.2013.771917

Hummers, W. S. Jr, & Offeman, R. E. (1958). Preparation of Graphene Oxide. *Journal of the American Chemical Society, 80*(6), 1339. doi:10.1021/ja01539a017

Humphreys, M., Nicol, F., & Roaf, S. (2016). *Adaptive Thermal Comfort: Foundations and Analysis* (1st ed.). Routledge.

Hunger, M., Entrop, A. G., Mandilaras, I., Brouwers, H. J. H., & Founti, M. (2009). The behavior of self-compacting concrete containing micro-encapsulated Phase Change Materials. *Cement and Concrete Composites, 31*(10), 731–743. doi:10.1016/j.cemconcomp.2009.08.002

Hunt, R. W. G., & Pointer, M. R. (2006). *Measuring Colour* (4th ed.). Wiley.

Hussain, A., & Khaled, H. (2020). Hybrid Technique for an Efficient PV System through Intelligent MPPT and Water Cooling Process. *International Journal of Power Electronics and Drive Systems, 11*(4), 1835-1843.

Hussain, A. (2017). Novel 9-Steps Automatic Voltage Regulator based on Two Step-Down Transformers. *Iranian Journal of Electrical and Computer Engineering, 7*(2), 576–583.

Hussain, A. (2019). Supplying DC Electricity to the Isolated Dwellings through MPP Tracked PV System Based on Artificial Neural Network. *6th International Conference of Control, Dynamic Systems, and Robotics (CDSR'19).*

Hussain, A. (2020). A New Intelligent Power Factor Corrector for Converter Applications. *6th International Conference on Control, Modeling and Computing (CMC 2020).*

Hussain, A. (2020). Artificial Neural Network Based Unity Power Factor Corrector for Single Phase DC-DC Converters. *Iranian Journal of Electrical and Computer Engineering, 10*(4), 4145–4154.

Hussain, A., & Ali, S. (2016). Novel Discrete Components Based Speed Controller for Induction Motor. *International Journal of Power Electronics and Drive Systems, 7*(4), 1075–1084.

Hussain, A., & Fernando, G. (2019). Stand-alone PV System with MPPT Function Based on Fuzzy Logic Control for Remote Building Applications. *International Journal of Power Electronics and Drive Systems, 10*(2), 842–851.

Hussain, A., Maissa, F., Mousa, M., Aaesha, A., Abdulrahman, S., Ahmad, C., Moath, A., Somaya, A., & Sumith, S. (2018). Novel Design of AURAK Hydrogen Car: Electrical Aspects. *The International Conference on Energy, Water & Environmental Sciences (ICEWES 2018).*

Hussain, A., Mousa, M., Basil, Q., Mohammed, A., Osman, A., & Zahidur, R. (2018). New Design and Implementation of AURAK Solar Car: Electrical Vision. *International Conference on Energy, Water & Environmental Sciences (ICEWES 2018).*

Hussain, A., Yousif, I., & Beza, N. (2014). *Design and Simulation of a High Performance Standalone Photovoltaic System. In Proceeding ICREGA'14.* Springer.

IDAE. (2009). *Condiciones de aceptación de Procedimientos alternativos a LIDER y CALENER.* Instituto para la Diversificación y Ahorro de la Energía.

IEA. (n.d.). http://www.ecbcs.org/

IES. (n.d.). *Tunable light could potentially improve student learning, increase teacher satisfaction and trim energy use. A recent pilot project explored these benefits.* https://www.ies.org/

IFC, EDGE. (2018). *EDGE User Guide _Version 2.1.* Author.

IFC. (2017-2018). *Terms of Reference-IFC. Cities Program Latin America and the Caribbean "Green Building and Municipalities-Peru".* IFC.

Illuminating Engineering Society of North America (IESNA). (n.d.). https://www.ies.org/standards/ies-standards-cross-reference/

INE. (2020). *Instituto Nacional de Estadística.* https://www.ine.es/covid/covid_inicio.htm

INEI. (2020). *Press Release: The population of Lima exceeds nine and a half million inhabitants.* Retrieved from http://m.inei.gob.pe/media/MenuRecursivo/noticias/notadeprensa006.pdf

Institute of Medicine. (2000). Clearing the Air: Asthma and Indoor Air Exposures. *Indoor Air.* doi: NBK224477

Instituto Nacional de Seguridad e Higiene en el Trabajo (INSHT). (2010). *Prevención de la exposición a formaldehído* (Vol. 830). Nota Técnica de Prevención.

International Well Building Institute. (n.d.). *WELL Building Standard v1*. Author.

Ioannou, A., Itard, L., & Agarwal, T. (2018). In-situ real time measurements of thermal comfort and comparison with the adaptive comfort theory in Dutch residential dwellings. *Energy and Building*, *170*, 229–241. doi:10.1016/j.enbuild.2018.04.006

IPCC. (2007). *Mitigation of climate change: Contribution of working group III to the fourth assessment report of the Intergovernmental Panel on Climate Change*. Intergovernmental Panel on Climate Change. https://www.ipcc.ch/publications_and_data/.htm

IPCC. (2013). *Climatic Change 2013*. Advance online publication. doi:10.1017/CBO9781107415324.Summary

IPCC. (2014). Climate Change 2014: Synthesis Report. Contribution of Working Groups I, II and III to the Fifth Assessment Report of the Intergovernmental Panel on Climate Change. IPCC.

ISO 13790. (2011). Thermal performance of buildings-calculation of energy use for space heating and cooling. ISO/DIS 13790:2008, International Organization for Standardization, Geneva.

ISO 8995-1 (2002): Lighting of indoor workplaces (CIE S008/2001)

Jacobs, D. E., Kelly, T., & Sobolewski, J. (2007). Linking Public Health, Housing, and Indoor Environmental Policy: Successes and Challenges at Local and Federal Agencies in the United States. *Environmental Health Perspectives*, *115*(6), 976–982. doi:10.1289/ehp.8990 PMID:17589610

James, M. (2012). *Smart Grid Fundamentals of Design and Analysis*. IEEE Press.

Javadi, F. S., Metselaar, H. S. C., & Ganesan, P. (2020). Performance improvement of solar thermal systems integrated with phase change materials (PCM), a review. *Solar Energy*, *206*, 330–352. doi:10.1016/j.solener.2020.05.106

Jiang, Kurihara, Torii, & Tsubota. (2018). Progress and Control of Myopia by Light Environments. *Eye Contact Lens, 44*(5), 273-278. . doi:10.1097/ICL.0000000000000548

Jie, W., Zheng, F., & Hao, J. (2013). Graphene/gallium arsenide-based Schottky junction solar cells. *Applied Physics Letters*, *103*(23), 233111. doi:10.1063/1.4839515

Jignesh, G. B. (2014). Smart Grid - An Evolutionary Version of Energy Infrastructure. *Smart World*, 32-35.

Ji, J., Luo, C., Chow, T. T., Sun, W., & He, W. (2011). Thermal characteristics of a building-integrated dual-function solar collector in water heating mode with natural circulation. *Energy*, *36*(1), 566–574. doi:10.1016/j.energy.2010.10.004

Jim, C. Y. (2015). Thermal performance of climber greenwalls: Effects of solar irradiance and orientation. *Applied Energy*, *154*, 631–643. doi:10.1016/j.apenergy.2015.05.077

Jotzo, F., & Betz, R. (2009). Australia's emissions trading scheme: Opportunities and obstacles for linking. *Climate Policy*, *9*(4), 402–414.

Jouhara, H., Żabnienska-Góra, A., Khordehgah, N., Ahmad, D., & Lipinski, T. (2020). Latent thermal energy storage technologies and applications: A review. *International Journal of Thermofluids*, *5-6*, 39–63. doi:10.1016/j.ijft.2020.100039

Jung, I., Dikin, D. A., Piner, R. D., & Ruoff, R. S. (2008). Tunable Electrical Conductivity of Individual Graphene Oxide Sheets Reduced at "Low" Temperatures. *Nano Letters*, *8*(12), 4283–4287. doi:10.1021/nl8019938 PMID:19367929

Kajtar, L. (2006). Influence of Carbon-dioxide Pollutant on Human Well-being and Work Intensity. In *H.B. 2006 - Healthy Buildings: Creating a Healthy Indoor Environment for People* (pp. 85–90). Available at https://www.scopus.com/inward/record.url?eid=2-s2.0-84871535246&partnerID=40&md5=0b0e1ad8a48325274153bf610d7761ea

Kalaiarasi, N., Paramasivam, S., & Sanchari, K. (2014). Control of Z-Source Inverter Based PV System with MPPT Using ANFIS. *International Review on Modelling and Simulations*, *7*(5), 797–806. doi:10.15866/iremos.v7i5.1970

Kalnæs, S. E., & Jelle, B. P. (2015). Phase change materials and products for building applications: A state-of-the-art review and future research opportunities. *Energy and Building*, *94*, 150–176. doi:10.1016/j.enbuild.2015.02.023

Karlsson, F., & Fahlén, P. (2008). Impact of design and thermal inertia on the energy saving potential of capacity controlled heat pump heating systems. *International Journal of Refrigeration*, *31*(6), 1094–1103. doi:10.1016/j.ijrefrig.2007.12.002

Karoly, V. (2002). Hybrid Combined Cycle Power Plant. *International Conference Nuclear Energy for New Europe 2002.*

Karunathilake, H., Hewage, K., Brinkerhoff, J., & Sadiq, R. (2019). Optimal renewable energy supply choices for net-zero ready buildings: A life cycle thinking approach under uncertainty. *Energy and Building*, *201*, 70–89. doi:10.1016/j.enbuild.2019.07.030

Katodrytis, G. (2015). Performative Urbanism: An Emerging Model of the Gulf. *Architectural Design: UAE and The Gulf Architecture and Urbanism Now,* 121–127.

Katsouyanni, K., Part, P., Andersson, I., Autrup, H., Baños de Guisasola, E., Bjerregaard, P., Mette, B., Bert, B., Calheiros, J., Casse, F. F., Castaño, A., Cochet, C. & Eisenreich, S. (2004) *Baseline Report on Needs in the framework of the European Environment and Helath Strategy ((COM 2003)338 final).* Academic Press.

Kawamura, H., Naka, K., Yonekura, N., Yamanaka, S., Kawamura, H., Ohno, H., & Naito, K. (2003). Simulation of I-V characteristics of a PV module with shaded PV cells. *Solar Energy Materials and Solar Cells*, *75*(3/4), 613–621. doi:10.1016/S0927-0248(02)00134-4

Keirstead, J., & Shah, N. (2011). Calculating minimum energy urban layouts with mathematical programming and Monte Carlo analysis techniques. *Computers, Environment and Urban Systems*, *35*(5), 368–377. doi:10.1016/j.compenvurbsys.2010.12.005

Kelly, S. (2011). *Do Homes that are More Energy Efficient Consume Less Energy?: A Structural Equation Model for England's Residential Sector*. EPRG Working Paper No. 1117. doi:10.17863/CAM.5540

Kelly, S., Crawford-brown, D., & Pollitt, M. G. (2012). Building performance evaluation and certification in the U.K.: Is SAP fit for purpose? *Renewable and Sustainable Energy Reviews. Elsevier*, *16*(9), 6861–6878. doi:10.1016/j.rser.2012.07.018

Kent, A., & Mercer, D. (2006). Australia's mandatory renewable energy target (MRET): An assessment. *Energy Policy*, *34*(9), 1046–1062. doi:10.1016/j.enpol.2004.10.009

Khademagha, P., Aries, M. B. C., Rosemann, A. L. P., & van Loenen, E. J. (2016). Implementing non-image-forming effects of light in the built environment: A review on what we need. *Building and Environment*, *108*, 263–272. doi:10.1016/j.buildenv.2016.08.035

Khan, Z. A., Labala, R. K., Yumnamcha, T., Devi, S. D., Mondal, G., Sanjita Devi, H., Rajiv, C., Bharali, R., & Chattoraj, A. (2018). Artificial Light at Night (ALAN), an alarm to ovarian physiology: A study of possible chronodisruption on zebrafish (Danio rerio). *The Science of the Total Environment*, *628–629*, 1407–1421. doi:10.1016/j.scitotenv.2018.02.101 PMID:30045561

Khedache, S., Makhlouf, S., Djefel, D., Lefebvre, G., & Royon, L. (2015). Preparation and thermal characterization of composite "Paraffin/Red Brick" as a novel form-stable of phase change material for thermal energy storage. *International Journal of Hydrogen Energy*, *40*(39), 13771–13776. doi:10.1016/j.ijhydene.2015.03.075

Khosravani, H. R., Castilla, M. M., Berenguel, M., Ruano, A. E., & Ferreira, P. M. (2016). A Comparison of Energy Consumption Prediction Models Based on Neural Networks of a Bioclimatic Building. *Energies*, *9*(1), 57. doi:10.3390/en9010057

Kim, I. T., Choi, A. S., & Sung, M. K. (2017). Development of a Colour Quality Assessment Tool for indoor luminous environments affecting the circadian rhythm of occupants. *Building and Environment*, *126*, 252–265. doi:10.1016/j.buildenv.2017.10.009

Kim, S., Kim, J. A., Kim, H. J., & Do Kim, S. (2006). Determination of formaldehyde and TVOC emission factor from wood-based composites by small chamber method. *Polymer Testing*, *25*(5), 605–614. https://doi.org/10.1016/j.polymertesting.2006.04.008

King, A. L. S., Valença, A. M., Silva, A. C., Sancassiani, F., Machado, S., & Nardi, A. E. (2014). "Nomophobia": Impact of Cell Phone Use Interfering with Symptoms and Emotions of Individuals with Panic Disorder Compared with a Control Group. *Clinical Practice and Epidemiology in Mental Health*, *10*(1), 28–35. doi:10.2174/1745017901410010028 PMID:24669231

Kish, L. (1965). Survey Sampling. John Wiley and Sons.

Klauson, D., Preis, S., Portjanskaja, E., Kachina, A., Krichevskayz, M., & Kallas, J. (2005). The influence of ferrous/ferric ions on the efficiency of photocatalytic oxidation of pollutants in groundwater. *Environmental Technology*, *26*(6), 653–661. doi:10.1080/09593330.2001.9619505 PMID:16035658

Klepeis, N. E., Nelson, W. C., & Ott, W. R. (2001). The national human activity pattern survey (NHAPS): A resource for assessing exposure to environmental pollutants. *Journal of Exposure Science & Environmental Epidemiology*, *11*, 231–252.

Knoppel, H., & Wolkoff, P. (1992). *Chemical, Microbiological, Health and Confort Aspects of Indoor Air Quality-State of the Art in SBS*. Kluwert Academic.

Koffi, B., Cerutti, A., Duerr, M., Iancu, A., Kona, A., & Janssens-Maenhout, G. (2017). *CoM Default Emission Factors for the Member States of the European Union - Version 2017, European Commission, Joint Research Centre (JRC)* [Dataset]. https://data.europa.eu/89h/jrc-com-ef-comw-ef-2017

Konis. (2017). A novel circadian daylight metric for building design and evaluation. *Building and Environment, 113*, 22-38.

Konya, A., & Vandenberg, M. (2011). *Design primer for hot climates*. Archimedia Press Limited.

Koo, J., So, H., Hong, S. W., & Hong, H. (2011). Effects of wallboard design parameters on the thermal storage in buildings. *Energy and Building*, *43*(8), 1947–1951. doi:10.1016/j.enbuild.2011.03.038

Kostiainen, R. (1995). Volatile organic compounds in the indoor air of normal and sick houses. *Atmospheric Environment*, *29*(6), 693–702. https://doi.org/10.1016/1352-2310(94)00309-9

Kottek, M., Grieser, J., Beck, C., Rudolf, B., & Rubel, F. (2006). World Map of the Köppen-Geiger climate classification updated. *Meteorologische Zeitschrift (Berlin)*, *15*(3), 259–263. doi:10.1127/0941-2948/2006/0130

Krieger, J. W., Takaro, T. K., Song, L., & Weaver, M. (2005, April). The Seattle-King County Healthy Homes Project: A randomized, controlled trial of a community health worker intervention to decrease exposure to indoor asthma triggers. *American Journal of Public Health*, *95*(4), 652–659. doi:10.2105/AJPH.2004.042994 PMID:15798126

Kuller, R., & Lindsten, C. (1992). Health and behavior of children in classrooms with and without windows. *Journal of Environmental Psychology, 12*(4), 305–317. doi:10.1016/S0272-4944(05)80079-9

Kumar, A., Sadanandhan, A. M., & Jain, S. L. (2019). Silver doped reduced graphene oxide as a promising plasmonic photocatalyst for oxidative coupling of benzylamines under visible light irradiation. *New Journal of Chemistry, 43*(23), 9116–9122. doi:10.1039/C9NJ00852G

Kumaran, V., Sudhagar, P., Konga, A. K., & Ponniah, G. (2020). Photocatalytic Degradation of Synthetic Organic Reactive Dye Wastewater Using GO-TiO2 Nanocomposite. *Polish Journal of Environmental Studies, 29*(2), 1683–1690. doi:10.15244/pjoes/109027

Kuznik, F., David, D., Johannes, K., & Roux, J. J. (2010). A review on phase change materials integrated in building walls. *Renewable & Sustainable Energy Reviews, 15*(1), 379–391. doi:10.1016/j.rser.2010.08.019

Kwon, K. S., Lee, I. B., Han, H. T., Shin, C. Y., Hwang, H. S., Hong, S. W., … Han, C. P. (2011). Analysing ventilation efficiency in a test chamber using age-of-air concept and CFD technology. *Biosystems Engineering, 110*(4), 421–433. doi:10.1016/j.biosystemseng.2011.08.013

Laamami, Benhamed, & Sbita. (2017). Analysis of shading effects on a photovoltaic array. *International Conference on Green Energy Conversion Systems (GECS).*

Larsen, M. A. D., Petrovic´, S., Radoszynski, A. M., McKenna, R., & Balyk, O. (2020). Climate change impacts on trends and extremes in future heating and cooling demands over Europe. *Energy and Building, 226*, 110397. doi:10.1016/j.enbuild.2020.110397

Laslett, D., Carter, C., Creagh, C., & Jennings, P. (2017). A large-scale renewable electricity supply system by 2030: Solar, wind, energy efficiency, storage and inertia for the South West Interconnected System (SWIS) in Western Australia. *Renewable Energy, 113*, 713–731. doi:10.1016/j.renene.2017.06.023

Lawrence, M. G., & Schäfer, S. (2019). Promises and perils of the Paris Agreement. *Science, 364*(6443), 829-830.

Laxen, D., Moorcroft, S., Marner, B., Laxen, K., Boulter, P., Barlow, T., Harrison, R., & Heal, M. (2010). *Final Report - PM 2.5 in the U.K.* SINFER.

Leaman, A., Thomas, L., & Vandenberg, M. (2007). 'Green'buildings: What Australian users are saying. *EcoLibrium.*

Lee, K. O., Medina, M. A., Sun, X., & Jin, X. (2018). Thermal performance of phase change materials (PCM)-enhanced cellulose insulation in passive solar residential building walls. *Solar Energy, 163*, 113–121. doi:10.1016/j.solener.2018.01.086

Lei, L., Xingong, C., Xiju, Z., Wen, W., & Chao, W. (2015). Research on Transmission Line Loss and Carrying Current Based on Temperature Power Flow Model. *3rd International Conference on Mechanical Engineering and Intelligent Systems (ICMEIS 2015).*

Leupen, B. (2005). Time-based Architecture (10th ed.). Academic Press.

Li, Y., Leung, G. M., Tang, J. W., Yang, X., Chao, C. Y. H., Lin, J. Z., … Yuen, P. L. (2007, February). Role of ventilation in airborne transmission of infectious agents in the built environment - A multidisciplinary systematic review. *Indoor Air.* doi:10.1111/j.1600-0668.2006.00445.x

Li, C., & Chow, T. T. (2011). Water-filled double reflective window and its year-round performance. *Procedia Environmental Sciences, 11*, 1039–1047. doi:10.1016/j.proenv.2011.12.158

Li, H. X., Edwards, D. J., Hosseini, M. R., & Costin, G. P. (2020). A review on renewable energy transition in Australia: An updated depiction. *Journal of Cleaner Production, 242*, 118475. doi:10.1016/j.jclepro.2019.118475

Li, J. (2019). Optimal sizing of grid-connected photovoltaic battery systems for residential houses in Australia. *Renewable Energy, 136*, 1245–1254. doi:10.1016/j.renene.2018.09.099

Li, L., Yu, H., & Liu, R. (2017). Research on composite-phase change materials (PCMs)-bricks in the west wall of room-scale cubicle: Mid-season and summer day cases. *Building and Environment, 123*, 494–503. doi:10.1016/j.buildenv.2017.07.019

Ling, T. C., & Poon, C. S. (2013). Use of phase change materials for thermal energy storage in concrete: An overview. *Construction & Building Materials, 46*, 55–62. doi:10.1016/j.conbuildmat.2013.04.031

Lin, Y., & Du, H. (2020). Graphene reinforced cement composites: A review. *Construction & Building Materials, 265*, 1–16. doi:10.1016/j.conbuildmat.2020.120312

Lipatov, A., Varezhnikov, A., Wilson, P., Sysoev, V., Kolmakov, A., & Sinitskii, A. (2013). Highly selective gas sensor arrays based on thermally reduced graphene oxide. *Nanoscale, 5*(12), 5426–5434. doi:10.1039/c3nr00747b PMID:23661278

Liu, D., Zhao, F. Y., & Tang, G. F. (2010). Active low-grade energy recovery potential for building energy conservation. *Renewable and Sustainable Energy Reviews. Elsevier Ltd, 14*(9), 2736–2747. doi:10.1016/j.rser.2010.06.005

Liu, J., Chen, X., Cao, S., & Yang, H. (2019). Overview on hybrid solar photovoltaic-electrical energy storage technologies for power supply to buildings. *Energy Conversion and Management, 187*, 103–121. doi:10.1016/j.enconman.2019.02.080

Liu, M., & Tilton, J. N. (2010). Spatial distributions of mean age and higher moments in steady continuous flows. *AIChE Journal. American Institute of Chemical Engineers, 56*(10), 2561–2572. https://doi.org/10.1002/aic.12151

Li, X., Pan, F., Sun, C., Zhang, M., Wang, Z., Du, J., ... Zhang, Z.-G. (2019). Simplified synthetic routes for low cost and high photovoltaic performance n-type organic semiconductor acceptors. *Nature Communications, 10*(1), 1–11. doi:10.103841467-019-08508-3 PMID:30705277

Li, X., Zhang, S., Wang, P., Zhong, H., Wu, Z., Chen, H., Liu, C., & Lin, S. (2015). High performance solar cells based on graphene-GaAs heterostructures. *Nano Energy, 16*, 310. doi:10.1016/j.nanoen.2015.07.003

Li, X., Zhao, Y., Wang, X., Wang, J., Gaskov, A. M., & Akbar, S. A. (2016). Reduced graphene oxide (rGO) decorated TiO2 microspheres for selective room-temperature gas sensors. *Sensors and Actuators. B, Chemical, 230*, 330–336. doi:10.1016/j.snb.2016.02.069

Li, X., Zhou, Y., Yu, S., Jia, G., Li, H., & Li, W. (2019). Urban heat island impacts on building energy consumption: A review of approaches and findings. *Energy, 174*, 407–419. doi:10.1016/j.energy.2019.02.183

Lixing, G. (2007). *Airflow Network Modeling in EnergyPlus | Simulation Research*. International Building Performance Simulation Association. Retrieved from https://simulationresearch.lbl.gov/publications/airflow-network-modeling-energyplus

Ln, Zhu, Guo, Zheng, Lu, Shih, & Wu. (2016). Multi-function indoor light sources based on light-emitting diodes–a solution for healthy lighting. *Optic Express*. doi:10.1364/OE.24.024401

Loch, S. (2006). Life transforms living transforms life. *Nordic Journal of Architectural Research, 19*(3), 33–44.

Loh, K. P., Bao, Q., Eda, G., & Chhowalla, M. (2010). Graphene oxide as a chemically tunable platform for optical applications. *Nature Chemistry, 2*(12), 1015–1024. doi:10.1038/nchem.907 PMID:21107364

López-Chao, V., Lorenzo, A. A., Saorín, J. L., De La Torre-Cantero, J., & Melián-Díaz, D. (2020). Classroom Indoor Environment Assessment through Architectural Analysis for the Design of Efficient Schools. *Sustainability, 12*(5), 2020. doi:10.3390u12052020

Lopez-Fernandez, O., Honrubia-Serrano, L., Freixa-Blanxart, M., & Gibson, W. (2014). Prevalence of problematic mobile phone use in british adolescents. *Cyberpsychology, Behavior, and Social Networking*, *17*(2), 91–98. doi:10.1089/cyber.2012.0260 PMID:23981147

Loryuenyong, V., Totepvimarn, K., Eimburanapravat, P., Boon-chompoo, W., & Buasri, A. (2013). Preparation and Characterization of Reduced Graphene Oxide Sheets via Water-Based Exfoliation and Reduction Methods. *Advances in Materials Science and Engineering*, *9*, 1–5. doi:10.1155/2013/923403

Losada, I. J., Toimil, A., Muñoz, A., Garcia-Fletcher, A. P., & Diaz-Simal, P. (2019). A planning strategy for the adaptation of coastal areas to climate change: The Spanish case. *Ocean and Coastal Management*, *182*, 104983. doi:10.1016/j.ocecoaman.2019.104983

Loukatou, A., Johnson, P., Howell, S., & Duck, P. (2021). Optimal valuation of wind energy projects co-located with battery storage. *Applied Energy*, *283*, 116247. doi:10.1016/j.apenergy.2020.116247

Lucas, Peirson, Berson, Brown, Cooper, Czeisler, Figueiro, Gamlin, Lockley, O'Hagan, Price, Provencio, Skene, & Brainard. (2014). Measuring and using light in the melanopsin age. *Trends in Neurosciences*, *37*(1), 1–9. doi:10.1016/j.tins.2013.10.004 PMID:24287308

Lu, G., Ocola, L. E., & Chen, J. (2009). Reduced graphene oxide for room-temperature gas sensors. *Nanotechnology*, *20*(44), 445502. doi:10.1088/0957-4484/20/44/445502 PMID:19809107

Lu, M., & Lai, J. (2020). Review on carbon emissions of commercial buildings. *Renewable & Sustainable Energy Reviews*, *119*, 109545. doi:10.1016/j.rser.2019.109545

Lu, M., & Lai, J. H. (2019). Building energy: A review on consumptions, policies, rating schemes and standards. *Energy Procedia*, *158*, 3633–3638. doi:10.1016/j.egypro.2019.01.899

Luo, M., de Dear, R., Ji, W., Bin, C., Lin, B., Ouyang, Q., & Zhu, Y. (2016). The dynamics of thermal comfort expectations: The problem, challenge and impication. *Building and Environment*, *95*, 322–329. doi:10.1016/j.buildenv.2015.07.015

Lu, Y., Huang, Z., & Zhang, T. (2013). Method and case study of quantitative uncertainty analysis in building energy consumption inventories. *Energy and Building*, *57*, 193–198. doi:10.1016/j.enbuild.2012.10.046

Lyons, M., Smuts, C., & Stephens, A. (2001). Participation, Empowerment and Sustainability: (How) Do the Links Work? *Urban Studies (Edinburgh, Scotland)*, *38*(8), 1233–1251. doi:10.1080/00420980125039

Lyu, Y. L., Chow, T. T., & Wang, J. L. (2018). Numerical prediction of thermal performance of liquid-flow window in different climates with anti-freeze. *Energy*, *157*, 412–423. doi:10.1016/j.energy.2018.05.140

Lyu, Y. L., Liu, W. J., Su, H., & Wu, X. (2019). Numerical analysis on the advantages of evacuated gap insulation of vacuum-water flow window in building energy saving under various climates. *Energy*, *175*, 353–364. doi:10.1016/j.energy.2019.03.101

Lyu, Y. L., Wu, X., Li, C., Su, H., & He, L. (2019). Numerical analysis on the effectiveness of warm water supply in water flow window for room heating. *Solar Energy*, *177*, 347–354. doi:10.1016/j.solener.2018.11.033

Mahdaoui, M., Hamdaoui, S., Msaad, A. A., Kousksou, T., El Rhafiki, T., Jamil, A., & Ahachad, M. (2021). Building bricks with phase change material (PCM): Thermal performances. *Construction & Building Materials*, *269*, 1–12. doi:10.1016/j.conbuildmat.2020.121315

Majid, N. H. A., Takagi, N., Hokoi, S., Ekasiwi, S. N. N., & Uno, T. (2014). Field survey of air conditioner temperature settings in a hot, dry climate (Oman). *HVAC & R Research*, *20*(7), 751–759. doi:10.1080/10789669.2014.953845

Manakshi, D., Sumita, D., & Champa, N. (2010). Maximum Photovoltaic Power Tracking Using Perturb & Observe Algorithm in MATLAB/Simulink Environment. *International Journal of Electrical Engineering and Technology*, *1*(1), 71–84.

Manisalidis, I., Stavropoulou, E., Stavropoulos, A., & Bezirtzoglou, E. (2020). Environmental and Health Impacts of Air Pollution: A Review. *Frontiers in Public Health*, *8*(14), 14. Advance online publication. doi:10.3389/fpubh.2020.00014 PMID:32154200

María, B. P. M., & Labarca, C. Y. (2015). Methodology for Assessing Daylighting Design Strategies in Classroom with a Climate-Based Method. *Sustainability*, *7*(1), 880–897. doi:10.3390u7010880

Maroni, M., Seifert, B., & Lindvall, T. (1995). *Indoor air quality a comprehensive reference book*. Amsterdam: Elsevier Science B.V. Available at: https://books.google.co.uk/books?hl=en&lr=&id=qsyLaKnn-nYC&oi=fnd&pg=PP2&dq=Indoor+Air+Quality+A+Comprehensive+Reference+Book,&ots=mDAkLMqddR&sig=RWCLF99Kto-MIWMGV8kLp2RUolM#v=onepage&q=Indoor Air Quality A Comprehensive Reference Book%2C&f=false

Martenies, S. E., & Batterman, S. A. (2018). Effectiveness of Using Enhanced Filters in Schools and Homes to Reduce Indoor Exposures to PM 2.5 from Outdoor Sources and Subsequent Health Benefits for Children with Asthma. *Environmental Science & Technology*, *52*(18), 10767–10776. doi:10.1021/acs.est.8b02053 PMID:30141330

Martin, N. J., & Rice, J. L. (2012). Developing renewable energy supply in Queensland, Australia: A study of the barriers, targets, policies and actions. *Renewable Energy*, *44*, 119–127. doi:10.1016/j.renene.2012.01.006

Maruthamani, D., Divakar, D., & Kumaravel, M. (2015). Enhanced photocatalytic activity of TiO2 by reduced graphene oxide in mineralization of Rhodamine B dye. *Journal of Industrial and Engineering Chemistry*, *30*, 33–43. doi:10.1016/j.jiec.2015.04.026

Masdías-Bonome, A. E., Orosa, J. A., & Vergara, D. (2020). A new methodology for decision-making in buildings energy optimization. *Applied Sciences (Basel, Switzerland)*, *10*(13), 4558. doi:10.3390/app10134558

Masoso, O. T., & Grobler, L. J. (2010). 'The dark side of occupants' behaviour on building energy use'. *Energy and Building*, *42*(2), 173–177. doi:10.1016/j.enbuild.2009.08.009

Mathur, D. (2014). Maximum Power Point Tracking with Artificial Neural Network. *International Journal of Emerging Science and Engineering*, *2*(3), 38–42.

Mavrotas, G., Florios, K., & Vlachou, D. (2010). Energy planning of a hospital using Mathematical Programming and Monte Carlo simulation for dealing with uncertainty in the economic parameters. *Energy Conversion and Management*, *51*(4), 722–731. doi:10.1016/j.enconman.2009.10.029

Mazarrón, F. R., Cid-Falceto, J., & Cañas, I. (2012). Ground thermal inertia for energy efficient building design: A case study on food industry. *Energies*, *5*(2), 227–242. doi:10.3390/en5020227

McGill, G. (2016). Meta-analysis of indoor temperatures in new-build housing. *Building Research & Information. Taylor & Francis*, *0*(0), 1–21. doi:10.1080/09613218.2016.1226610

McGill, G., Oyedele, L. O., & McAllister, K. (2015). Case study investigation of indoor air quality in mechanically ventilated and naturally ventilated U.K. social housing. *International Journal of Sustainable Built Environment*, *4*(1), 58–77. doi:10.1016/j.ijsbe.2015.03.002

Mcnutt, M. (2013). *Climate Change Impacts*. Science., doi:10.1126cience.1243256

Meadowcroft, J. (2004). Participation and Sustainable Development: Modes of Citizen, Community and Organisational Involvement. In W. M. Lafferty (Ed.), *Governance for Sustainable Development. The Challenge of Adapting Form and Function*. Edward Elgar Publishing, Inc. doi:10.4337/9781845421700.00014

Meiss, A., & Feijó-Muñoz, J. (2014). The energy impact of infiltration: A study on buildings located in north central Spain. *Energy Efficiency*, *8*(1), 51–64. doi:10.100712053-014-9270-x

Meiss, A., & Muñoz, F. (2013). La Necesidad De Estudiar Las Infiltraciones : Estudio De Caso. In *Viviendas Sociales De La Urbanización Zabalgana*. Boletín Académico, Revista de Investiga., doi:10.17979/bac.2013.3.0.997

Memon, S. A. (2013). Phase change materials integrated in building walls: A state of theart review. *Renewable & Sustainable Energy Reviews*, *31*, 870–906. doi:10.1016/j.rser.2013.12.042

Meng, W., & Khayat, K. H. (2016). Mechanical properties of ultra-high-performance concrete enhanced with graphite nanoplatelets and carbon nanofibers. *Composites. Part B, Engineering*, *107*, 113–122. doi:10.1016/j.compositesb.2016.09.069

MeteoGalicia. (2020). *Anuario climatolóxico de Galicia. Consellería de Medio Ambiente*. Xunta de Galicia. https://www.meteogalicia.gal/observacion/informesclima/informesIndex.action

Miao, X., Tongay, S., Petterson, M., Berke, K., Rinzler, A., Appleton, B., & Hebard, A. (2012). High Efficiency Graphene Solar Cells by Chemical Doping. *Nano Letters*, *12*(6), 2745–2750. doi:10.1021/nl204414u PMID:22554195

Michael, H., Damir, N., & Mariesa, W. (2017). Electric Power Grid Modernization Trends, Challenges, and Opportunities. *IEEE Advancing Technology for Humanity.* https://www.ieee.org/content/dam/ieee-org/ieee/web/org/about/corporate/ieee-industry-advisory-board/electric-power-grid-modernization.pdf

Miguez, J. L., Porteiro, J., López-González, L. M., Vicuña, J. E., Murillo, S., Morán, J. C., & Granada, E. (2006). Review of the energy rating of dwellings in the European Union as a mechanism for sustainable energy. *Renewable & Sustainable Energy Reviews*, *10*(1), 24–45. doi:10.1016/j.rser.2004.08.003

Ministerio de Fomento del Gobierno de España. (2017). Código técnico de la Edificación (CTE). Author.

Ministerio de la Presidencia, Gobierno de España. (2013). Real Decreto 235/2013, de 5 de abril, por el que se aprueba el procedimiento básico para la certificación de la eficiencia energética de los edificios. Boletín Ofical del Estado. Author.

Ministry of Housing, Construction, and Housing. (2021). *Peru GBC & MVCS - Technical Code of Sustainable Construction Peru 2021*. Retrieved 2021, from https://issuu.com/perugbc/docs/codigo_-_ctcs

Ministry of Housing, Construction, and Sanitation. (2014, May 13). *Standard EM.110 Thermal and Light Comfort with Energy Efficiency*. Retrieved from http://www3.vivienda.gob.pe/dnc/archivos/Estudios_Normalizacion/Normalizacion/normas/DS-006-2014-VIVIENDA.pdf

Min-Joong, K., & Huei, P. (2007). Power Management and Design Optimization of Fuel Cell/Battery Hybrid Vehicles. *Journal of Power Sources*, *165*(2), 819–832. doi:10.1016/j.jpowsour.2006.12.038

Miranda, A. (2018, May 28). *Water or Energy?* Academic Press.

Mishra, A. K., & Ramgopal, M. (2013). Field studies on human thermal comfort— An overview. *Building and Environment*, *64*, 94–106. doi:10.1016/j.buildenv.2013.02.015

Mishra, A., Mehta, A., Basu, S., Shetti, N. P., Reddy, K. R., & Aminabhavi, T. M. (2019). Graphitic carbon nitride (g–C3N4)–based metal-free photocatalysts for water splitting. *Carbon*, *149*, 693–721. doi:10.1016/j.carbon.2019.04.104

Mishra, V., & Smyth, R. (2017). Conditional convergence in Australia's energy consumption at the sector level. *Energy Economics*, *62*, 396–403. doi:10.1016/j.eneco.2016.07.009

Mivivienda, F. (2020). *Certification Procedure for projects of the Sustainable Housing Program*. Retrieved from https://www.mivivienda.com.pe/PORTALCMS/archivos/documentos/8586096294727711719.PDF

Modeste, K. N., Mempouo, B., René, T., Costa, A. M., Orosa, J. A., Raminosoa, C. R. R., & Mamiharijaona, R. (2015). Resource potential and energy efficiency in the buildings of Cameroon: A review. *Renewable & Sustainable Energy Reviews*, *50*, 835–846. doi:10.1016/j.rser.2015.05.052

Mohamed, L., Ali, K. A., & Mazen, A., & S., Ookawara (2016). Performance Enhancement of Constant Voltage Based MPPT for Photovoltaic Applications Using Genetic Algorithm. *3rd International Conference on Power and Energy Systems Engineering (CPESE 2016).*

Molhave, L. (1991). Volatile organic compounds, indoor air quality and health. *Indoor Air*, *1*(4), 357–376. doi:10.1111/j.1600-0668.1991.00001.x

Molhave, L., Jensen, J. G., & Larsen, S. (1991). Subjective Reactions To Volatile Organic Compounds As Air Pollutants. *Atmospheric Environment*, *25*(7), 1283–1293. doi:10.1016/0960-1686(91)90240-8

Monsalve, F. (2011). *Influencia del tiempo y de la contaminación atmosférica sobre enfermedades de los sistemas circulatorio y respiratorio en Castilla-La Mancha*. Academic Press.

Montaner, J. M., & Muxí, Z. (2010). Reflexiones para proyectar viviendas del siglo XXI. *Dearq*, *06*(6), 82–89. doi:10.18389/dearq6.2010.09

Moreno Rangel, A. B. J., Roh, T., Xu, X., &, Carrillo, G. (2020). Assessing Impact of Household Intervention on Indoor Air Quality and Health of Children with Asthma in the US-Mexico Border: A Pilot Study. Texas A&M University, School of Public Health.

Moreno Santamaria, B., Ama Gonzalo, F., Lauret Aguirregabiria, B., & Hernandez Ramos, J. A. (2020). Evaluation of Thermal Comfort and Energy Consumption of Water Flow Glazing as a Radiant Heating and Cooling System: A Case Study of an Office Space. *Sustainability*, *12*, 7596.

Moreno, B., & Hernández, J. A. (2018). Analytical solutions to evaluate solar radiation overheating in simplified glazed rooms. *Building and Environment*, *140*, 162–172.

Morse, R., Cooper, P., & Proctor, D. (1974). *The status of solar energy utilization in Australia for industrial commercial and domestic purposes.* CSIRO Solar Energy Studies Report No. 74/1.

Moyano, Moyano, López, Aznal, & González Lezcano. (2020). Photometric and colorimetric analysis of LED luminaires for interior lighting design. *Optik*. doi:10.1016/j.ijleo.2020.165599

Moyano, D. B., Mónica, S. J. F., & Roberto, A. G. L. (2020). Towards a Sustainable Indoor Lighting Design: Effects of Artificial Light on the Emotional State of Adolescents in the Classroom. *Sustainability*, *12*(10), 4623.

Moyano, D. B., Sola, Y., & Roberto, A. G. L. (2020). Blue-light levels emitted from portable electronic devices compared to sunlight. *Energies*, *13*(16), 4276. doi:10.3390/en13164276

MUGAK. (2020a). *Noticias: MUGAK cierra su segunda edición, con una bienal más internacional y más cercana a la ciudadanía*. Accessed March 26, 2020. https://www.bienalmugak.eus/es/noticias

MUGAK. (2020b). *La Bienal: ¿Puntualizar? ¿Adaptar? ¿Repensar? ¿Volver a empezar?* Accessed March 26, 2020 https://www.bienalmugak.eus/es/bienal

Nabil, A., & Mardaljevic, J. (2006). Useful daylight illuminances: A replacement for daylight factors. *Energy and Building*, *38*(7), 905–913. doi:10.1016/j.enbuild.2006.03.013

Namiesnik, J., Gorecki, T., Krosdon-Zabiegala, B., & Lusiak, J. (1992). Indoor air quality pollutants, their sources and concentration levels. *Building and Environment*, *27*(3), 339–356.

Nanz, P., & Fritsche, M. (2014). La partecipazione dei cittadini: un manuale. Metodi partecipativi: protagonisti, opportunità e limiti. Assemblea legislativa della Regione Emilia-Romagna, Italia.

Naspi, F., Arnesano, M., Zampetti, L., Stazi, F., Revel, G. M., & D'Orazio, M. (2018). Experimental study on occupants' interaction with windows and lights in Mediterranean offices during the non-heating season. *Building and Environment, 127*(September), 221–238. doi:10.1016/j.buildenv.2017.11.009

Nastasi, B. (2015). Renewable hydrogen potential for low-carbon retrofit of the building stocks. *Energy Procedia, 82*, 944–949. doi:10.1016/j.egypro.2015.11.847

Nathanson, T. (1993). *Indoor Air Quality in Office Buildings: A Technical Guide*. Health Canadá.

National Environmental Protection Council. (2003). *National Environment Protection (Ambient Air Quality)*. Measure.

National Institute of Environmental Health Sciences. (2020). *Air Pollution and Your Health*. NIEHS. Retrieved November 16 from https://www.niehs.nih.gov/health/topics/agents/air-pollution/index.cfm

Neill, M. S. O., & Ebi, K. L. (2009). Temperature Extremes and Health: Impacts of Climate Variability and Change in the United States. *Journal of Occupational and Environmental Medicine, 51*(1), 13–25. doi:10.1097/JOM.0b013e318173e122 PMID:19136869

Nelson, T., Nelson, J., Ariyaratnam, J., & Camroux, S. (2013). An analysis of Australia's large scale renewable energy target: Restoring market confidence. *Energy Policy, 62*, 386–400. doi:10.1016/j.enpol.2013.07.096

Nematchoua, M. K., Ricciardi, P., Orosa, J. A., & Buratti, C. (2018, August). A detailed study of climate change and some vulnerabilities in Indian Ocean: A case of Madagascar island. *Sustainable Cities and Society, 41*, 886–898. doi:10.1016/j.scs.2018.05.040

Nematchoua, M. K., Tchinda, R., & Orosa, J. A. (2014). Thermal comfort and energy consumption in modern versus traditional buildings in Cameroon: A questionnaire-based statistical study. *Applied Energy, 114*, 687–699. doi:10.1016/j.apenergy.2013.10.036

Nematchoua, M. K., Tchinda, R., Orosa, J. A., & Andreasi, W. A. (2015). Effect of wall construction materials over indoor air quality in humid and hot climate. *Journal of Building Engineering, 3*, 16–23. doi:10.1016/j.jobe.2015.05.002

Nguyen, D., & Lehman, B. (2006). *Modeling and simulation of solar PV arrays under changing illumination conditions*. IEEE Workshops on Computers in Power Electronics. doi:10.1109/COMPEL.2006.305629

Nicol, F. (2004). Adaptive thermal comfort standards in the hot-humid tropics. *Energy and Building, 36*(7), 628–637. doi:10.1016/j.enbuild.2004.01.016

Nicol, F., Humphreys, M., & Roaf, S. (2012). *Adaptive thermal comfort: Principles and practice* (1st ed.). Routledge. doi:10.4324/9780203123010

Nicol, J. F., & Humphreys, M. (2002). Adaptive thermal comfort and sustainable thermal standards for buildings. *Energy and Building, 34*(6), 563–572. doi:10.1016/S0378-7788(02)00006-3

Nikolopoulou, M., Baker, N., & Steemers, K. (2001). Thermal comfort in outdoor urban spaces: Understanding the human parameter. *Solar Energy, 70*(3), 227–235. doi:10.1016/S0038-092X(00)00093-1

Nikolopoulou, M., & Steemers, K. (2003). Thermal comfort and psychological adaptation as a guide for designing urban spaces. *Energy and Building, 35*(1), 95–101. doi:10.1016/S0378-7788(02)00084-1

Ning, H., Wang, Z., & Ji, Y. (2016). Thermal history and adaptation: Does a long-term indoor thermal exposure impact human thermal adaptability? *Applied Energy, 183*, 22–30. doi:10.1016/j.apenergy.2016.08.157

Nocera, Faro, Costanzo, & Raciti. (2018). Daylight Performance of Classrooms in a Mediterranean School Heritage Building. *Sustainability, 10*, 3705. doi:10.3390u10103705

OMS. (1987). *Air Quality Guidelines for Europe.* European Series, no. 23. Copenhague: Publicaciones Regionales de la OMS.

OMS. (2005). *Guías de calidad del aire de la OMS relativas al material particulado, el ozono, el dióxido de nitrógeno y el dióxido de azufre.* OMS.

OMS. (2012). *Atlas de la salud y del clima.* OMS.

Oñativia, D. A. (2011, May 10). *The Geographical Regions of Peru.* Retrieved from Blog HISTORY 2DO YEAR IEM 2011: http://historiasegundoiemunsa2011.blogspot.com/2011/05/las-regiones-geograficas-del-peru.html

Orosa, J. A. (2012). Thermal inertia and ISO 13790 - Have its effects on future energy consumption been fully considered? *Energy Education Science and Technology, 29*(1), 339–352.

Orosa, J. A., Costa, A. M., Rodríguez-Fernández, A., & Roshan, G. (2014). Effect of climate change on outdoor thermal comfort in humid climates. *Journal of Environmental Health Science & Engineering, 12*(1), 46. doi:10.1186/2052-336X-12-46 PMID:24517127

Orosa, J. A., Nematchoua, M. K., & Reiter, S. (2020). Air changes for healthy indoor ambiences under pandemic conditions and its energetic implications: A Galician case study. *Applied Sciences (Basel, Switzerland), 10*(20), 7169. doi:10.3390/app10207169

Orosa, J. A., & Oliveira, A. C. (2010). Impact of climate change on cooling energy consumption. *Journal of the Energy Institute, 83*(3), 171–177. doi:10.1179/014426010X12759937396876

Orosa, J. A., & Oliveira, A. C. (2012). A field study on building inertia and its effects on indoor thermal environment. *Renewable Energy, 37*(1), 89–96. doi:10.1016/j.renene.2011.06.009

Østergaard, P. A., Duic, N., Noorollahi, Y., Mikulcic, H., & Kalogirou, S. (2020). *Sustainable development using renewable energy technology.* Elsevier. doi:10.1016/j.renene.2019.08.094

OXFAM. (n.d.). *OXFAM in Peru. Retrieved from Between 7 and 8 million Peruvians do not have access to drinking water.* https://peru.oxfam.org/qu%C3%A9-hacemos-ayuda-humanitaria/entre-7-y-8-millones-de-Peruvians don't have access to drinking water

Pal, J. S., & Eltahir, E. A. B. (2016). Future temperature in southwest Asia projected to exceed a threshold for human adaptability. *Nature Climate Change, 6*(2), 197–200. doi:10.1038/nclimate2833

Parameshwaran, R., Naresh, R., Ram, V.V., & Srinivas, P.V. (2021). Microencapsulated bio-based phase change material-micro concrete composite for thermal energy storage. *Journal of Building Engineering*, 1-39. doi:10.1016/j.jobe.2021.102247

Paricio, I., & Sust, X. (2004). *La vivienda contemporánea.* Barcelona: Institut de Tecnología de la Construcció de Catalunya-ITeC.

Parra, J., Guardo, A., Egusquiza, E., & Alavedra, P. (2015). Thermal performance of ventilated double skin façades with venetian blinds. *Energies, 8*(6), 4882–4898. doi:10.3390/en8064882

Pasupathy, A., Velraj, R., & Seeniraj, R. V. (2008). Phase change material-based building architecture for thermal management in residential and commercial establishments. *Renewable & Sustainable Energy Reviews, 12*(1), 39–64. doi:10.1016/j.rser.2006.05.010

Peacock, A. D., Jenkins, D. P., & Kane, D. (2010). Investigating the potential of overheating in U.K. dwellings as a consequence of extant climate change. *Energy Policy. Elsevier*, *38*(7), 3277–3288. doi:10.1016/j.enpol.2010.01.021

Peeters, L., de Dear, R., Hensen, J., & D'haeseleer, W. (2009). Thermal comfort in residential buildings: Comfort values and scales for building energy simulation. *Applied Energy*, *86*(5), 772–780. doi:10.1016/j.apenergy.2008.07.011

Peng, L. L. H., & Jim, C. Y. (2013). Green-roof effects on neighborhood microclimate and human thermal sensation. *Energies*, *6*(2), 598–618. doi:10.3390/en6020598

Pérez-Lombard, L., Ortiz, J., & Pout, C. (2008). A review on buildings energy consumption information. *Energy and Building*, *40*(3), 394–398. doi:10.1016/j.enbuild.2007.03.007

Persily, A. K. (2014). *Indoor Air Quality in High Performance Buildings: What is and isn't in ASHRAE/IES/USGBC Standard 189.1. In 13th International Conference on Indoor Air Quality and Climate*. Indoor Air. Available at https://www.isiaq.org/docs/paper/HP1014.pdf

Petersdorff, C., Boermans, T., & Harnisch, J. (2006). Mitigation of CO2 Emissions from the EU-15 Building Stock Beyond the E.U. Directive on the Energy Performance of Buildings. *Environmental Science and Pollution Research International*, *13*(5), 350–358. doi:10.1065/espr2005.12.289 PMID:17067030

Peterson, K., Torcellini, P., & Grant, R. (2015). *A common definition for Zero Energy Buildings. U.S. Department of energy*. Energy Efficiency and Renewable Energy.

Pilehvar, S., Sanfelix, S. G., Szczotok, A. M., Rodríguez, J. F., Valentini, L., Lanzón, M., Pamies, R., & Kjøniksen, A. L. (2020). Effect of temperature on geopolymer and Portland cement composites modified with Micro-encapsulated Phase Change materials. *Construction & Building Materials*, *252*, 55–66. doi:10.1016/j.conbuildmat.2020.119055

Pirkl James, J. (2008). Transgenerational Design: A Heart Transplant for Housing. In F. Kohlbacher & C. Herstatt (Eds.), *The Silver Market Phenomenon*. Springer. doi:10.1007/978-3-540-75331-5_10

Pirkl, J. J. (1994). *Transgenerational Design: Products for an Aging Population*. John Wiley & Sons.

Piselli, C., Prabhakar, M., de Gracia, A., Saffari, M., Pisello, A. L., & Cabeza, L. F. (2020). Optimal control of natural ventilation as passive cooling strategy for improving the energy performance of building envelope with PCM integration. *Renewable Energy*, *162*, 171–181. doi:10.1016/j.renene.2020.07.043

Porteous, C. D. A. (2011). Sensing a Historic Low-CO 2 Future. In D. N. Mazzeo (Ed.), *Chemistry, Emission Control, Radioactive Pollution and Indoor Air Quality* (pp. 213–246). InTech. doi:10.5772/16918

Portney, K. E., & Berry, J. M. (2010). Participation and the Pursuit of Sustainability in U.S. Cities. *Urban Affairs Review*, *46*(1), 119–139. doi:10.1177/1078087410366122

Poza-Casado, I., Meiss, A., Padilla-Marcos, M. Á., & Feijó-Muñoz, J. (2020). Airtightness and energy impact of air infiltration in residential buildings in Spain. *International Journal of Ventilation*, 1–7. doi:10.1080/14733315.2020.1777029

Prabhakar-Murthy, D. N., Bulmer, M., & Eccleston, J. A. (2004). Weibull model selection for reliability modelling. *Reliability Engineering & System Safety*, *86*(3), 257–267. doi:10.1016/j.ress.2004.01.014

Priarone, A., Silenzi, F., & Fossa, M. (2020). Modelling Heat Pumps with Variable EER and COP in EnergyPlus: A Case Study Applied to Ground Source and Heat Recovery Heat Pump Systems. *Energies*, *13*, 794.

Priyadarsini, S., Mohanty, S., Mukherjee, S., Basu, S., & Mishra, M. (2018). Graphene and graphene oxide as nanomaterials for medicine and biology. *Journal of Nanostructure in Chemistry*, *8*(2), 123–137. doi:10.100740097-018-0265-6

PUPC. (2018). *Puntoedu report on water stress in Peru*. Retrieved from https://puntoedu.pucp.edu.pe/noticia/informe-puntoedu-sobre-estres-hidrico-en-el-peru/

Qader, M. R. (2009). Electricity Consumption and GHG Emissions in GCC Countries. *Energies*, *2*(4), 1201–1213. doi:10.3390/en20401201

Qiu, J., Zhang, P., Ling, M., Li, S., Liu, P., Zhao, H., & Zhang, S. (2012). Photocatalytic Synthesis of TiO2 and Reduced Graphene Oxide Nanocomposite for Lithium Ion Battery. *Interfaces*, *4*, 3636. PMID:22738305

Querol, X., Viana, M., Moreno, T., & Alastuy, A. (2012). *Bases científicos-técnicas para un Plan Nacional de mejora de la Calidad del Aire. Informes CSIC*. Consejo Superior de Investigaciones Científicas.

Raftery, P., Keane, M., & O'Donnell, J. (2011). Calibrating whole building energy models: An evidence-based methodology. *Energy and Building*, *43*(9), 2356–2364. doi:10.1016/j.enbuild.2011.05.020

Ragette, F. (2012). *Traditional Domestic Architecture of the Arab Region* (1st ed.). American Univeristy of Sharjah.

Ramanathan, V., & Feng, Y. (2009). Air pollution, greenhouse gases and climate change: Global and regional perspectives. *Atmospheric Environment. Elsevier Ltd*, *43*(1), 37–50. doi:10.1016/j.atmosenv.2008.09.063

Rao, A. S. R. S. (2005). On joint Weibull probability density functions. *Applied Mathematics Letters*, *18*(11), 1224–1227. doi:10.1016/j.aml.2005.02.011

Rasmussen, M., Foldbjerg, P., Christoffersen, J., Daniell, J., Bang, U., Galiatto, N., & Bjerre, K. (2017). *Healthy Homes Barometer 2017 - Buildings and Their Impact on the Health of Europeans. VELUX Group.*

Rathore, P. K. S., & Shukla, S. K. (2021). Enhanced thermophysical properties of organic PCM through shape stabilization for thermal energy storage in buildings: A state of the art review. *Energy and Building*, *236*, 1–29. doi:10.1016/j.enbuild.2021.110799

Real Decreto 102/2011. (2011). *Relativo a la mejora de la calidad del aire*. https://www.boe.es/buscar/act.php?id=BOE-A-2011-1645

Rea, M. S., & Figueiro, M. G. (2018). Light as a circadian stimulus for architectural lighting. *Lighting Research & Technology*, *50*(4), 497–510. doi:10.1177/1477153516682368

Rea, M. S., Figueiro, M. G., Bierman, A., & Hamner, R. (2012). Modelling the spectral sensitivity of the human circadian system. *Lighting Research & Technology*, *44*(4), 386–396. doi:10.1177/1477153511430474

Rea, M. S., Figueiro, M. G., Bullough, J. D., & Bierman, A. (2005). A model of phototransduction by the human circadian system. *Brain Research. Brain Research Reviews*, *50*(2), 213–228. doi:10.1016/j.brainresrev.2005.07.002 PMID:16216333

Reda, F., & Fatima, Z. (2019). Northern European nearly zero energy building concepts for apartment buildings using integrated solar technologies and dynamic occupancy profile: Focus on Finland and other Northern European countries. *Applied Energy*, *237*, 598–617. doi:10.1016/j.apenergy.2019.01.029

Reddy, T. A. (2006). Literature Review on Calibration of Building Energy Simulation Programs. *ASHRAE Transactions*, *112*(1), 226–240.

Rehmani, M. H., Reisslein, M., Rachedi, A., Erol-Kantarci, M., & Radenkovic, M. (2018). Integrating Renewable Energy Resources into the Smart Grid: Recent Developments in Information and Communication Technologies. *IEEE Transactions on Industrial Informatics*, *14*(7), 2814–2825. doi:10.1109/TII.2018.2819169

Rehman, S. K. U., Kumarova, S., Memon, S. A., Javed, M. F., & Jameel, M. (2020). A Review of Microscale, Rheological, Mechanical, Thermoelectrical and Piezoresistive Properties of Graphene Based Cement Composite. *Nanomaterials (Basel, Switzerland)*, *10*(2076), 1–42. PMID:33096675

Reinhart, C. F., Mardaljevic, J., & Rogers, Z. (2006). Dynamic daylight performance metrics for sustainable building design. *Leukos*, *3*(1), 7–31. doi:10.1582/LEUKOS.2006.03.01.001

Ren, M., Wen, X., Gao, X., & Liu, Y. (2021). Thermal and mechanical properties of ultra-high performance concrete incorporated with microencapsulated phase change material. *Construction & Building Materials*, *273*, 1–10. doi:10.1016/j.conbuildmat.2020.121714

Ren, X., Yan, D., & Wang, C. (2014). Air-conditioning usage conditional probability model for residential buildings. *Building and Environment*, *81*, 172–182. doi:10.1016/j.buildenv.2014.06.022

Rey, E., Aguayo, F., & Peralta, M. (2015). Integración de métodos escalares y vectoriales en la evaluación del riesgo de incendio en el ciclo de vida de una construcción. *Informes de la Construcción*, *67*(539), e110. doi:10.3989/ic.14.154

Ritchie, H., & Roser, M. (2020). CO_2 and greenhouse gas emissions. *Our World in Data.*

RITE. (2020). *Reglamento de Instalaciones Térmicas de los Edificios*. Academic Press.

Roberts, M., Bruce, A., & MacGill, I. (2019). Opportunities and barriers for photovoltaics on multi-unit residential buildings: Reviewing the Australian experience. *Renewable & Sustainable Energy Reviews*, *104*, 95–110. doi:10.1016/j.rser.2018.12.013

Roshan, G., Ghanghermeh, A., & Orosa, J. A. (2014). Thermal comfort and forecast of energy consumption in northwest Iran. *Arabian Journal of Geosciences*, *7*(9), 3657–3674. doi:10.100712517-013-0973-7

Roshan, G., Orosa, J. A., & Nasrabadi, T. (2012). Simulation of climate change impact on energy consumption in buildings, case study of Iran. *Energy Policy*, *49*, 731–739. doi:10.1016/j.enpol.2012.07.020

Rosseau, D., Bowser, D., & Mattock, C. (2001). *A Guide to Mechanical Equipment for Healthy Indoor Environments*. Available at: https://www.cmhc-schl.gc.ca/odpub/pdf/62015.pdf

Rupp, R. F., Andersen, R. K., Toftum, J., & Ghisi, E. (2021). Occupant behaviour in mixed-mode office buildings in a subtropical climate: Beyond typical models of adaptive actions. *Building and Environment, 190*(December). doi:10.1016/j.buildenv.2020.107541

Šadauskienė, J., Paukštys, V., Šeduikytė, L., & Banionis, K. (2014). Impact of air tightness on the evaluation of building energy performance in Lithuania. *Energies*, *7*(8), 4972–4987. doi:10.3390/en7084972

Sadineni, S. B., Madala, S., & Boehm, R. F. (2011). Passive building energy savings: A review of building envelope components. *Renewable and Sustainable Energy Reviews. Elsevier Ltd*, *15*(8), 3617–3631. doi:10.1016/j.rser.2011.07.014

Saéz, E. (2017). Análisis de la calidad del aire en función de la tipología de ventilación. Aplicación al frototipo. Universidad Politécnica de Valencia.

Sagmeister, S. (2013). *The Happy Show*. The Museum of Contemporary Art. Accessed April 18, 2020. https://www.moca.org/exhibition/stefan-sagmeister-the-happy-show

Saidi, K., & Omri, A. (2020). The impact of renewable energy on carbon emissions and economic growth in 15 major renewable energy-consuming countries. *Environmental Research*, *186*, 109567. doi:10.1016/j.envres.2020.109567 PMID:32361260

Salgado, M. A. (2006). Antes y después de la vivienda cinematográfica. *Revista Arquitectura*, *345*, 98–103.

Saltelli, A., Tarantola, S., Campolongo, F., & Ratto, M. (2004). *Sensitivity Analysis in Practice: A Guide to Assessing Scientific Models*. Wiley. Retrieved from https://www.wiley.com/en-us/Sensitivity+Analysis+in+Practice%3A+A+Guide+to+Assessing+Scientific+Models-p-9780470870945

Salvia, G., Morello, E., Rotondo, F., Sangalli, A., Causone, F., Erba, S., & Pagliano, L. (2020). Performance gap and occupant behavior in building retrofit: Focus on dynamics of change and continuity in the practice of indoor heating. *Sustainability (Switzerland)*, *12*(14), 1–25. doi:10.3390u12145820

Samuel, R. B. (2015). A Comparative Analysis of Integrated Boost Fly back Converter using PID and Fuzzy Controller. *International Journal of Power Electronics and Drive System, 5*(4), 486-501.

Samy, M., Barakat, S., & Ramadan, H. (2020). Techno-economic analysis for rustic electrification in Egypt using multi-source renewable energy based on PV/wind/FC. *International Journal of Hydrogen Energy*, *45*(20), 11471–11483. doi:10.1016/j.ijhydene.2019.04.038

Sanglier, G., Robas, M., & Jiménez, P. A. (2020b). Gamma Radiation in Aid of the Population in Covid-19 Type Pandemics. *Contemporary Engineering Sciences, 13*(1), 113-129. doi:10.12988/ces.2020.91456

Sanglier, G., González, R. A., Cesteros, S., & López, E. J. (2020c). Validation of the Mathematical Model Applied to Four Autonomous Communities in Spain to Determine the Number of People Infected by Covid-19. *Modern Applied Science*, *14*(6). https://doi.org/10.5539/mas.v14n6p110

Sanglier, G., Robas, M., & Jiménez, P. A. (2020a). Application of Forecasting to a Mathematical Models for adjusting the Determination of the Number of People Infected by Covid-19 in the Comunity of Madrid (Spain). *Contemporary Engineering Sciences*, *13*(1), 89–101. https://doi.org/10.12988/ces.2020.91453

Santin, O. G., Itard, L., & Visscher, H. (2009). The effect of occupancy and building characteristics on energy use for space and water heating in Dutch residential stock. *Energy and Building*, *41*(11), 1223–1232. doi:10.1016/j.enbuild.2009.07.002

Sartori, I., Napolitano, A., & Voss, K. (2012). Net zero energy buildings: A consistent definition framework. *Energy and Building*, *139*, 220–232.

Satish, U., Mendell, M. J., Shekhar, K., Hotchi, T., Sullivan, D., Streufert, S., & Fisk, W. J. (2012). Is CO2 an indoor pollutant? direct effects of low-to-moderate CO2 concentrations on human decision-making performance. *Environmental Health Perspectives*, *120*(12), 1671–1677. doi:10.1289/ehp.1104789 PMID:23008272

Saxena, R., Rakshit, D., & Kaushik, S. C. (2019). Phase change material (PCM) incorporated bricks for energy conservation in composite climate: A sustainable building solution. *Solar Energy*, *183*, 276–284. doi:10.1016/j.solener.2019.03.035

Schakib-Ekbatan, K., Çakici, F. Z., Schweiker, M., & Wagner, A. (2015). Does the occupant behavior match the energy concept of the building? - Analysis of a German naturally ventilated office building. *Building and Environment*, *84*, 142–150. doi:10.1016/j.buildenv.2014.10.018

Schellnhuber, H. J., Schellnhuber, H. J., Cramer, W., Nakicenovic, N., Wigley, T., & Yohe, G. (2006). Avoiding Dangerous Climate Change. Cambridge University Press.

SCHER. (2007). *Opinion on risk assessment on indoor air quality*. SCHER.

Scherelis, C., Penesis, I., Hemer, M. A., Cossu, R., & Wright, J. T. (2020). Dataset for concurrent echosounder and ADCP measurements at a tidal energy candidate site in Australia. *Data in Brief*, *31*, 105873. doi:10.1016/j.dib.2020.105873 PMID:32642508

Schieweck, A., Uhde, E., Salthammer, T., Salthammer, L. C., Morawska, L., & Mazaheri, M. (2018). Smart homes and the control of indoor air quality. Renewable and Sustainable Energy Reviews, 94, 705–718. doi:10.1016/j.rser.2018.05.057

Schweiker, M., & Wagner, A. (2016). The effect of occupancy on perceived control, neutral temperature, and behavioral patterns. *Energy and Building*, *117*, 246–259. doi:10.1016/j.enbuild.2015.10.051

Sedaghat, A., Ram, M. K., Zayed, A., Kamal, R., & Shanahan, N. (2014). Investigation of Physical Properties of Graphene-Cement Composite for Structural Applications. *Journal of Composite Materials*, *4*(01), 12–21. doi:10.4236/ojcm.2014.41002

Senamhi. (n.d.). *Climatic Events - Phenomenon of the Child*. Retrieved from Senamhi: http://senamhi.gob.pe/?p=fenomeno-el-nino

Sendra, J. J., Domínguez-Amarillo, S., Bustamante Rojas, P., & Leon, A. L. (2013). Energy Intervention in the residential sector in the south of Spain: Current challenges. *Informes de la Construcción*, *65*(532), 457–464. doi:10.3989/ic.13.074

Setiawan, A. F., Tzu-Ling, H., Chun-Ta, T., & Chi-Ming, L. (2015). The effects of envelope design alternatives on the energy consumption of residential houses in Indonesia. *Energies*, *8*(4), 2788–2802. doi:10.3390/en8042788

Shafiullah, G., Amanullah, M., Ali, A. S., Jarvis, D., & Wolfs, P. (2012). Prospects of renewable energy–a feasibility study in the Australian context. *Renewable Energy*, *39*(1), 183–197. doi:10.1016/j.renene.2011.08.016

Shahzad, S., Umair, U., & Shan, U. (2015). Electric Power Transmission and Distribution Losses Overview and Minimization in Pakistan. *International Journal of Scientific and Engineering Research*, *6*(4), 1108–1112.

Shang, Y., Zhang, D., Yang, C., Liu, Y., & Liu, Y. (2015). Effect of graphene oxide on the rheological properties of cement pastes. *Construction & Building Materials*, *96*, 20–28. doi:10.1016/j.conbuildmat.2015.07.181

Sharma, B., & Rabinal, M. K. (2013). Ambient synthesis and optoelectronic properties of copper iodide semiconductor nanoparticles. *Journal of Alloys and Compounds*, *556*, 198–202. doi:10.1016/j.jallcom.2012.12.120

Sharpe, T., McGill, G., Gregg, M. & Mawditt, I. (2016). *Characteristics and performance of MVHR systems: A meta study of MVHR systems used in the Innovate U.K. Building Performance Evaluation*. Academic Press.

Shaw-Williams, D., Susilawati, C., Walker, G., & Varendorff, J. (2019). Valuing the impact of residential photovoltaics and batteries on network electricity losses: An Australian case study. *Utilities Policy*, *60*, 100955. doi:10.1016/j.jup.2019.100955

Sherman, M. (1995). The Use of Blower-Door Data. *Indoor Air*, *5*(3), 215–224. doi:10.1111/j.1600-0668.1995.t01-1-00008.x

Sherman, M. H. (1987). Estimation of infiltration from leakage and climate indicators. *Energy and Building*, *10*(1), 81–86. doi:10.1016/0378-7788(87)90008-9

Shetti, N. P., Malode, S. J., Malladi, R. S., Nargund, S. L., & Shukla, S. S. (2019). Electroanalysis of paracetamol at nanoclay modified graphite electrode. *Microchemical.*, *146*, 387–392. doi:10.1016/j.microc.2019.01.033

Shoaib, M., Siddiqui, I., Rehman, S., Khan, S., & Alhems, L. M. (2019). Assessment of wind energy potential using wind energy conversion system. *Journal of Cleaner Production*, *216*, 346–360. doi:10.1016/j.jclepro.2019.01.128

Shukla, S., & Saxena, S. (2011). Spectroscopic investigation of confinement effects on optical properties of graphene oxide. *Applied Physics Letters*, *98*(7), 073104. doi:10.1063/1.3555438

Sierra, P., & Hernandez, J. A. (2017). Solar heat gain coefficient of water flow glazing. *Energy and Building*, *139*, 133–145.

Singh, E., & Nalwa, H. (2015). Graphene-based bulk-heterojunction solar cells: A Review. *Journal of Nanoscience and Nanotechnology*, *15*(9), 6237–6278. doi:10.1166/jnn.2015.11654 PMID:26716184

Singh, M. K., Mahapatra, S., & Atreya, S. K. (2011). Adaptive thermal comfort model for different climatic zones of North-East India. *Applied Energy*, *88*(7), 2420–2428. doi:10.1016/j.apenergy.2011.01.019

Siong, V. L. E., Lee, K. M., Juan, J. C., Lai, C. W., Tai, X. H., & Khe, C. S. (2019). Removal of methylene blue dye by solvothermally reduced graphene oxide: A metal-free adsorption and photodegradation method. *RSC Advances*, *9*(64), 37686–37695. doi:10.1039/C9RA05793E

Smart Buildings © ACEEE. (2017). *Smart Buildings: Using Smart Technology to Save Energy in Existing Buildings*. Author.

Soares, N., Costa, J. J., Gaspar, A. R., & Santos, P. (2013). Review of passive PCM latent heat thermal energy storage systems towards buildings' energy efficiency. *Energy and Building*, *59*, 82–103. doi:10.1016/j.enbuild.2012.12.042

Souayfane, F., Fardoun, F., & Biwole, P. H. (2016). Phase change materials (PCM) for cooling applications in buildings: A review. *Energy and Building*, *129*, 396–431. doi:10.1016/j.enbuild.2016.04.006

Spanish Regulation of Thermal Installations in Buildings RITE. (2016). *Factores de Emisión de CO_2 y Coeficientes de paso a Energía Primaria de Diferentes Fuentes de Energía Final Consumidas en el Sector de Edificios en España. Instituto para la Diversificación y Ahorro de la Energía*. IDAE.

Spengler, J. D., & Chen, Q. (2000). Indoor air quality factors in designing a healthy building. *Annual Review of Energy and the Environment*, *25*, 567–600. https://doi.org/10.1146/annurev.energy.25.1.567

Spengler, J., & Sexton, K. (1987). Indoor air pollution: A public health perspective. *Science*, *221*(4605), 9–17. doi:10.1126cience.6857273 PMID:6857273

Spiekman, M. (2010). *ASIEPI. The final recommendations of the ASIEPI project: How to make EPB-regulations more effective?* Retrieved from https://ec.europa.eu/energy/intelligent/projects/sites/ieeprojects/%0Afiles/projects/documents/asiepi_access_the_results._en.pdf

Sposito, G. (1998). On Points of Zero Charge. *Environmental Science & Technology*, *32*(19), 2815–2819. doi:10.1021/es9802347

Steiger, S., Hellwing, R., & Junker, E. (2008). Distribution of carbon dioxide in a naturally ventilated room with high internal heat load. In C. Rode (Ed.), *Symposium on Building Physics in the Nordic Countries* (pp. 377–384). Available at http://publica.fraunhofer.de/documents/N-97349.html

Stevens, R. G. (2009). Light-at-night, circadian disruption and breast cancer: Assessment of existing evidence. *International Journal of Epidemiology*, *38*(4), 963–970. doi:10.1093/ije/dyp178 PMID:19380369

Stymne, H., Axel Boman, C., & Kronvall, J. (1994). Measuring ventilation rates in the Swedish housing stock. *Building and Environment*, *29*(3), 373–379. https://doi.org/10.1016/0360-1323(94)90037-X

Subiyanto & Hannan. (2009), Maximum Power Point Tracking in Grid Connected PV System Using A Novel Fuzzy Logic Controller. *Proceedings of SCOReD*.

Sudman, S. (1976). *Applied Sampling*. A. Press.

Sundell, J. (2004). On the history of indoor air quality and health. *Indoor Air*, *14*(s7, Suppl. 7), 51–58. doi:10.1111/j.1600-0668.2004.00273.x PMID:15330772

Sundell, J., Levin, H., Nazaroff, W. W., Cain, W. S., Fisk, W. J., Grimsrud, D. T., Gyntelberg, F., Li, Y., Persily, K., Pickering, C., Samet, J. M., Spengler, J. D., Taylor, S. T., & Weschler, C. J. (2011). Ventilation rates and health: Multidisciplinary review of the scientific literature. *Indoor Air*, *21*(3), 191–204. doi:10.1111/j.1600-0668.2010.00703.x PMID:21204989

Tamburrini, M., Palmer, D., & Macdonald, I. A. (2003). *Calibrating models for simulation use in design practices.* Retrieved from http://www.sesg.strath.ac.uk

Tanner, C. K. (2008). Explaining relationships among student outcomes and the school's physical environment. *Journal of Advanced Academics*, *19*(3), 444–471. doi:10.4219/jaa-2008-812

Tarjomannejad, A., Farzi, A., Niaei, A., & Salari, D. (2016). An experimental and kinetic study of toluene oxidation over LaMn1−x B x O3 and La0.8A0.2Mn0.3B0.7O3 (A=Sr, Ce and B=Cu, Fe) nano-perovskite catalysts. *Korean Journal of Chemical Engineering*, *33*(9), 2628–2637. doi:10.100711814-016-0108-4

Tatsidjodoung, P., Le Pierrès, N., & Luo, L. (2013). A review of potential materials for thermal energy storage in building applications. *Renewable & Sustainable Energy Reviews*, *18*, 327–349. doi:10.1016/j.rser.2012.10.025

Teichman, K., & Howard-reed, C. (2016). *NIST Technical Note 1868 Characterizing Indoor Air Quality Performance Using a Graphical Approach.* doi:10.6028/NIST.TN.1868

Tham, K. W. (2016). Indoor air quality and its effects on humans—A review of challenges and developments in the last 30 years. *Energy and Buildings, 130*, 637–650. doi:10.1016/j.enbuild.2016.08.071

The International Institute for Sustainable Development. (2014). *The Impact of Fossil-Fuel Subsidies on Renewable Electricity Generation.* https://www.iisd.org/system/files/publications/impact-fossil-fuel-subsidies-renewable-electricity-generation.pdf

Torii, H., Kurihara, T., & Yuko, S. K. N. (2017). Violet Light Exposure Can Be a Preventive Strategy Against Myopia Progression. *EBioMedicine*, *15*, 210–219. doi:10.1016/j.ebiom.2016.12.007 PMID:28063778

Traditional geographical regions of Peru. (n.d.). Obtained from Wikipedia: https://es.wikipedia.org/wiki/Regiones_geogr%C3%A1ficas_tradicionales_del_Per%C3%BA

Treasury, H. M. (2015). Fixing the Foundations: Creating a more prosperous nation. *The Cambridge Law Journal.* Advance online publication. doi:10.1017/S0008197300099359

Treut, L. (2007). Historical Overview of Climate Change Science. IPCC WG1 Fourth Assessment Report, 93–127. doi:10.1016/j.soilbio.2010.04.001

Tsalikis, G., & Martinopoulos, G. (2015). Solar energy systems potential for nearly net zero energy residential buildings. *Solar Energy*, *115*, 743–756. doi:10.1016/j.solener.2015.03.037

Tunçbilek, E., Arıcı, M., Bouadila, S., & Wonorahardjo, S. (2020). Seasonal and annual performance analysis of PCM-integrated building brick under the climatic conditions of Marmara region. *Journal of Thermal Analysis and Calorimetry*, *141*(1), 613–624. doi:10.100710973-020-09320-8

Turiel, I. (1986). *Indoor Air Quality and Human Health.* Stanford Univ. Press.

Tushar, Q., Bhuiyan, M., Sandanayake, M., & Zhang, G. (2019). Optimizing the energy consumption in a residential building at different climate zones: Towards sustainable decision making. *Journal of Cleaner Production*, *233*, 634–649. doi:10.1016/j.jclepro.2019.06.093

Tyagi, V. V., Kaushik, S. C., Tyagi, S. K., & Akiyama, K. (2011). Development of phase change materials based microencapsulated technology for buildings: A review. *Renewable & Sustainable Energy Reviews*, *15*(2), 1373–1391. doi:10.1016/j.rser.2010.10.006

U.S. Green Building Council. (2019). *LEED v4 for Building Design Construction.* Author.

UNE EN ISO 7730:2006. Determinación analítica e interpretación del bienestar térmico mediante el cálculo de los índices PMV y PPD y los criterios de bienestar térmico local.

UNE-EN 13829:2002. (n.d.). *Thermal performance of buildings - Determination of air permeability of buildings - Fan pressurization method. (ISO 9972:1996, modified).*

United States E. P. A. (2011). *National Ambient Air Quality Standards.* https://www.epa.gov/ttn/naaqs

United States Environmental Protection Agency. (2020a). *Indoor Sources of Contaminants.* Retrieved November 27 from https://www.epa.gov/expobox/exposure-assessment-tools-media-air#indoorair

United States Environmental Protection Agency. (2020b). *Introduction to Air Quality.* Retrieved November 16 from https://www.epa.gov/indoor-air-quality-iaq/introduction-indoor-air-quality

Upadhyay, R. K., Soin, N., & Roy, S. S. (2014). Role of graphene/metal oxide composites as photocatalysts, adsorbents and disinfectants in water treatment. *RSC Advances*, *4*(8), 3823–3851. doi:10.1039/C3RA45013A

UPC (Universidad Politécnica de Cataluña). (2019). *Una conversa sobre el manjar i la ciutat* (Película). Habitar-UPC.

US Department of Housing and Urban Development. (2020). *A Guide to Healthy Homes.* https://www.hud.gov/program_offices/healthy_homes/healthyhomes

US Green Building Council. (2017). *LEED v4 for Homes design and Construction.* LEED.

Van Buggenhout, S., Zerihun Desta, T., Van Brecht, A., Vranken, E., Quanten, S., Van Malcot, W., & Berckmans, D. (2006). Data-based mechanistic modelling approach to determine the age of air in a ventilated space. *Building and Environment*, *41*(5), 557–567. https://doi.org/10.1016/j.buildenv.2005.02.029

Van Holsbeeck, S., & Srivastava, S. K. (2020). Feasibility of locating biomass-to-bioenergy conversion facilities using spatial information technologies: A case study on forest biomass in Queensland, Australia. *Biomass and Bioenergy*, *139*, 105620. doi:10.1016/j.biombioe.2020.105620

Vila-Traver, J., Aguilera, E., Infante-Amate, J., & González de Molina, M. (2021). Climate change and industrialization as the main drivers of Spanish agriculture water stress. *The Science of the Total Environment*, *760*, 143399. doi:10.1016/j.scitotenv.2020.143399 PMID:33293093

Villalba, D., Fajardo, E. J., & Romero, H. (2018). *Relación entre el material particulado PM10 y variables meteorológicas en la ciudad de Bucaramanga – Colombia: Una aplicación del análisis de datos longitudinal.* Simposio Internacional de Estadística.

Virote, J., & Neves-Silva, R. (2012). Stochastic models for building energy prediction based on occupant behavior assessment. *Energy and Building*, *53*, 183–193. doi:10.1016/j.enbuild.2012.06.001

Wadden, R. A., & Schelf, P. A. (1983). *Indoor Air Pollution. Characterization, Prediction and Control.* Wiley.

Wagner, H.U., Verhoeff, A.P., Colombi, A., Flannigan, B., Gravesen, S., Moulliseux, A., Nevelainen, A., Papadakis, J., & Sebel, K. (1993). *Biological Particles in In-door Environments. Indoor Air Quality and Its Impact.* Comisión Europea.

Walker, G. R., & Sernia, P. C. (2004). Cascaded dc-dc converter connection of photovoltaic modules. *IEEE Transactions on Power Electronics*, *19*(4), 1130–1139. doi:10.1109/TPEL.2004.830090

Wang, B., Jiang, R., & Wu, Z. (2016, November 04). Investigation of the mechanical properties and microstructure of graphene nanoplatelet-cement composite. *Nanomaterials (Basel, Switzerland)*, *6*(11), 200–208. doi:10.3390/nano6110200 PMID:28335328

Wang, X., Chen, D., & Ren, Z. (2010). Assessment of climate change impact on residential building heating and cooling energy requirement in Australia. *Building and Environment*, *45*(7), 1663–1682. doi:10.1016/j.buildenv.2010.01.022

Wang, X., Shen, Y., Xie, A., Qiu, L., Li, S., & Wang, Y. (2011). Novel structure CuI/PANI nanocomposites with bifunctions: Superhydrophobicity and photocatalytic activity. *Journal of Materials Chemistry*, *21*(26), 9641–9646. doi:10.1039/c0jm04558f

Wargocki, P. (2015, August 25). *What are indoor air quality priorities for energy-efficient buildings? Indoor and Built Environment*. SAGE Publications Ltd. doi:10.1177/1420326X15587824

Wargocki, P. (2000). The effects of outdoor air supply rate in an office on perceived air quality, sick building syndrome (SBS) symptoms and productivity. *Indoor Air*, *10*(4), 222–236. https://doi.org/10.1034/j.1600-0668.2000.010004222.x

Wargocki, P. (2013). The Effects of Ventilation in Homes on Health. *International Journal of Ventilation*, *12*(2), 101–118. doi:10.1080/14733315.2013.11684005

Wargocki, P. (2013). The Effects of Ventilation in Homes on Health. *International Journal of Ventilation*, *12*(2), 101–118. https://doi.org/10.1080/14733315.2013.11684005

Wargocki, P. (2016). Ventilation, indoor air quality, health, and productivity. In P. Wargocki (Ed.), *Ergonomic Workplace Design for Health, Wellness, and Productivity* (pp. 39–72). CRC Press.

Weatherbase. (2018). *Muscat, Oman: Monthly - Weather averages summary*. Author.

Wells, L., Rismanchi, B., & Aye, L. (2018). A review of Net Zero Energy Buildings with reflections on the Australian context. *Energy and Building*, *158*, 616–628. doi:10.1016/j.enbuild.2017.10.055

Whiffen, T. R., & Riffat, S. B. (2013). A review of PCM technology for thermalenergy storage in the built environment:Part I. *The International Journal of Low Carbon Technologies*, *8*(3), 147–158. doi:10.1093/ijlct/cts021

WHO. (1991). *Indoor Environment: Health Aspects of Air Quality, thermal environment, light and noise*. Available at: http://whqlibdoc.who.int/hq/1990/WHO_EHE_RUD_90.2.pdf

WHO. (1997). Assessment of exposure to indoor air pollutants. *WHO Regional Publications. European Series no. 78*, 139.

WHO. (2000). *Air quality guidelines for Europe. WHO Regional Publications, European Series; No. 91* (F. Theakston, Ed.; 2nd ed.). WHO Regional Publications. Available at http://link.springer.com/10.1007/BF02986808%5Cnhttp://www.springerlink.com/index/10.1007/BF02986808

WHO. (2010). WHO guidelines for indoor air quality: selected pollutants. World Health Organization Regional Office for Europe. doi:10.1186/2041-1480-2-S2-I1

Wilfried van Sark. (1826). Photovoltaic System Design and Performance. *Energies*, 12.

Wilson, B., Trieu, L. H., & Bowen, B. (1994). Energy efficiency trends in Australia. *Energy Policy*, *22*(4), 287–295. doi:10.1016/0301-4215(94)90003-5

Winterbottoma, M., & Wilkinsb, A. (2009). Lighting and discomfort in the classroom. *Journal of Environmental Psychology*, *29*(1), 63–75. doi:10.1016/j.jenvp.2008.11.007

World Allergy Organization. (2013). *White Book on Allergy*. https://www.sanofi.com/en/about-us/indoor-air-pollution-the-invisible-enemy-of-allergy-sufferers

World Health Organization. (2010). *Guidelines for Indoor Air Quality: Selected Pollutants*. WHO Regional Office for Europe.

Xi, F., Satyajayant, M., Guoliang, X., & Dejun, Y. (2012). Smart Grid – The New and Improved Power Grid: A Survey. *IEEE Communications Surveys and Tutorials, 14*(2), 944–980.

Xing, Y. F., Xu, Y. H., Shi, M. H., & Lian, Y. X. (2016). The impact of PM2.5 on the human respiratory system. *Journal of Thoracic Disease, 8*, E69–E74.

Xu, C., Luo, X., Yu, C., & Cao, S. J. (2020). The 2019-nCoV epidemic control strategies and future challenges of building healthy smart cities. In *Indoor and Built Environment*. SAGE Publications Ltd. doi:10.1177/1420326X20910408

Xu, B., & Li, Z. (2014). Performance of novel thermal energy storage engineered cementitious composites incorporating a paraffin/diatomite composite phase change material. *Applied Energy, 121*, 114–122. doi:10.1016/j.apenergy.2014.02.007

Xu, X., Xiao, B., & Li, C. Z. (2020). Critical factors of electricity consumption in residential buildings: An analysis from the point of occupant characteristics view. *Journal of Cleaner Production, 256*, 120423. doi:10.1016/j.jclepro.2020.120423

Yang, L., Qiao, Y., Liu, Y., Zhang, X., Zhang, C., & Liu, J. (2018). A kind of PCMs-based lightweight wallboards: Artificial controlled condition experiments and thermal design method investigation. *Building and Environment, 144*, 194–207. doi:10.1016/j.buildenv.2018.08.020

Yang, L., Yan, H., & Lam, J. C. (2014). Thermal comfort and building energy consumption implications–a review. *Applied Energy, 115*, 164–173. doi:10.1016/j.apenergy.2013.10.062

Yang, Y., Wu, W., Fu, S., & Zhang, H. (2020). Study of a novel ceramsite-based shape-stabilized composite phase change material (PCM) for energy conservation in buildings. *Construction & Building Materials, 246*, 1–9. doi:10.1016/j.conbuildmat.2020.118479

Ye, B., Jiang, J., Liu, J., Zheng, Y., & Zhou, N. (2021). Research on quantitative assessment of climate change risk at an urban scale: Review of recent progress and outlook of future direction. *Renewable & Sustainable Energy Reviews, 135*, 110415. doi:10.1016/j.rser.2020.110415

Yen-Jung, M. T., Aiguo, P. H., & Nirmal, K. N. (2006). Evaluation of Micro Controller Based Maximum Power Point Tracking Methods Using dSPACE Platform. *Australian University Power Engineering Conference*. https://citeseerx.ist.psu.edu/viewdoc/download?doi=10.1.1.526.1472&rep=rep1&type=pdf

Yocom, J. E., & McCarthy, S. M. (1991). *Measuring Indoor Air Quality: A Practical Guide*. Wiley.

Younes, C., Shdid, C. A., & Bitsuamlak, G. (2012). Air infiltration through building envelopes: A review. *Journal of Building Physics, 35*(3), 267–302. doi:10.1177/1744259111423085

Yu, H., Li, C., Zhang, K., Tang, Y., Song, Y., & Wang, M. (2020). Preparation and thermophysical performance of diatomite-based composite PCM wallboard for thermal energy storage in buildings. *Journal of Building Engineering, 32*, 1–11. doi:10.1016/j.jobe.2020.101753

Zarrabi, M., Haghighi, M., Alizadeh, R., & Mahboob, S. (2019). State of the art: Synthesis and characterization of functionalized graphene nanomaterials. *IOP Publishing Ltd., 211*, 738–752.

Zero Carbon Hub. (2012). *Understanding overheating - where to start: An introduction for house builders and designers.* Author.

Zero Carbon Hub. (2015). *Overheating in Homes: the big picture*. doi:10.1017/CBO9781107415324.004

Zhang, H., Lv, X., Li, Y., Wang, Y., & Li, J. (2009). P25-Graphene Composite as a High Performance Photocatalyst. *ACS Publications., 4*, 380–386. PMID:20041631

Zhang, J., & Smith, K. R. (2003). Indoor air pollution: A global health concern. *British Medical Bulletin*, *68*(1), 209–225. doi:10.1093/bmb/ldg029 PMID:14757719

Zhou, D., Zhao, C. Y., & Tian, Y. (2012). Review on thermal energy storage with phase change materials (PCMs) in building applications. *Applied Energy*, *92*, 593–605. doi:10.1016/j.apenergy.2011.08.025

Zhu, X., & Liu, Y. (2014). Characterization and Risk Assessment of Exposure to Volatile Organic Compounds in Apartment Buildings in Harbin, China. *Bulletin of Environmental Contamination and Toxicology*, *92*(1), 96–102. doi:10.100700128-013-1129-x PMID:24158356

Zurigat, Y. H., Al-Hinai, H., Jubran, B. A., & Al-Masoudi, Y. S. (2003). Energy efficient building strategies for school buildings in Oman. *International Journal of Energy Research*, *27*(3), 241–253. doi:10.1002/er.871

About the Contributors

Roberto A. González-Lezcano is a PhD in Industrial Engineering, Postgraduate Diploma in Plant Engineering, Postgraduate Diploma in University Didactics, Master in Mathematical Engineering Extraordinary Doctorate Award. Assistant Professor of the Department of Architecture and Design, in the area of Facilities at the Higher Polytechnic School of the CEU San Pablo University. He has been teaching at Building Installations, Electromechanics and Materials, Physics, Mathematics, Mechanical Engineering and Mechanics of Continuous Media and Theory of Structures. Accredited Professor by the ACAP in the figures of Professor Hired Doctor, Professor Doctor of Private University and Tenured Professor. A six-year research period recognized by the CNEAI. He has taught in undergraduate and master's degrees at the Carlos III University of Madrid, Camilo José Cela University, European University of Madrid, Antonio de Nebrija University, Technical University of Madrid and Alfonso X El Sabio University. He has published several books and articles concerning the areas of Energy, Installations and Continuous Media Mechanics.

* * *

Hussain Attia earned his Ph.D. in Power Electronics from the University of Malaya, Kuala Lumpur, Malaysia in March 2019 and his M.Sc. in Electronic Engineering and B.Sc. in Electronics and Communications Engineering from the University of Technology, Baghdad, Iraq in 1999, and 1991 respectively. Dr. Hussain Attia is a faculty member in the School of Engineering SoE. Dr. Attia is a member of the IEEE Society, and an active reviewer for IEEE Transaction on Power Electronics, IET-Power Electronics, IET-Electric Power Applications, IEEE Conferences. His research interests are including Power Electronics, AC/DC Drives, Solar PV Converters and Inverters, Pulse Width Modulation (PWM) Techniques, Improving Inverter Performance, Harmonics Elimination/Reduction Techniques, and Maximum Power Point Tracking Techniques. Since 2000, he has been working on many research projects in power electronics and renewable energy generations and conversions. During the last five years, Dr. Hussain published more than 60 research papers in the fields of research interests.

Genny Carrillo is an Associate Professor of Environmental and Occupational Health at Texas A&M University. She is also Director of the Program of Asthma Research and Education (P.A.R.E.) at the Texas A&M School of Public Health. Through these positions, she conducts applied research focusing on children's environmental health issues in the Lower Rio Grande Valley of South Texas. Concentrating on the effects of asthma and diabetes on children's and families' health, she studies how environmental pollutants and air quality most affect low-income communities of color. Dr. Carrillo also serves as a

Co-Facilitator for the Environmental Education Committee of the U.S.-Mexico Border 2020 Environmental Program. She works directly with EPA Region 6 and the Texas Commission of Environmental Quality to identify high-priority environmental issues for both countries. Her work has been published in journals such as BMC Public Health; Environmental Research; and Pediatric Allergy, Immunology, and Pulmonology.

Fernando del Ama is an Assistant Professor of Architecture at Keene State College. Fernando de Ama possesses a Ph.D. in Architectural Acoustics from the School of Architecture of Madrid, Spain. He studied spatial perception of an architectural project and empirical parameters to qualify a Concert Hall. He has a seventeen year teaching experience related to construction technology, building physics, acoustics, energy management in buildings, building construction and architectural design. His research interests are related to energy simulation and monitoring systems in net zero energy buildings; new materials for regenerative buildings; Active Noise Control and acoustical components and geometric considerations for spatial perception of an architectural project.

Samuel Domínguez-Amarillo is a Vice Dean in the College of Architecture and Professor in the Department of Construction at the University of Seville, Spain. He is a member of the research group TEP-130: Architecture, Heritage and Sustainability of the Andalusian Scientific Research. In his professional career he acts as Expert Consultant on Hi-performance systems for buildings and in the assessment of sustainability and energy efficiency of buildings and their urban environment. He has extensive background as adviser on environmental control for Cultural Heritage, with a wide portfolio of national and international projects.

Savaş Erdem, after receiving his PhD from the Department of Civil Engineering at the University of Nottingham-England-UK, joined the Department of Civil Engineering at the University of Istanbul in 2012. He progressed rapidly through the academic ranks, in 2016 becoming Associate Professor of Civil Engineering, one of the youngest associate professors at that time. His doctoral research has created a new understanding of the micro- and nanomechanical damage and fracture behaviour of sustainable concrete under short-duration extreme dynamic loads. The results of his study have been published in the most prestigious journals in civil engineering.

Jesica Fernández-Agüera is a Postdoctoral researcher at Seville University, Department of Construction, member of the research group TEP-130: Architecture, Heritage and Sustainability of the Andalusian Scientific Research, Spain and the General Manager of Dimarq SL. Her research focuses on energy efficiency, low-energy buildings, airtightness, indoor air quality and ventilation, with special interest in the interaction between environmental parameters, health, and thermal comfort in dwellings.

Feliciano Fraguela is Master in Marine Engineering and also Master in Maintenance. His PhD was focused on the statistical procedure for wind energy maintenance. Nowadays, he lectures in the University of A Coruña (Spain) where he develops his research activity in several engineering fields, especially in statistic tools and energy.

Iosu Gabilondo is a graduate in Architecture from the University of the Basque Country (UPV), Donostia-San Sebastian, Spain (2.007). His work focuses primarily in environmental and social issues, regardless of scale, setting or budget. His practice is based in a highly collaborative design method, to ensure the dialogue between all the parts involved in the process.

Matthew Griffin is currently a college undergrad studying architecture. Matthew hopes to pursue graduate study in architecture and building science, and in the future work in an architectural firm promoting the study of building science and sustainability.

Ezgi Gurbuz is a PhD student at the Department of Civil Engineering of the Istanbul University-Cerrahpasa.

Juan Antonio Hernandez Ramos is an associate professor of Aerospace Engineering at the Polytechnic University of Madrid, Spain. He possesses a PhD. in Numeric Simulation from the School of Aeronautics and Space Engineering, UPM. He has been the director of the CES CEU Architecture (UPM) and co-founder of the Technology Based Company, InTelliGlass S.L. His research experience is related to the development of high-energy performance of water-flow glazed façades for buildings and numerical simulations and high order numerical schemes for thermo-fluid-dynamic problems. As a result of his research he has published papers in high impact factor journals, along with patents.

Omar Ibrahim is a PhD student at the Department of Civil Engineering in Istanbul University-Cerrahpasa.

Antonio E. Masdías is Master in Science and graduated in Electrical Engineering. His PhD was focused on energy saving in building construction and its related decisions based on real data. Nowadays, he is a professor at the University of A Coruña and a member of the Energy and Marine Propulsion Research Unit. He develops research activities in different energy-saving topics like the Internet of things (IOT), Big data and new sensors for energy saving in buildings.

Enrique Masdías-Bonome is graduated in Electrical Engineering and Master in Engineering. His PhD was focused on statistical procedures for energy saving in buildings. Nowadays he is a lecturer in the University of A Coruña and develops his research activities in building energy-saving procedures.

Eray Meric is a MSc student at the Department of Civil Engineering of the Istanbul University-Cerrahpasa.

Belen Moreno is an architect specialist in nearly zero energy building (nZEB). She is a lecturer of the Master of Technological Facades and Sustainable Envelopes at the Technical School of Architecture of Madrid. Over the last ten years she has been working as a Senior Researcher at the School of Aeronautics and Space Engineering in the Department of Applied Mathematics at the UPM. Founder of her own architecture office, her professional career has been enriched in the field of research and development of new technologies focused on industrialized construction and energy efficiency in building at National and International level.

Alejandro Moreno-Rangel's main research interests are sustainable architecture and the indoor environment – indoor air quality (IAQ) and thermal comfort – and its connections to health, urban and human behaviours to create healthy homes. Recently, Alejandro has developed an interest in using low-cost sensors as research tools and the effect that they could have on residential behaviour, design and human health & well-being. Alejandro is also interested in developing low-energy homes, particularly through the Passivhaus Standard, in developing countries such as Latin American Countries.

José A. Orosa is Master in Marine Engineering, graduated in Naval architecture and graduated in Mechanical Engineering. Nowadays, he is full professor at the University of A Coruña. In last years he was identified as the second more relevant researcher in sustainability and energy saving. His research activities are centred in mathematical methods, energy-saving and engineering applications.

Claudia Pennese is a Doctor of Architecture. Director of the university strat-ups specialized in urban regeneration, citizen participation and gender perspective in urbanism.

José Rizzo Sierra has a Physics Degree by Universidad Industrial de Santander (Bucaramanga, Santander, Colombia) with a thesis on scientific epistemology. Magister Degree in Physics by the same institution with a thesis on Computational Plasma Physics. Doctor of Technology Degree by Instituto Politécnico Nacional (Santiago de Querétaro, Querétaro, México) with an applied physics thesis on Alternative (magnetohydrodynamic) Power Generation. I currently work as Automotive Technology Engineering Researcher/Professor at Universidad Politécnica de Querétaro (El Marqués, Querétaro, México). My research interests include theoretical/applied physics/engineering towards efficiency, computational physics/engineering, interdisciplinary research, alternative energies, and philosophy of science.

Estela Karem Samamé Zegarra is an architect with a master's degree in Environmental Management and Building Design at Navarra University in Pamplona, Spain. Currently a teacher in Peruvian universities like UNASAM.

Juan José Sendra is a Professor at the University of Seville, was Director for 15 years (2000-2015) of the University Institute of Architecture and Construction Sciences of the University of Seville. He is currently the Head Researcher of the research group TEP130 Architecture, Heritage and Sustainability: Acoustics, Lighting, Optics and Energy. He is Principal Investigator (PI) of various research projects and contracts, at regional, national, and international level.

Diego Vergara is graduated in Civil Engineering and also graduated in Materials Engineering. His Ph.D. was focused on hydrogen embrittlement of pearlitic steel wires. Nowadays, he is a lecturer at the Catholic University of Ávila (Spain) where he develops his research activity in several engineering fields, especially in materials science, energy, and virtual reality.

Index

A

air conditioning 72, 123-125, 130, 146-148, 150-151, 158-163, 196, 213, 220, 222, 265, 270, 286
Air Exchange Rate 95
air infiltration 73-74, 77, 83-84, 93-95, 141
airtightness 51, 57, 72-75, 77-78, 82-84, 86, 88, 91-95
Arabian Gulf 146, 148, 150-152, 158-159, 161-163, 168
asthma 49-51, 57, 61-65, 67-68, 122-123, 125

B

Behavioural Adaptation 146-147, 168
BREEAM 52, 54, 275-276, 280, 282-290
Building Comfort 1
building energy consumption 51, 195, 198-199, 201, 203, 210, 213, 215-217, 220, 272, 274
building envelope 8, 16, 18, 22-24, 44, 46-47, 55, 69, 82, 95

C

carbon emissions 22, 49, 51-55, 263, 272-273
circadian lighting 239, 251, 253
circulating system 31-32, 47
climate change 42, 49-51, 66-70, 150, 166, 191, 195-199, 201-204, 209-210, 212-213, 215-220, 261, 263, 266-267, 271, 274-276, 278
COAVN 96-98
construction materials 1, 15, 60, 139, 218, 281
cooling energy 2, 8, 24, 146, 148, 151, 163, 168, 197, 202, 204-209, 215-216, 218, 274
Covid-19 118-120, 140, 144
cultural attitude 146, 148, 168

D

data analysis 118
daylighting 239, 241, 243-244, 252-255
DC-DC Converter design 169
DGNB 275, 282, 290
dynamic g-factor 47
Dynamic U-Value 47

E

EDGE BONO VERDE 275
energy efficiency 1-2, 8, 11, 15, 19, 21-23, 43, 45-47, 52-53, 55, 65, 94, 118, 122, 141, 149-150, 218-219, 222, 244, 253-254, 265, 271, 274-276, 291
energy generation 21, 23, 257-262, 264, 268-269
Energy Performance Monitoring 21
Energy Rating Software 213, 220
Energy Smart Homes 65
energy utilization 257-260, 262, 264-266, 269, 272
Epidemiology 118, 143, 255-256
Extreme Climate 146, 168

G

graphene oxide 221-227, 230-238

H

Healthy Homes 49-50, 62-63, 66-67, 71, 144
housing components 96
HQE 275, 280, 282-283, 285, 287-288, 290-291

I

IAQ 49-51, 54-57, 61, 66, 73-74, 119, 122, 151
illuminance 239, 242-243, 245, 247-253
indoor air quality 49-50, 52, 54, 56, 63-72, 93, 95, 118-119, 121-122, 130, 137, 139-140, 142-145, 151, 212-214, 218
indoor ambience 213-214, 220
indoor lighting 239, 241-242, 247-249, 251-252, 255
ipRGCs 239-241, 243, 251-252

ISO 95, 123, 144, 195, 198-199, 201-202, 213, 215, 217-218, 240, 247-248, 252, 254, 289

L

LEED 52, 54, 256, 275-277, 280-281, 283-291

M

mass flow rate 24, 28-29, 31, 33, 35, 38, 43-44, 47
mechanical ventilation 50-51, 57, 95, 130, 146, 148
Mediterranean area 72, 93
melatonin 239, 241, 243, 246-247, 249, 254
microgrid system 169, 173-174, 182, 184, 190
Model Prediction 72
Monte Carlo 195, 198, 201-204, 213, 215, 217
MUGAK Architecture Biennial 96
Muscat 146, 150, 152, 155-156, 164, 167

N

Nanocomposite 221, 232, 236-237
natural ventilation 8, 18, 52, 79, 95, 130, 142, 146, 148, 151, 158-163
nursing homes 118, 120, 124, 128, 130-132, 137-138, 141

O

Oman 146, 150-152, 156, 164, 166-168, 190, 257
optimization 8, 16, 101, 193, 195-196, 214, 216-217, 220, 252

P

particulate matter 49, 59-61, 64, 123, 126, 130, 133, 136, 138-140
pavilion 96-98, 100-106, 108, 111, 115
Peru 275-279, 281-282, 285-288, 290-291
phase change materials 1-3, 8, 13, 15-20, 43, 215
photo catalyst 221, 224-226, 232-234
Photovoltaic System 182, 191-192, 194
PM2.5 59-61, 71, 123, 126, 128, 130, 139
Predictive Modeling 95
public participation 97
PV matrix 176, 182, 184-186, 190
PV panel 172, 175-180, 186, 190

Q

quality air 118
questionnaire 146-147, 151, 156, 163, 168

R

Reduced Graphene Oxide 221-222, 224-227, 234-238
renewable energy 18, 21, 24, 36, 41-42, 44-45, 47, 164, 169, 172-173, 175, 187, 191, 193, 218, 257-263, 265-273
renewable energy sources 169, 172-173, 175, 258-259, 261, 263, 266-267, 269-270
residential 1, 18, 43, 49-51, 59, 65-67, 69, 73-74, 79, 83-84, 86, 93-94, 119, 122-123, 125, 139-141, 146, 148-151, 157-158, 162, 165-168, 170, 199, 203, 219, 257-261, 264-274, 276, 278, 286, 288, 290-291

S

San Sebastian 96-97
Setpoint Temperature 146, 168
smart power grid 169, 172, 190-191
social housing 68, 72-73, 84, 91, 93, 165, 277
structured workshops 96, 110
sustainable homes 49, 276

T

thermal comfort 8, 10, 15-16, 47, 52, 55-56, 69, 93, 146-152, 158, 162-168, 212-213, 218-220, 241, 274
thermal energy storage 1-4, 7-8, 11-12, 16-20, 26
traditional power grid 169-172, 190
Transgenerational House 96

U

ultrafine particulate 49, 61

V

ventilation 8, 18, 22, 41, 44, 49-52, 55, 57, 62-66, 70-74, 79-80, 83, 94-95, 119, 122-123, 126, 130, 140-146, 148-149, 151, 158-163, 195, 198, 201-202, 213-215, 220, 222, 258, 265
volatile organic compounds 57, 60-61, 64-65, 68, 71, 122, 127, 142-143

W

Water Flow Glazing 21-22, 24-27, 42-44, 47-48
Water Heat Gain Efficiency 36, 48
Weibull 195, 198-199, 202-203, 209-215, 218, 220

Weibull model 195, 198-199, 202, 211-215, 218, 220

Z

Zero Energy Buildings 21-22, 47, 67, 196, 274

www.ingramcontent.com/pod-product-compliance
Lightning Source LLC
LaVergne TN
LVHW060416300926
849162LV00006B/1445